PARTICLE PHYSICS
at the Year of Astronomy

Faculty of Physics of Moscow State University

INTERREGIONAL CENTRE
FOR ADVANCED STUDIES

Proceedings of the Fourteenth Lomonosov
Conference on Elementary Particle Physics

PARTICLE PHYSICS at the Year of Astronomy

Dedicated to the late Academician Alexey Sissakian

Moscow, Russia 19 – 25 August 2009

Editor

Alexander I. Studenikin

Department of Theoretical Physics
Moscow State University, Russia
and
Joint Institute for Nuclear Research (Dubna), Russia

World Scientific

NEW JERSEY • LONDON • SINGAPORE • BEIJING • SHANGHAI • HONG KONG • TAIPEI • CHENNAI

Published by

World Scientific Publishing Co. Pte. Ltd.

5 Toh Tuck Link, Singapore 596224

USA office: 27 Warren Street, Suite 401-402, Hackensack, NJ 07601

UK office: 57 Shelton Street, Covent Garden, London WC2H 9HE

British Library Cataloguing-in-Publication Data
A catalogue record for this book is available from the British Library.

PARTICLE PHYSICS AT THE YEAR OF ASTRONOMY
Proceedings of the Fourteenth Lomonosov Conference on Elementary Particle Physics

ISBN-13 978-981-4329-67-5
ISBN-10 981-4329-67-3

Printed in Singapore by Mainland Press Pte Ltd.

Moscow State University
Faculty of Physics
Interregional Centre for Advanced Studies

FOURTEENTH LOMONOSOV CONFERENCE ON ELEMENTARY PARTICLE PHYSICS

Moscow, August 19-25, 2009

Mikhail Lomonosov
1711-1765

Sponsors

Russian Foundation for Basic Research
Russian Agency for Science and Innovation
Russian Academy of Sciences
Russian Agency for Atomic Energy
Dmitry Zimin "Dynasty" Foundation

Supporting Institutions

Faculty of Physics of Moscow State University
Skobeltsyn Institute of Nuclear Physics, Moscow State University
Interregional Centre for Advanced Studies
Joint Institute for Nuclear Physics (Dubna)
Institute for Nuclear Research (Moscow)
Institute of Theoretical and Experimental Physics (Moscow)
Budker Institute of Nuclear Physics (Novosibirsk)

International Advisory Committee

Organizing Committee

Moscow State University
Interregional Centre for Advanced Studies

EIGHTH INTERNATIONAL MEETING ON PROBLEMS OF INTELLIGENTSIA

"INTELLIGENTSIA and WORLD CRISIS"

Moscow, August 25, 2009

Presidium of the Meeting

V.A.Sadovnichy (MSU) - Chairman
V.V.Belokurov (MSU)
J.Bleimaier (Princeton)
G.Diambrini-Palazzi (University of Rome)
V.G.Kadyshevsky (JINR)
S.P.Kapitza (Russian Academy of Sciences)
V.V.Mikhailin (MSU)
A.I.Studenikin (MSU & ICAS) - Vice Chairman
V.I.Trukhin (MSU)

FOREWORD

The 14th Lomonosov Conference on Elementary Particle Physics was held at the Moscow State University (Moscow, Russia) on August 19-25, 2009 under the Patronage of the Rector of the Moscow State University Victor Sadovnichy.

The conference was organized by the Faculty of Physics and Skobeltsyn Institute of Nuclear Physics of the Moscow State University in cooperation with the Interregional Centre for Advanced Studies and supported by the Joint Institute for Nuclear Research (Dubna), the Institute for Nuclear Research (Moscow), the Budker Institute of Nuclear Physics (Novosibirsk) and the Institute of Theoretical and Experimental Physics (Moscow). The Russian Foundation for Basic Research, the Russian Agency for Science and Innovation, the Russian Academy of Sciences, the Dmitry Zimin "Dynasty" Foundation and the Russian Agency for Atomic Energy sponsored the conference.

It was more than twenty years ago when the first of the series of conferences (from 1993 called the "Lomonosov Conferences"), was held at the Department of Theoretical Physics of the Moscow State University (June 1983, Moscow). The second conference was held in Kishinev, Republic of Moldavia, USSR (May 1985).

After the four years break this series was resumed on a new conceptual basis for the conference programme focus. During the preparation of the third conference (that was held in Maykop, Russia, 1989) a desire to broaden the programme to include more general issues in particle physics became apparent. During the conference of the year 1992 held in Yaroslavl it was proposed by myself and approved by numerous participants that these irregularly held meetings should be transformed into regular events under the title "Lomonosov Conferences on Elementary Particle Physics". Since then at subsequent meetings of this series a wide variety of interesting things both in theory and experiment of particle physics, field theory, astrophysics, gravitation and cosmology were included into the programmes. It was also decided to enlarge the number of institutions that would take part in preparation of future conferences.

Mikhail Lomonosov (1711-1765), a brilliant Russian encyclopaedias of the era of the Russian Empress Catherine the 2nd, was world renowned for his distinguished contributions in the fields of science and art. He also helped establish the high school educational system in Russia. The Moscow State University was founded in 1755 based on his plan and initiative, and the University now bears the name of Lomonosov.

The 6th Lomonosov Conference on Elementary Particle Physics (1993) and all of the subsequent conferences of this series were held at the Moscow State University on each of the odd years. Publication of the volume "Particle Physics, Gauge Fields and Astrophysics" containing articles written on the basis of presentations at the 5th and 6th Lomonosov Conferences was supported by the Accademia Nazionale dei Lincei (Rome, 1994). Proceedings of the 7th and 8th Lomonosov Conference (entitled "Problems of Fundamental Physics" and "Elementary Particle Physics") were published by the Interregional Centre for

Advanced Studies (Moscow, 1997 and 1999). Proceedings of the 9th, 10th, 11th, 12th and 13th Lomonosov Conferences (entitled "Particle Physics at the Start of the New Millennium", "Frontiers of Particle Physics", "Particle Physics in Laboratory, Space and Universe", "Particle Physics at the Year of 250th Anniversary of Moscow University" and "Particle Physics on the Eve of LHC") were published by World Scientific Publishing Co. (Singapore) in 2001, 2003, 2005, 2006 and 2009, correspondently.

The physics programme of the 14th Lomonosov Conference on Elementary Particle Physics (August 19-25, 2009) included review and original talks on wide range of items such as neutrino and astroparticle physics, electroweak theory, fundamental symmetries, tests of standard model and beyond, heavy quark physics, non-perturbative QCD, quantum gravity effects, physics at the future accelerators. Totally there were more than 350 participants with 143 talks including 43 plenary (25 min) talks, 38 session (20 min) talks, 47 brief (15-10 min) reports and 15 poster presentations. One of the goals of the conference was to bring together scientists, both theoreticians and experimentalists, working in different fields, so that no parallel sessions were organized at the conference. The Round table discussion on "The Year of Astronomy: Astroparticle Physics and Frontiers of the Universe" was held during the last day of the 14th Lomonosov Conference.

Following the tradition that has started in 1995, each of the Lomonosov Conferences on particle physics has been accompanied by a conference on problems of intellectuals. The 8th International Meeting on Problems of Intelligentsia held during the 13th Lomonosov Conference (August 29, 2007) was dedicated to discussions on the issue "Intelligentsia and World Crisis".

The success of the 14th Lomonosov Conference was due in a large part to contributions of the International Advisory Committee and Organizing Committee. On behalf of these Committees I would like to warmly thank the session chairpersons, the speakers and all of the participants of the 14th Lomonosov Conference and the 8th International Meeting on Problems of Intelligentsia.

We are grateful to the Rector of the Moscow State University, Victor Sadovnichy, the Vice Rector of the Moscow State University, Vladimir Belokurov, the Dean of the Faculty of Physics, Vladimir Trukhin, the Directors of the Joint Institute for Nuclear Research, Victor Matveev, the Director of the Budker Institute of Nuclear Physics, Alexander Skrinsky, and the Vice Dean of the Faculty of Physics of the Moscow State University, Anatoly Kozar for the support in organizing these two conferences.

This volume is dedicated to the memory of the late Alexey Sisakian, the Director of the Institute for Nuclear Research, who for many years provided very important help in organizing of the Lomonosov conferences.

I would like to thank Tom Browder, Giorgio Chiarelli, Dmitri Denisov, Guido Drexlin, Richard Dubois, Antonio Ereditato, Bellisario Esposito, Paolo Franzini, Peter Jenni, Takuya Hasegawa, Cristina Lazzeroni, Tsuyoshi Nakaya, Monica Pepe-Altarelli, Luc Poggioli, Frank Porter, David Reitze, Uwe Schneekloth, Stefan Schoenert, Christian Spiering, Christian Stegmann, Horst Wahl, Gaston

Wilquet, Bolek Wyslouch for their help in planning of the scientific programme of the meeting and inviting speakers for the topical sessions of the conference.

Special thanks are due to Alexander Suvorinov (the Russian Agency for Science and Innovations) and Valery Rachkov (the Russian State Corporation for Atomic Energy) for their very valuable help.

Furthermore, I am very pleased to mention Ilya Balantsev, Alexander Grigoriev and Alexey Lokhov, the Scientific Secretaries of the conference, Andrey Egorov, Hamid Gadimi, Mila Polyakova, Konstantin Kiselyev, Ilya Tokarev, Svetlana Esakova, Tanyusha Pogorelova and Maria Skrigolovskaya for their very efficient work in preparing and running the meeting.

These Proceedings were prepared for publication at the Interregional Centre for Advanced Studies with support by the Russian Foundation for Basic Research, the Russian Agency for Science and Innovations and the Russian Agency for Atomic Energy.

Alexander Studenikin

CONTENTS

Neutrino Physics

Physics at Accelerators and Studies in SM and Beyond

STATUS OF THE ATLAS EXPERIMENT

P. S. Wells [a], on behalf of the ATLAS Collaboration
Dept. PH, CERN, CH-1211 Geneva 23, SWITZERLAND

Abstract. The ATLAS experiment is a general purpose detector to observe proton-proton collisions at the Large Hadron Collider. The experiment has been commissioned with a large sample of cosmic ray events recorded in 2008 and 2009, and the first proton-proton collisions were observed at the end of 2009. The very good state of readiness of the detector, and some first results of collisions, are reported.

1 Introduction

The ATLAS experiment [1] is a general purpose detector, designed to measure the products of high energy proton-proton (pp) and heavy ion collisions from the Large Hadron Collider (LHC). The detector was built and is operated by a collaboration of about 2,800 physicists, including 800 PhD students, drawn from 169 institutions, coming from 37 countries, spanning 5 continents. This paper discusses the readiness of the ATLAS detector to observe LHC collisions, based on calibration work and studies of a large sample of cosmic ray events.

A total of 216 million cosmic ray events was collected in the autumn of 2008, and a further 93 million in summer 2009. The overall event rate was up to 700 Hz, but with less than 1 Hz passing through the pixel detector at the centre of the experiment. In December 2009, about one million pp collision interactions were recorded at the LHC injection energy, corresponding to 900 GeV in the centre-of-mass system. The beams were declared stable for about 460 thousand of these collisions, and for these events the detector could be fully switched on. In addition, 34 thousand events were recorded at a higher centre-of-mass energy of 2.36 TeV, but without the full tracking detector at operational voltage.

2 The ATLAS detector

The overall apparatus is 25 m high, 44 m long, and weighs 7,000 tonnes. A cut-away diagramme is shown in Figure 1. The layout of the detector is driven by the magnet systems, in particular the barrel and end-cap toroid magnets of the muon spectrometer.

2.1 The ATLAS magnets and the muon spectrometer

A 2.5 m long, 2.5 m diameter superconducting solenoid provides a uniform 2 T field for the inner detector, with an excitation current of 7.7 kA. The barrel and two end-cap toroids each have eight coils, with an excitation current of 20.5 kA.

[a] e-mail: pippa.wells@cern.ch

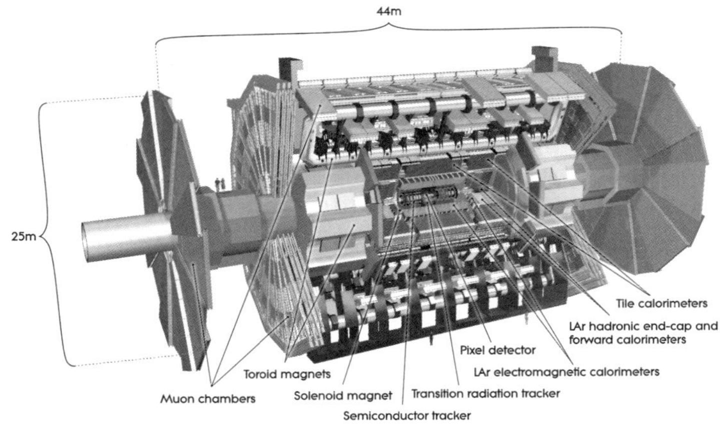

Figure 1: Cut-away view of the ATLAS detector.

The 8 barrel coils are each 25 m long, 5 m wide and weigh 100 tonnes. The typical field is 0.5 T. In contrast, each end-cap toroid has 8 coils in a common cryostat of 11 m diameter. At 240 tonnes each, these were the heaviest indidual pieces to be lowered into the ATLAS experimental cavern. The typical end-cap field is 1 T. The full magnet system has operated reliably for long periods.

The muon spectrometer provides a muon trigger and momentum measurement, with a momentum resolution better than 10% up to about 1 TeV. The barrel toroid field alone covers the range $|\eta| < 1.4$, and then overlaps the end-cap field for $1.4 < |\eta| < 1.6$. There are 700 precision chambers made from monitored drift tubes (MDT) in the barrel, complemented by 600 resistive plate trigger chambers (RPC) for $|\eta| < 1.05$. When the detector is in its operating configuration, the most dramatic sight is that of the end-cap big wheels of the forward muon system. These add an additional 400 MDT chambers with $|\eta| < 2.7$. The precision chamber coverage is completed by 32 cathode strip chambers (CSC) of the innermost "small wheel" ($2.0 < |\eta| < 2.7$). The forward trigger is provided by 3,600 thin gap chambers (TGC), which instrument the range $1.05 < |\eta| < 2.7$, and extend the trigger to $|\eta| < 2.4$.

Projecting RPC stand-alone cosmic ray muon tracks to ground level, 81m above the centre of ATLAS, the rate dependence reveals the two main access shafts down which the parts of the detector were lowered, and the two smaller personnel elevator shafts, as shown in Figure 2.

The alignment of the MDT chambers was tested with cosmic rays, with the toroid magnet switched off(giving a sample of approximately straight tracks). The sagitta in the precision plane of the middle barrel chamber was measured before alignment, after alignment with the optical alignment system, and after

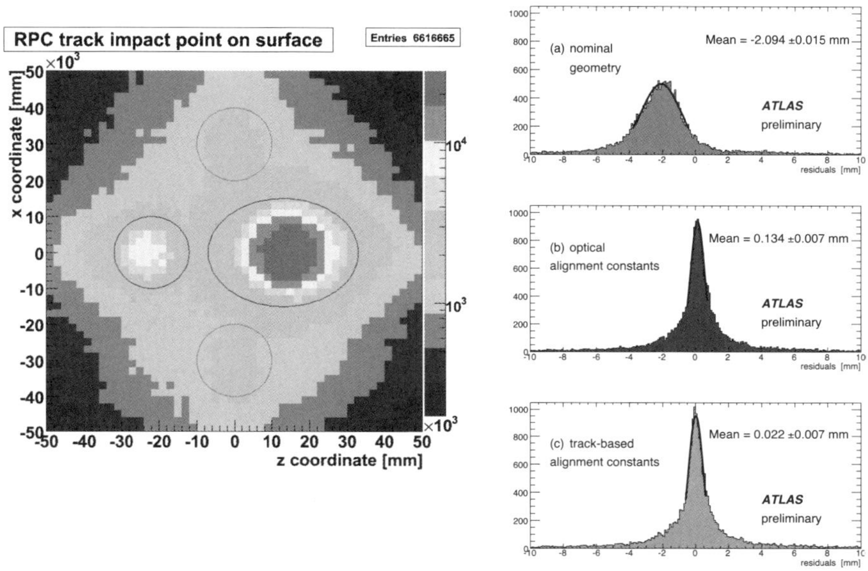

Figure 2: Map of cosmic ray tracks reaching the detector, revealing the access shafts (left) and MDT sagitta for straight tracks at three stages in the alignment process (right).

applying a track-based alignment correction. The sagitta for a 1 TeV muon with field on would be typically 500 μm, and the target is to achieve a mis-measurement less than 30 μm for a momentum resolution better than 10% at 1 TeV. With the track based alignment, the average sagitta is 22 μm. The alignment residuals are illustrated in Figure 2.

2.2 The calorimeters

The electromagnetic calorimeter covers $|\eta| < 3.2$, while the hadronic calorimetery extends to $|\eta| < 4.9$. The lead/liquid-Argon electromagnetic calorimeter has an accordion geometry, and provides triggering, identification and energy measurement of electrons and photons. The energy resolution is $\sigma/E \approx 10\%/\sqrt{E} \oplus 0.7\%$. The liquid Argon cryostats also include a copper and tungsten layers to provide the forward hadronic calorimeter. The central hadronic calorimeter is iron with scintillator tiles. The calorimeter can trigger on and measure jets and missing transverse energy, with a resolution of $\sigma/E \approx 50\%/\sqrt{E}$ $\oplus 3\%$. The calorimeter noise and response have been shown to be uniform across the detector, and stable in time. The calorimeter pulse-shape modelling has also been studied. Two examples are given in Figure 3.

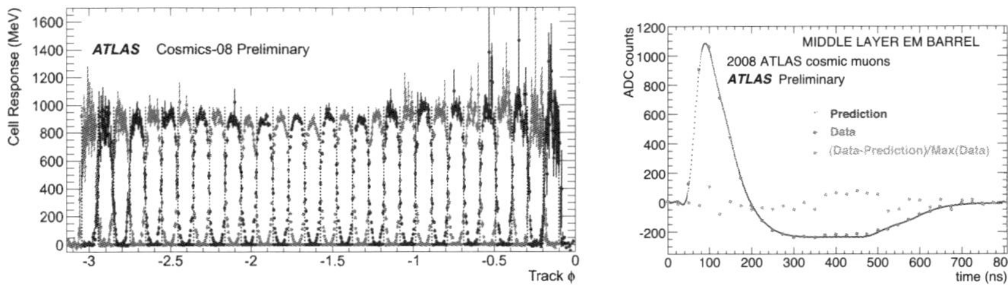

Figure 3: Energy deposited in cells of the tile calorimeter as a function of the ϕ position of extrapolated tracks (left) and the pulse shape of the liquid argon EM calorimeter (right).

2.3 The inner detector

The inner detector (ID) measures the tracks of charged particles. A typical barrel track would pass three layers of pixel detectors, four layers of double sided silicon micro-strip detectors (SCT) and more than thirty straws of the transition radiation tracker (TRT). Including the barrel and both end-caps, the overall ID covers $|\eta| < 2.5$, with 80 million 50x400 μm pixels on 1,744 modules, 6.3 million SCT strips on 4,088 modules, with a 40 mrad stereo angle between the two sides, and 400 thousand 4mm TRT straws. The expected momentum resolution satisfies $\sigma/p_{\rm T} \approx 3.4 \times 10^{-4} p_{\rm T}(\text{GeV}) \oplus 0.015$. Out of 216 million cosmic ray events, 7.6 million pass through the ID, and 420 thousand have at least one pixel hit.

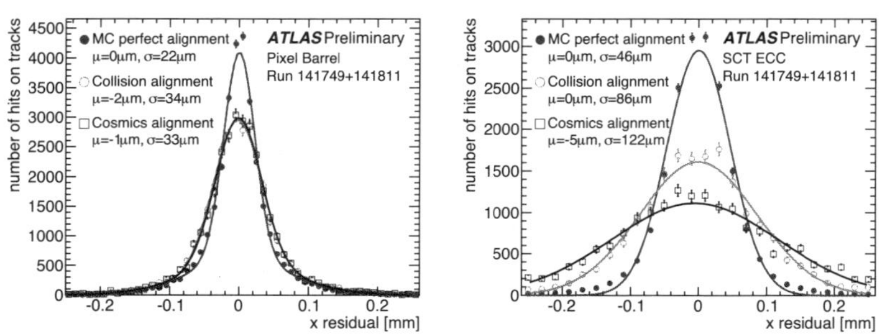

Figure 4: Examples of the ID alignment resolution for the most precise coordinates of the pixel barrel (left) and the SCT end-cap (right).

The silicon pixel and strip noise occupancies are below the design specification: less than 10^{-9} for the pixels and less than 5×10^{-4} for the SCT. For the TRT, the thresholds are adjusted such that the noise occupancy is 2%.

The intrinsic sensor efficiencies have been measured with cosmic ray tracks to be much greater than 99% for the silicon detectors, and above 97% for an individual TRT straw. Furthermore, the cosmic ray sample allowed the ID alignment procedure to be exercised, especially for the barrel regions which are well illuminated with cosmic ray tracks. Using the 2008 alignment results on the Summer 2009 cosmic ray sample already showed that the detector was remarkably stable. The good quality of the barrel alignment was confirmed with collision data at the end of the year. The collision data also allowed a rapid improvement in the quality of the end-cap alignment, as illustrated in Figure 4.

3 Detector performance

3.1 Operational efficiency and trigger performance

Sub-detector	Number of channels	Operational fraction (%)
Pixel	80 M	97.5
SCT silicon strips	6.3 M	99.3
TRT transition radiation tracker	350 k	98.2
LAr EB calorimeter	170 k	98.6
Tile calorimeter	9800	98.0
Hadronic end-cap LAr calorimeter	5600	99.9
Forward LAr calorimeter	3500	100
LVL1 calorimeter trigger	7200	99.5
MDT muon drift tubes	350 k	99.7
CSC muon end-cap	31 k	98.5
RPC barrel muon trigger	370 k	99.5
TGC end-cap muon trigger	320 k	100

Table 1: Fraction of operational channels for the main ATLAS sub-detectors.

The operational fraction of the main subsystems at the time of December 2009 are listed in Table 1. The typical operational efficiency for simulated LHC fills in Summer 2009 was about 83%. The data-taking efficiency with colliding beams was around 90%.

The ATLAS trigger has three levels, designed to reduce the event rate of pp collision data from 40 MHz to about 200 Hz. The large samples of cosmic ray tracks have allowed detailed studies of the trigger performance. Two examples are shown in Figure 5.

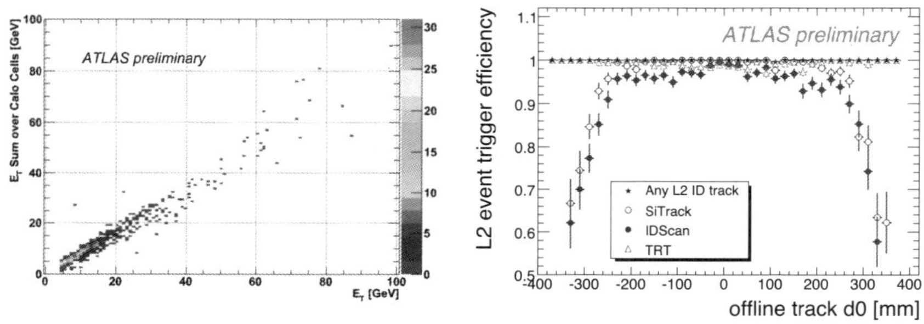

Figure 5: Comparision of the fast energy measurement made by the level-1 calorimeter trigger and the more precise measurement from the full calorimeter readout (left), and the high efficiency of level-2 tracking algorithms after tuning for tracks from cosmic rays (right).

3.2 Reconstruction of particles

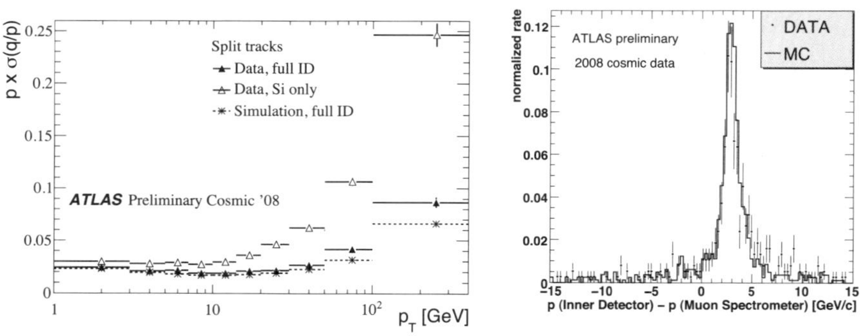

Figure 6: The momentum resolution derived by the cosmic ray split-track method (left), and the evaluation of energy deposited in the calorimeter by comparing the momentum measured in the ID and muon spectrometer (right).

Splitting cosmic ray tracks which pass through the whole ID into the upper and lower hemisphere, and fitting these as two separate tracks, allows a direct measurement of track resolutions. At low momentum, the resolution is dominated by multiple scattering, so the good agreement between MC and data shown in Figure 6 indicates that the material in the detector is well modelled. The cosmic ray sample also allowed checks of the agreement between track parameters measured in the ID and the muon spectrometer, and comparisons between data and simulation of τ-lepton identfication variables. For example, in Figure 6, the difference in momentum measured in the ID and in the muon spectrometer is shown to be in good agreement between simulation and data, consistent with a typical energy loss of 3 GeV as the muons pass through the material of the calorimeters.

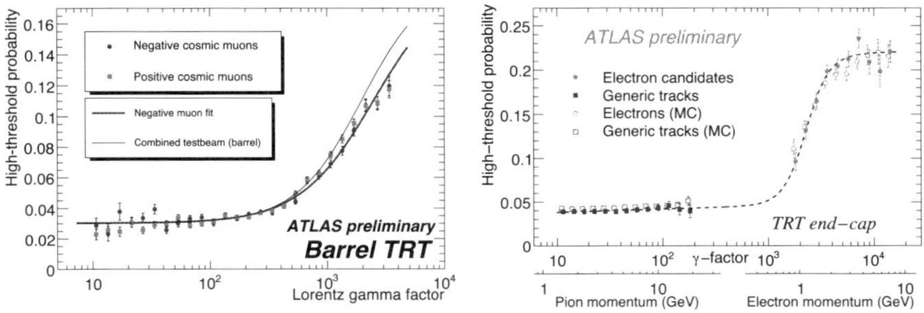

Figure 7: TRT transition radiation for cosmic-ray muons (left), and comparing hadrons (generic tracks) and electron candidates in collision data (right).

Transition radiation in the TRT is intended to be used for electron-pion separation. However, the highest energy cosmic rays in ATLAS had a sufficiently high Lorentz gamma factor to be able to see the turn on of transition radiation, resulting in a larger fraction of high threshold hits, as shown in Figure 7. In the same figure, a transition radiation signal from electron candidates from converted photons in pp collisions is also shown.

Combining transition radiation information with the ratio between the track momentum and the energy of an associated electromagnetic calorimeter cluster allowed a sample of delta rays from cosmic rays to be identified.

Neutral kaons and lambdas have been identified by combining oppositely charged tracks with a common vertex, under either a pion-pion or pion-proton hypothesis, to construct the invariant mass. The K_S^0 mass peak is shown in Figure 8. In the same figure, the result of combining photon pairs to reconstruct

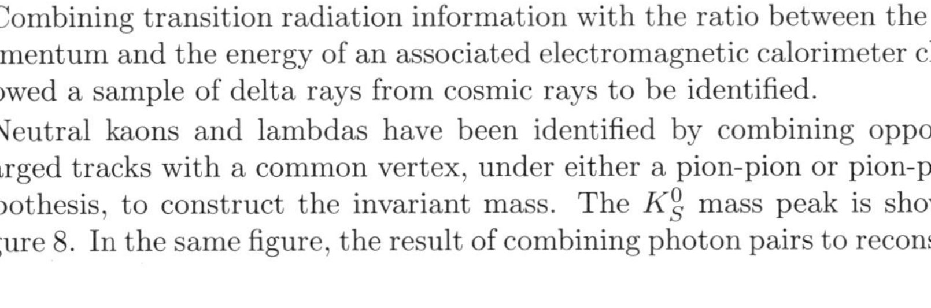

Figure 8: Mass peaks formed from two charged tracks (K_S^0, left) or two photons (π^0, right).

neutral pions is also shown.

Algorithms to identify and measure jet energies and missing transverse energy were studied in cosmic ray events, randomly trigged noise events and collision events. Examples from collision data of the very good agreement between data and simulation seen in all the variables studied is given in Figure 9.

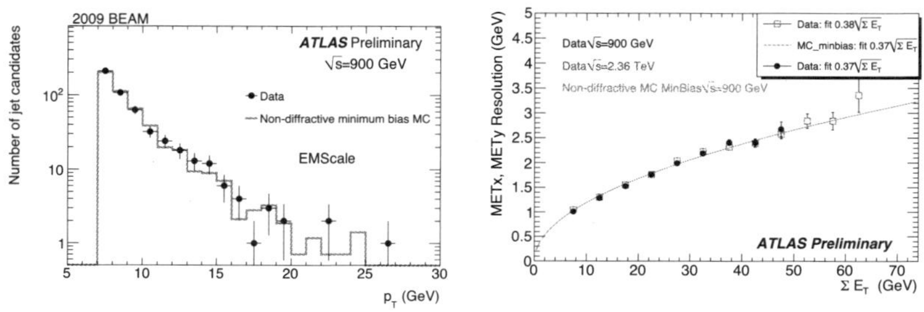

Figure 9: Transverse momentum spectrum of jets (left), and the resolution of the transverse momentum in collision events as a function of the total transverse energy (right).

4 Conclusion

The detailed preparation of the ATLAS experiment with cosmic rays has resulted in a very well understood detector, equally well modelled by Monte Carlo simulation. The first measurements of LHC pp collision events at 900 GeV and 2.36 TeV at the end of 2009 proved this remarkable state of readiness. The experiment took data with high efficiency, and has demonstrated that the variables to identify charged leptons, photons, jets and missing transverse energy are all under control. The collaboration is ready and eager to collect a large data sample of 7 TeV centre-of-mass energy collisions in 2010.

Acknowledgments

I would like to thank the conference organisers for making us all so welcome in Moscow, and for a very enjoyable and informative conference.

References

[1] The ATLAS Collaboration, G. Aad et al., The ATLAS Experiment at the CERN Large Hadron Collider, JINST 3 (2008) S08003.

EARLY PHYSICS WITH ATLAS AT LHC

Bellisario Esposito [a] - On behalf of the ATLAS Collaboration
Istituto Nazionale di Fisica Nucleare, Laboratori Nazionali di Frascati, 00044 Frascati (Italy)

Abstract. The ATLAS experiment is aimed at the exploitation of the full physics potential of LHC. The 2009 - 2010 LHC run will provide the first collision data at an energy above the Tevatron energy. The detector calibration with physics processes and the main physics measurements possible with early data are presented.

1 Introduction

The LHC collider is designed for a proton-proton collision energy of 7+7 TeV and a luminosity of $10^{34}\text{cm}^{-2}\text{s}^{-1}$ in order to provide access to the experimentation at an energy scale high enough to make it possible to clarify the mechanism of the spontaneous gauge symmetry breaking and the origin of the particle masses. In addition, new physics expected by models beyond the Standard Model (SM), or unexpected, can take place at such high energy and intensity. The ATLAS experiment aims at covering the full physics potential offered by LHC [1,2]. The ATLAS detector is fully operational [3] and ready to start the campaign of measurements foreseen in its physics program.
The physics perspectives of the first LHC run depend on the performance of the LHC, namely on which is the collision energy and what integrated luminosity is delivered. At the moment of writing these proceedings, data has been collected in the Nov.-Dec. 2009 run ($\sim$ 920k minimum bias events at $\sqrt{s} = 900\,\text{GeV}$ and $\sim$ 34k minimum bias events at $\sqrt{s} = 2.36\,\text{TeV}$) and the plan is to have a run in 2010-2011 at 3.5+3.5 TeV and reach an integrated luminosity of $1\,\text{fb}^{-1}$. The scenario discussed here corresponds to what was expected at the time of the conference: beam energy between 3.5 and 5 TeV, luminosity $\sim 10^{31}$-$10^{32}\,\text{cm}^{-2}\text{s}^{-1}$, integrated luminosity $\sim 100\,\text{pb}^{-1}$.
Based on the integrated luminosity expected, a discussion of the physics processes observable is made in Section 2. In Section 3 the use of physics processes for the in-situ calibration is discussed. Section 4 is devoted to report on the main early physics measurements.

2 Physics processes observable with early data

The main known processes (minimum bias, high p_T jet production, $b\bar{b}$, $c\bar{c}$, W, Z and top) have large cross section. With $100\,\text{pb}^{-1}$ of integrated luminosity large enough statistics can be collected to perform interesting studies of these processes in a new energy regime, to improve the knowledge of the PDF using jets, W and Z events, and to tune the Monte Carlo generator.

[a] e-mail: bellisario.esposito@lnf.infn.it

Most of the interesting discovery processes, like Higgs production, have cross section very low and for their discovery a much larger integrated luminosity is required. There are however also new processes like Z′ production or gluino or squark production in supersymmetry for which the expected cross section is such that with $100\,\mathrm{pb}^{-1}$ there are chances that they are observed, or a significant limit can be established on their mass.

3 Detector performance and in-situ calibration

The ATLAS detector has been designed to provide excellent performance in the measurement of the charged particle tracks in the Inner Detector, the muon tracks in the Muon Detector and in the calorimetric energy measurement of electron, photon and jets [1,2].

	Initial	Ultimate	Samples for calibration
ECAL uniformity	~2.5%	0.7%	Isolated e, Z →ee
e/γ Energy-scale	2 – 3%	≤0.1%	J/ψ, Z →ee, E/p for e
Jet Energy-scale	5 – 6%	1%	γ/Z+1j, W →jj in $t\bar{t}$ events
ID alignment	20 – 200 μm	5 μm	Tracks, isolated μ, Z → $\mu\mu$
Muon alignment	40 – 1000 μm	40 μm	Straight μ, Z → $\mu\mu$

Table 1: Initial and ultimate detector performances.

With the calibration and performance studies of the various subdetectors with test beam and cosmic rays, the so-called "day-1" performance, which is already very good, has been obtained. In order to obtain the ultimate design performance it is necessary to perform a calibration of the detector using collision events such as Z → $\mu\mu$, Z →e e, J/ψ, γ/Z +jet,W →jj in top events etc.. Table 1 summarizes the initial, obtained, and the ultimate, expected, detector performance and the calibration processes to be used.

4 Early physics results

With the expected integrated luminosity many studies can be performed both on standard processes and for the search of new physics. The main expected physics results are discussed in the following subsections.

4.1 Minimum Bias

The goals of the study of the minimum bias events at LHC are to measure the properties of the inelastic pp interaction processes in a new energy regime, and

to determine the characteristics of the background at high luminosity, due to pile-up events. The detector performances required for those studies are an unbiased trigger and tracking efficiency at low p_T.
The results expected for the measurement of the charged particle pseudorapidity η in minimum bias events using the low p_T tracking algorithm are reported in Figure 1. The systematic uncertainty coming from the trigger and selection efficiency correction and from the tracking efficiency correction is about 6-8%. The results of these measurements will allow to disentangle between different expectations based on different extrapolations from low energy data. These expectations have indeed a large uncertainty, as shown in Figure 2.

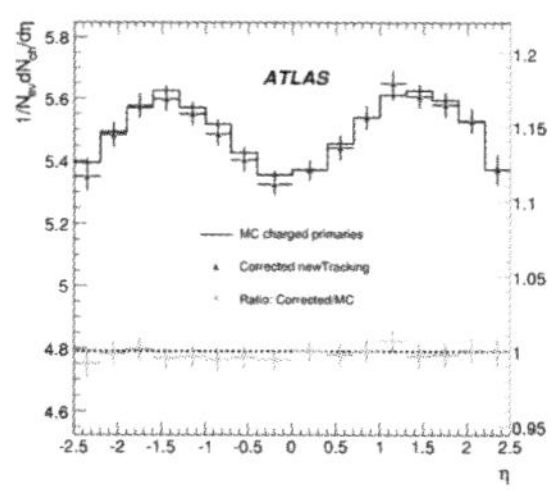

Figure 1: Charged particles pseudorapidity distributions. MC truth (full line) and the ratio of the analysis result over the MC (at the bottom) are also shown.

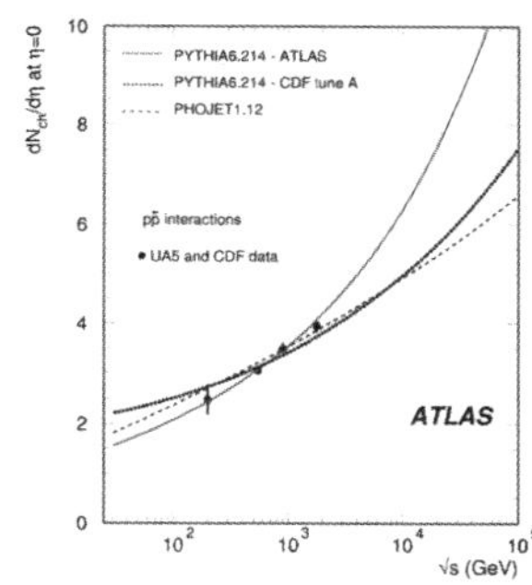

Figure 2: Central charged particle density for non-single diffractive inelastic events as a function of energy.

4.2 QCD Jet Physics

The LHC energy opens up a much higher p_T region with respect to the Tevatron (Figure 3). The goals of the study of the high p_T jet events are to measure the properties of the very hard pp interaction processes, to look for deviations from QCD predictions due to new physics (quark substructure, resonant production, large extra dimensions) and to determine the characteristics of the background from QCD events for the observation of other processes.
The use of a jet algorithm appropriate for comparison with theoretical calculations (colinear and infrared safe) is required, and the jet energy scale (JES) has to be calibrated accurately for detector effects (non compensation, noise, cracks) and for physics effects (clustering, fragmentation, Initial State Radiation (ISR), Final State Radiation (FSR), Underlying Event (UE)).
In-situ calibration with physics processes (di-jet, γ / Z + jet, multi-jet, W $\rightarrow$jj) is used. The calibration with γ-jet events is illustrated by the plots in Figure 4.

With $100\,\text{pb}^{-1}$ a statistical uncertainty on JES of 1-2 % for $100\text{-}200 < p_T < 500\,\text{GeV}$ is obtained. The systematics from physics is 1-2 % (ISR/FSR, UE).

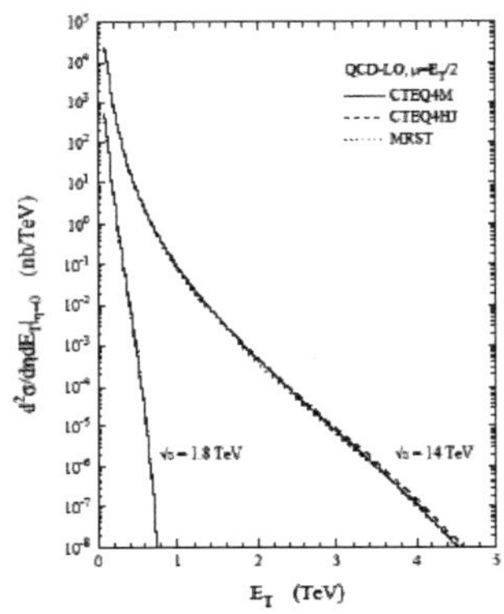

Figure 3: Inclusive differential jet cross section as a function of E_T.

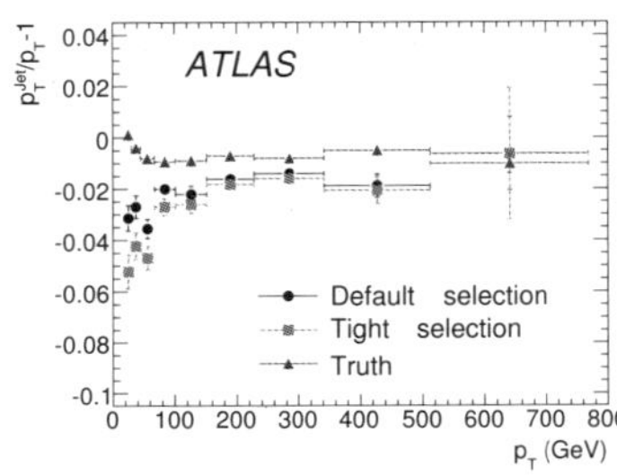

Figure 4: The most probable value of γ/jet p_T balance as a function of p_T^{γ}.

4.3 W *and* Z *Physics*

The goals of the study of the W and Z events are to measure their production cross-sections known theoretically with uncertainty at the level of 1 %, to measure p_T distribution to probe QCD initial parton radiation, and to measure rapidity distribution to probe parton density functions (PDF).

The results of an analysis to measure the W and Z cross sections with $50\,\text{pb}^{-1}$

	N(x10^4)	B(x10^4)	σ (pb)
$W \to e\nu$	22.67 ± 0.04	0.61 ± 0.92	$20520 \pm 40 \pm 1060$
$W \to \mu\nu$	30.04 ± 0.05	2.01 ± 0.12	$20530 \pm 40 \pm 630$
$Z \to ee$	2.71 ± 0.02	0.23 ± 0.04	$2016 \pm 16 \pm 83$
$Z \to \mu\mu$	2.57 ± 0.02	0.010 ± 0.002	$2016 \pm 16 \pm 76$

Table 2: Signal (N), Background (B) events and cross section measurement with integrated luminosity of $50\,\text{pb}^{-1}$.

are reported in Table 2. W and Z events can also be used to perform in-situ detector calibration (absolute energy and momentum scale, resolution, trigger and reconstruction efficiency).

4.4 *Top Physics*

The goals of the study of top-quark physics are to measure $t\bar{t}$ cross-section and to study top properties and decay. The $t\bar{t}$ events can also provide samples of b-jet and W $\rightarrow$j j , very useful for the in-situ detector calibration (b-tagging efficiency, light jet energy scale).
The $t\bar{t}$ signal can be observed in the single lepton channel. The three jet top mass peak and the two jet W mass peak can be observed with a simple cut selection and no b-tagging (Figure 5 and 6). The $t\bar{t}$ cross section measurement with 10-15% accuracy is possible with 100 pb^{-1}. The $t\bar{t}$ signal can also be observed in the di-lepton channel with lower statistics but systematic uncertainties smaller than for the single lepton channel.

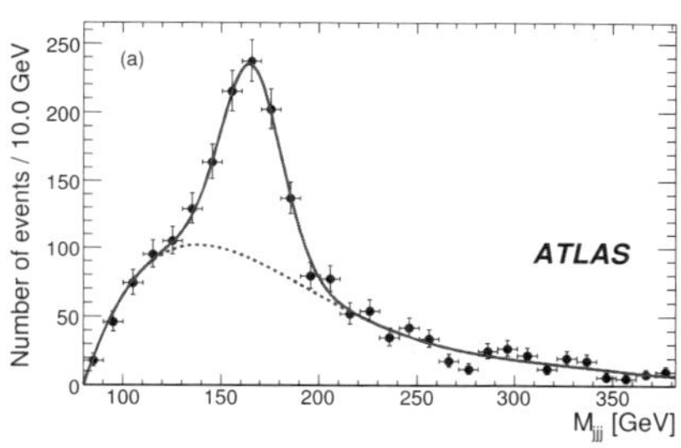

Figure 5: Three jet mass distribution. The fit to the top signal (gaussian) and to the background (Chebychev polynomial) are indicated.

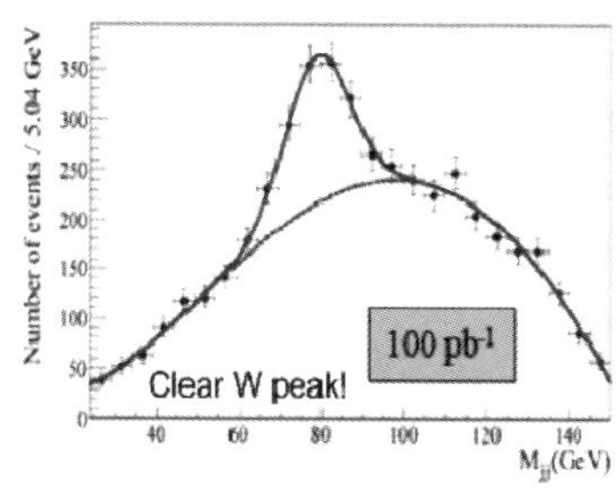

Figure 6: Two jet mass distribution. The fit to the W signal (gaussian) and to the background (Chebychev polynomial) are indicated.

4.5 *Early discoveries of New Physics*

The LHC run will not only provide data to study known processes in a new energy regime but will also give the chance of new discoveries even with the initial expected luminosity. In fact with integrated luminosity of 100 pb^{-1} early discoveries are possible.

Z′ searches

A Z′ signal would show up as a (narrow) mass peak above small and smooth SM background . The discovery for mass up to 1 TeV is possible with 100 pb^{-1} at $\sqrt{s}$=10 TeV (Figure 7). The observation of Z $\rightarrow$ll signal does not require ultimate detector performance.

Supersymmetry (SUSY) searches

Squarks and gluinos are produced via strong interactions and have large cross-section. Their decay produces spectacular final states with many jets, leptons, missing transverse energy. The results of a simulation study at $\sqrt{s} = 10\,\text{TeV}$ are summarized in Figure 8. A discovery up to squark and gluino masses of 600-700 GeV is possible with $200\,\text{pb}^{-1}$. This however requires a very good understanding of the detector effects and of the physics background. The $E_{\text{T}}^{\text{miss}}$ can be checked with known processes, and data-driven methods to estimate the background from the SM processes are to be used.

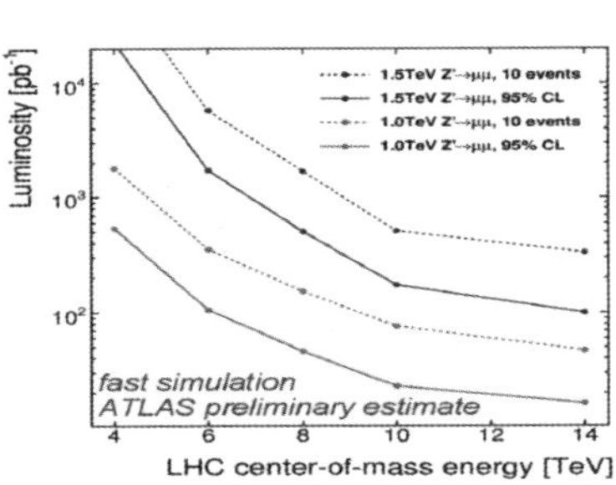

Figure 7: Integrated luminosity needed for Z′ discovery as a function of center-of-mass energies.

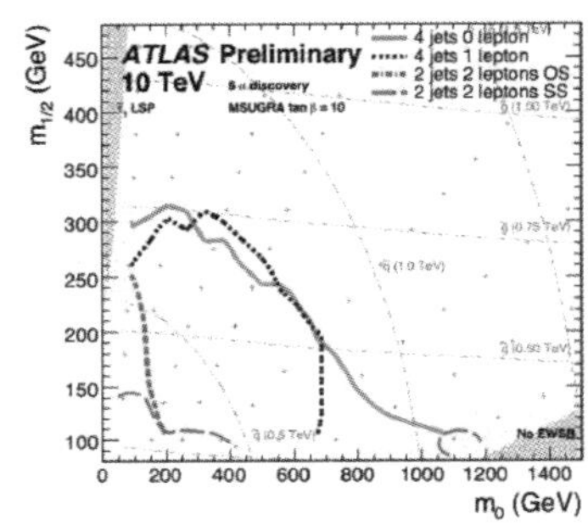

Figure 8: SUSY discovery reach in the plan of universal scalar mass m_0 and universal gaugino mass $m_{1/2}$.

5 Conclusions

The study of a variety of SM processes in a new energy regime and the search for some of the new particles foreseen by the models beyond the SM are the physics prospects of the first LHC run. The analysis of the data collected will also provide the verification and the tuning of the ATLAS detector calibration, necessary to improve the performances and reduce the systematics.
With a well understood and calibrated detector, unexpected effects possibly leading to surprising discoveries can be looked upon.

References

[1] ATLAS Collaboration, JINST **3** (2008) S08003.

[2] ATLAS Collaboration, arXiv:0901.0512, CERN-OPEN-2008-020.

[3] P. Wells, these proceedings.

THERMAL EFFECTS FOR QUARK AND GLUON DISTRIBUTIONS IN HEAVY-ION COLLISIONS AT NICA

G.I.Lykasov [a], A.N.Sissakian [b], A.S.Sorin [c], O.V.Teryaev [d]

Joint Institute for Nuclear Research, 141980 Dubna, Russia

Abstract. In-medium effects for distributions of quarks and gluons in central A+A collisions are considered. We suggest a duality principle, which means similarity of thermal spectra of hadrons produced in heavy-ion collisions and inclusive spectra which can be obtained within the dynamic quantum scattering theory. Within the suggested approach we show that the mean square of the transverse momentum for these partons grows and then saturates when the initial energy increases. It leads to the energy dependence of hadron transverse mass spectra which is similar to that observed in heavy ion collisions.

1 Introduction

Searching for a new physics in heavy-ion collisions at NICA, AGS, SPS and RHIC energies has led to intense theoretical and experimental activities in this field of research [1]. In this respect the search for signals of a possible transition of hadrons into the QCD predicted phase of deconfined quarks and gluons, quark-gluon plasma (QGP), is of particular interest. One of these signals can be the recent experimental observation of the transverse-mass spectra of kaons and pions from central Au+Au and Pb+Pb collisions which revealed "anomalous" dependence on the incident energy. The inverse slope parameter of the transverse mass distribution (the so called effective transverse temperature) at the mid-rapidity rather fast increases with incident energy in the AGS domain [2], then saturates at the SPS [3] and RHIC energies [4].

In this paper we would like to discuss the physical meaning of the so-called thermal spectra of hadrons produced in heavy-ion collisions, see for example [5, 6], and try to understand the dynamic reason of such inclusive spectra. Then we focus on a possible theoretical interpretation of the nontrivial energy dependence for the inverse slope parameter of the transverse mass spectra of mesons produced in central heavy-ion collisions.

2 Duality principle

According to many experimental data, inclusive spectra of hadrons produced in heavy-ion collisions can be fitted by the Fermi-Dirac distribution, corresponding to the thermodynamic equilibrium (**LE**) for the system of final hadrons, see

[a] e-mail: lykasov@jinr.ru
[b] e-mail: sisakian@jinr.ru
[c] e-mail: sorin@theor.jinr.ru
[d] e-mail: teryaev@theor.jinr.ru

for example [5, 6]

$$f_h^A = C_T^A \left\{\exp((\epsilon_h - \mu_h)/T) \pm 1\right\}^{-1} , \qquad (1)$$

where + is for fermions and − is for bosons, ϵ_h and μ_h are the kinetic energy and the chemical potential of the hadron h, T is the temperature, C_T^A is the normalization coefficient depending on T. Actually, the parameter T depends on the incident energy $\sqrt{s}$ in the $N-N$ c.m.s. For mesons simplifying this case we can assume that $\mu_h \simeq 0$, (in fact, it generally cannot be strictly zero [7]); then Eq.(1) is usually presented in the form

$$f_h^A \simeq C_T^A \exp(-\epsilon_h/T) . \qquad (2)$$

On the other hand, according to the Regge theory and the $1/N$ expansion in QCD, the inclusive spectrum of hadrons produced, for example in $N-N$ collisions at high energies, has the scaling form, e.g., it depends only on M_X^2/s, where M_X is the missing mass of produced hadrons, s is the initial energy squared in the $N-N$ c.m.s., and $M_X^2/s = 1 - x_r$, where $x_r = 2E_h^*/\sqrt{s}$ is the radial Feynman variable, E_h^* is the energy of the hadron h in the $N-N$ c.m.s. For example, the quantum scattering theory and the fit of the experimental data for inclusive meson spectra at low x_r results in

$$\rho_m^{NN}(x_r) \sim C_N(1-x_r)^{d_N} \qquad (3)$$

If $x_r << 1$, Eq.(3) can be presented in the exponential form

$$\rho_m^{NN} \sim C_N exp(-d_N x_r) \qquad (4)$$

Inserting the form for x_r in Eq.(4), we get the inclusive spectrum of mesons in the form similar to that of the thermal spectrum given by Eq.(2)

$$\rho_m^{NN} = C_N \exp(-d_N x_r) \equiv C_N \exp(-\frac{E_h^*}{T_s^N}) , \qquad (5)$$

where $T_s^N = \sqrt{s}/2d_N$. However, in contrast to Eq.(2), the form of the inclusive spectrum of mesons produced in $N-N$ collisions given by Eq.(5) does not assume introduction of any temperature of mesons like T. Figure 1 illustrates the approximate equivalence between ρ_m^{NN} given by Eq.(3) and ρ_m^{NN} given by Eq.(5). One can see from Fig.1 that at high energies these two forms for the meson spectrum are very similar to each other to the meson energies about a few GeV. Therefore, ρ_m^{NN} can be presented in the exponential form at low and even moderate energies E_h^*.

Let us assume that in central $A-A$ collisions at high energies in the first $N-N$ interaction at some time the mini-jet consisting of $(q\bar{q})$ pairs is created and then pions are produced in the central rapidity region. We also suggest

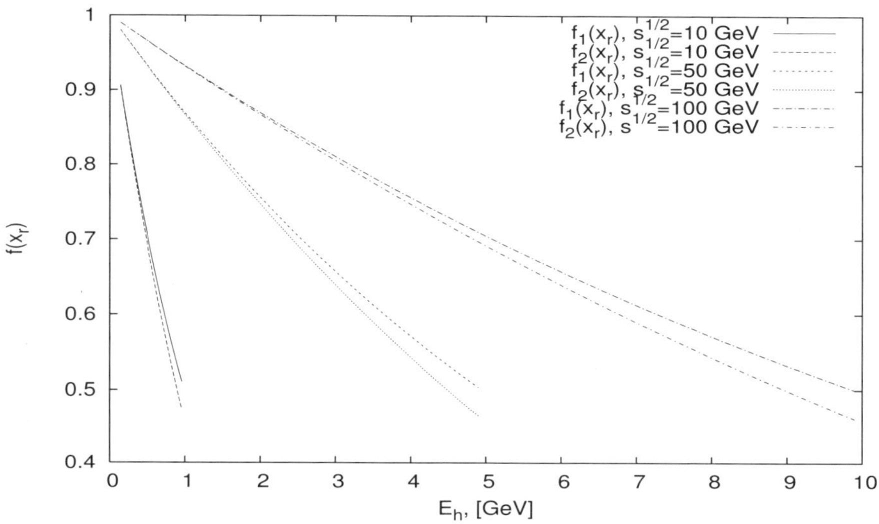

Figure 1: $f_1(x_r) = exp(-dx_r)$ and $f_2(x_r) = (1 - x_r)^d$ as functions of E_h.

that the distribution of these pairs in the $(q\bar{q})$ mini-jet has the form similar to the one given by Eq.(5)

$$f_{q\bar{q}}^{jet} = C_A(1 - x_r)^{d_A} \simeq C_A \exp(-\frac{E_h^*}{T_s^A}) , \tag{6}$$

where $T_s^A = \sqrt{s}/2d_A$. In the general case, the parameter d_A is not the same as d_N which can be found from the quantum scattering theory or fitting the experimental data on inclusive spectra of mesons produced in $N - N$ collisions. Let us call the assumption corresponding to Eq.(6) the **Dynamic ansatz (DA)**. One can suggest the **duality principle** which is the similarity of thermal spectra given by Eq.(2) and the dynamical spectra given by Eq.(6). To find the form for d_A one can use the approach suggested by Kuti, Weiskopf [8] for the calculation of parton distribution in a nucleon. One can show that d_A is proportional to the number n of $q\bar{q}$ pairs in the mini-jet, e.g., $d_A \sim n$. To estimate the in-medium effects we replace n by the mean multiplicity $< n >_\pi^{NN}$ of pions produced in NN collision, e.g., $d_A \simeq d_0 < n >_\pi^{NN}$.

3 Parton distribution in medium

Recently the parton distribution in a medium was analyzed on the assumption of the local thermodynamic equilibrium for quark objects like hadrons produced in heavy ion central collisions [9]. It was shown that, for example, the valence quark distribution in the quark object like the hadron h, which is in local thermodynamic equilibrium with surrounding nuclear matter, can be calculated by the following equation:

$$f_{q_v}^A(x, \mathbf{p}_t) = \int_0^1 dx_1 \int_0^1 dx_h \int d^2p_{1t}d^2p_{ht}q_v^h(x, \mathbf{p}_t)q_r^h(x_1, \mathbf{p}_{1t}) \times \\ f_h^A(x_h, \mathbf{p}_{ht}) \times \delta(x + x_1 - x_h)\delta^{(2)}(\mathbf{p}_t + \mathbf{p}_{1t} - \mathbf{p}_{ht}) \; , \tag{7}$$

where $f_h^A(x_h, \mathbf{p}_{ht})$ is the distribution of quark objects like hadrons locally equilibrated in a medium (**LE**); q_v^h, q_r^h are the probabilities to find the valance quark and other partons (valence, sea quarks (antiquarks) and gluons) in h; x_1, x_h, x are the Feynman variables, $\mathbf{p}_t, \mathbf{p}_{1t}, \mathbf{p}_{ht}$ are the transverse momenta. The thermodynamic distribution like Eq.(2) was assumed in [9] for $f_h^A(x_h, \mathbf{p}_{ht})$. The same form for $f_{q_v}^A$ can be obtained suggesting the dynamic distribution (**DA**) for f_h^A given by Eq.(6) instead of Eq.(2). Assuming the factorized form for $f_q^h(x, p_t) = f_q(x)g_q(p_t)$ we have approximately [9] the following form for the mean transverse momentum squared of the valence quark in a medium:

$$< p_{q,t}^2(x \simeq 0) >_{q,appr.}^A \quad \simeq \quad \frac{< p_t^2 >_q^h + \tilde{T}\sqrt{m_h^2 + s/4}}{1 + \tilde{T}\sqrt{m_h^2 + s \; / \; 4}/(2 < p_t^2 >_q^h)} \; , \tag{8}$$

where $\tilde{T} = T$ for Eq.(2) (**LE**) and $\tilde{T} = T_s = \sqrt{s}/2d_A$ for Eq.(6) (**DA**); $< p_t^2 >_q^h$ is the mean value for the transverse momentum squared of the quark in the free hadron h. As is seen from Eq.(8) $< p_{q,t}^2 >_{q,appr.}^A$ grows when $\sqrt{s}$ increases and then saturates, its more careful calculation is presented in [9]. Note that in the **LE** case $\sqrt{s}$ is some scale energy which cannot be equal to the initial energy $\sqrt{s_{NN}}$ [9], whereas in the **DA** case it is the same as $\sqrt{s_{NN}}$. For mesons produced in central $A-A$ collisions we have similar broadening for the hadron p_t-spectrum [9]

$$< p_{h_1t}^2 >_{appr.}^{AA} \quad \simeq \quad \frac{< p_{h_1t}^2 >^{NN} /(1 + r) + \tilde{T}\sqrt{m_m^2 + s/4}}{1 + \tilde{T}\sqrt{(m_m^2 + s/4)}(1 + r) \; / \; (2 < p_{h_1t}^2 >^{NN})} \\ + \frac{< p_{h_1t}^2 >^{NN}}{r} \; , \tag{9}$$

where $< p_{h_1t}^2 >$ is the mean value for the transverse momentum squared of the meson h_1 produced in the central heavy-ion collision, $< p_t^2 >_q^m$ is the same quantity for a quark in a medium, $r = \gamma_c/\gamma_q$, $\tilde{T} = T$ (**LE**) or $\tilde{T} = \sqrt{s}/2d_A$. Here γ_q and γ_c are the slopes in the Gauss form of the p_t dependence for the quark distribution in the hadron h and its fragmentation function, see details in [9]. As is seen from Eqs.(8,9), the saturation properties for $< p_{q,t}^2 >_q^A$ and

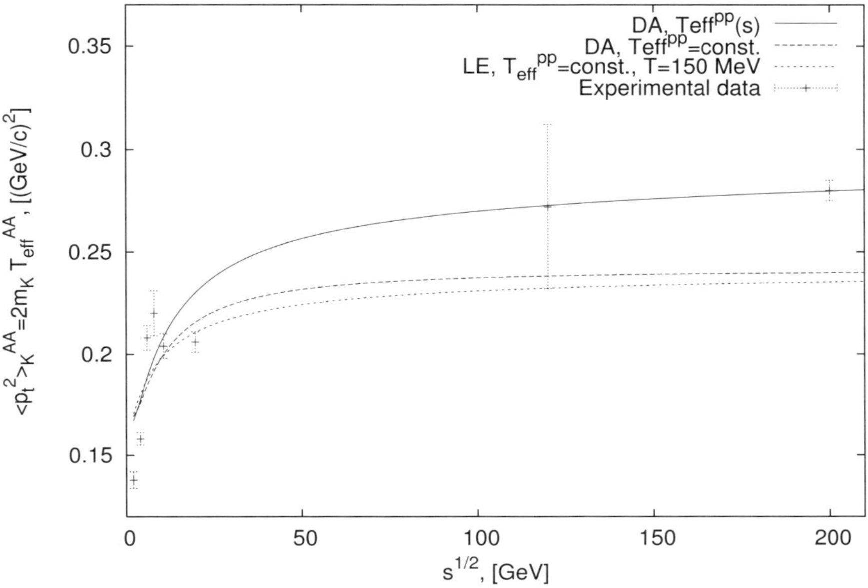

Figure 2: The mean transverse momentum squared of the K^--meson produced in the central A-A collision. The solid line corresponds to the DA, when the effective slope for the transverse mass spectrum of kaons produced in the $p-p$ collision T^{pp}_{eff} depends on $\sqrt{s}$ [11]. The long dashed line corresponds to the DA, when $T^{pp}_{eff} = constant$, whereas the short dashed line corresponds to the LE, when $T^{pp}_{eff} = constant$ and the temperature $T = 150$ MeV. Experimental data were taken from [2–4].

$< p^2_{h_1 t} >^{AA}$ at high $\sqrt{s}$ do not depend on the values of $\tilde{T}$, whereas the growth of these quantities at $\sqrt{s} \leq 20 - 30$ GeV is very sensitive to the value of d_A. To describe the experimental data on the transverse momentum squared of K-mesons produced in central $A - A$ collisions we took $d_0 = 0.5$ and the energy dependence for $< n >^{NN}_{\pi}$ from [10]. We included also the energy dependence for the mean squared of the transverse momentum of kaons produced in the $N - N$ collision, see [11] and references there in.

In Fig.2 we present our estimation for the mean transverse momentum squared of the K^--meson produced in the A-A collision. The suggested approach results in the saturation of the effective slope T_{eff} for the transverse mass spectrum of mesons produced in central heavy-ion collisions that is directly related to the quantity presented in Fig.2. In contrast to this the thermodynamic models predict the increase in T_{eff} when $\sqrt{s_{NN}}$ increases even to very high energies [12]. Therefore, the presented results can be verified by more careful measurements at the SPS energy and future experiments at the LHC and NICA.

Acknowledgments

The authors are grateful to A.Andronic, P.Braun-Munzinger, A.V.Efremov, L.L.Frankfurt, M.Gazdzicki, S.B.Gerasimov, J.Cleymans, A.B.Kaidalov, H.Satz, Yu.Sinyukov and V.D.Toneev for very useful discussions. This work was supported in part by RFBR project No 08-02-01003 and by the special program of the Ministry of Education and Science of the Russian Federation (grant RNP.2.1.1.5409).

References

[1] Proceedings of Quark Matter '2005', edited by T. Csörgö, D. Gabor, P.Lévai, and G. Papp, Nucl.Phys. A**774** (2006); Proc.'QM 2006', ed. by Y. G. Ma, E. K. Wang, X. Cai, H. Z. Huang, X. N. Wang and Z. Y. Zhu, J.Phys. G**34** (2007).

[2] L. Ahle *et. al.*, E866 and E917 Collaboration, Phys. Let. B**476**, 1 (2000); B**490**, 53 (2000).

[3] S. V. Afanasiev *et al.* (NA49 Collab.), Phys.Rev. C**66**, 054902(2002); C. Alt et al., J. Phys. G**30**, S119 (2004); M. Gazdzicki, *et al.*, J. Phys. G**30**, S701 (2004).

[4] C. Adler*et al.*, STAR Collaboration, nucl-ex/0206008; O. Barannikova *et al.*, Nucl. Phys. A**715**, 458 (2003); K. Filimonov *et al.*, hep-ex/0306056; D. Ouerdane *et al*, BRAHMS Collaboration, Nucl. Phys. A**715**,478 (2003); J. H. Lee *et al.*, J. Phys. G**30**, S85 (2004);
S. S. Adler *et al.*, PHENIX Collaboration, nucl-ex/0307010 nucl-ex/0307022.

[5] A. Andronic, P. Braun-Munzinger, J. Stachel, Nucl.Phys.A**772**, 167(2006). e-Print: nucl-th/0511071

[6] J.Cleymans, J.Phys.G35:044017,2008; J.Cleymans, R.Sahoo, D.K.Strivastava, S.Wheaton, Eur.Phys.J.ST 155:13-18,2008; J.Cleymans *et al.*, hep-ph/0803.3940

[7] S.V.Akkelin, P.Braun-Munzinger, Yu.M.Sinyukov, Nucl.Phys.A**710**, 439 (2002); nucl-th/0111050

[8] J.Kuti and V.F.Weiskopf, Phys.Rev. D**4**, 3418 (1971).

[9] G.I.Lykasov, A.N.Sissakian, A.S.Sorin and V.D.Toneev Phys.Atom.Nucl **71**, 1600 (2008).

[10] B.B.Back, et al., Phys.Rev.C**69**,054901 (2004).

[11] A.Andronic, P.Braun-Munzinger, J.Stachel, Nucl.Phys. A**772**, 167 (2006).

[12] Yu.M.Sinyukov, Acta Phys.Pol. B**37**, 3343 (2006).

CMS: STATUS AND PHYSICS PROSPECTS

Cristina Biino [a]

Istituto Nazionale di Fisica Nucleare, via P. Giuria 1, 10125 Torino, Italy

Abstract. CMS is one of the two general purpose particle detector experiments located at the Large Hadron Collider (LHC) at CERN. After a brief overview of the CMS experiment, the final detector installation and its commissioning is reviewed. Some of the achievements of the past year are highlighted. Preliminary subdetectors performance results from the large data sample of cosmic ray collected during 2008 and 2009 and events produced by the first LHC beams are presented. The early physics potential is illustrated.

1 Introduction

The CMS experiment [1] was completely assembled in the fall of 2008 after more than a decade of design, construction and installation. During the last two years, cosmic ray data were collected on a regular basis. These data have enabled CMS to align the detector components, both spatially and temporally and to establish the relative alignment of the CMS tracking and muon systems. In addition, the CMS calorimetry has been crosschecked with test beam data, thus providing an initial energy calibration of CMS calorimetry to about 5%. The CMS magnet has been powered and field mapped. The trigger and data acquisition systems have been run at full speed. The data analysis system has been exercised at full design bandwidth for Tier0, Tier1 and Tier2 sites. Monte Carlo simulation of the CMS detector at a detailed geometric level has been tuned using test-beam and other production data to provide a realistic model of the CMS detector prior to first collisions. A few preliminary results from the subdetectors commissioning during the long COSMIC Muon Run at Four Tesla (CRAFT) field are presented in this paper.

In 2008 the LHC beam was commissioned and successfully steered through the CMS detector. First commissioning results with real beam data are shown focusing on detector performance.

The LHC collider run 2009-2010 will restart in October 2009 after a short beam commissioning phase. Beam intensity, bunch spacing and instantaneous luminosity will be gradually increased. The run is expected to deliver $\simeq$ 300 pb^{-1} integrated luminosity at $\sqrt{s} \simeq 7$ TeV. The early physics potential and a roadmap toward discoveries is highlighted. [2]

2 The CMS Detector

A detailed description of the Compact Muon Solenoid (CMS) experiment can be found elsewhere Ref. [1] and the overall layout is shown in Fig.1, left side. At design luminosity ($10^{34} cm^{-2} s^{-1}$) the 7 TeV proton beams will cross inside

[a] e-mail: cristina.biino@to.infn.it

CMS every 25 ns providing an average of 20 collisions per crossing, leading to up to 1000 tracks in the detector. In order to sustain such a rate the detector has extremely high granularity. The overall size of the detector is

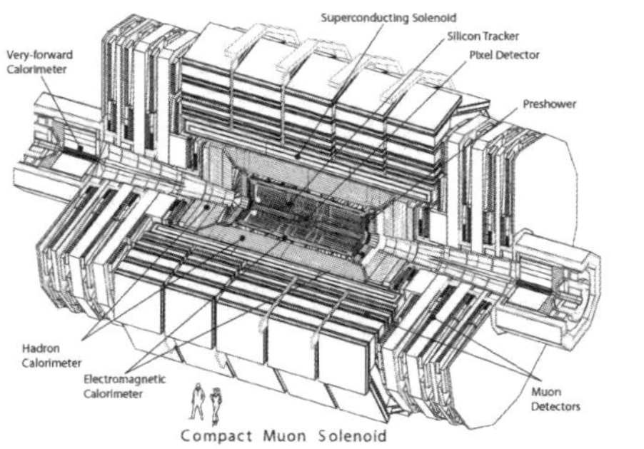

Figure 1: (left) A perspective view of the CMS detector. (right) The CMS detector during the installation of the silicon strip tracker, December 2007.

set by the muon tracking system that in turn makes use of the steel return yoke of the magnet. The central feature of the CMS apparatus is therefore the superconducting solenoid of 6 m internal diameter delivering a uniform magnetic field of 3.8 T. Immersed in the magnetic field are the inner tracker, the crystal electromagnetic calorimeter (ECAL) and the brass-scintillator hadron calorimeter (HCAL). The tracker consists of 66M silicon pixel channels and 10M silicon microstrip channels (SST) and measures charged particles within the pseudorapidity range $|\eta| < 2.5$. It provides an impact parameter resolution of about 100 μm and a transverse momentum resolution of about 0.7% for 1 GeV/c charged particles. During the early beam data taking period 98.4% of the Pixel and 97.2% of the SST channels were operational. The fraction of noisy Pixel channels was less than 10^{-5}. The signal-to-noise ratio in the SST depends on the sensor thickness and was measured to be between 28 and 36, consistent with the design expectations. The tracker was aligned using cosmic ray data in advance of the LHC commissioning. The precision achieved for the positions of the detector modules with respect to particle trajectories is 3-4 μm in the barrel for the coordinate in the bending plane. ECAL, made of about 76,000 lead tungstate scintillating crystals, has been designed to obtain an energy resolution of about 0.5% for high energy electromagnetic showers, with the Higgs decay to two photons as a benchmark channel. Muons are measured in different gas-ionization detectors (Drift Tubes, Cathode Strip Chambers and Resistive Plate Chambers) and the momentum resolution is 9% at 200 GeV and 15-40% at 1 TeV and, combined with the tracker, it becomes 1% at low p_t and 5% at 1 TeV. The overall detector is segmented in wheels and disks. It was assembled on surface and lowered in the collision hall slice by slice (see Fig. 1, right side).

Cosmic ray runs started early in 2007, first with only the data acquisition system itself, and then grew up to include almost all CMS at the end of August 2008. CMS ran in global mode as a single detector a few days per week and a full week per month logging more than 350 million cosmic triggers in the period March-August 2008.

3 Performance in LHC Beam Runs

On September 2008, LHC injected beam in the accelerator and in the following days the experiment accumulated many beam splash and beam halo events from circulating beams. Single beam shots of 2×10^9 protons at injection energy were sent onto a closed collimator about 150m upstream of CMS. Hundreds of thousand of scattered muons deposited several hundred TeV of energy in the calorimeters. We used these splash events to synchronize the trigger, to improve the calorimeters timing (to better than 1 ns) and validate the intercalibration of the ECAl endcaps channels. The first RF capture of the beam resulted in millions of orbits through CMS. Halo muons, mainly parallel to the beam, were reconstructed and were used to improve on alignment precision in the muon endcaps.

On September 19th an interconnection between two magnets broke leading to a premature shutdown of the LHC machine.

4 Cosmic Run at Four Tesla (CRAFT)

After the LHC incident CMS went back to cosmic operation. The aim of CRAFT was to run CMS for four weeks during 2008 fall with all subsystems participating, collecting data continuously to further gain operational experience. (see Fig.2, left side). We collected more than 370 million cosmic events.

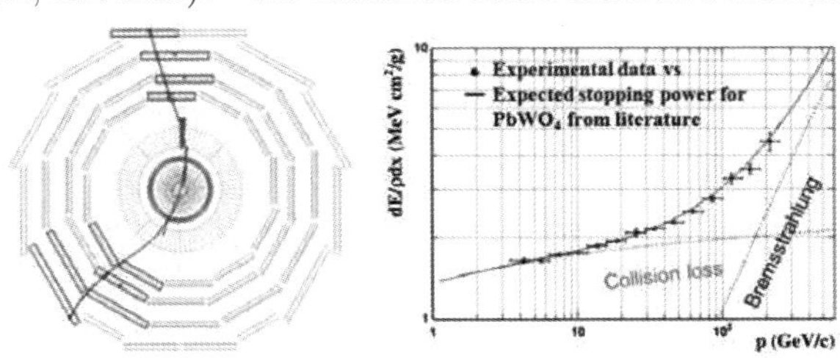

Figure 2: (left) Event display of a typical cosmic muon passing through the CMS detector. (right) ECAL stopping power for muons as a function of the muon momentum.

This data has been particularly useful to evaluate efficiencies, resolutions, improve on calibration and alignment constants. A lot of information has been extracted from these data and a series of paper is being prepared for publication. The study of energy deposition in ECAL crystals is one of such analysis.

We studied the stopping power of the CMS electromagnetic crystal calorimeter as a function of the muon momentum as measured in the tracker (see Fig.2, right side). The good agreement between the data and the theoretical curve proves the correctness of the tracker momentum scale and the ECAL energy scale, calibrated at test beams.

5 Early Physics Program

As we have seen, much of the detector commissioning has been done using cosmics and test-beams. In 2009 and 2010, with early beam, splash events, muons halos, we will repeat the first in-situ alignment and calibration, we will study the detector synchronization, and continue commissioning the trigger.

With first collisions it will be possible to study the detector and reconstruction performance for physics objets: muons, electrons, lepton rates, minimum bias events, jets, b-tags, τ, missing transverse energy (MET).

Everything other than the hard scattering process in a collision is referred as the underlying event. Soft interactions are not calculable from first principles. We will compare phenomenological models: different models consistent with Tevatron results show significant differences at LHC energies in the number of charged particles as a function of leading charged jet transverse momentum.

Soft hadronic interactions, minimum bias events will be the most common events. The measurement of their properties will be top priority in order to study reconstruction efficiency and understand background contribution to physics studies.

In general the understanding of Standard Model (SM) processes will be crucial in the early searches for new physics, therefore a considerable effort should be invested to measure them precisely. Even at low integrated luminosity such as 10-20 pb^{-1} and up to several 100 pb^{-1} we will continue measuring SM, we will record many di-jets, W, Z and top quark events. We will improve the understanding of physics objects, we will start to study the jet energy scale from $W \to jetjet$ and start an extensive use of b-tagging. We will understand and measure the background to SUSY and Higgs searches and start looking for possible extraordinary signatures, i.e. Z', W' resonances.With collisions at higher energy we will extend searches, explore large part of SUSY and start searches for resonances at about a few TeV. With luminosity in the 1000 pb^{-1} range we will enter the Higgs discovery era.

Leptons and photons play a crucial role at LHC in several fields. They are essential in Higgs decay channels, they can be signature of the decay of new heavy bosons, they play a role in SUSY and are, of course, central in the reconstruction of electroweak and QCD processes. CMS, featuring a finely grained, high resolution e.m. calorimeter and excellent muon tracking performances, is well equipped for the task of measuring these particles with high resolution.

No new physics will be possible before the rediscover of SM, that can be used as the ultimate detector commissioning tool. In the roadmap towards discoveries with leptons at LHC as a function of increasing luminosity (see Fig.3),

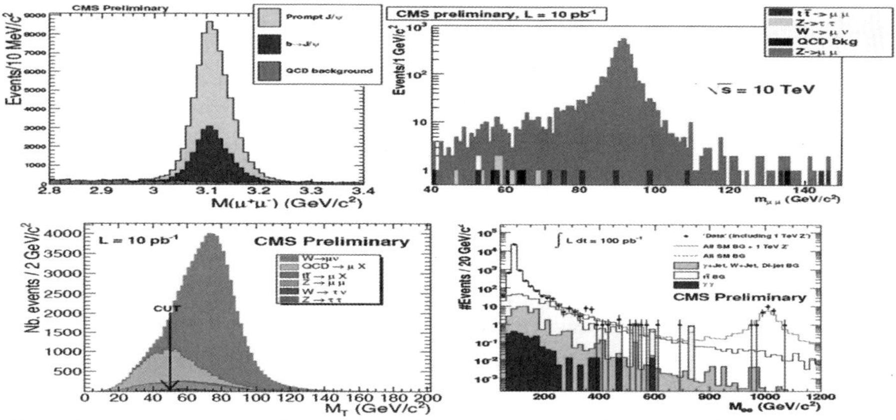

Figure 3: The roadmap to discoveries with leptons. From left to right: $J\psi$ Resonance Mass Distribution (few pb^{-1}). Z Boson Peak (10 pb^{-1}). W Cross Section Measurement (10 pb^{-1}) Drell-Yan Spectrum.

we can first measure low mass resonances (J/Ψ and Υ should be available with just a few pb^{-1} of integrated luminosity). We should establish important SM reference signals like the Z peak (order of 10 pb^{-1}) and W cross-section measurement. We will look for Drell-Yan spectrum beyond $M_{leptlept} > 700$ GeV^2. These SM measurements, at the beginning of the data taking, can be used for a detailed understanding of instrumental effects, detector performance, optimization of reconstruction algorithms, tune MonteCarlo detector simulation, understand the underlying events and the MET resolution.

The production of Z boson is foreseen to be used to estimate lepton reconstruction efficiencies and trigger efficiency from the data using a tag and probe method. This is done by selecting events with a lepton satisfying very stringent cuts (tag lepton) and then looking for a second lepton in the event which is consistent with originating from Z decay and passing the set of cuts for which one wants to determine the efficiency (probe lepton). W and Z production have several advantages: the cross section is large, of the order of nb, they have charged leptons in the final state which can be triggered on and their decay properties have been measured at previous accelerators to high precision. In Z events the background can be estimated from off resonance regions in the dilepton invariant mass spectrum (side bands). Since cross section theoretical calculations are very precise (1%) one can turn this measurement into a luminosity measurement.

One of the most promising signals for new physics in the first data are high mass lepton pairs. If these particles have SM like couplings to leptons and quarks then one expects sizable production cross section and BR to e^+e^- and $\mu^+\mu^-$ resulting in very clean signatures above a low and well defined Drell-Yan continuum background. Extending the knowledge beyond the Z boson reach relies on MonteCarlo extrapolation. Good understanding of the Drell-Yan spectrum from data will be a key check-point on the roadmap to discoveries.

The t-tbar production cross section at LHC at NLO is about a factor 100 higher at LHC compared to the Tevatron. LHC will be therefore a top quark factory and should allow to study the properties of top in great details, not only mass and cross section but also BR, coupling and rare decays. Final states of top pair events are determined by the fact that the decays are almost exclusively into W and a b-quark and by the branching fraction of the W boson. The b-tagging performance can be studied in these events.

If SUSY exists at the electroweak scale should be discovered at LHC but first we will have to overcome MET challenges.

A key question in particle physics is the origin of electroweak symmetry breaking. The most elegant explanation within the SM is a Higgs field with at least one scalar particle: the Higgs Boson. At high luminosity ($\simeq$ fb^{-1}) we will start to hunt for the Higgs but the potential of discovering at relatively early days of LHC operation strongly depends on its mass. As the luminosity starts reaching a few hundreds pb^{-1} we will reach 150-160 GeV sensitivity, comparable to the Tevatron.

6 Conclusions

The year 2008 saw the final achievement of almost 20 years of development and building of CMS. Since 2008 all subdetectors have been routinely in global readout and the stability of the running has been proven. A long cosmic data taking confirmed that all components are operational and close to their performance design specifications. The first physics analyses are only preparations for the physics goals for which CMS has been designed, but the understanding of the detector demonstrate its readiness for data taking. CMS has shown that can deliver high quality data, and CMS is more than ready for first LHC collisions data in 2009.

References

[1] R.Adolphi et al. (CMS Collaboration), *JINST* **3** S08004 (2008).

[2] CMS Collaboration, CMS Physics Technical Design Report, Volume 1: Detector Performance and Software, *CERN/LHCC* 2006-001 (2008); CMS Collaboration, CMS Physics Technical Design Report, Volume 2: Physics Performance, *J.Phys. G: Nucl.Part.Phys.* **34** 995 (2007).

Z' DISCOVERY POTENTIAL AT THE LHC FOR MODELS WITH CONTINUOUSLY DISTRIBUTED MASS

N.V.Krasnikov
INR RAN, Moscow 117312

Abstract. We study the Large Hadron Collider (LHC) discovery potential for Z' models with continuously distributed mass for $\sqrt{s} = 7,\ 10$ and $14\ TeV$ centre-of-mass energies. One of possible LHC signatures for such models is the existence of broad resonance structure in Drell-Yan reaction $pp \to Z' + ... \to l^+l^- + ...$.

In this note we study the Large Hadron Collider (LHC) discovery potential for Z' models [1] with large invisible decay width. [a] One of possible LHC signatures is the existence of broad resonance structure in Drell-Yan reaction $pp \to Z' + ... \to l^+l^- + ...$. The Z' models with broad resonance structure naturally arise in the context of Z' models with continuously distributed masses [4,5].[b] Also scenario with large invisible decay width can be realized if Z' boson couples to a hidden sector and Z' boson decays mainly into heavy neutral leptons (massive neutrino). We present the LHC discovery potential for Z' models with continuously distributed mass for $\sqrt{s} = 7,\ 10$ and $14\ TeV$ centre-of-mass energies [c]. In most standard Z' models [1] the total decay width Γ_{tot} is rather narrow. Typically $\Gamma_{tot} \leq O(10^{-2}) \cdot m_{Z'}$. The interaction of Z' boson with additional neutral fermions increases invisible Z' decay width and can lead to broad total decay width of the Z' boson.

In this paper we consider three Z' models. In model A the Z' boson interaction with quarks and leptons is the same as for standard Z boson. In model B the interaction of Z' boson with quark and lepton fields is determined by the current

$$J_\mu = \frac{g}{2}[\sum_{q,l}(\bar{q}\gamma_\mu(1+\gamma_5)q + \bar{l}\gamma_\mu(1+\gamma_5)l], \qquad (1)$$

where $g = 0.65$ is the Standard Model $SU(2)_L$ coupling constant. In model C the Z' boson interacts with (B-L) current

$$L_{mu} = g_{B-L}\sum_{q,l}[\frac{1}{3}\bar{q}\gamma_\mu q - \bar{l}\gamma_\mu l]\,. \qquad (2)$$

with coupling constant $g_{B-L} = 0.9$. To make Z' resonances broad we modified the Z' boson interaction with the first flavour of neutrino, namely we made the replacement

[a] The LHC Z' boson discovery potential at CMS and ATLAS detectors has been studied for narrow Z' resonances in refs. [2,3].

[b] Note that recent notion of an unparticle, introduced by Georgi [6] can be interpreted as a particular case of a field with continuously distributed mass [4,5,7–10].

[c] In recent papers [11] the LHC Z' boson discovery potential was studied for energies $\sqrt{s} = 7,\ 10$ and $14\ TeV$.

$$Z^{'}_{\mu} g_{\nu} \bar{\nu} (\gamma_{\mu} + \gamma_{\mu}\gamma_5)\nu \to Z^{'}_{\mu} g_{\nu} \bar{\nu} (\gamma_{\mu}(1+u) + \gamma_{\mu}\gamma_5(1-u)),$$

where u is some parameter The parameter $u = 0$ corresponds to the case of standard interaction and the invisible decay width into righthanded neutrino is proportional to $|u]^2$ and for large u it dominates. In other words, in considered models we introduced additional interaction of $Z^{'}$ boson with righthanded neutrino that allows to increase full decay width of the $Z^{'}$ boson and make $Z^{'}$ boson rather wide. The values of total decay width Γ_{tot} for some parameters u for models A, B and C are given in Table 1.

The reaction we are interested in is di-lepton production

$$pp \to \gamma^*, Z^*, Z^{'*} + ... \to l^+ l^- + ...$$

Here $l = e, \mu$. The main background is the Drell-Yan production $PP \to \gamma^*, Z^* + ... \to l^+l^- + ...$. Other backgrounds like WW, ZZ, WZ or $t\bar{t}$ are small. For leptons standard acceptance cuts are the following:

$$p_T^l > 10 \ GeV \,, \tag{3}$$

$$|\eta^l| < 2.4 \,. \tag{4}$$

For di-lepton invariant mass M_{ll} we apply cut

$$M_{ll} > M_{Z^{'}} \,. \tag{5}$$

This cut differs from standard cut

$$|M_{ll} - M_{Z^{'}}| < \frac{\Gamma_{Z^{'}}}{2} \tag{6}$$

often used for estimation of the LHC $Z^{'}$ boson discovery potential for the case of narrow $Z^{'}$ boson. The reason is that the Drell Yan cross section is sharply decreasing function on M_{ll} and for wide $Z^{'}$ boson the Drell-Yan background dominates in mass region $M_{Z^{'}} - \frac{\Gamma_{Z^{'}}}{2} < M_{ll} < M_{Z^{'}}$. In our calculations we use PYTHIA 6.3 code [12] with STEQ6L [13] parton distribution functions evaluated at the scale $Q^2 = M^2_{Z^{'}}$. We did not take into account K-factors. An account of K-factor which is similar for both background and signal leads to the increase of significance. Also we don't take into account detector effects like effectiveness of electron or muon registration but at the CMS and ATLAS detectors the effectivness of lepton registration is typically bigger than 80 percent [2, 3] that can decreases significance by factor ≥ 0.8. An account of nonzero K-factor partly compensates this decrease. For the LHC total energy $\sqrt{s} = 14 \ TeV$ and for $Z^{'}$ boson masses $1 \ TeV \leq M_{Z^{'}} \leq 4 \ TeV$ K-factor for both signal and background is approximately 1.3 - 1.4 [2]. An account of K-factor leads to the increase of the significance by factor $\sqrt{K} \approx 1.1 - 1.2$. For the estimation of the significance we we use the method [14] based on frequentist approach which for the case of large statistics ($N_S \gg 1, N_B \gg 1$) leads to the approximate formula $S = 2(\sqrt{N_S + N_B} - \sqrt{N_B})$ for significance.

The first two years LHC will work with center-of-mass energy $\sqrt{s} = 7\ TeV$ and with total integrated luminosity $L_{tot} = O(1)\ fb^{-1}$. Finally the energy will be increased up to $\sqrt{s} = 14\ TeV$. In our estimates we use total integrated luminosity $L_{tot} = 1\ fb^{-1}$ for energies $\sqrt{s} = 7\ TeV$, $10\ TeV$ and $L_{tot} = 100\ fb^{-1}$ for the LHC energy $\sqrt{s} = 14\ TeV$. For models A, B and C we used $Z^{'}$ boson resonances with total decay widths shown in Table 1. Our results are summarized in Tables 2-4.

Table 1. The ratio $\frac{\Gamma_{Z^{'},tot}}{M_{Z^{'}}}$ for models A,B,C and parameters $u = 0$, 3, 5, 10, 15, 20.

u-parameter	$\frac{\Gamma_{Z^{'},tot}}{M_Z}$(mod.A)	$\frac{\Gamma_{Z^{'},tot}}{M_Z}$(mod.B)	$\frac{\Gamma_{Z^{'},tot}}{M_Z}$(mod.C)
$u = 0$	0.031	0.046	$6.2 \cdot 10^{-3}$
$u = 3$	0.048	0.062	0.028
$u = 5$	0.078	0.092	0.061
$u = 10$	0.22	0.23	0.21
$u = 15$	0.45	0.47	0.45
$u = 20$	0.78	0.79	0.79

Table 2. The LHC cross sections for signal plus background in fb and significances (in brackets) for total energy $\sqrt{s} = 7\ TeV$ and $M_{Z^{'}} = 1\ TeV$. The background cross section is $\sigma_B = 1.9\ fb$ and total integrated luminosity is $L_{tot} = 1\ fb^{-1}$.

u-parameter	model A	model B	model C
$u = 0$	94 (17)	166 (23)	24 (7.3)
$u = 3$	58 (13)	114 (18)	5.8 (2.7)
$u = 5$	36 (6.2)	72 (14)	3.2 (1.7)
$u = 10$	12 (4.7)	20 (6.5)	nondetectable
$u = 15$	5.6 (2.7)	7.2 (3.2)	nondetectable
$u = 20$	3.6 (1.8)	5.2 (2.5)	nondetectable

Table 3. The LHC cross sections for signal plus background in fb and significances (in brackets) for total energy $\sqrt{s} = 10\ TeV$ and $M_{Z^{'}} = 1.5\ TeV$. The background cross section is $\sigma_B = 0.66\ fb$ and total integrated luminosity is $L_{tot} = 1\ fb^{-1}$.

u-parameter	model A	model B	model C
$u = 0$	34 (10)	58 (14)	8.8 (4.5)
$u = 3$	22 (7.8)	40 (11)	2.1 (1.6)
$u = 5$	13 (5.6)	26 (8.7)	nondetectable
$u = 10$	4.2 (2.8)	7.8 (4.2)	nondetectable
$u = 15$	2.0 (1.5)	2.6 (2.0)	nondetectable
$u = 20$	nondtectable	nondetactable	nondetectable

Table 4 .The LHC cross sections for signal plus background in fb and significances (in brackets) for total energy $\sqrt{s} = 14\ TeV$ and $M_{Z^{`}} = 3\ TeV$. The background cross section is $\sigma_B = 0.040\ fb$ and total integrated luminosity is $L_{tot} = 100\ fb^{-1}$.

u-parameter	model A	model B	model C
$u = 0$	2.0 (24)	3.6 (32)	0.54 (11)
$u = 3$	1.9 (23)	3.4 (31)	0.12 (4.0)
$u = 5$	1.4 (20)	2.6 (28)	0.064 (2.4)
$u = 10$	0.54 (11)	1.0 (17)	nondetectable
$u = 15$	0.24 (5.6)	0.46 (9.4)	nondetectable
$u = 20$	0.13 (3.2)	0.22 (5.4)	nondetectable

As we can see from Tables 2-4 the LHC discovery potential depends rather strongly on the total decay width and it decreases with the increase of the total decay width. The reason is trivial and it is due to the dilution factor $k = \frac{\Gamma_{tot}(u=o)}{\Gamma_{tot}(u=0)+\Gamma_{inv}(u)}$. Note that direct Tevatron experimental bound on the mass of $Z^{`}$ boson depends on the $Z^{`}$ model and for $Z^{`}$ boson with standard couplings $M_{Z^{`}} > 923\ GeV$ [15]. For considered in this paper $Z^{`}$ models with large invisible decay width the Tevaton bound will be more weak for wide $Z^{`}$ bosons.

To conclude in this note we studied the LHC discovery potential for $Z^{'}$ models with continuously distributed mass. One of the possible LHC signatures for such models is the existence of broad resonance structure in Drell-Yan reaction $pp \to Z^{'} + ... \to l^+l^- + ...$. We made our estimates at the parton level and did not take into accont detector level Rough estimates show that detector effects are not very essential and will lead to the decrease of the significance by factor $0.8-0.9$. An account of K-factor will compensate this decrease. The LHC discovery potential for wide $Z^{`}$ bosons is more weak than for the case of narrow $Z^{`}$ bosons due to the dilution factor for $Br(Z^{`} \to l^+l^-)$. Nevertheless for some models it is possible to detect wide $Z^{`}$ bosons for masses up to $3-4\ TeV$ at final LHC energy $\sqrt{s} = 14\ TeV$ and total integrated luminosity $L_{tot} = 100\ fb^{-1}$. For the 2010-2011 LHC energy $\sqrt{s} = 7\ TeV$ and total integrated luminosity $L_{tot} = 1\ fb^{-1}$ it is possible to discover wide $Z^{`}$ boson with masses up to 1.2 TeV.

I am indebted to A.N.Toropin for useful discussions. This work was supported by the Grant RFBR 08-02-91007-CERN.

References

[1] As a review, see for example:
P.Langacker, Rev.Mod.Phys. **81**, 1199 (2008);
P.Langacker, arXiv:0911.4294[hep-ph] (2009);
J.Hewett and T.Rizzo, Phys.Rep. **183**. 193 (1989).

[2] G.L.Baytian et al., J.Phys. **G 34**, 995 (2007).

[3] ATLAS collaboration, Detector and Physics Performance, Technical Desigh Report ,
CERN/LHCC/99-14.
[4] N.V.Krasnikov, Phys.Lett. **B 325**, 430 (1994).
[5] N.V.Krasnikov, Int.J.Mod.Phys. **A 22**, 5117 (2007).
[6] H.Georgi, Phys.Rev.Lett. **98**, 221601 (2007);
H.Georgi, Phys.Lett. **B 650**, 275 (2007).
[7] M.A.Stephanov, Phys.Rev. **D 76**, 035008 (2007).
[8] Patricio Gaete and Euro Spallucci, Phys.Lett. **B 661**, 319 (2008).
[9] H.Nikolic, Mod.Phys.Lett. **A23**, 2645 (2008).
[10] J.Lukerski, In Erice 1978 Proceedings, Hadronic matter at Extreme Energy density, p.343.
[11] L.Basso et al., arXiv:1002.3586[hep-ph] (2010);
E.Salvioni, G.Villadoro anf F.Zwirner, JHEP **1003**:012 (2009);
arXiv:0909.1320[hep-ph];
W.Emam and P.Mine, J.Phys. **G 35**, 11508 (2009).
[12] T.Sjostrand, L.Lonnblad, S.Mrenna and P.Skands, arXiv:hep-ph/0308153 (2003).
[13] J.Pumplin et al., Phys.Rev. **D65**, 014011 (2002);
http://durpdg.dur.ac.uk/hepdata/pdf.html.
[14] S.I.Bityukov and N.V.Krasnikov, Mod.Phys.Lett. **A 13**, 3235 (1998);
S.I.Bityukov and N.V.Krasnikov, Nucl.Instr. and Meth. **A 452** (2000).
[15] C.Amsler et al., Phys.Lett. **B 667** (2008).

BLACK HOLES AT LHC?

D. V. Galt'sov [a], P. A. Spirin [b]

Faculty of Physics, Moscow State University, 119991 Moscow, Russia

*Abstract.*We critically review the scenario of black hole production by colliders in view of the recent calculations of the emission of Kaluza-Klein gravitons in the transplanckian collisions within the TeV-scale gravity models. These results indicate that strong radiative damping may substantially decrease the rate of black hole production at LHC.

TeV-scale gravity [1] assumes that the "true" gravitational constant corresponds to the mass scale M_* of the order of one or several TeV, at the expense of making gravity multidimensional. The weakness of the usual four-dimensional gravity results from the fact that only a small fraction of the gravitational lines of force is responsible for gravitational interaction in our world which can be thought of as four-dimensional hypersurface (brane) in the $D = n + 4$ dimensional bulk. The standard model particles live only on the brane, while the bulk is inhabited only by gravitons which participate in interactions with matter with equal strength.

This (and similar) scenario opens a fascinating possibility of production of black holes at colliders. This prediction [2] made about ten years ago, has been the subject of intense theoretical study and numerical simulations (for a review see [3]). The prediction is based on the assumption that for impact parameters of the order of the horizon radius corresponding to the center of mass collision energy $2E = \sqrt{s}$ (s being the Mandelstam variable)

$$r_S = \frac{1}{\sqrt{\pi}} \left[\frac{8\Gamma\left(\frac{d+3}{2}\right)}{d+2} \right]^{\frac{1}{d+1}} \left(\frac{G_D\sqrt{s}}{c^4} \right)^{\frac{1}{d+1}}$$

an event horizon should form due to the non-linear nature of gravity. The $D = 4 + d$ dimensional gravitational constant can be replaced by the D-dimensional Planck mass M_* through $G_D = \hbar^{d+1}/(M_*^{d+2}c^{d-1})$. It is related to the four-dimensional Planck mass as $M_{\rm Pl} \equiv M_4$ via

$$M_{\rm Pl}^2 = M_*^{d+2}V, \qquad V = (2\pi R)^d,$$

where R is the compactification radius assumed to be of mesoscopic scale up to micrometers.

This classical, essentially, picture of BH formation is justified for transplanckian energies $s \gg G_D^{-2/(d+2)} = M_*^2$. Indeed [4], in this case the D-dimensional Planck length

$$l_* = \left(\hbar G_D/c^3\right)^{1/(d+2)} = \hbar/M_*c$$

[a] e-mail: galtsov@physics.msu.ru

[b] e-mail: salotop@list.ru

and the de Broglie length of the collision $\lambda_B = \hbar c/\sqrt{s}$ satisfy the *classicality* condition

$$\lambda_B \ll l_* \ll r_S.$$

Furthermore, gravity is believed to be the dominant force in the transplanckian region. Thus, for BH masses large compared to M_*, the use of classical Einstein theory is well justified. Moreover, it seems that formation of BHs in four dimensions is predicted by string theory Thus a consensus has been reached that the prediction of BHs in ultra-high energy collisions is robust and is summarized in the widely accepted four-stage process of formation and evaporation of BHs in colliders [2], namely (i) formation of a closed trapped surface (CTS) in the collision of shock waves modeling the head-on particle collision, (ii) the balding phase, during which the BH emits gravitational waves and relaxes to the Myers-Perry BH, (iii) Hawking evaporation and superradiance phase in which the experimental signatures are supposed to be produced, and (iv) the quantum gravity stage, where more fundamental theory like superstrings is important. This scenario was implemented in computer codes [5] to simulate the BH events in LHC, where they are expected to be produced at a rate of several per second, and in ultra high energy cosmic rays.

Our concern here is the stage (i). Replacing the field of an ultrarelativistic particle by that of a black hole in the infinite momentum frame corresponds to the linearized theory picture, while the full non-linear gravitational dynamics is more complicated. First of all, it is intrinsically dissipative due to gravitational radiation. To see the difference, it is enough to compare the Lorentz contracted Coulomb field of the uniformly moving charge and the retarded field of an accelerated charge which has quite different fall-off law at infinity. Continuing this analogy with electrodynamics, one can argue that strong radiation damping may essentially change the particle trajectory and decrease the particle energy so that conditions for the black hole creation will not be encountered at all. These conditions are quite likely to be met in the transplanckian region in TeV-scale gravity, similarly to the case of string theory [6].

Full quantum treatment of the bremsstrahlung in the transplanckian region is not available for the moment neither in the usual Einstein gravity nor in the TeV-scale models. Lack of better, we can use the resent results [7] of purely classical treatment, which is partially justified in the classicality region mentioned above if, in addition, the emitted quanta are soft. Classical calculation of bremsstrahlung at ultrarelativistic collisions of point masses m, m' at an impact parameter b gives the energy loss in the rest frame of one of the particles [7]

$$E_{\rm rad} = C_D \frac{m^2 m'^2 \kappa_D^6}{b^{3d+3}} \gamma^{d+3},$$

where $\kappa_D^2 = 16\pi G_D$, γ is the Lorentz-factor of the collision and C_D is some dimension-dependent coefficient. To pass to the center of mass frame, we calculate the relative energy loss (radiation efficiency) $\epsilon \equiv E_{\rm rad}/E$, and express the result in terms of the Lorentz factor in the CM frame via (for $m = m'$):

$$\epsilon = C_d \left(\frac{r_S}{b}\right)^{3(d+1)} \gamma_{\rm cm}^{2d+1}, \quad \gamma_{\rm cm}^2 = (1+\gamma)/2. \tag{1}$$

This expression contains the large factor $\gamma_{\rm cm}^{2d+1}$ which rapidly increases with the number of extra dimensions. Physically it is due to the large number of light KK modes involved both in the gravitational force and in the radiation. It is also worth mentioning the growing with d coefficient C_d (see [7] for details).

The classical description of small angle ultrarelativistic scattering is strictly speaking valid for impact parameter in the region

$$\gamma^\nu r_S \ll b \ll b_c, \quad \nu = 1/2(d+1),$$

where b_c is a special length parameter arising if $d \geq 1$

$$b_c \equiv \pi^{-1/2} \left[\Gamma(d/2) G_D s/\hbar c^5\right]^{1/d} \sim r_S \left(r_S/\lambda_B\right)^{1/d},$$

marking the scale beyond which the classical notion of particle trajectory is lost [4]. These two restrictions come from the validity of the perturbation theory in terms of the gravitational constant, and classicality of motion in ADD. Another restriction comes from the quantum bound on the radiation frequency $\hbar\omega_{\rm cr} < m\gamma$, which is equivalent to $b > \lambda_C \equiv \hbar/(mc)$. For $d \neq 0$ the two conditions overlap provided $\lambda_C < b_c$. A rough estimate of ϵ can be obtained in the case $\gamma^\nu r_S \ll \lambda_C \ll b_c$ by setting $b = \lambda_C$ in (1), which becomes

$$\epsilon = B_d (sm/M_*^3)^{d+2},$$

with $B_d = 7.4, 0.6, 0.36, 0.45, 0.81, 2.1, 7.0$ for $d = 0, 1, \ldots, 6$. Thus, a simple condition for strong radiation damping is

$$sm \geq M_*^3 \,,$$

which may well hold for heavy point particles with LHC energies. For example, in collisions of particles with $m \sim \mathcal{O}(100GeV)$ and energy $\sqrt{s} \sim \mathcal{O}(10TeV)$ all conditions for extreme gravitational bremsstrahlung are satisfied for $d = 2$.

However, even it seems physically reasonable, one cannot strictly speaking apply (1) to the case of proton-proton collisions in LHC. Protons are not point-particles, while their constituents are too light and lie outside the region of validity of our approximation. For the same reason formula (1) cannot be applied to massless particle collisions, which seem to require a different treatment [6]. Therefore, even though quantum gravity effects can be ignored for some parameters in the transplanckian region, non-gravitational quantum features have to be taken into account.

Nevertheless, (a) one may have to include the reaction force in the study of processes such as BH production, which might even exclude the formation of a CTS. Note that there are indications that gravitational collapse of an oscillating macroscopic string does not take place, once gravitational radiation is taken into account [8]. Also, (b) bremsstrahlung, is a strong process leading to missing energy signatures in transplanckian collisions, which may further constrain the ADD parameters.

We thank Th.Tomaras and G.Kofinas for collaboration, and G. Veneziano for useful discussions and correspondence. The work was supported by the RFBR project 08-02-01398-a.

References

[1] N. Arkani-Hamed, S. Dimopoulos and G. Dvali, Phys. Lett. B **429**, 263 (1998) [hep-ph/9803315]; I. Antoniadis, N. Arkani-Hamed, S. Dimopoulos and G. Dvali, Phys. Lett. B **436**, 257 (1998) [hep-ph/9804398]; G. F. Giudice, R. Rattazzi and J. D. Wells, Nucl. Phys. B **544**, 3 (1999) [arXiv:hep-ph/9811291]; T. Han, J. D. Lykken and R. J. Zhang, Phys. Rev. D **59**, 105006 (1999) [arXiv:hep-ph/9811350].

[2] P. C. Argyres, S. Dimopoulos and J. March-Russell, Phys. Lett. B **441**, 96 (1998) [arXiv:hep-th/9808138]; S. B. Giddings and S. D. Thomas, Phys. Rev. D **65**, 056010 (2002) [arXiv:hep-ph/0106219]; D. M. Eardley and S. B. Giddings, Phys. Rev. D **66**, 044011 (2002) [arXiv:gr-qc/0201034]; J. L. Feng and A. D. Shapere, Phys. Rev. Lett. **88**, 021303 (2002) [arXiv:hep-ph/0109106].

[3] M. Cavaglia, Int. J. Mod. Phys. A **18**, 1843 (2003); P. Kanti, Int. J. Mod. Phys. A **19**, 4899 (2004); D. M. Gingrich, Int. J. Mod. Phys. A **21**, 6653 (2006); G. L. Landsberg, J. Phys. G **32**, R337 (2006); B. Koch, M. Bleicher and H. Stoecker, J. Phys. G **34**, S535 (2007).

[4] G. F. Giudice, R. Rattazzi and J. D. Wells, Nucl. Phys. B **630**, 293 (2002)

[5] C. M. Harris, P. Richardson and B. R. Webber, JHEP **0308**, 033 (2003) [arXiv:hep-ph/0307305]; M. Cavaglia *et al.*, Comput. Phys. Commun. **177**, 506 (2007); [arXiv:hep-ph/0609001]. J. A. Frost *et al.*, arXiv:0904.0979 [hep-ph]; D. C. Dai *et al.*, arXiv:0902.3577 [hep-ph].

[6] D. Amati, M. Ciafaloni and G. Veneziano, Nucl. Phys. B **347**, 550 (1990); Phys. Lett. B **289**, 87 (1992); Nucl. Phys. B **403**, 707 (1993).

[7] D. V. Gal'tsov, G. Kofinas, P. Spirin and T. N. Tomaras, **JHEP** 0905:074, 2009 [arXiv:0903.3019 [hep-ph]]; Phys. Lett. B **683** (2010) 331 [arXiv:0908.0675 [hep-ph]]; **JHEP** (2010) [arXiv:1003.2982 [hep-th]].

[8] R. Iengo, hep-th/0702087.

RECENT ELECTROWEAK MEASUREMENTS AT THE TEVATRON

Andrew Askew [a]

Physics Department, Florida State University, Tallahassee FL, U.S.A.

*Abstract.*The CDF and D0 experiments at the Fermilab Tevatron have now collected in excess of 6 fb^{-1} in integrated luminosity. These large datasets make available a wide range of electroweak measurements. Electroweak cross sections provide a great deal of vital information. Large inclusive cross sections (and asymmetries within those processes) are sensitive to parton distribution functions. Measurement of the W boson mass, a key parameter in the Standard Model (SM), can be made to high precision. Much smaller diboson cross sections are sensitive to trilinear coupling vertices, and are also common backgrounds to new physics searches. An overview of the most recent measurements from the CDF and D0 experiments will be presented.

1 Introduction

The Fermilab Tevatron has delivered more than 6 fb^{-1} of integrated luminosity to the CDF and D0 experiments. This provides an opportunity to study W and Z production properties with high statistics, as well as low cross section diboson processes. In this paper, the newest of these results are reviewed, which utilize 1 to 4.9 fb^{-1}.

2 W boson analyses

2.1 W Charge Asymmetry

The W boson charge asymmetry arises from the differing momentum fractions carried by the initial state partons. Thus, a measurement of this asymmetry as a function of the W-rapidity can be used to further constrain the momentum distribution of partons within the proton (and anti-proton). This is complicated by the decay of the W ($W \to \ell\nu$), since the neutrino escapes, the W-rapidity cannot be exactly calculated. CDF weights events and then unfolds the W charge asymmetry [1] (the first time this method has been used), while D0 measures the charge asymmetry of the decay leptons [2] in both electron and muon channels (see Figure 1). This data can be used to further constrain the parton distribution function fits.

2.2 W Mass

A full discussion of the measurement of the W-mass and width far exceeds the scope of these proceedings. Briefly, a parametrized detector simulation is created to reflect the detector reponse, resolution and efficiency. Monte

[a] e-mail: askew@hep.fsu.edu

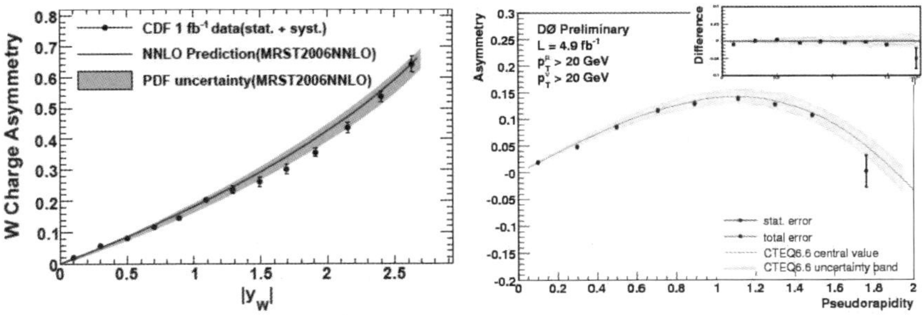

Figure 1: Left, unfolded W charge asymmetry distribution from CDF, using 1 fb^{-1}. Right, muon charge asymmetry distribution from D0, using 4.9 fb^{-1}.

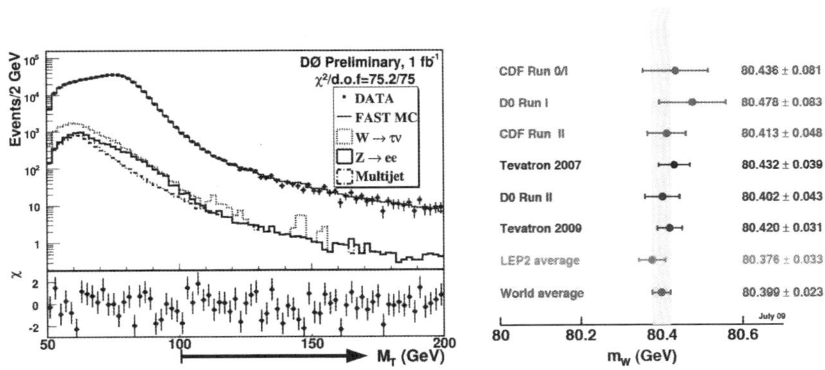

Figure 2: Left, W-width measurement from D0 (arrow denotes the fit range). Right, world combination of W-mass measurements, the precision now driven by the Tevatron measurements.

Carlo mass templates are then generated using this simulation, which are then compared with the data distribution [3], [4]. For the measurement of the W-mass, the position of the peak of the transverse mass distribution is used, while for the W-width the tail of the distribution is used(see Figure 2). The D0 W-mass measurement is currently the most precise single measurement, and in the combination, the Tevatron value now drives the precision on the world average.

3 Z/γ^* boson analyses

The Z/γ^* boson analyses mentioned here only make use of $Z/\gamma^* \rightarrow ee$ events, due to the larger detector acceptance for electrons (for both CDF and D0).

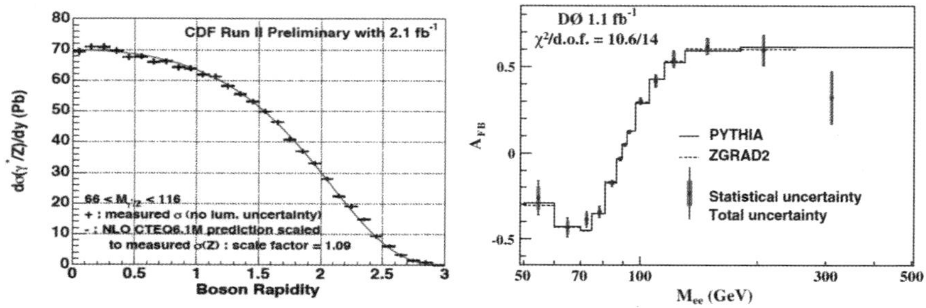

Figure 3: Left, Z-rapidity measurement from CDF. Right, forward-backward asymmetry from D0.

The differential cross section in Z/γ^* rapidity, like the W charge asymmetry, is used as a probe of the initial momentum fractions of the colliding partons. Using 2.1 fb^{-1} of integrated luminosity, the CDF experiment measures the cross section versus rapidity in the invariant mass range of $66 < M_{ee} < 116$ GeV, shown in Figure 3 [5].

The forward-backward asymmetry probes the coupling of the neutral current to fermions, and is sensitive to $\sin^2\theta_W$. If additional structure is present at high invariant masses, it could be indicative of interference with new gauge bosons. Using 1.1 fb^{-1} of data, the D0 experiment measures the forward-backward backward asymmetry to be in agreement with the SM, and measures $\sin^2\theta_W^{eff}=0.2326\pm0.0018$(stat)$\pm0.0006$(syst) [6].

4 Diboson analyses

The study of the production of pairs of vector bosons probes the underlying gauge structure of the SM. The boson trilinear couplings, whether required or forbidden, are predicted exactly by the SM. Any deviation would be evidence for new physics beyond the SM.

WW production simultaneously probes both the $WW\gamma$ and WWZ couplings. SM WW production is also the major background to searches for $H \to WW$, one of the most sensitive channels at the Tevatron. The lowest background decay process is $WW \to \ell\nu\ell\nu$. Using 1 fb^{-1} of integrated luminosity, D0 measures a cross section of $\sigma_{WW}=11.5\pm2.2$ pb [7]. Using 3.6 fb^{-1} of integrated luminosity, CDF measures a cross section of $\sigma_{WW}=12.1^{+1.8}_{-1.7}$(stat+syst) pb [8].

The semileptonic decay of WW production ($WW \to \ell\nu jj$) cannot be distinguished from the similar final state in WZ. It is thus typical to measure the combined cross section, which suffers from very large backgrounds of W bosons produced in association with jets. Using 1.1 fb^{-1} of integrated luminosity, D0 measures a cross section of $\sigma(WW+WZ)$=20.2$\pm$4.5 pb, at 4.4 σ

Parameter	Minimum	68%C.L.	95%C.L.
$\Delta\kappa_\gamma$	0.07	[-0.13,0.23]	[-0.29,0.38]
Δg_1^Z	0.05	[-0.01,0.11]	[-0.07,0.16]
λ	0.0	[-0.04,0.05]	[-0.08,0.08]

Table 1: Combined anomalous coupling limits from D0.

significance [9]. CDF uses 3.9 fb^{-1} of integrated luminosity, and measures a cross section $\sigma(WW+WZ)$=14.4±3.8 pb, at 4.6 σ significance [10].

In a similar vein, CDF searches for events in which two vector bosons are produced, where one decays hadronically, and the other decay involves a neutrino. Using 3.5 fb^{-1} of integrated luminosity, CDF measures a cross section $\sigma(VV+X)$=18.0±2.8(stat)±2.4(syst)±1.1(lumi) pb, at 5.3 sigma significance [11].

D0 has produced anomalous vector boson coupling limits in several analyses ($W\gamma \to \ell\nu\gamma$, $WW \to \ell\nu\ell\nu$, $WZ \to \ell\ell\ell\nu$ and $WW+WZ \to \ell\nu jj$), and combines them under the assumption of $\mathrm{SU}(2)_L \times \mathrm{U}(1)_Y$ symmetry. The result of this combination is shown in Table 1 [12].

5 Summary

The Tevatron continues to deliver vast quantities of data to the CDF and D0 experiments. With over 6 fb^{-1} of integrated luminosity delivered, these results are still only a small portion of the electroweak measurements that can be performed with the full data sample.

References

[1] T. Aaltonen *et al.* Phys. Rev. Lett. **102**, 181801 (2009).
[2] V. M. Abazov *et al.* Phys. Rev. Lett. **101**, 211801 (2008).
[3] V. M. Abazov*et al.* Phys. Rev. Lett. **103**, 141801 (2009).
[4] V. M. Abazov*et al.* Phys. Rev. Lett. **103**, 231802 (2009).
[5] CDF Collaboration, arXiv: 0908.3914.
[6] V. M. Abazov *et al.* Phys. Rev. Lett. **101**, 191801 (2008).
[7] V. M. Abazov *et al.* Phys. Rev. Lett. **103**, 191801 (2009).
[8] CDF Collaboration, arXiv: 0912.4691.
[9] V. M. Abazov *et al.* Phys. Rev. Lett. **102**, 161801 (2009).
[10] CDF Collaboration, arXiv: 0911.4449.
[11] CDF Collaboration, arXiv: 0905.4714.
[12] D0 Collaboration, arXiv:0907.4952.

TOP PHYSICS WITH THE ATLAS DETECTOR AT LHC

Dominique Pallin, on behalf of the ATLAS collaboration [a]
Laboratoire de Physique Corpusculaire, Clermont Université, Université Blaise Pascal, CNRS/IN2P3, FR - 63177 Aubière Cedex, France

Abstract. Studies performed at 14 TeV and 10 TeV centre-of-mass energies to estimate the potential for top quark physics of the ATLAS experiment at the LHC are reviewed.

1 Introduction

Since its discovery in 1995 [1,2] the top quark properties have been extensively explored at Tevatron. Indeed, the top quark has quite interesting features. It has a very large mass compared to the other quarks, its measured mass being about 173 GeV [3] from CDF and D0 experiments. It interacts heavily with the Higgs sector. This suggest that the top quark plays a specific role in the electro-weak symmetry breaking mechanism and this has been exploited in many models beyond the standard model (SM) [4, 5]. Another interesting property is that due to its short lifetime the top quark decays before hadronisation. It is the only quark which can be studied as a bare quark. Consequently, the physics content of top physics is quite broad and a large number of precision measurements could be performed to see if the top quark behaves as expected from the SM and if some new physics could be revealed in the top quark production or decays.

In 2009, the Large Hadron Collider (LHC) provided first proton-proton collisions. Even at low center-of-mass energies and luminosities, top quark events are expected to be produced copiously at the LHC, opening a new opportunity for precision measurements of the top quark properties. Prior to any measurements, the top quark will be a reference point and top events will be used as a tool for detector commissioning. Top quark rediscovery and top quark pair production cross section will be important early measurements. With increasing luminosity and centre-of-mass energy, measurements will gain in precision improving our knowledge in the top quark sector and possibly opening a window for physics beyond the standard model (BSM).

A brief overview of the potential of the ATLAS detector [6] for measuring the top quark properties at LHC is given here. Studies performed at 10 TeV and 14 TeV [7–10] to determine the top quark pair production cross section, to measure the top quark mass, the top quark charge, to search for rare top quark decays and $t\bar{t}$ resonance production are reviewed.

Top quarks are mainly produced via the pair production mechanism. Top quark pairs can be produced through gluon-gluon fusion and quark-antiquark

[a] e-mail: pallin@clermont.in2p3.fr

annihilation. In pp collisions at the LHC $t\bar{t}$ production via gluon-gluon dominates, the theoretical $t\bar{t}$ cross section at 10 TeV being 400 pb with an uncertainty of about 6% [11] . The single top production is far from being negligible with a total cross section of about one third of the $t\bar{t}$ production.

In the SM, a top quark decays almost exclusively through the $t \rightarrow Wb$ decay mode. The W decays in about one third of the cases into $l\nu$, all lepton flavours being produced at an approximately equal rate. In the remaining two thirds the W decays into $q\bar{q}$. From an experimental point of view, the $t\bar{t}$ decay mode is defined according to the W decay mode: the dilepton mode where both W boson decay leptonically , the semi-leptonic mode where only one W decays leptonically and the full hadronic mode where both W decay hadronically.

2 Top quark measurements at the LHC

The gold plated channel to investigate top properties is the semi-leptonic channel accounting for 30% of the $t\bar{t}$ decays considering only the electronic and muonic W decays. The dilepton channel has a very clean signature but represents only 5% of the decays and cant be fully reconstructed due to the precence of two neutrinos. The presence of lepton(s) is used to trigger the di-leptonic and semileptonic modes. The full hadronic channel suffers from a large QCD background and will be difficult to exploit in early data. The measurement of most of the top properties implies the detection and identification by the ATLAS detector of all top decay products. Two of the produced jets being b jets, good b-tagging capabilities are a real bonus to choose the right assignment of the jets to the two top quarks.

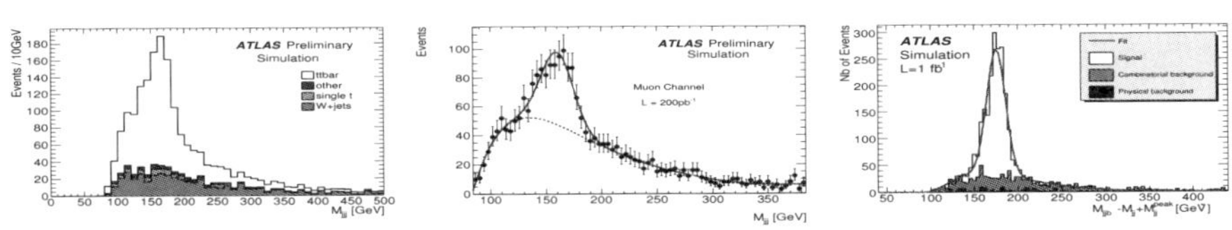

Figure 1: Left: Expected distribution of the three-jet invariant mass in the electron channel mass, normalised to 200 pb^{-1}. Middle: The likelihood fit in the three-jet invariant mass for the muon channel. The statistical uncertainties correspond to an integrated luminosity of 200 pb^{-1}. Right: Hadronic top mass for semileptonic events after purification cuts.

Since $t\bar{t}$ events are a major background for many processes in the SM and BSM where many models of physics predict events with similar signatures, the measurement of the $t\bar{t}$ production is an essential stepping stone to new physics. The latest ATLAS prospects for the $t\bar{t}$ cross section measurement [8,9] have been determined at $\sqrt{s} = 10\,\text{TeV}$ assuming an integrated luminosity of $200\,pb^{-1}$ from early data. The studies of the semi-leptonic and di-leptonic

modes rely on very simple and robust selections, and do not make use of b tagging capabilities. The baseline analysis for the semi-leptonic mode consists of selecting events which pass the lepton trigger (15 GeV) and contain one isolated lepton with p_T >20 GeV, at least 4 jets with $p_T > 20$ GeV, 3 of them having a $p_T > 40$ GeV. The expected transverse missing energy from the escaping neutrino has to be larger than 20 GeV. Only the hadronic top is fully reconstructed. It is defined as the 3 jet combination which gives the highest transverse momentum sum. Among these selected 3 jets, one jet pair should originate from the W boson. A further selection cut reducing a large fraction of the W+jets background requires that one jet pair is within 10 GeV of the W boson mass. The hadronic top invariant mass distribution is shown in figure 1(left). Two methods have been used to measure the cross section. The first uses a very simple approach based on a Cut and Count method, which counts the total number of events passing the cuts, substracts the number of background events expected from simulation and data driven background determination (mainly W+jets events), and finally corrects for signal efficiency and luminosity. The second method uses a fitting procedure where the top mass distrution is modeled by a Gaussian function for the signal and a Chebychev distribution for the background (figure 1 middle). The two methods have different systematics uncertainties, the first one being more sensitive to background normalisation (from the uncertainty on the jet enegy scale), the second one to signal and background shapes. The precision for the $t\bar{t}$ cross section measurement in the semileptonic channel is expected to be better than 20%, excluding the luminosity uncertainty. A Cut and Count method is also applied to measure the $t\bar{t}$ cross section in the di-leptonic channel. The obtained $t\bar{t}$ cross section sensitivity is of the same order as in the semi-leptonic case.
ATLAS used the semileptonic decay of the $t\bar{t}$ process to determine the reachable precision on the top mass using a sample of 1 fb^{-1} of data at 14 TeV [7]. The strategy followed was to remove as much as possible the combinatorial background contribution making use of b-tagging capabilities (events with two b-tagged jets) and to rely only on well measured objects. The top mass estimator is built from the invariant mass of the hadronic top candidate which is reconstructed in two steps: first a W candidate is selected as the 2 jet pair which is the closest in space and the top candidate is built by adding the closest b-jet to this pair. The mass distribution is shown in figure 1(right). The main contribution to the systematic error on the mass measurement is coming from the b-jet energy scale uncertainy, the top mass being expected to be measured at the level of 1 (3.5) GeV if the b-jet scale is known with a 1 (5)% precision. The full reconstruction of the event then allows the $t\bar{t}$ invariant mass to be reconstructed. The leptonic W can be reconstructed from a W mass constraint assuming that the neutrino transverse momentum is identified with the missing transverse energy. The remaining b jet is added to the leptonic W to form the

leptonic top candidate. ATLAS established from the $t\bar{t}$ invariant mass distribution the discovery potential for a generic resonance which could appear on top of this distribution. With $1\,fb^{-1}$ at 14 TeV, ATLAS will be able to discover a $t\bar{t}$ resonance of mass 700 GeV if $\sigma\times$ BR is greater than 11 pb [7]. This standard top reconstruction is not very efficient for highly boosted top quarks since top decay products tend to overlap and produce top monojets. ATLAS developed new reconstruction schemes for highly boosted tops leading to a much improved efficiency for large $M_{t\bar{t}}$. 95% CL expected limits for $1\,fb^{-1}$ at 14 TeV were set for 2 and 3 TeV $t\bar{t}$ resonance masses to be 550 and 160 fb [10].

The top quark charge is one of the top fundamental properties which will be probed at LHC. ATLAS expects to perform the measurement of the electric charge of the top quark using the semi- and di-leptonic channels where both b jets are tagged. This measurement requires the determination of the charge of the top decay products. The W and b originating from the same top have to be correctly assigned. The charge of the W boson is determined through its leptonic decay while for the charge of the b jet the best method uses a weighted technique where the b jet charge is obtained from the charges of the tracks belonging to the b-jet. Using this technique ATLAS will be able to establish with $0.1\,fb^{-1}$ if the top quark has the SM charge or if it is an exotic particle with charge -4/3, with 5σ significance [7].

Top quark rare decays through FCNC processes are strongly suppressed in the SM but may appear at tree level in SUSY and others models. ATLAS investigated such top rare decays where top quarks decay into a quark and a particle which could be either a photon, a Z or a gluon. One of the top quark is assumed to decay according to the SM while the other procceeds through FCNC decay. For $1\,fb^{-1}$at 14 TeV in the absence of a FCNC top quark decay signal, BR limits at 95% CL are expected to be $6.8\times10^{-4}, 2.8\times10^{-3}$ and 1.2×10^{-2} for the $t\rightarrow q\gamma, t\rightarrow qZ$ and $t\rightarrow qg$ respectively [7].

References

[1] F.Abe et al, *Phys. Rev. Lett* **74** 2626, 2631 (1995).

[2] S Abachi et al, *Phys. Rev. Lett* **74** 2632, 2637 (1995).

[3] Tevatron Electroweak Working Group , *arXiv:0811.4682[hep-ex]* (2009).

[4] M. Beneke at al, Top quark Physics *hep-ph/0003033* (2000).

[5] C.T. Hill and E.H Simmons, *Phys. Rep.* **381** 235 (2003).

[6] The ATLAS Collaboration, *JINST* **0803** S08004 (2008).

[7] The ATLAS Collaboration, *CERN-OPEN-2008-020.*

[8] The ATLAS Collaboration, *ATLAS-PHYS-PUB-2009-086* (2009).

[9] The ATLAS Collaboration, *ATLAS-PHYS-PUB-2009-087* (2009).

[10] The ATLAS Collaboration, *ATLAS-PHYS-PUB-2009-081* (2009).

[11] S Moch and P. Uwer, *Phys. Rev.* **78** (2008).

QCD STUDIES AT THE LHC WITH THE ATLAS DETECTOR

Sebastian Eckweiler [a] on behalf of the ATLAS Collaboration
University of Mainz, Staudinger Weg 7, 55126 Mainz, Germany

*Abstract.*This paper describes a selection of early QCD analyses, planned to be performed with the ATLAS experiment. Plans for measurements of underlying event properties and minimum bias events in early data are discussed. Selected analyses including jets are presented.

1 Introduction

This paper describes a selection of QCD analyses, planned to be performed with early data at $\sqrt{s} = 14\,\text{TeV}$ of the ATLAS experiment [1]. We present plans for a measurement of properties of inelastic proton proton collisions, using minimum bias events. These analyses will provide a better understanding of underlying event modelling. Measurements including jets are discussed: They will enable better tuning of Monte Carlo (MC) generators and decrease the uncertainties of the high-x gluon parton density function (pdf).

2 Minimum bias

We distinguish two definitions for *minimum bias* events: One referring to minimum bias events as all inelastic events and one restricted to non-single-diffractive (NSD) events. Both sorts of events are dominated by particles with low transverse momentum. Their multiplicity and momentum spectrum are quantities, which will help constrain the modelling of soft QCD processes in MC generators. This is especially important in LHC scenarios, where more than one proton proton interaction per bunch crossing takes place. Though several such measurements have been made at lower energies, the extrapolation to the LHC design energy is uncertain. A comparison in [2] of simulations using PYTHIA [3] and PHOJET [4] illustrates that these uncertainties cannot be neglected.

A detailed preparation study of this measurement is given in [5]. An important detail in a minimum bias analysis is the trigger, whose bias naturally has to be minimised and well known. The above cited analysis uses the minimum bias trigger scintillators (MBTS), which are, as a part of the first trigger level, implemented in hardware. The system covers the pseudorapidity region of $2.1 < |\eta| < 3.8$. One of the triggers used is MBTS_2, which requires in total two hits anywhere in the MBTS system and thus also incorporates asymmetric topologies. Information from PYTHIA and PHOJET is used to correct for the included single-diffractive events in case of the measurement on a NSD event sample, as it is shown in Figure 1. The statistical uncertainties will be

[a]e-mail: Sebastian.Eckweiler@Uni-Mainz.de

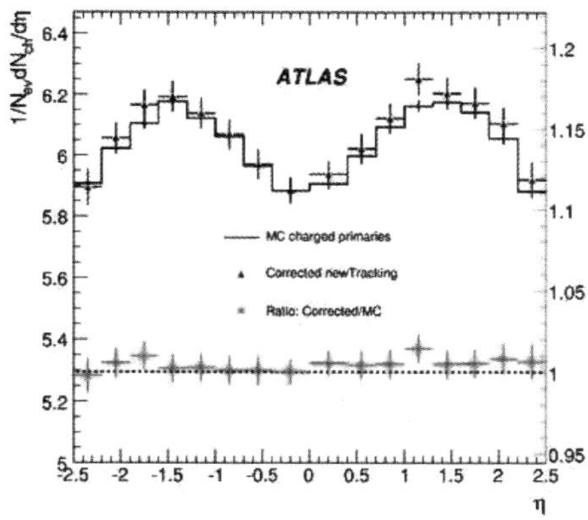

Figure 1: Pseudorapidity distribution of charged particles with $p_T > 150\,\mathrm{MeV}$ in non-single-diffractive events at $\sqrt{s} = 14\,\mathrm{TeV}$, which were selected using the MBTS_2 trigger. Figure taken from [5].

negligible, compared to the systematic uncertainties. The dominant systematic uncertainties are the the mis-alignment of the tracking system and the predictions for diffractive cross sections at LHC energies, which differ significantly between PHOJET and PYTHIA. In total, the systematic uncertainties are evaluated to be 8%, which will allow to incorporate this measurement in tunes of MC generators for the LHC.

3 Underlying event

We refer to the underlying event (UE) as everything but the particles from the primary hard scattering. Thus, this includes the beam remnants, their connection to the hard scattering, initial and final state radiation and possible interactions of secondary pairs of partons. The effect of the underlying event is important as it contaminates the primary event and even adds physics objects to it, such as low p_T jets. Thus a proper modelling of its properties is crucial to be able to correct for its effects.

Though the UE has been thoroughly studied at the Tevatron and MC generators have been tuned appropriately, the extrapolation to LHC energies is uncertain and differs significantly between MC generators. A commonly used quantity to characterise the UE is the charged particle distribution in regions transverse to the leading jet in an event. The study in [6] compares the UE properties predicted by PYTHIA and HERWIG [7] in combination with JIMMY [8].

Once PYTHIA and JIMMY are tuned to describe the available CDF data

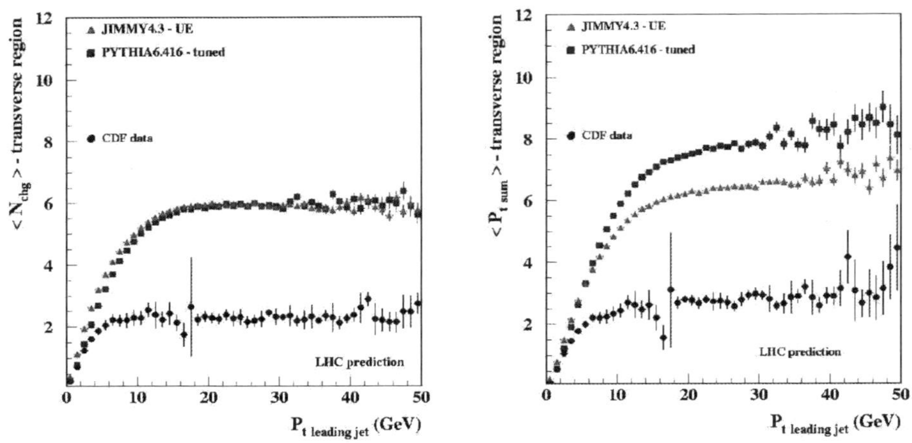

Figure 2: Predictions from JIMMY and PYTHIA for multiplicity and transverse momenta of charged particles in the region transverse to the leading jet at $\sqrt{s}$ =14 TeV. Figure taken from [6].

the UE description is extrapolated to $\sqrt{s} = 14$ TeV. This is illustrated in Figure 2. While both generators agree in the description of the charged particle multiplicity, the average transverse momentum sum $p_{T,Sum}$ that is predicted by PYTHIA is roughly 20% higher in the plateau region. These studies imply that with an integrated luminosity of $10\,\mathrm{pb}^{-1}$, ATLAS will be able to provide important input to generator tunes and a significant decrease in the uncertainty of the UE description.

4 Jet physics

Jet physics very directly gives a handle to the momentum fractions x of the scattered partons. Thus, observables like the jet cross section as function of transverse momentum and rapidity can give a handle on the gluon pdf. Especially constraints on this pdf at high values of Q^2 and x are expected.

In [9] a toy study was made, using ATLAS pseudo jet data: The inclusive jet cross section in three bins of pseudorapidity was added to a pdf fit and the effect on the gluon uncertainty was studied. The pseudo data was simulated to correspond to an integrated luminosity of $10\,\mathrm{fb}^{-1}$. The result is shown in Figure 3: the uncertainty at the high-x end is reduced to from 18% to 7%.

5 Summary and conclusions

With the early data QCD analyses presented, we will be able to improve our knowledge of minimum bias events and the underlying event. This will become

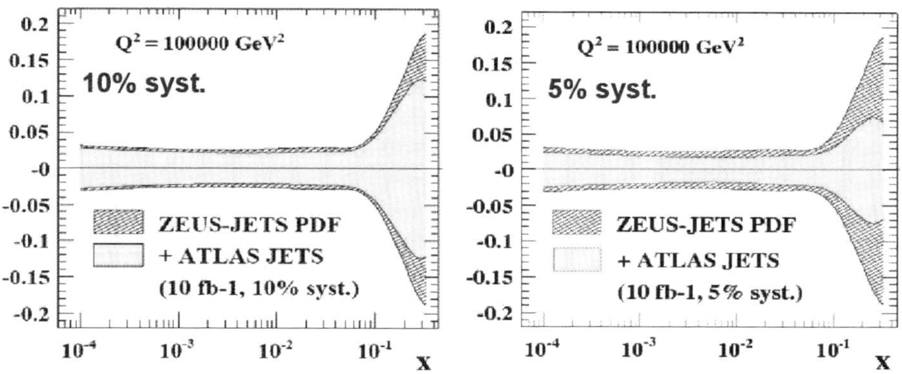

Figure 3: Fractional uncertainty on the gluon density function for jet energy scale uncertainties assumed to be 10% (left) and 5% (right). Figure taken from [9].

even more important once the LHC will run in configurations with more than one proton-proton-collision per bunch crossing.

Acknowledgements

I am grateful for very useful discussion with Arthur Moraes. I acknowledge the support of the BMBF, Germany under the contract number 05H09UM1.

References

[1] The ATLAS Collaboration, G Aad *et al*, *JINST* **3** S08003 (2008).
[2] A. Moraes, C. Buttar and I. Dawson, *Eur. J. Phys.* **C50** 435 (2007).
[3] T. Sjostrand, S. Mrenna and P. Skands, *JHEP* **05** 26 (2006).
[4] R. Engel, *Z. Phys. C.* **66** 203 (1995).
[5] The ATLAS Collaboration, G. Aad *et al.*, *arXiv*:0901.0512, (2009).
[6] A. Moraes, *Modeling the Underlying Event: MC tunes for LHC predictions* at 4th HERA and the LHC Workshop, CERN (2008): `http://indico.cern.ch/getFile.py/access?contribId=44&sessionId=1&resId=0&materialId=slides&confId=27458`
[7] G. Corcella *et al.*, *JHEP* **01** 10 (2001).
[8] J.M. Butterworth, J.R. Forshaw and M.H. Seymour, *Z. Phys. C.* **72** 637 (1996).
[9] K. Rabbertz, *Expectations for QCD with first LHC Data* at 4th HERA and the LHC Workshop, CERN (2008): `http://indico.cern.ch/getFile.py/access?contribId=34&sessionId=5&resId=6&materialId=slides&confId=27458`

ATLAS DISCOVERY POTENTIAL OF THE STANDARD MODEL HIGGS BOSON

Christian Weiser [a] on behalf of the ATLAS collaboration
Physikalisches Institut, Albert-Ludwig-Universität Freiburg, Hermann-Herder-Str. 3, D-79104 Freiburg

*Abstract.*The Standard Model of elementary particles is remarkably succesful in describing experimental data. The Higgs mechanism as the origin of electroweak symmetry breaking and mass generation, however, has not yet been confirmed experimentally. The search for the Higgs boson is thus one of the most important tasks of the ATLAS experiment at the Large Hadron Collider (LHC). This talk will present an overview of the potential of the ATLAS detector for the discovery of the Standard Model Higgs boson. Different production processes and decay channels -to cover a wide mass range- will be discussed.

1 Introduction

The discovery of the Standard Model Higgs boson [1] is one of the most important outstanding issues in elementary particle physics. The current experimental knowledge on the Higgs boson mass, the only free parameter of the theory, comes from direct searches at LEP, setting a lower limit of $m_H >$114.4 GeV [2], direct searches at the Tevatron, excluding the mass range 162 GeV $< m_H <$ 166 GeV [3], and the combination of precision measurements of electroweak observables, suggesting a relatively light Higgs boson ($m_H <$157 GeV) [4][b]. Figure 1 shows the Higgs boson production cross sections for the different pro-

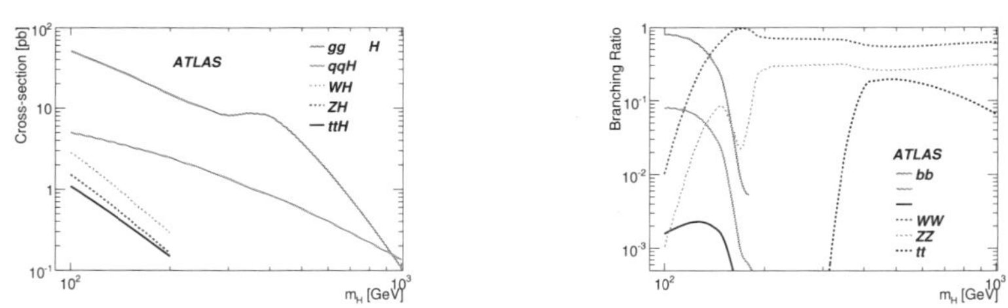

Figure 1: The cross sections at the LHC of different Higgs boson production processes (left) and the branching ratios of the Higgs boson (right) as function of the Higgs boson mass.

duction processes (gluon fusion, vector boson fusion and associated production with W and Z bosons or a $t\bar{t}$ pair) and the decay branching ratios as function of the Higgs boson mass. In the following, studies performed by the ATLAS collaboration with realistic full simulations of the ATLAS detector, including simulations of the trigger systems, and real reconstruction algorithms are

[a] e-mail: Christian.Weiser@cern.ch

[b] all limits given here at a confidence level of 95%

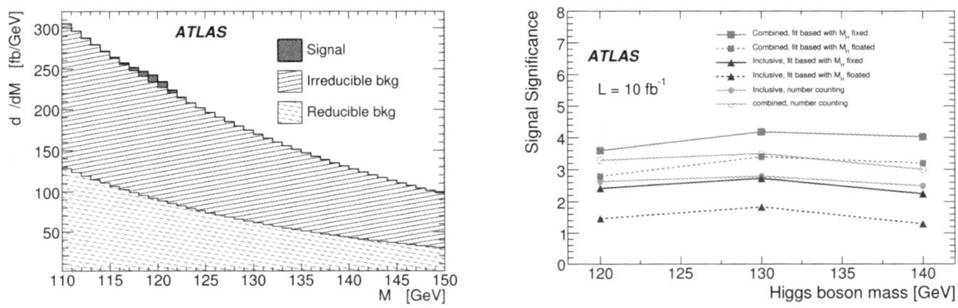

Figure 2: The expected signal and background components (left) for a Higgs boson mass of m_H =120 GeV and the expected signal significance using several methods for its estimation for an integrated luminosity of $\mathcal{L}^{int.} = 10$ fb^{-1} (right).

presented. The focus lies on studies of the discovery potential in the low and medium mass range (up to $m_H \approx 200$ GeV) at the design centre-of-mass energy of the LHC of $\sqrt{s}$=14 TeV.

2 $H \rightarrow \gamma\gamma$

Despite its very small branching ratio at the permille level (see Fig. 1), the decay of the Higgs boson into two photons, $H \rightarrow \gamma\gamma$, offers good discovery potential for low masses of the Higgs boson because of its clear experimental signature of two isolated photons in the detector. The main background processes in this channel are the irreducible $\gamma\gamma$ processes and the reducible $\gamma + jet$ and $jet + jet$ processes. Excellent photon identification and photon-jet separation are crucial to suppress the reducible backgrounds. An excellent di-photon mass resolution is needed to be able to establish a signal over the huge combinatorial background (see Fig. 2, left). The expected signal significance as function of the Higgs boson mass using different techniques is shown in Fig. 2, right.

3 $qqH, H \rightarrow \tau^+\tau^-$

The decay of the Higgs boson into a pair of τ-leptons, $H \rightarrow \tau^+\tau^-$, has the second largest branching ratio for low Higgs boson masses. The distinct experimental signature of the vector boson fusion production process with two *tag jets* in the forward and backward regions of the detector with a large rapidity gap (Fig. 3, left) is exploited to suppress background processes. The *collinear approximation* allows the computation of the $\tau^+\tau^-$ invariant mass (Fig. 3, middle). To be able to estimate the contribution of the dominating $Z \rightarrow \tau^+\tau^-$ background from the experimental data themselves, $Z \rightarrow \mu^+\mu^-$ decays are studied, with a proper transformation of the muons to τ leptons. The expected

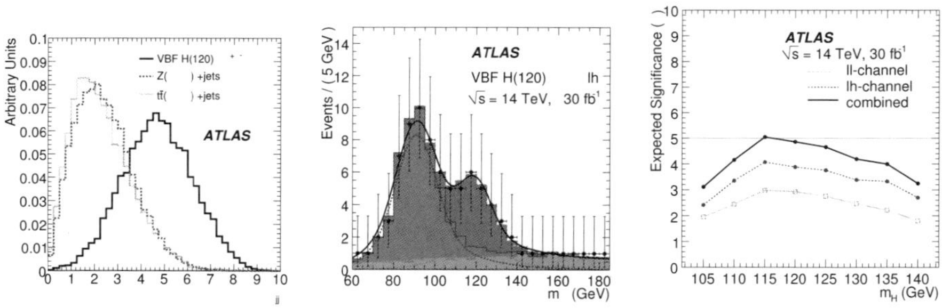

Figure 3: The rapidity difference between the tag jets for signal and background processes (left), the expected invariant $\tau^+\tau^-$ mass distribution for a Higgs boson mass of m_H =120 GeV (middle) and the expected signal significance (right) for an integrated luminosity of $\mathcal{L}^{int.} = 30$ fb^{-1}.

signal significance for the decay channels of the τ leptons decaying either both leptonically or one of them decaying into a hadronic final state is shown in Fig. 3, right.

4 $WH/ZH, H \to b\bar{b}$

As can be seen in Fig. 1, for masses of the Higgs boson up to about 135 GeV, the dominant decay mode is into a pair of b quarks. The dominant Higgs boson production process of gluon-gluon fusion, however, is not accessible experimentally because of the huge QCD production of jets, including real b jets.

Some time ago the associated production with a $t\bar{t}$ pair was thought to be the channel with the best discovery potential in the $b\bar{b}$ final state and being able to contribute significantly to a discovery of the Higgs boson for low masses [5, 6]. More recent studies, using realistic simulations of the detectors and reconstruction algorithms, have shown that this channel might only be able to marginally contribute to a discovery if systematic uncertainties are taken into account [7,8], as can be seen in Fig. 4, left. The production modes of the Higgs boson in association with a W or Z boson and the subsequent decay of the Higgs boson in a pair of b quarks had been studied extensively in the past at the LHC experiments (see e.g. [9]). However, it turned out that these channels offer only a very modest discovery potential because of the overwhelming backgrounds, mainly from $t\bar{t}$ and $W + jets$ events.

Recently, a new analysis on the WH and ZH channels has been performed in ATLAS [10], following an idea described in [11]. In this analysis, one selects events where both the W and Higgs bosons have a large transverse momentum, typically above 200 GeV. This results in the two b jets from the Higgs boson decay to end up in one *fat* jet. Despite the fact that the requirement

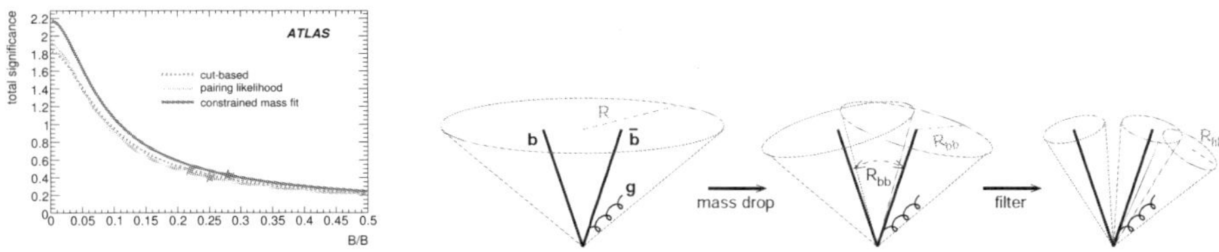

Figure 4: The dependence of the significance for the $t\bar{t}H, H \to b\bar{b}$ channel on the relative uncertainty of the backgrounds (left). Sketch of the jet decomposition technique as used for the $WH/ZH, H \to b\bar{b}$ analyses (right).

of such large transverse momenta selects only about 5% of the signal events, background processes such as $t\bar{t}$ and $W+jets$ are strongly suppressed. Furthermore, these events populate preferably the central part of the detector leading to increased acceptances and performances of many reconstruction algorithms. The key element of the analysis is the decomposition of the broad jet by iteratively undoing the last steps of the jet clustering, thus breaking up the jets into smaller components (see Fig. 4, right), until a configuration is obtained that contains the two b jets from the Higgs boson decay and possibly an additional jet from final state radiation.

The analysis has been carried out in three final states as motivated by the (semi-)leptonic decays of the W and Z bosons: (1) lepton[c]-$E_{\mathrm{T}}^{\mathrm{miss}}$, (2) lepton-lepton and (3) $E_{\mathrm{T}}^{\mathrm{miss}}$, where mainly (1) $W \to l\bar{\nu}$, (2) $Z \to l^+l^-$ and (3) $Z \to \nu\bar{\nu}$ contribute to these topologies. Figure 5 shows the mass distributions expected for a Higgs boson mass of m_H =120 GeV and an integrated luminosity of $\mathcal{L}^{int.} = 30$ fb^{-1} for these topologies. The statististical significances in terms of standard deviations obtained are (1) 3.0, (2) 1.5 and (3) 1.6, respectively. A combination yields a statistical significance of 3.7 standard deviations. The signal to background ratio in this analysis is significantly better than in the case of the $t\bar{t}H$ channel. This has the advantage that the dependence on the background uncertainties is strongly reduced, e.g. assuming an uncertainty of 15% on all background production cross sections still results in a significance of 3.0 standard deviations.

5 $H \to ZZ^{(*)}$

The decay of the Higgs boson into two Z bosons, $H \to ZZ^{(*)}$, which are then both decaying into leptons (electrons or muons), offers a very clear experimental signature with typically at least one reconstructed Z boson close to its nominal mass and a narrow Higgs boson mass peak. The main background processes are

[c] A lepton denotes an electron or muon here.

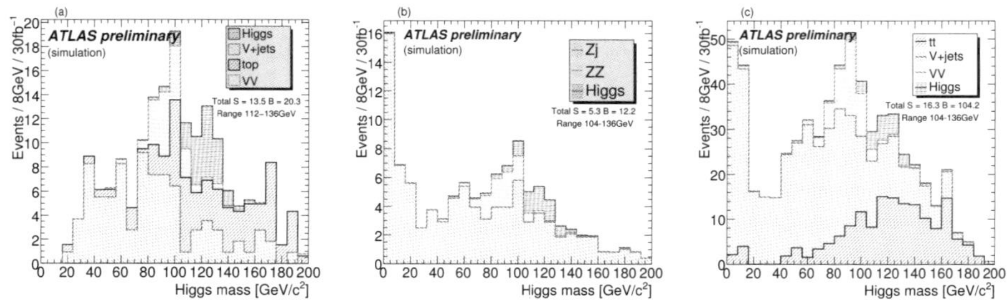

Figure 5: The expected $b\bar{b}$ mass distributions for signal and background events for the lepton-$E_{\mathrm{T}}^{\mathrm{miss}}$ (left), lepton-lepton (middle) and $E_{\mathrm{T}}^{\mathrm{miss}}$ (right) channels, for a Higgs boson mass of m_H =120 GeV and an integrated luminosity of $\mathcal{L}^{int.} = 30$ fb^{-1}.

ZZ (irreducible), $Zb\bar{b}$ and $t\bar{t}$ production, the latter two being reduced efficiently by lepton isolation and impact parameter requirements, since these leptons stem mainly from decays of heavy flavoured hadrons. The expected invariant mass distribution of the four leptons for a Higgs boson mass of m_H =150 GeV, as well as the expected signal significance for the muons, electrons and mixed Z boson decays are shown in Fig. 6, left and middle. It can be seen, that this channel can cover a wide range of Higgs boson masses. The dip in the expected signal significance around $m_H \approx 2 \cdot m_W \approx 160$ GeV is caused by the fact that the Higgs boson decays almost exclusively into a pair of W bosons in this mass range.

6 $H \to WW^{(*)}$

The decay of the Higgs boson into a pair of W bosons, $H \to WW^{(*)}$, also shows a very clear experimental signature of two leptons and large missing transverse energy caused by the undetectable neutrinos. However, in contrast to the

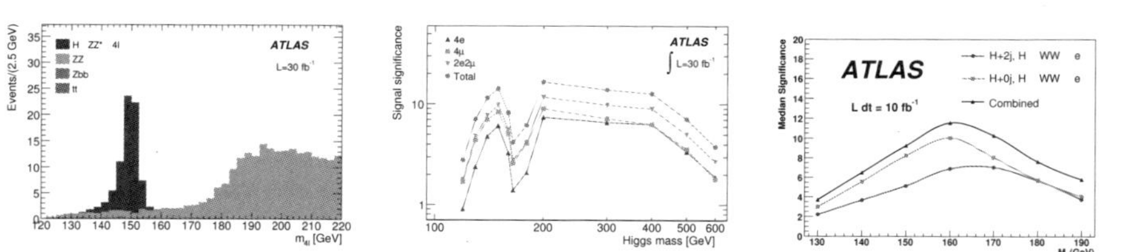

Figure 6: The invariant mass of four leptons in the decay mode $H \to ZZ^{(*)}$ for a Higgs boson mass of m_H =150 GeV (left) and the expected signal significances separated for the different Z boson decay modes and after combination (middle) for an integrated luminosity of $\mathcal{L}^{int.} = 30$ fb^{-1}. The expected signal significance for the channel $H \to WW^{(*)}$ for both final state topologies studied for an integrated luminosity of $\mathcal{L}^{int.} = 10$ fb^{-1} (right).

ZZ channel, no mass peak and intermediate resonances can be reconstructed making the interpretation of this channel more challenging. This channel has been analysed in the *+0 jet* and *+2 jets* modes, which are dominated by the gluon-gluon and vector boson fusion production processes, respectively. The spins of the particles and helicity structure of the decays involved allows the use of the opening angle between the two leptons in the transverse plane as efficient discriminating variable, especially against the irreducible WW background. This channel has an excellent discovery potential, especially around $m_H \approx 2 \cdot m_W \approx 160$ GeV, already with a modest integrated luminosity as can be seen in Fig. 6, right.

7 Combined Discovery Potential

To obtain the best possible sensitivity for either a discovery or exclusion of the Standard Model Higgs boson, the channels discussed before have been combined following the methods described in [7] and the results are shown in Fig. 7. This indicates that the ATLAS experiment has the potential to reach a signal significance corresponding to more than five standard deviations for an integrated luminosity of $\mathcal{L}^{int.} = 10$ fb^{-1} for Higgs boson masses $m_H > 127$ GeV. If no evidence for a Higgs boson is found, Higgs boson masses $m_H > 115$ GeV are expected to be excluded at 95% confidence level with an integrated luminosity of $\mathcal{L}^{int.} = 2$ fb^{-1}. Such a combination is particularly important for low Higgs boson masses where no single channel alone offers sufficient discovery potential. The inclusion of the channels $WH/ZH, H \to b\bar{b}$ (see Sec. 4) will improve the situation in this mass range.

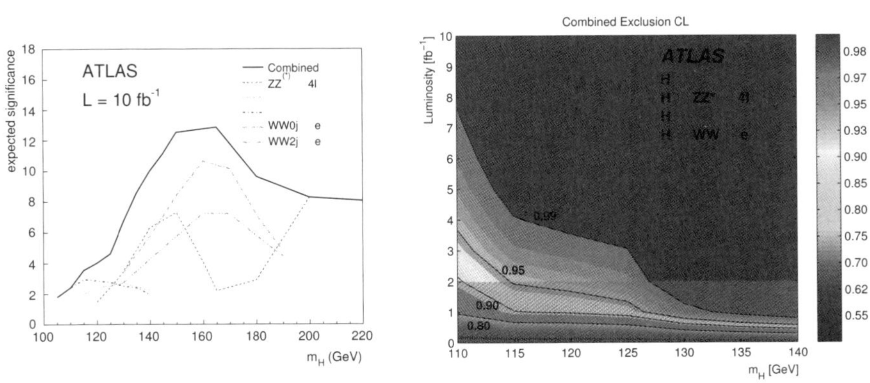

Figure 7: The potential of the ATLAS experiment for adiscovery (left, for an integrated luminosity of $\mathcal{L}^{int.} = 10$ fb^{-1}) and exclusion (right) of the Standard Model Higgs boson in dependence of its mass.

References

[1] P.W. Higgs, Phys. Lett. **12** (1964) 132 and Phys. Rev. **145** (1966) 1156; F. Englert and R. Brout, Phys. Rev. Lett. **13** (1964) 321; G.S. Guralnik, C.R. Hagen and T.W. Kibble, Phys. Rev. Lett. **13** (1964) 585.
[2] ALEPH, DELPHI, L3 and OPAL Collaborations, Phys. Lett. **B565** (2003) 61.
[3] CDF and D0 Collaborations, Phys. Rev. Lett. **104**, 061802 (2010).
[4] The LEP Electroweak Working Group, The Tevatron Electroweak Working Group, The SLD Electroweak and Heavy Flavour Working Groups, arXiv:0911.2604 [hep-ex].
[5] J. Cammin and M. Schumacher, ATL-PHYS-2003-024 .
[6] V. Drollinger et al., CMS Note 2001/054.
[7] ATLAS Collaboration, arXiv:0901.0512 [hep-ex].
[8] D. Benedetti et al., J. Phys. G: Nucl. Part. Phys. **34** (2007) N221.
[9] E. Richter-Was, ATL-PHYS-2000-024.
[10] ATLAS Collaboration, ATL-PHYS-PUB-2009-088.
[11] J. Butterworth et al., Phys. Rev. Lett. **100**,242001 (2008).

SIGNATURES OF ADS/CFT USING THE CMS EXPERIMENT AT THE LHC

David Krofcheck [a]

Dept of Physics, University of Auckland, Private Bag 92019, Auckland, New Zealand

Abstract. Measurements of jet suppression via heavy ion collisions at RHIC lead to an interpretation that the produced, high temperature nuclear matter is "color opaque". Observation of large collective motion, known as elliptic flow in this strongly interacting nuclear matter, led to the suggestion that ideal hydrodynamic behavior was observed. We will look at what could be measured with the CMS detector, and how the higher centre-of-mass energy at the LHC will permit access to new observables and laboratory tests of AdS/CFT versus pQCD.

1 Introduction

The study of relativistic nuclear collisions at RHIC has generated data that supports the idea that Quantum Chromodynamics(QCD) may share a deep mathematical link with a form of string theory via anti-de Sitter/conformal field theory (the AdS/CFT conjecture [1]). Ideal hydrodynamic fluid behavior observed at RHIC from the strongly interacting quark gluon plasma (sQGP) has prompted new theoretical work (see [2]) on comparing measurable quantities like: nuclear matter flow; jet propagation through hot nuclear matter; jet energy loss via gauge boson tagging; p_T dependent quarkonia melting; and nuclear modification factors (R_{AA}), with perturbative QCD (pQCD) calculations and string theory inspired predictions for these effects.

2 Experimental Signatures

2.1 Elliptic Flow

Elliptic flow data [3] was obtained at RHIC using the STAR detector. Ideal fluid hydrodynamical calculations [4] were shown to reproduce the data. An analysis [5] using the AdS/CFT conjecture also explained the data by requiring a minimum viscosity-to-entropy ratio $\geq 1/4\pi$ for the hot, flowing, nuclear matter. The CMS detector [6] at the Large Hadron Collider (LHC) should be capable of measuring low-p_T elliptic flow for $p_T \leq 1 GeV/c$. Particle identification for protons, pions, and kaons in the CMS detector will be done using the pixel barrel and silicon strip detectors. [7] If the same ideal hydrodynamic behavior is observed at the LHC beam energy, then the LHC will also provide a testing ground for the AdS/CFT conjecture and would permit comparisons with pQCD.

[a] e-mail: d.krofcheck@auckland.ac.nz

2.2 *Gauge Boson Tagging*

The CMS will primarily reconstruct the Z^0 in the dimuon channel The large p_T of its decay muons allows for a clean identification of this gauge boson over the underlying heavy-ion event background. Energy loss of the jet on the away side can be compared against the gauge boson energy to determine jet energy loss in the colored medium. It is important to note that contributions from the γ^* to dimuons has only been simulated at the generator level in the CMS acceptance. [7] In principle, the high resolution CMS electromagnetic calorimeter with large coverage and segmentation may make it possible to reconstruct γ-jet events. The γ escapes the hot colored nuclear matter without depositing much energy.

2.3 *Shock Waves*

A heavy quark moving through a sQGP will generate a diffusion wake and a "sonic" boom (Mach cone) like pattern of emitted particles. Thermal AdS/CFT calculations in the rest frame of the sQGP lead to the suggestion that the energy flow from the sonic boom may dissipate approximately $\sim (1+v^2)$ more energy than is fed in by the diffusion wake energy flow. [8] The CMS detector, with its ability to measure energy flow may allow an empirical check of this predicted velocity dependence in the supression.

2.4 *Quarkonia Melting*

The dissociation of charmonium bound states was suggested as a signal of the production of hot and deconfined quark gluon plasma by Matsui and Satz. [9] It is expected that the melting of bottomonium bound states should occur in sQGP matter for center-of-mass energies expected at the LHC.

The total production of high-p_T charmonium and bottomium is expected to be greatly enhanced at the LHC when compared to RHIC. In Fig. 1 the production and detection range of high-p_T quarkonia are simulated for the CMS detector. An observation of changes in the melting pattern of the quarkonia resonances (ψ', Υ' and Υ''), predicted using AdS/CFT, can be used to explore this conjecture. It is predicted [10] that the melting temperatures for the various J/ψ and the Υ families should show $T(v)_{Melt} \sim T(0)_{Melt}(1+v^2)^{1/4}$ for heavy mesons as a function of their transverse momenta. Here, $T(0)_{Melt}$ is the dissociation temperature for a heavy quark meson in its rest frame

2.5 *Nuclear Modification Factor*

Recently, there was a proposal [11] to use nuclear modification factors for a possibly more sensitive high-p_T test comparing AdS/CFT with pQCD. The idea is to use the momentum independence and heavy quark inverse mass de-

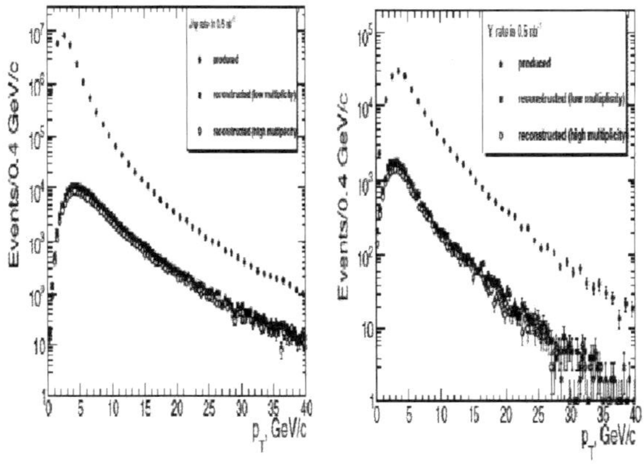

Figure 1: Simulated production and generated CMS detection of high-p_T J/ψ (left) and Υ (right) for PbPb at 5.5 TeV. [7]

pendence of the AdS/CFT drag coefficient. This drag coefficient is completely different to the $log(p_T/M)/p_T$ variation of the drag coefficient of pQCD.

Acknowledgments

This work was partly supported by the New Zealand Performance Based Research Fund(PBRF) at the University of Auckland Department of Physics.

References

[1] J. Maldacena, *Adv. Theor. Math. Phys.* **2** (1998) 231.
[2] W. Horowitz, Y. Kovchegov, *Physics Letters* **B680** (2009) 56.
[3] R. Snellings *AIP Conf. Proc.* **756** (2005) 390.
[4] P. Huovinen, P. Ruuskanen *Ann. Rev. Nucl. Part. Sci.* **56** (2005) 163.
[5] P. Kovtun, D. Son, A. Starinets, *Phys. Rev. Lett.* **94**(2005) 111601.
[6] The CMS Collaboration, *JINST* **3** (2008) S08004.
[7] The CMS Collaboration, *J. Phys. G: Nucl. Part. Phys.* **34** (2007) 2307.
[8] S.Gubser, S. Pufu, A. Yarom, *Phys. Rev. Lett.* **100** (2008) 012301.
[9] T. Matsui, H. Satz, *Phys. Lett.* **B178** (1986) 416.
[10] H. Liu, *J. Phys. G: Nucl. Part. Phys.* **34** (2007) S361-368.
[11] W. Horowitz, M. Gyulassy, *Phys. Lett.* **B666** (2008) 320.

STUDY OF JET TRANSVERSE STRUCTURE WITH CMS EXPERIMENT AT 10 TEV

Natalia Ilina [a] for the CMS collaboration

Institute for Theoretical and Experimental Physics, 117218, Moscow, Russia

*Abstract.*We present a study of the jet transverse structure which could be performed at CMS start up conditions. The analysis corresponds to the first $10pb^{-1}$ of proton-proton collisions at $\sqrt{s} = 10$ TeV. Jet transverse structure is studied using the second moment of the jet profile in transverse momentum which reflects the distribution of transverse energy inside the jet cone. Different sources of systematics were considered and a comparison of predictions from different generators (HERWIG++ and PYTHIA) was made. The jet structure differs for quark and gluon initiated jets and the possibility of estimating the fractions of quark and gluon jets from the second moment of the jet profile in transverse momentum is discussed.

1 Introduction

In hadron-hadron collisions the jet shape provides a good test for different models of parton cascades and the hadronization [1, 2]. The comparison of the jet transverse structure between CMS data and Monte Carlo events can be a powerful tool for tuning Monte Carlo generators used for hadron jet simulations. This analysis is devoted to the analysis of the jet transverse structure studied with the CMS experiment at $\sqrt{s} = 10$ TeV, using the new observable, the second moment of the jet profile in transverse momentum, $\langle \delta R^2_{jet} \rangle$.

We use this new variable for testing the two main hadronization models (implemented in PYTHIA [3] and HERWIG++ [4] Monte Carlo generators) at the level of fully simulated CMS data. The possibility to measure the fraction of quark-induced jets using the second moment is shown.

2 Jet transverse structure variables

The variable used in the CMS collaboration for the study of jet transverse structure was proposed in [5]:

$$\langle \delta R^2_{jet} \rangle (p_{\rm T}) = \frac{\sum\limits_{i \in \rm jet} \Delta R^2(i, jet) \cdot p^i_{\rm T}}{\sum\limits_{i \in \rm jet} p^i_{\rm T}} \qquad (1)$$

where $\Delta R^2(i, j) = (\phi_i - \phi_j)^2 + (\eta_i - \eta_j)^2$. The summation can be performed either over stable particles, or over calorimeter towers or over reconstructed tracks in the $\eta - \phi$ cone around the jet axis.

[a]e-mail: Natalia.Ilina@cern.ch

In order to avoid the impact of the jet angular resolution, we use the 2nd central moments of ϕ and η. Then, we combine the central moments to calculate the trace of the $\eta - \phi$ correlation matrix $\langle \delta R_c^2 \rangle$.

While all jets become narrower as the transverse momentum increases, gluon initiated jets are found to be on average broader than quark initiated jets. Both generators predict for both quark and gluon initiated jets separately that $\langle \delta R_{jet}^2 \rangle$ does not depend on η for fixed jet transverse momentum. The quark jet fraction rises with p_{T} and higher quark fraction values are occurred for higher $|\eta|$.

In order to avoid the magnetic field shifts and the non-linearity effects of the calorimeter energy response we analyze the shape of the charged component of the jet, as reconstructed from charged particle tracks. The tracker provides more accurate measurements of the kinematics of low-p_{T} charged particles.

We have considered three sources of systematic uncertainty [5]: jet energy scale, tracker p_{T} resolution and simulation bias correction.

3 Testing of hadronization models in PYTHIA and HERWIG++

The comparison of the particle level and detector level observables shows out that the size of the difference is similar for both generators. The correction for this deviation was done.

Predictions of generators on the particle level are shown in Fig. 1(left) with data points which were produced by HERWIG++ samples passed through the full CMS detector simulation and reconstruction. The figure demonstrates that a CMS data sample corresponding to 10 pb^{-1} of integrated luminosity should be sufficient to distinguish between the predictions of the PYTHIA and HERWIG++.

Figure 1 (right) shows the variable $\langle \delta R_c^2 \rangle$ as a function of the jet p_{T} for quark, gluon and all jets for particle level Monte Carlo events and for all jets extracted from the data for $|\eta| < 1$. The different $\langle \delta R_c^2 \rangle$ distributions correspond to a variation of the quark induced jet fraction from 30% at $p_{\mathrm{T}} = 100$ GeV to 70% at $p_{\mathrm{T}} = 900$ GeV.

4 Conclusions

The study of the jet transverse structure was performed using the second central moment of the jet profile in transverse momentum, $\langle \delta R^2 \rangle$. We show that with MC generated data corresponding to the integrated luminosity of 10 pb^{-1}, the $\langle \delta R_c^2 \rangle_{tr}$ measured with future data from the CMS detector will be enough to distinguish between HERWIG++ and PYTHIA generators.

The results of this study can be used for tuning the MC models in a central η range. The fraction of quark- and gluon initiated jets does not depend on

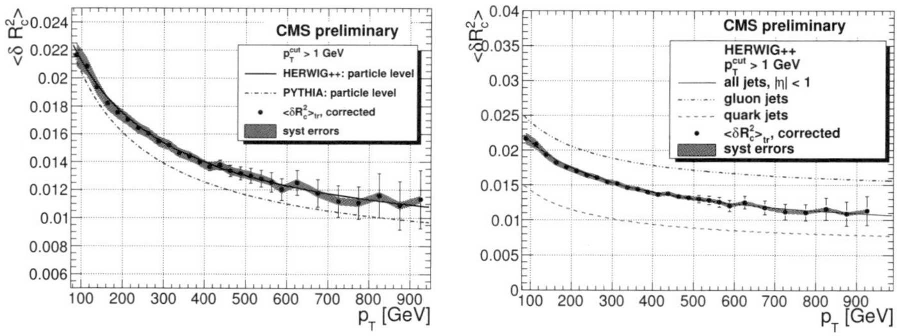

Figure 1: Comparison between PYTHIA and HERWIG++ predictions for the jet transverse structure for $|\eta| < 1$(left) and HERWIG++ prediction for the jet transverse structure for quark jets and gluon jets in $|\eta| < 1$(right). The data points are calculated from the reconstructed tracks as a function of the reconstructed jet transverse momentum and the lines show $\langle \delta R_c^2 \rangle$ calculated from the charged particles as a function of the generated jet transverse momentum. The error bars correspond to the statistical uncertainty for 10 pb^{-1}. The total systematic uncertainties are indicated by the shaded region.

parameters of the fragmentation models, therefore the main tuning that should be done is that of transverse structure of quark- and gluon jets. Once the tuning for central η-region is done and the data are well described, the fraction of quark induced jets as a function of jet p_T and η can be extracted from experimental CMS data for the other η-regions and compared with the Monte Carlo predictions.

5 Acknowledgements

This work was developed in the QCD physics analysis working group of CMS. We thank our CMS colleagues and especially the QCD conveners for comments and discussions.

References

[1] S.D. Ellis, Z. Kunszt and D.E. Soper, Phys.Rev.Lett. 69, 3615 (1992).
[2] J. Pumplin, Phys.Rev. D 44, 7(1991).
[3] T. Sjostrand, S. Mrenna and P. Skands, hep-ph/0603175, *"PYTHIA 6.4 physics and manual"*.
[4] M. Bahr et al., *"Herwig++ 2.2 Release Note"*, arXiv:0804.3053.
[5] CMS Collaboration, CMS PAS **QCD-08-002** (2009).

ELLIPTIC FLOW STUDIES IN HEAVY-ION COLLISIONS USING THE CMS DETECTOR AT THE LHC

Sergey Petrushanko [a] (for the CMS collaboration)
Skobeltsyn Institute of Nuclear Physics, Moscow State University, Leninskiye Gory, 119991 Moscow, Russia

Abstract. The azimuthal anisotropy of charged particles in heavy ion collisions is an important probe of quark-gluon plasma evolution at early stages. The nuclear reaction plane can be determined independently by different CMS detector subsystems and using different analysis methods. The capabilities of the CMS detector at the LHC to reconstruct the reaction plane of the collision and to measure elliptic flow with calorimetry and a tracking system are discussed.

1 Introduction

In non-central collisions between two nuclei the beam direction and the impact parameter vector define a reaction plane for each event. A measurement of the azimuthal anisotropy of particle production with respect to the reaction plane is one of the important tools for studying the properties of the dense matter created in ultra relativistic heavy-ion collisions [1,2]. This report is dedicated to studying the capability of the CMS detector [3] at the LHC to reconstruct the reaction plane and to measure elliptic flow, using calorimetry and the tracking system. The high tracking efficiency and low rate of fake tracks at the CMS, together with a large calorimetric coverage, provide a precise measurement of global event characteristics, event by event [4,5].

2 Methods

The elliptic flow parameter, v_2, is defined as the second harmonic coefficient in the Fourier expansion of the particle azimuthal distribution with respect to the reaction plane:

$$\frac{dN}{d\varphi} = \frac{N_0}{2\pi}[1 + 2v_1\cos(\varphi - \Psi_R) + 2v_2\cos 2(\varphi - \Psi_R) + ...], \quad (1)$$

where Ψ_R is the true reaction plane angle and N_0 stands for full multiplicity. Then v_2 is the average over particles of $\cos(2(\varphi - \Psi_R))$.

There exists a wealth of anisotropic flow measurement methods, each of which has its advantages and limitations. Here we have used a v_2 determination method based on plane angle measurement. Usually the true elliptic flow coefficient in the event plane (EP) method is evaluated by dividing the observed v_2 value by a factor, R [6], which accounts for the event plane resolution:

$$v_2\{EP\} = \frac{v_2^{\rm obs}\{EP\}}{R} = \frac{\langle\cos 2(\varphi - \Psi_2)\rangle}{\langle\cos 2(\Psi_2 - \Psi_R)\rangle}. \quad (2)$$

[a] e-mail: Serguei.Petrouchanko@cern.ch

Here the event plane angle Ψ_2 is the estimate of the true reaction plane angle Ψ_R. The mean was taken over all charged particles in a given event and then over all events. In order to avoid the trivial autocorrelation of particles, the event plane angle Ψ_2 and, hence, R are calculated from the angular distribution of a sample of events, and v_2 from another event sample with the same multiplicity.

The event plane angle, Ψ_n, can be determined from the measured n-th harmonics via the standard method [6, 7]:

$$\tan n\Psi_n = \frac{\sum\limits_i w_i \sin(n\varphi_i)}{\sum\limits_i w_i \cos(n\varphi_i)}, \qquad n \geq 1, \quad 0 \leq \Psi_n < 2\pi/n, \tag{3}$$

where φ_i is the azimuthal angle of the i-th particle and w_i is a weight. The sum runs over all particles in each given event.

3 Analysis and discussion

For the estimation of the azimuthal anisotropy of particles in heavy ion collisions, the HYDJET event generator [8] was used with full GEANT simulation of the CMS detector responses.

Study of event plane resolution with CMS calorimeters. It was found that the CMS calorimetric system can be used for the determination of the event plane, using $\omega_i = p_{Ti}$ in Eq. (3) [4]. Although the anisotropic flow is maximal at midrapidity, the much larger total energy deposition in the CMS calorimeter endcaps ($1.5 < |\eta| < 3.$) results in reduced relative fluctuations and, accordingly, in a much better event plane resolution. Moreover, energy flow measurements in the endcaps are less sensitive to the magnetic field than in the barrel region ($|\eta| < 1.5$).

Study of v_2 reconstruction with the CMS tracker. A sample of 10^5 Pb+Pb events at impact parameter $b = 9$ fm within the pseudorapidity window $|\eta| < 2.4$ (the CMS tracker acceptance) was utilized. The standard settings were used to reconstruct tracks (i.e. more than 12 hits per track and a track fit probability above 1 %). A cut on $p_T > 0.9$ GeV/c was set in both simulated and reconstructed events. The number of reconstructed tracks per event is about 170 at this centrality.

The differential p_T dependence of the elliptic flow in Pb+Pb collisions for impact parameter $b = 9$ fm is shown in Fig. 1. Here we have calculated the dependence using the event plane angle determination from Eq. (3). Two sub-event sets were used, with $\eta > 0$ and $\eta < 0$.

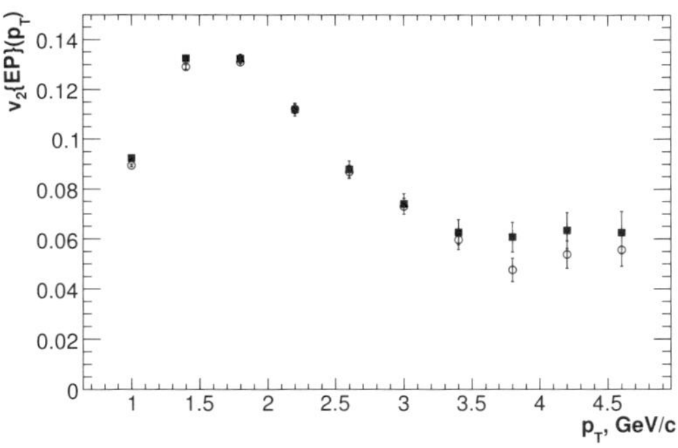

Figure 1: The p_T dependence of $v_2\{EP\}$ in Pb+Pb collisions for impact parameter $b = 9$ fm, calculated with the simulated (open circles) and reconstructed events (closed squares). Statistical errors are shown for 10^5 events. Non-flow systematic uncertainties not included.

4 Summary

We have shown that the CMS detector at the LHC will be able to determine the reaction plane using the calorimeters and the tracker. The transverse momentum dependence of the elliptic flow coefficient v_2 can be reconstructed in the CMS tracker with high accuracy. The CMS track reconstruction performance induces a systematic error estimated to be about 3% on the v_2 determination.

Acknowledgments

I would like to thank the members of the CMS collaboration for providing the materials. I wish to express the gratitude to the organizers of the 14th Lomonosov Conference on Elementary Particle Physics for the possibility to give my talk. This work was partly supported by Russian Foundation for Basic Research (grants No 08-02-91001 and No 08-02-92496) and Russian Ministry for Education and Science (contracts 01.164.1-2.NB05 and 02.740.11.0244).

References

[1] J.-Y.Ollitrault, *Phys.Rev.* **D** 46, 229 (1992).
[2] H.Sorge, *Phys. Rev.Lett.* 82, 2048 (1999).
[3] CMS Collaboration, *JINST* 3, S08004 (2008).
[4] D.d'Enterria et al., *J.Phys.* **G** 34, 2307 (2007).
[5] V.L.Korotkikh et al., *Phys.Atom.Nucl.* 71, 2142 (2008).
[6] A.M.Poskanzer, S.A.Voloshin, *Phys.Rev.* **C** 58, 1671 (1998).
[7] S.A.Voloshin, Y.Zhang, *Z.Phys.* **C** 70, 665, (1996).
[8] I.P.Lokhtin, A.M.Snigirev, *Eur.Phys.J.* **C** 46, 211 (2006).

ION PHYSICS IN CMS EXPERIMENT AT LHC

L.I.Sarycheva (on behalf the CMS Collaboration)
Faculty of Physics, Moscow State University, 119991 Moscow, Russia

Abstract. We review the capabilities of the CMS experiment to measure various observables in Pb-Pb collisions at the LHC, including elliptic flow, jet quenching, J/ψ and Υ production. The results are based on full simulation studies using the Monte-Carlo event generator HYDJET.

1 CMS detector and Heavy Ion program

The Compact Muon Solenoid (CMS) at LHC is a detector optimised for measurements of the characteristics of high energy leptons, photons, hadrons and hadronic jets, which provide the unique possibility to make "hard probes" in both pp and AA collisions. The detailed description of the CMS apparatus can be found in [1]– [4]. CMS features high magnetic field of 4T in 13 m long, $\sim$ 3 m dia volume inside. The tracking system and muon chambers cover the pseudorapidity region $|\eta| < 2.4$, the electromagnetic (ECal) and hadron (HCal) calorimeters reach the values $\eta = \pm 3$ while the very forward (HF) calorimeters — $\eta = \pm 5.2$. In addition, a quartz-fibre calorimeter "CASTOR" covers the region $5.2 < |\eta| < 6.6$, and another "zero degree" calorimeter (ZDC) 140 m distant from the interaction point — the region $|\eta| \geq 8.3$. CASTOR and ZDC extend the area of observations to forward physics including ultraperipheral interaction, limiting fragmentation, saturation, color glass condensate, exotics. The tracking system of CMS apparatus allows track reconstruction and momentum esimation with resolution better than 2% in the region P_T between 0.5 GeV/c and few tens GeV/c. CMS is an excelent device for the study of quark-gluon plasma using hard probes: quarkonia (J/ψ, Υ), heavy quarks ($b\bar{b}$) and Z^0, high P_T jets, as well as jet-jet, jet-γ, jet-γ^*/Z^0 correlations [5, 6]. Heavy Ion (HI) Physics program at CMS is presented in detail in [7].

2 Quarkonia and heavy quarks

Quark-gluon plasma becomes hotter and lives longer as energy increases from $\sqrt{s} = 200$ GeV/n-n (RHIC) to 5500 GeV/n-n (LHC). Quarkonia (J/ψ, ψ', Υ, Υ', Υ'') and Z^0 should be observed with high statistics under the conditions of LHC. Large cross-section for heavy quark (b, c) production allows evaluation of medium-induced energy loss of partons basing on the spectra of large mass $\mu\mu$ pairs and secondary J/ψ. SPS and RHIC experiments report the J/ψ suppression, and now we have a question: will recombination or suppression be observed at LHC? At LHC we have the new probes Υ: Υ' and Υ'', so the question is: should we expect such regeneration or suppression for these particles? From $B \to J/\psi$ and $BB \to \mu^+\mu^-$ events one can obtain information about b-quark rescattering and energy loss in the medium. Fig. 1 shows the mass distributions of dimuons with opposite charge assuming the multiplicity 2500 and the full acceptance ($|\eta| < 2.4$). The different quarkonia resonances appear on top of the continuum due to the various decay modes: π^+K, charm and bottom decays.

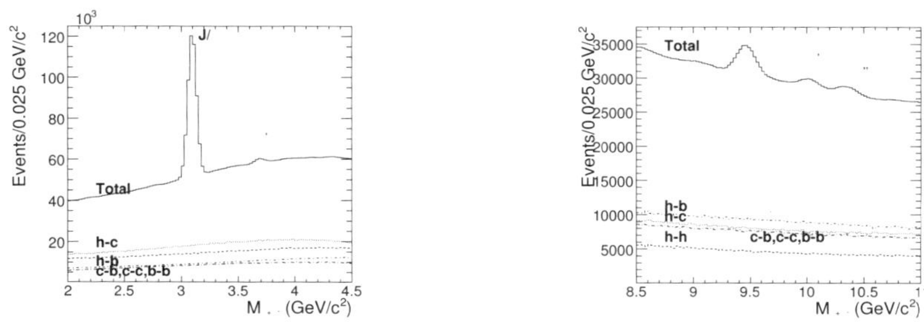

Figure 1: Dimuon mass distributions within $|\eta| < 2.4$ for PbPb events with $dN_{\rm ch}/d\eta|_{\eta=0} =$ 2500 in the J/ψ and Υ mass regions. The main background contributions are also shown: h, c and b stand for $\pi + K$, charm, and bottom decay muons, respectively.

3 Jet Quenching and azimuthal anysotropy

The algorithm of jet reconstruction in HI collisions is based on the event-by-event η dependent background substraction. This work started in 1994 [5] and has been in progress up to now [6]. The possibility of hard QCD jet reconstruction with initial parton energies in the range $50 \div 300$ GeV was investigated (using HIJING model) in central PbPb collisions. An important tool for studies of the properties of quark-gluon plasma (QGP) in ultrarelativistic HI collisions is the analysis of QCD jet production: medium-induced energy loss of energetic partons (jet quenching) is very different in cold nuclear matter and in QGP. The energy lost by partons in nuclear matter appears to be 10^2 times lager than that in a hadron gas. The jet quenching should manifest itself in P_T distribution, elliptic flow, jet fragmentation function, distribution of azimuthal angles in *non-central* HI collisions, and other observables. Fig. 2 shows the ratio of jet fragmentation functions $D^{\rm PbPb}(Z)/D^{pp}(Z) \simeq 0.25$. The jet quenching is observed in the nuclear modification factor for charged hadrons: $R_{AA}(P_T) = d^2N^{AA}/dP_Td\eta/T_{AA}d^2\sigma^{NN}/dP_Td\eta$. Fig. 3 shows the factor of nuclear modification for charged hadrons up to $P_T \sim 300$ GeV/c.

Non-central HI collisions (impact parameter $b \neq 0$) result in the elliptic volume of interacting nuclear matter. The energy flow from this elliptic volume may be non-isotropic. Fig. 4 shows the behaviour of parameter $v_2(P_T, y) = \langle \cos[2(\phi - \Psi_R)] \rangle$ as a function of P_T and η, obtained using CMS Tracker.

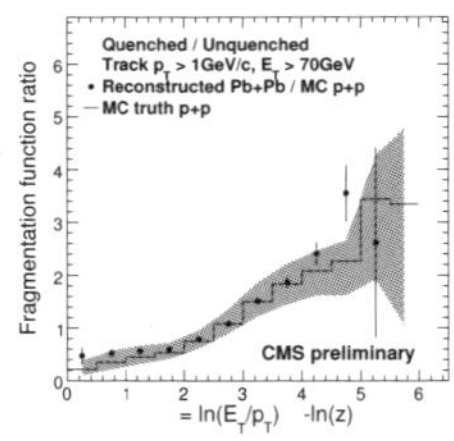

Figure 2: Generated (histogram) and reconstructed (points) ratio of quenched $D^{\rm PbPb}(z)$ and unquenched $D^{pp}(z)$ fragmentation functions as a function of ξ.

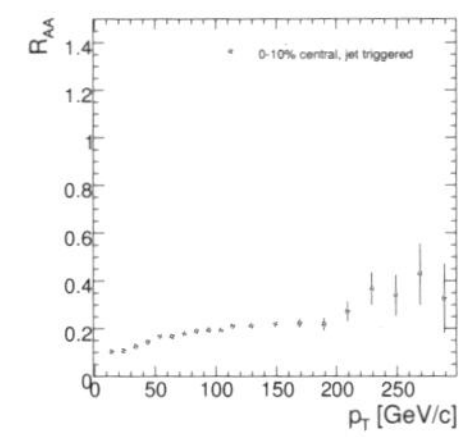

Figure 3: Nuclear modification factor $R_{AA}(P_T)$ for the charged hadrons (HYDJET).

Figure 4: Azimuthal anisotropy in HI Collisions: $v_2(P_T, y) = \langle\cos[2(\phi - \Psi_R)]\rangle$, where azimuthal angle $\phi = \arctan(P_y/P_x)$. Open circles — simulated, closed squares — reconstructed events(HYDJET, 10000 PbPb events, $b = 9$ fm).

4 Summary and Acknowledgement

With its large acceptance, nearly fine granularity hadronic and electromagnetic calorimetry, precision muon and tracking system, the CMS detector is an excellent device for the study of hard probes (such as quarkonia, jets, photons and high P_T hadrons) in heavy ion collisions at the LHC.

The author is grateful to the members CMS Collaboration for providing the material for this talk.

References

[1] CMS HCal Technical Design Report 1997 CERN/LHCC 97-31.
[2] CMS ECal Technical Design Report 1997 CERN/LHCC 97-32.
[3] CMS MUON Technical Design Report 1997 CERN/LHCC 97-33.
[4] CMS Tracker Technical Design Report 1998 CERN/LHCC 98-6.
[5] CMS TN-1995/052: O.L. Kodolova *et al.* "Transverse energy and jets in the S-S collisions in CMS".
[6] CMS NOTE-1999/016: M.Bedjidian, O.L. Kodolova, *et al.* "Jet Physics in CMS Heavy Ion Programme".
[7] J. Phys. G: Nucl. Part. Phys. **34** (2007) 2307–2455. CMS Collaboration "High Density QCD with Heavy Ions".

SYSTEMATIC UNCERTAINTIES IN EXPERIMENTS AT LHC

Sergei I. Bityukov [a]
Institute for High Energy Physics, Protvino, 142281, Russia
Nikolai V. Krasnikov
Institute for Nuclear Research RAS, Moscow, 117312, Russia

Abstract. We discuss the issues related to systematics in experimental data. We consider the possibilities both for classifying and estimating of them and for implementing the estimators of systematic uncertainties in the data analysis of experiments at LHC.

1 Introduction

Most measurements of physical quantities involve both a statistical uncertainty and a systematic uncertainty. Typically, systematic uncertainties are not well defined and are not straightforward to determine. They are also usually not Gaussian distributed, and combining systematic uncertainties from different sources is problematic.

The consideration of issues of uncertainty propagation and combining is a very important part of planning of experiments at the LHC (see, for example, PTDR CMS v.II [1]). Such type studies [2–5] often include proposals to combine statistical and systematic uncertainties, and techniques to estimate the magnitude of systematic uncertainties.

2 Classification of systematic uncertainties

Statistical uncertainties are the result of stochastic fluctuations arising from the fact that a measurement is based on a finite set of observations.

Systematic uncertainties arise from uncertainties associated with the nature of the measurement apparatus, assumptions made by the experimenter, or the model used to make inferences based on the observed data. Examples of systematic uncertainties include uncertainties that arise from the calibration of the measurement device, the probability of detection of a given type of interaction, and parameters of model used to make inferences that themselves are not precisely known.

Let us consider systematic uncertainties which are usually taken into account in experiment. In ref.[4] is proposed the classification of systematic uncertainties.

– Class 1 systematics are uncertainties that can be constrained by ancillary measurements and can therefore be treated as statistical uncertainties.

[a] e-mail: Serguei.Bitioukov@cern.ch

– Class 2 systematics arise from model assumptions in the measurement or from poorly understood features of the data or analysis technique that introduce a potential bias in the experimental outcome.

– Class 3 systematics from uncertainties in the underlying theoretical paradigm used to make inferences using the data.

3 Typical sources of systematic uncertainties in LHC

The part of systematic uncertainties depends on the integrated luminosity of experiment. The another part depends on statistics in ancillary measurements during experiment and, correspondingly, has internal limitations. Third part has not dependence from any experimental data. The list of typical sources of systematics is presented below.

– Luminosity from detector: the uncertainty on the luminosity detector acceptance and efficiency.

– Luminosity from cross section: the uncertainty of the inelastic and diffractive cross sections.

– Signal modeling: the systematic uncertainty arising from uncertainties in the modeling of the signal.

– Detector modeling: the systematic uncertainty arising from the uncertainty on the event detection efficiencies for object identification and MC mismodeling of data.

– Background from MC: the systematic uncertainty arising from uncertainties in modeling of the different background sources.

– Background from data: the systematic uncertainty arising from uncertainties in modeling of the different background sources that are obtained using data-driven methods.

– Particle identification and misstagging: the systematic uncertainty coming from the uncertainty on the identification of particles and mistag rate and shape modeling.

– JES and E_T^{miss} scale: the JES uncertainty which originates from limitations in the calibration data samples used.

4 Treatment of systematic uncertainties

Class 1 systematics are statistical in nature and will therefore naturally scale with the sample (special or regular) size.

A common technique for estimating the magnitude of systematic uncertainty of Class 2 is to determine the maximum variation in the measurement, associated with the given source of systematic uncertainty.

Class 3 systematics do not depend on how well we understand the measurement per se, but are fundamentally tied to the theoretical model or hypothesis being tested.

Often the distribution of errors is supposed to be normal distribution. It allows to combine the error estimates of several uncertainties into the overall uncertainty using a root sum-of-the-squares approach.

The systematic uncertainties usually not Gaussian distributed. For example, distributions of efficiencies are asymmetric. In this case the combination of different systematic uncertainties by standard methods is problematic.

If we measure the random variable to estimate the parameter of model which describe the behavior of this random variable, we can describe the systematic uncertainties via confidence densities [6, 7]. In this case the uncertainty in determination of parameter is associated with probability.

More commonly, if we know the probability density of uncertainties during planning of experiment we can use MC experiments to imitate the possible result and to construct the confidence densities of parameters in frame of the model of experiment. It is usual practice in experiment preparation and it is done for experiments at LHC.

By the using of notion confidence distributions (and, correspondingly, confidence densities) all sources of uncertainties can be included in a logically consistent way to description of overall uncertainty [6]. We use this notion in statistical studies during preparing experiments at LHC [8, 9].

Acknowledgments

This work was supported by the RFBR Grant 08-02-91007.

References

[1] CMS PTDR v.2 (Appendix B), CERN/LHCC 2006-021, 2006

[2] G.D'Agostini and M.Raso, e-Print: hep-ex/0002056, 2000

[3] R.Barlow, e-Print: hep-ex/0207026, 2002

[4] P.K. Sinervo, in *"PHYSTAT2003"*, (Proceedings of the Conference, September 8-11 2003, Stanford, USA), eds. by L. Lyons, R. Mount, R. Reitmeyer, SLAC, 122, 2003; eConf: C030908.

[5] J.Heinrich and L.Lyons, *Annu.Rev.Nucl.Part.Sci.* 57, 145 (2007).

[6] K.Singh, M.Xie, W.E.Strawderman, *The Annals of Statistics* 33, 159 (2005).

[7] S.Bityukov, N.Krasnikov, IHEP Preprint 2008-10, Protvino, 2008 (in russian).

[8] S.Bityukov, *JHEP* 09, 060 (2002).

[9] S.Bityukov, N.Krasnikov, A.Nikitenko, V.Smirnova, *Proc.of Sci.* (**ACAT08**) 118, 2009.

ELECTROWEAK PHYSICS AND SEARCHES FOR NEW PHYSICS AT HERA

U.Schneekloth [a]

Deutsches Elektronen-Synchrotron DESY
Hamburg, Germany

Abstract. Recent results from the H1 and ZEUS experiments are reported on electroweak physics and on searches for new physics. All results are in good agreement with the Standard Model.

1 Introduction

High energy electron- (positron-) proton collisions at the HERA collider, colliding 27.5 GeV electrons (or positrons) on 920 GeV protons, provide a unique opportunity for studying electroweak physics and for searches for physics beyond the Standard Model. Extensive studies have been performed by the H1 and ZEUS collaborations with the final data sets, corresponding to an integrated luminosity of about 0.5 fb^{-1} per experiment. Recently, significant improvements have been achieved by combining the results of both experiments. A summary of electroweak studies and searches for new physics is presented in this paper.

2 Evidence for Electroweak Unification

One of the main goals of the HERA physics program has been the precise measurement of the differential cross sections in neutral (NC) and charged current (CC) deep inelastic scattering (DIS). These measurements have not only been of great importance for the understanding of the quark and gluon content of the proton, but have also provided a basis for electroweak (EW) studies [1].

Figure 1 shows the single differential NC and CC e^+p and e^-p cross sections measured by H1 [2, 3] and ZEUS [4] for $Q^2 > 200\,\mathrm{GeV}^2$ as a function of the momentum transfer Q^2. The NC data show a $1/Q^4$ behavior due to the electromagnetic current, whereas the CC cross section is proportional to $1/(Q^2 + M_W^2)^2$, which gives a much less steeper cross section decrease as a function of Q^2, since the propagator term includes the W mass. At high Q^2, $Q^2 > M_W^2$, both NC and CC are mediated by a unified electroweak current and both cross sections are of comparable size. There is excellent agreement with the SM predictions over 7 (4) orders of magnitude for NC (CC) scattering.

[a] e-mail: uwe.schneekloth@desy.de

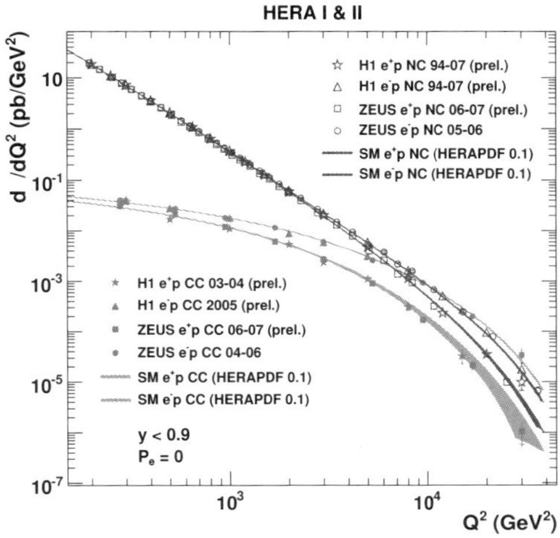

Figure 1: *Differential cross section for NC and CC ep scattering, as measured by H1 and ZEUS, compared to the SM expectations using the HERAPDF parametrization of the proton parton distribution functions.*

3 Neutral Current Cross Sections

The unpolarized NC cross section has been measured over a large range of Q^2 (200 - 30 000 GeV^2) for e^-p and e^+p DIS. At high Q^2, the e^-p cross section is significantly larger than the e^+p cross section. This charge asymmetry can be exploited to measure the interference structure function $xF_3^{\gamma Z}$:

$$xF_3^{\gamma Z} \simeq x\tilde{F}_3 \frac{(Q^2 + M_Z^2)}{\alpha_e \kappa Q^2}, \qquad x\tilde{F}_3 = \frac{Y_+}{2Y_-}(\tilde{\sigma}^{e^-p} - \tilde{\sigma}^{e^+p}).$$

Since $xF_3^{\gamma Z}$ has little dependence on Q^2, the measurements from $1\,500 < Q^2 < 30\,000\,\mathrm{GeV}^2$ were extrapolated to $5\,000\,\mathrm{GeV}^2$ and then averaged to obtain higher statistical significance. Figure 2 shows $xF_3^{\gamma Z}$ measured at $Q^2 = 5\,000\,\mathrm{GeV}^2$ [6]. It is well described by the SM prediction.

A direct measure of EW effects are the charge dependent polarization asymmetries of the NC cross sections, which are now accessible using the HERA II data. The cross section asymmetries $A^{\pm}$, as defined below, can be used to measure to a good approximation the structure function ratio, which is proportional to the product $a_e v_q$, where $a_e (v_q)$ is the axial (vector) coupling of the electron (quark q) to the Z boson, and thus gives a direct measure of parity violation.

$$A^{\pm} = \frac{2}{P_R - P_L} \cdot \frac{\sigma^{\pm}(P_R) - \sigma^{\pm}(P_L)}{\sigma^{\pm}(P_R) + \sigma^{\pm}(P_L)} \simeq \mp \kappa a_e \frac{F_2^{\gamma Z}}{F_2} \propto a_e v_q,$$

where P_R (P_L) is the right (left) handed lepton beam polarization.

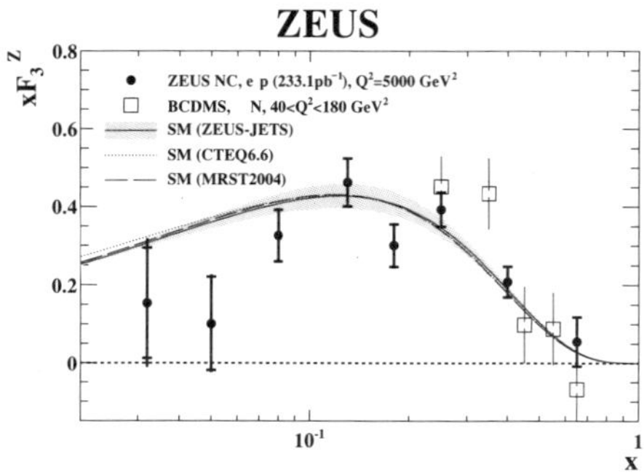

Figure 2: *The structure function $xF_3^{\gamma Z}$ extrapolated to a single Q^2 value of 5 000 GeV^2, plotted as a function of x.*

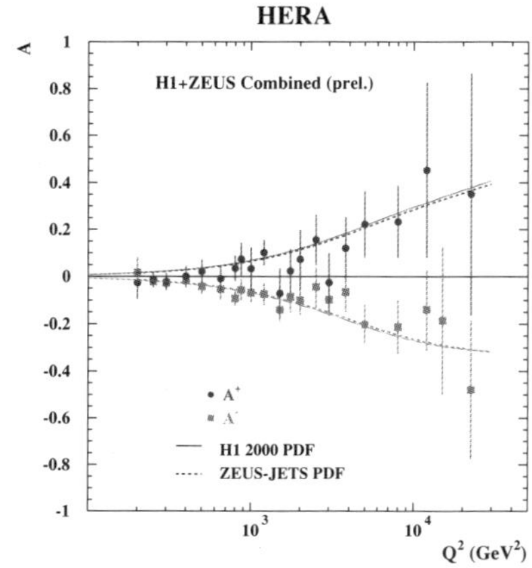

Figure 3: *The polarization asymmetry A plotted as a function Q^2.*

The asymmetries obtained from the combined H1 and ZEUS data are shown in Fig. 3 [7] and are well described by the SM predictions as obtained from the H1 and ZEUS QCD fits. The data demonstrate parity violation at very small distances, down to 10^{-18} m.

4 Polarized Charged Current Cross Sections

The total CC cross sections have been measured by H1 [3] and ZEUS [4] as a function of the lepton beam polarization P_e in the common phase space $Q^2 > 400\,\text{GeV}^2$ and $y < 0.9$. In Fig. 4 the results are compared with SM predictions based on CTEQ6D, MRST 2004 and HERAPDF0.1 fits. The linear dependence of the CC cross sections on P_e is expected as the W boson interacts only with e_L^- and e_R^+. A straight line fit to these cross sections is sensitive to exotic right-handed and left-handed current additions to the SM Lagrangian. Assuming SM couplings and a massless right-handed neutrino, the existence of charged currents involving right-handed fermions mediated by a boson of mass below 208 GeV is excluded at 95% C.L. [5].

5 Combined Electroweak-QCD Fits

The NC cross sections provide information on the quark couplings to the Z^0 boson. For the HERA kinematic regime, the axial (a_q) and vector (v_q) coupling constants are dominant in the unpolarized xF_3^0 and polarized F_2^P structure functions, respectively. These electroweak parameters can be fitted simultaneously with the PDF parameters to perform a model independent extraction.

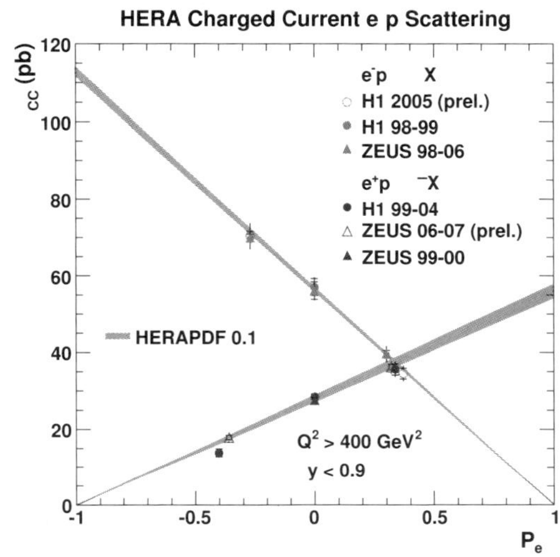

Figure 4: *Total cross sections for e^-p and e^+p CC DIS as a function of longitudinal electron (positron) beam polarization.*

The HERA [8, 9] results are shown in Fig. 5 and compared to LEP and CDF results [10, 11].

6 Isolated Lepton Events with Missing P_T

A search for events with high transverse energy isolated leptons (electrons or muons) and missing transverse momentum has been performed by the H1 and ZEUS collaborations [12] using the full data sets. In general, the observed events yields are in good agreement with the SM model predictions, which is dominated by W production. An excess at high P_T^X, $P_T^X > 25\,\mathrm{GeV}$, is observed by H1 in the e^+p data sample, which is not observed by ZEUS. A small excess remains in the common analysis [12]: the number of observed events with $P_T^X > 25\,\mathrm{GeV}$ is 23, compared to 14.0 ± 1.9 expected.

The measured single W production cross section is shown in Fig. 6. The total cross section of this process is measured as $1.07 \pm 0.16(\mathrm{stat.}) \pm 0.08(\mathrm{sys.})\,\mathrm{pb}$, in agreement with the SM prediction of $1.26 \pm 0.19\,\mathrm{pb}$.

7 Multi-Lepton Production

The production of multileptons (electron or muon) at high transverse momenta has been studied by the H1 and and ZEUS collaborations [13] using the full $e^{\pm}p$ data sample. The yields of di- and tri-lepton events are in good agreement with

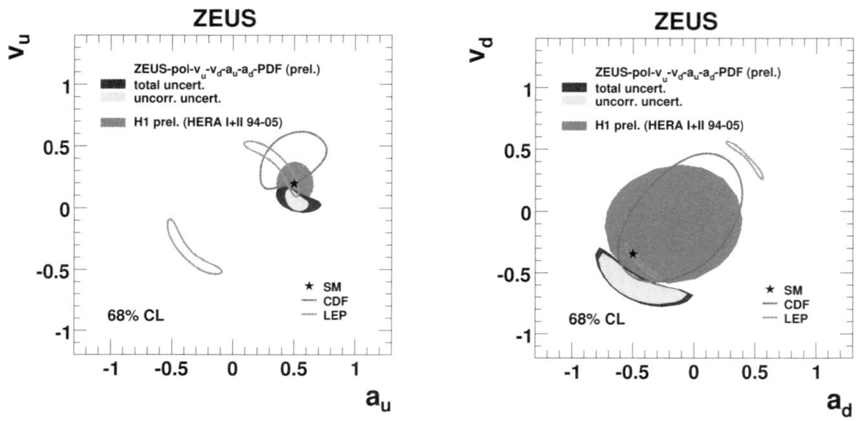

Figure 5: *Contour plots of the 68% C.L. limits on the electroweak couplings of the quarks to the Z^0 boson. Left side v_u vs. a_u (u quarks). Right side v_d vs. a_d (d quarks).*

SM predictions. Distributions of the invariant mass M_{12} of the two highest P_T leptons and of the scalar sum of the lepton transverse momenta $\sum P_T$ are in good agreement with the SM expectations. Events are observed in $ee, e\mu, eee$ and $e\mu\mu$ topologies with invariant masses M_{12} above 100 GeV, where the SM prediction is low. Both experiments observe high mass and high $\sum P_T$ events in e^+p collisions only, while, for comparable SM expectations, none are observed in e^-p collisions. In the combined analysis seven events have a $\sum P_T > 100\,\text{GeV}$, whereas the corresponding SM expectation for e^+p collisions is 1.94 ± 0.17 events [13] .

The total and differential cross sections for electron and muon pair photoproduction are measured in a restricted phase space dominated by photon-photon interactions and are found in good agreement with the SM expectations.

8 Single-top Production

Observables sensitive to flavor-changing neutral current (FCNC) interactions are particularly useful as probes for physics beyond the SM, since SM rates are very small due to the GIM mechanism. At the HERA collider, single-top production is a prime reaction to search for FCNC, where the incoming lepton exchanges a γ or Z with an up-type quark in the proton, producing a top quark in the final state. Deviations from the SM can be parameterized in terms of the coupling constants $\kappa_{tu\gamma}$, v_{tuZ} [14].

The studies performed by H1 and ZEUS considered top quark decays into a b quark and a W boson with subsequent leptonic or hadronic decay of the W.

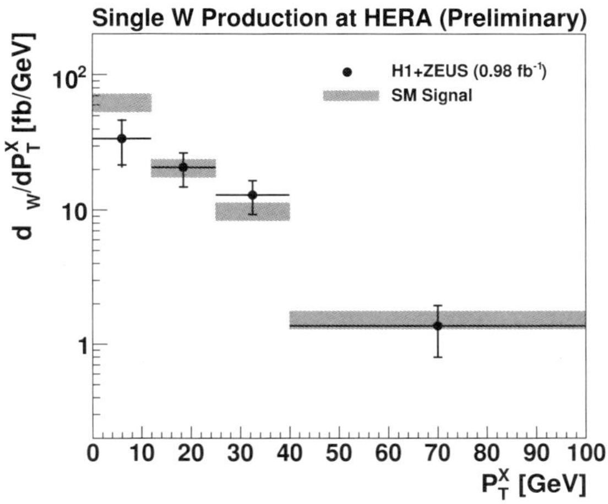

Figure 6: *The differential single W cross section plotted as a function* P_T^X.

The search is therefore based on a sample of events with isolated leptons and missing transverse momentum and a sample of multi-jet events. No evidence for single top quark production is observed. A 95% C.L. limit on the anomalous coupling $\kappa_{tu\gamma}$, $\kappa_{tu\gamma} < 0.18$ for H1 [15] and $\kappa_{tu\gamma} < 0.13$ for ZEUS [16] is set for the scale of new physics of $\Lambda \equiv m_{top} \equiv 175\,\mathrm{GeV}$.

9 Are Quarks elementary?

A possible quark sub-structure can be detected by measuring the spatial distribution of the quark charge. If the quark has a finite radius, the cross section will decrease as the probes penetrates into it. Deviations from the SM cross section are described by:

$$\frac{d\sigma}{dQ^2} = \frac{d\sigma^{\mathrm{SM}}}{dQ^2}\left(1 - \frac{R_q^2}{6}Q^2\right)^2 \left(1 - \frac{R_e^2}{6}Q^2\right)^2,$$

where R_e and R_q are the root-mean-square radii of the electroweak charge of the electron and the quark, respectively.

The high Q^2 ($Q^2 > 1\,000\,\mathrm{GeV}^2$) neutral current data sample has been used for this analysis. Assuming the electron to be point-like, the 95% C.L. limit on the quark radius is obtained as: $R_q < 0.74 \cdot 10^{-18}\,\mathrm{m}$ (H1 [2]) and $R_q < 0.63 \cdot 10^{-18}\,\mathrm{m}$ (ZEUS [17]).

The ZEUS data set has also been used to derive limits on the mass scale parameter in models with large extra dimensions and on the effective mass scale limits for contact-interaction model ranging from 3.8 to 8.9 TeV [17].

10 Search for Excited Fermions

Excited fermions ($e^\star, \nu^\star$ and $q^\star$) would be a signature of compositeness at the compositeness scale Λ. The cross section is proportional to the coupling constants f and f' [18]. All electroweak decays of excited fermions have been considered, including final states from Z or W hadronic decays. No evidence for excited fermion production is found. Exclusion limits on f/Λ at 95% C.L. are determined by H1 as a function of the mass of the excited fermions. Assuming $f/\Lambda = 1/M_{f^\star}$, the following mass limits are derived at 95% C.L.: $M_{e^\star} > 272\,\text{GeV}$, $M_{\nu^\star} > 213\,\text{GeV}$ and $M_{q^\star} > 252\,\text{GeV}$ (for $f_s = 0$) [18, 19].

11 Leptoquarks

A search for scalar and vector leptoquarks (LQ) coupling to first generation fermions has been performed by the H1 collaboration using the full HERA data set [20]. Leptoquark decays into eq and νq were considered, where q represents both quarks and anti-quarks. Such LQ decays lead to final states similar to those of DIS NC and CC interactions at very high Q^2. No evidence for direct or indirect production of leptoquarks is found in data samples with a large transverse momentum final state electron or with large missing transverse momentum. For each of the LQ species defined in the Buchmüller-Rückl-Wyler (BRW) model [21], the present analysis excludes a previously unexplored domain in the plane spanned by the mass of the leptoquark and its coupling to fermions.

As an example limits on the coupling λ for $\tilde{S}_{1/2,L}$ and $S_{0,L}$ leptoquarks are shown in Fig. 7 as function of the LQ mass.

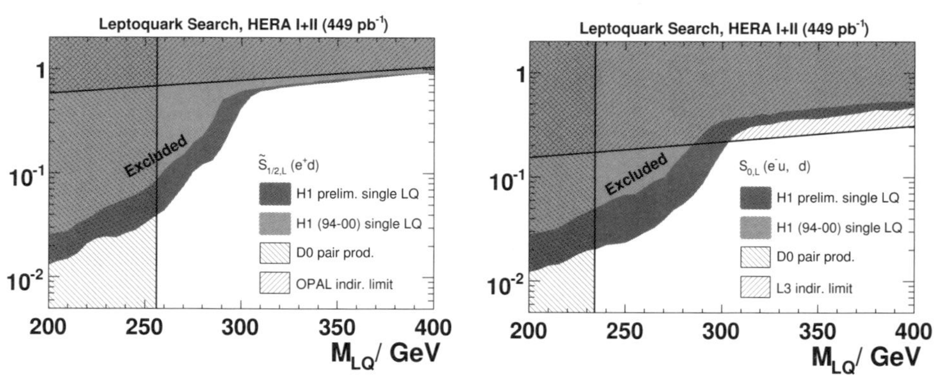

Figure 7: *Limits of the coupling constant λ for $\tilde{S}_{1/2,L}$ and $S_{0,L}$ leptoquarks as a function of their mass.*

12 General Searches for High - P_T Phenomena

H1 performed a model independent, generic search in final states with at least two high-P_T objects: electrons, muons, jets, photons or neutrinos [22]. The transverse momentum of these objects is required to be larger than 20 GeV. The events were classified according to their final states. Forty different final states were considered. In general, the events yields are in good agreement with Standard Model expectations. No statistically significant deviation is observed.

Acknowledgments

I would like to thank the organizers, in particular A.Studenikin, for the invitation to the conference. I am grateful to M.Turcato and S.Schmitt for a critical reading of the manuscript.

References

[1] A.M. Cooper-Sarkar, R.C.E. Devenish and A. De Roeck, *Int. J. Mod. Phys.* **A 13**, 3385 (1998).
[2] H1 Collaboration, prelim-07-141.
[3] H1 Collaboration, prelim-06-041.
[4] ZEUS Collaboration, ZEUS-prel-09-001.
[5] A. Aktas et al., H1 Collaboration, *Phys. Lett.* **B634**, 173 (2006).
[6] S.Chekanov et al., ZEUS Collaboration, *Eur. Phys. J.* **C 62**, 625 (2009).
[7] H1 and ZEUS Collaborations, H1 prelim-06-142 and ZEUS prel-06-022.
[8] A. Aktas et al., H1 Collaboration, *Phys. Lett.* **B632**, 35 (2006).
[9] S.Chekanov et al., ZEUS Collaboration, ZEUS-prel-07-027.
[10] D. Acosta et al., CDF Collaboration, *Phys. Lett.* **B634**, 173 (2006).
[11] LEP Electroweak Working Group, *Phys. Rep.* **427**, 257 (2006).
[12] H1 and ZEUS Collaborations, H1 prelim-09-161, ZEUS-prel-09-044.
[13] H1 and ZEUS Collaborations, *JHEP10* 013 (2009).
[14] T. Han, J.L. Hewett, *Phys. Rev.* **D60**,074015 (1999).
[15] F.D.Aaron et al., H1 Collaboration, DESY 09-050 (2009).
[16] ZEUS Collaboration, ZEUS-prel-09-009.
[17] ZEUS Collaboration, ZEUS-prel-09-013.
[18] F.D.Aaron et al., H1 Collaboration, *Phys. Lett.* **B678**, 335 (2009).
[19] F.D.Aaron et al., H1 Collaboration, DESY 09-040, arXiv:0904.3392v2 (2009).
[20] H1 Collaboration, H1 prelim-07-164, (2007)
[21] W.Buchmüller, R.Rückl and D.Wyler, Phys.Lett. **B191**, 442 (1987). Erratum in Phys.Lett. **B448**, 320 (1999).
[22] F.D.Aaron et al., H1 Collaboration, *Phys. Lett.* **B 674**, 257 (2009).

PROTON STRUCTURE FUNCTION MEASUREMENTS AT HERA

V. Chekelian (Shekelyan) [a]

Max Planck Institute for Physics, Foehringer Ring 6, 80805 Munich, Germany

Abstract. Recent results from the electron-proton collider HERA on the structure of the proton and understanding of the data in terms of QCD are presented.

1 Introduction

At the first ep collider HERA, electrons of 27.5 GeV collided with protons of 920 GeV (820 GeV until 1997). This corresponds to an ep centre of mass energy of 320 GeV. The maximum negative four-momentum-transfer squared from the lepton to the proton, Q^2, accessible with this machine was as high as 100000 GeV^2. The two ep interaction regions were instrumented with the multi-purpose detectors of the H1 and ZEUS collider experiments. Over 15 years of data taking from 1992 to 2007, these two experiments together collected a total integrated luminosity of $\approx 1fb^{-1}$, about equally shared between positive and negative polarities and positive and negative longitudinal polarisations of the lepton beam.

This paper concentrates on recent developments related to the central topic of the HERA physics program - the measurements of the inclusive neutral current (NC) and charged current (CC) deep inelastic scattering (DIS) cross sections, determination of the proton structure and understanding of the data in terms of Quantum Chromodynamics (QCD). The ultimate goal for these measurements is to obtain one unique HERA data set, which averages the H1 and ZEUS results produced in different years with emphasis on different features of the apparatus. Recently the combination is performed for HERA I, the first phase of the HERA project from 1992 to 2000, for which both experiments completed and published final results. These combined data are analysed in the QCD framework, and the parton distribution functions (PDFs) in the proton are determined. At the end of the HERA data taking, special runs were performed with reduced energies of the proton beam of 460 and 575 GeV. Using these data the longitudinal structure function was measured in a model independent way.

2 Deep inelastic NC and CC ep scattering

The deep inelastic NC $e^{\pm}p$ scattering cross section with unpolarised beams can be expressed in the reduced form as

$$\tilde{\sigma}^{\pm}_{NC}(x,Q^2) = \frac{\mathrm{d}^2\sigma^{\pm}_{NC}}{\mathrm{d}x\mathrm{d}Q^2}\frac{xQ^4}{2\pi\alpha^2}\frac{1}{Y_+} = F_2(x,Q^2) \mp \frac{Y_-}{Y_+}xF_3(x,Q^2) - \frac{y^2}{Y_+}F_L(x,Q^2), \quad (1)$$

[a] e-mail: shekeln@mail.desy.de

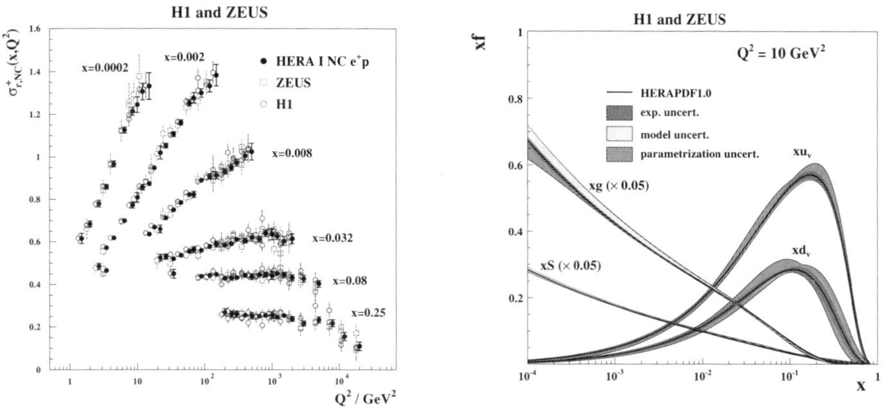

Figure 1: Left: Selection of combined NC e^+p reduced cross sections from HERA I compared to the separate H1 and ZEUS data input to the averaging procedure. Right: The parton distribution functions from HERAPDF1.0, $xu_v, xd_v, xS = 2x(\bar{U} + \bar{D})$ and xg, at $Q^2 = 10$ GeV^2. The gluon and sea distributions are scaled down by a factor 20. The experimental, model and parametrisation uncertainties are shown separately.

where α is the fine structure constant, x is the Bjorken scaling variable, y characterises the inelasticity of the interaction and $Y_\pm = 1 \pm (1 - y^2)$. The NC cross section is dominated by the contribution of the proton structure function $F_2(x, Q^2)$. At leading order in QCD, F_2 is related to a linear combination of sums of the quark and anti-quark momentum distributions in the proton, and the structure function $xF_3(x, Q^2)$ is related to a linear combination of their differences. The longitudinal structure function F_L is equal to zero in the quark-parton model due to the spin one half nature of the quarks. Non-zero values of the longitudinal structure function appear in QCD due to gluon radiation.

The deep inelastic CC cross section can be expressed in a similar manner. The $W^\pm$-boson exchange in the charged current process distinguishes between the charges of the constituent quarks. So, for the electron (positron) beam the cross section depends on the $u(d)$ quark content of the proton.

3 Combination of the H1 and ZEUS inclusive data

The published H1 and ZEUS measurements on inclusive NC and CC reactions cover a wide range of x and Q^2. For NC the kinematic range is $6 \cdot 10^{-7} \leq x \leq 0.65$ and $0.045 \leq Q^2 \leq 30000\,\text{GeV}^2$, and for CC it is $1.3 \cdot 10^{-2} \leq x \leq 0.40$ and $300 \leq Q^2 \leq 30000\,\text{GeV}^2$. All published NC and CC cross section measurements from H1 and ZEUS obtained using data collected in the years 1994-2000 are

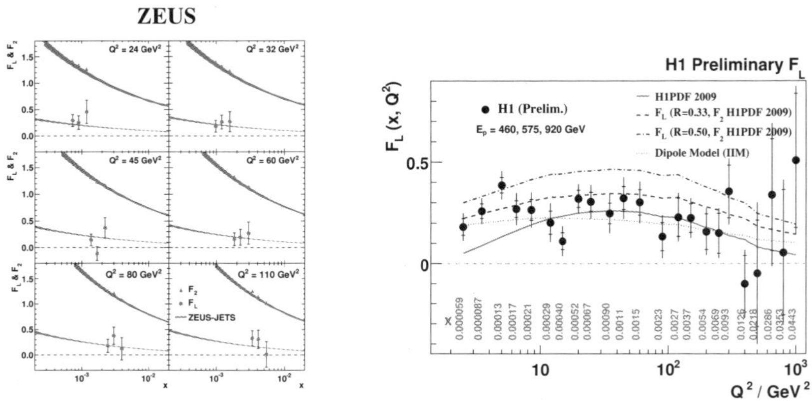

Figure 2: Left: F_2 and F_L as measured by ZEUS as a function of x at fixed values of Q^2. Right: The H1 measurement of $F_L(Q^2)$ averaged over x at fixed values of Q^2. The resulting x values of the averaged F_L are given in the figure for each point in Q^2.

combined [1] in one simultaneous minimisation. The resulting shifts of the correlated systematic uncertainties are propagated to both CC and NC data such that one coherent data set is obtained. This combined data set contains complete information on inclusive DIS cross sections measured by H1 and ZEUS at HERA I.

The data sets considered for the combination were taken with a proton beam energy of 820 and 920 GeV. Therefore, the data are corrected to a common centre of mass energy corresponding to $E_p = 920$ GeV and then averaged. The NC data for $y \geq 0.35$ are kept separate for the two proton beam energies.

The total integrated luminosity of the combined data set corresponds to about 200 pb^{-1} for e^+p and 30 pb^{-1} for e^-p. In total 1402 data points are combined to 741 cross section measurements. The data show good consistency, with $\chi^2/n_{\text{dof}} = 636.5/656$. There are in total 110 sources of correlated systematic uncertainty. None of these systematic sources shifts by more than $2\,\sigma$ of the nominal value in the averaging procedure. The total uncertainty of the combined data set reaches 1% for NC scattering in the best measured region, $20 < Q^2 < 100$ GeV^2.

In Figure 1 (left) averaged NC data are compared to the input H1 and ZEUS data. Since H1 and ZEUS have employed different experimental techniques, using different detectors and methods of kinematic reconstruction, the combination leads to a significantly reduced correlated systematic uncertainty. This reduction propagates to all average points, including also those which are based solely on the measurement from one experiment.

4 QCD analysis of the combined data

The combined data set on the NC and CC e^+p and e^-p inclusive cross sections from HERA I is used as the sole input to a next-to-leading order (NLO) QCD PDF fit, called HERAPDF1.0 [1]. The consistency of the present input data justifies the use of the conventional χ^2 tolerance, $\Delta\chi^2 = 1$, when determining the experimental uncertainties of the fit.

The QCD predictions for the structure functions are obtained by solving the DGLAP evolution equations at NLO in the $\overline{\text{MS}}$ scheme with the renormalisation and factorization scales chosen to be Q^2. The DGLAP equations yield the PDFs at all values of Q^2, if they are provided as functions of x at some input scale Q_0^2. This scale is chosen to be $Q_0^2 = 1.9$ GeV^2 such that the starting scale is below the charm mass threshold, $Q_0^2 < m_c^2$. The heavy quark coefficient functions are calculated in the general-mass variable-flavour-number scheme. The heavy quark masses are chosen to be $m_c = 1.4$ GeV and $m_b = 4.75$ GeV. The strong coupling constant is fixed to $\alpha_s = 0.1176$. A minimum Q^2 cut on the data of $Q^2_{min} = 3.5$ GeV^2 is imposed to remain in the kinematic region, where perturbative QCD should be applicable. PDFs are parametrised at the input scale by the generic form

$$xf(x) = Ax^B(1-x)^C(1+\epsilon\sqrt{x}+Dx+Ex^2). \tag{2}$$

The parametrised PDFs are the gluon distribution xg, the valence quark distributions xu_v, xd_v, and the u-type and d-type anti-quark distributions $x\bar{U}$, $x\bar{D}$.

Figure 1 (right) shows a summary plot of the HERAPDF1.0 distributions at $Q^2 = 10$ GeV^2. The analysis yields small experimental uncertainties and makes an assessment of uncertainties introduced both by model assumptions and by assumptions about the form of the parametrisation. Due to the high precision of the combined data set, the parametrisation HERAPDF1.0 has total uncertainties at the level of a few percent at low x.

5 HERA results for $F_L(x, Q^2)$

A model independent measurement of the longitudinal structure function $F_L(x, Q^2)$ requires several sets of NC cross sections at fixed x and Q^2 but different y. This was achieved at HERA by variation of the proton beam energy. The measurements are performed using e^+p data collected in 2007 with a positron beam energy of 27.5 GeV and with three proton beam energies: the nominal energy of 920 GeV, the smallest energy of 460 GeV and an intermediate energy of 575 GeV.

According to eq. 1, the $F_L(x, Q^2)$ contribution to the reduced cross section is proportional to $f(y) = y^2/[1+(1-y)^2]$. Therefore, the F_L values are determined as the slopes of straight-line fits of the measured $\tilde{\sigma}_{NC}(x, Q^2, y)$ values as a function of the y-dependent factor $f(y)$. The ZEUS result for $F_L(x, Q^2)$ is shown in Figure 2 (left). The result is consistent with the expectation derived from the ZEUS-JETS QCD fit [3] to previous data. The H1 measurements of $F_L(x, Q^2)$ are averaged over x at fixed Q^2, and the resulting $F_L(Q^2)$ is shown in Figure 2 (right). The averaged F_L is compared with calculations based on the H1PDF 2009 fit [5]. For $Q^2 \geq 10\,\mathrm{GeV}^2$, the data are well described by the QCD predictions, whereas at lower Q^2 the QCD calculations underestimate the F_L data. Dipole models are found to describe the measurements. The F_L data are consistent with $R = F_L/(F_2 - F_L) = 0.25$.

6 Summary

Inclusive cross sections of NC and CC $e^{\pm}p$ scattering, measured by H1 and ZEUS at HERA I, are combined providing the most accurate data set with a total uncertainty which reaches 1% for NC in the bulk of the phase space.

A NLO QCD analysis is performed based exclusively on these combined cross section data. A new set of parton distribution functions, HERAPDF1.0, is obtained using a variable-flavour-number scheme. The parametrisation HERAPDF1.0 has total uncertainties at the level of a few percent at low x, which include experimental as well as model and parametrisation uncertainties.

The H1 and ZEUS collaborations measured the longitudinal proton structure function in DIS at low x. The $F_L(x, Q^2)$ values are determined from three sets of cross section measurements at fixed x and Q^2, but different inelasticity y, obtained using three different proton beam energies at HERA. For the $Q^2 \geq 10\,\mathrm{GeV}^2$, the F_L results are consistent with the QCD predictions. At lower Q^2 values the predictions underestimate the F_L measurements.

References

[1] H1 and ZEUS Collab., F.D. Aaron *et al.*, *JHEP01* 109 (2010).
[2] ZEUS Collab., S. Chekanov *et al.*, *Phys. Lett.* **B682** 8 (2009).
[3] ZEUS Collab., S. Chekanov *et al.*, *Eur. Phys. J.* **C42** 1 (2005).
[4] H1 Collab., F.D. Aaron *et al.*, *Phys. Lett.* **B665** 139 (2008);
H1 Collaboration, H1prelim-08-042, H1prelim-09-044,
`http://www-h1.desy.de/publications/H1preliminary.short_list.html`
[5] H1 Collab., C. Adloff *et al.*, *Eur. Phys. J.* **C30** 1 (2003).

SEARCHES FOR PHYSICS BEYOND THE STANDARD MODEL AT TEVATRON

Andrey Shchukin [a], on behalf of the CDF and D0 Collaborations
Institute for High Energy Physics, 142281 Protvino, Russia

Abstract. Recent searches beyond the Standard Model in $p\bar{p}$ collisions at center-of-mass energy of 1.96 TeV at Tevatron RunII are presented. The analyses use 1.1-4.1 fb^{-1} of data collected with CDF and D0 detectors.

1 Introduction

The Standard Model (SM) of elementary particle physics describes the structure of fundamental particles and their interactions via gauge bosons. To date most of experimental tests have agreed with the predictions of the SM. However, many questions can be raised, indicating that SM is not complete model. Therefore many theoretical extensions of the SM have been proposed. Together with supersymmetry as most popular extension of SM there are other well motivated models like extra dimensions, 4th generation, compositeness, etc. In this article we report most recent results of signature based searches obtained using 1.1-4.1 fb^{-1} of data collected at CDF and D0 detectors.

2 Searches for high mass resonances

Attempting to unify SM forces or explain the large difference between the SM and gravitational scales some theories predict new particles which decay to lepton-lepton or photon-photon pair, such as Z' and Randall-Sundrum (RS) graviton. The Z' is predicted be E_6 model where six mass eigen states Z'_ψ, Z'_χ, Z'_η, Z'_I, Z'_{sq}, Z'_N are built with specific mixing angles. The RS graviton is the lightest neutral spin-2 particle predicted by RS model. The CDF has reported two searches [1, 2] in $\mu^+\mu^-$ and e^+e^- channels based on 2.5 fb^{-1} and 2.3 fb^{-1} data correspondingly. The upper limits on $\sigma \cdot \Gamma(X \to ee)$ at 95% CL are set as function of mass. The X's are spin-1 or spin-2 particles. The SM coupling Z' and the RS graviton with $k/\bar{M}_{Pl} = 0.1$ are excluded for masses below 966 and 850 GeV/c^2 respectively. The similar studies are performed in $\mu^+\mu^-$ channel and the SM coupling Z' is excluded for masses bellow 1030 GeV/c^2 and RS graviton with $k/\bar{M}_{Pl} = 0.1$ is excluded for masses below 921 GeV/c^2.

3 Searches for heavy quarks

The SM of particle physics defines three generations of quarks but do not prohibit a fourth. CDF has performed the search [3] for pair production of heavy bottom-type quarks decaying to bW which leads to same sign dilepton

[a] e-mail: schukin@ihep.ru

final state with missing transverse energy, two jets and positive b-tag. Also the search for exotic top partners $(B, T_{5/3})$ which have similar decays has been conducted. The results obtained using 2.7 fb^{-1} of data exclude a b' (or B) with mass less than 338 GeV/c^2, and a $T_{5/3}$ with mass less than 365 GeV/c^2 at 95% CL.

4 Searches for leptoquarks

The leptoquarks are hypothetical bosons that allow lepton-quark transitions are predicted by numerous extensions of standard model. The leptoquarks would come in three generation and can be either scalar of vector particles. At the Tevatron leptoquark pairs would be produced mainly through $q\bar{q}$ annihilation or through gluon-gluon fusion. The D0 has performed search [4] for the final state with two acoplanar b-jets from the third generation leptoquarks and missing energy due to escaping neutrinos or neutralinos. The preliminary results are obtained with 4.0 fb^{-1} data sample. At 95% CL D0 excludes third generation leptoquark with mass below 252 GeV/c^2 and sets new cross section limit in the $(m_{\tilde{b}_1}, m_{\tilde{\chi}_1^0})$ plane.

5 Searches for extra dimensions

Some theoretical models propose the existence for extra spatial dimensions to resolve the SM hierarchy problem. One of such extensions is Arkani-Hamed, Dimopulos, Dvali (ADD) model which assumes that SM forces are confined to a four-dimensional brane, but the gravity propagates in several additional (up to 7) dimensions. ADD model gives massive stable Kaluza-Klein gravitons G_{KK} and can explain why gravity is weak. The CDF has looked [5] for indication of large extra dimensions in $q\bar{q} \to G\gamma$ channel using 2.0 fb^{-1} data sample and set the new limit on M_D as function of the number of extra dimensions, Fig. 1.

The D0 has probed [6] extra dimensions by searching the channels $G_{KK} \to ee$ and $G_{KK} \to \gamma\gamma$ using 1.1 fb^{-1} of collected data. The limit on effective Planck scale M_S has been set at 95% CL from 2.09 TeV to 1.29 TeV for two to seven dimensions.

6 Searches for supersymmetry

Supersymmetry (SUSY) is a popular extension of the SM. It postulates the symmetry between fermionic and bosonic degrees of freedom. Since no SUSY particles has been observed yet, we can conclude that supersymmetry must be broken and supersymmetric particles must be heavier than corresponding SM partners. Both CDF and D0 experiments have performed the searches for many different channels with different SUSY scenarios.

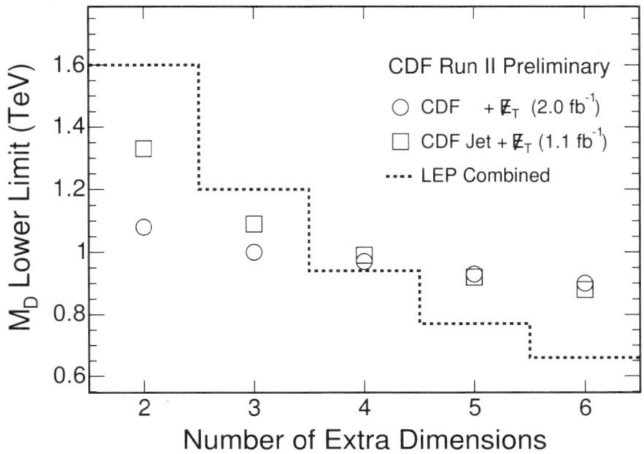

Figure 1: CDF search for LED. 95% CL lower limits in the ADD model on M_D as function of the number of extra dimensions.

6.1 Searches for stop quark

As possible exception stop quark might be lighter then top quark. In the framework of minimal supersymmetric standard model with R-parity conservation the final state $\tilde{t}_1\bar{\tilde{t}}_1 \to b\bar{b}ll'\tilde{\nu}\tilde{n}\tilde{u}$ might be observed. The D0 has performed the search [7] for the final states where ll' is $e^{\pm}\mu^{\mp}$ or e^+e^-. The integrated luminosity of analyzed data sample is 1.0 fb^{-1}. The 95% CL exclusion contour as function of scalar top quark and sneutrino masses is shown on the Fig 2. Assuming that χ_1^0 is the lightest supersymmetric particle and $m_{\tilde{t}_1} < m_t$, stop quarks will decay into final state similar to that of pair produced top quarks. The CDF has updated [8] their search for stop quark in $\tilde{t}_1 \to b\tilde{\chi}_1^{\pm} \to b\tilde{\chi}_1^0 l\nu$ channel using 2.7 fb^{-1} of data. The 95% CL exclusion contour as function of scalar top quark and sneutrino masses is shown on the Fig. 3.

6.2 Search for SUSY in trilepton final state

In $p\bar{p}$ collisions charginos and neutralinos can be produced in pairs via off-shell W boson or the exchange of squarks. These processes can lead to the signatures with three isolated leptons and large missing transverse energy. The search for this channel is performed by D0 [9] in four different signatures: eel, $\mu\mu l$, $e\mu l$, $\mu\tau l$ using 2.3 fb^{-1} of data. New cross section limit as function of chargino mass have been set. To explore broader parameters space m_0 and $m_{1/2}$ were scanned simultaneously. The exclusion contour in $m_0 - m_{1/2}$ plane are shown on the Fig. 4.

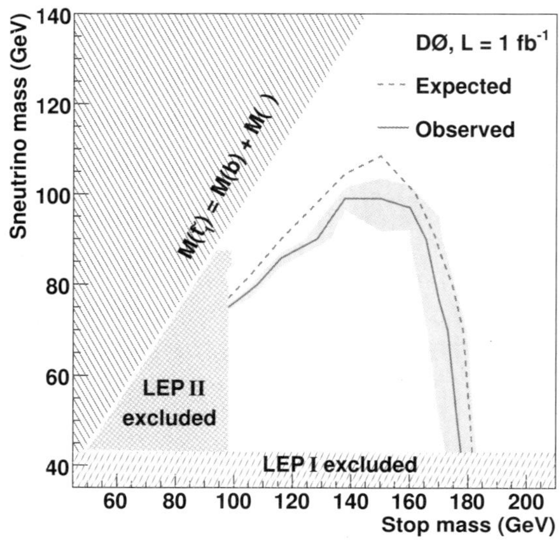

Figure 2: D0 search for stop. The 95% CL exclusion contour in the sneutrino mass versus stop mass plane.

6.3 Search for SUSY with Hidden Valleys

Hidden valley models introduce a new hidden sector which is very weakly coupled to the SM particles, and therefore can easily escape detection. An important subset of hidden valley models also contain supersymmetry. Supersymmetric partners are pair produced and cascade to lightest neutralinos that can decay into the hidden sector state plus either a photon or a dark photon. The dark photon decays through its mixing with a photon into fermion pairs. Such scenario leads to unexplored before final state with missing transverse energy from darkino, photon and two spatially close leptons. The D0 has performed this search [10] using 4.1 fb^{-1} of data. No evidence for dark photon events is found. For dark photon masses of 0.2, 0.782, and 1.5 GeV/c^2 chargino masses of 230, 142, and 200 GeV/c^2, respectively, are excluded.

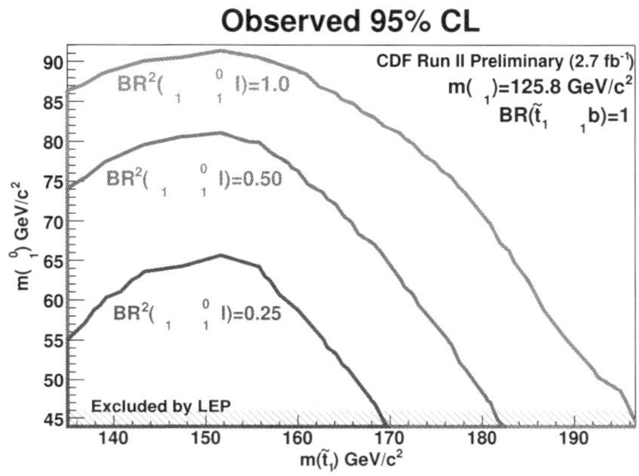

Figure 3: CDF search for stop. The observed 95% CL in the sneutrino vs stop mass plane at chargino mass of 125.8 GeV/c^2 for various dilepton branching ratios.

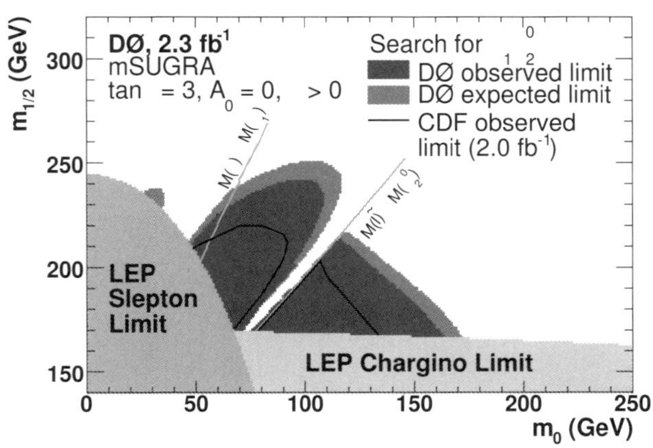

Figure 4: Region excluded by D0 in $m_0 - m_{1/2}$ plane in comparison with the limits from the searches at LEP and CDF.

7 Conclusion

Many searches for new physics are performed by CDF and D0 experiments using up to 4.1 fb^{-1} of RunII Tevatron data. No evidence for the physics beyond th Standard Model is found and new exclusion limits have been set. By the end of 2009 CDF and D0 experiments expect to collect up to 7 fb^{-1} of data. With growing integrated luminosity there can be many promising updates.

References

[1] T.Aaltonen, *et al.*, *Phys.Rev.Lett.* **102**, 091805 (2009).
[2] T.Aaltonen, *et al.*, *Phys.Rev.Lett.* **102**, 031801 (2009).
[3] T.Aaltonen, *et al.*, arXiv:0912.1057v1 [hep-ex]
[4] V.M.Abazov, *et al.*, www-d0.fnal.gov/Run2Physics/WWW/results/prelim/NP/N68
[5] T.Aaltonen *et al.*, Phys.Rev.Lett. **101**, 181602 (2008).
[6] V.M.Abazov *et al.*, Phys.Rev.Lett. **102**, 051601 (2009).
[7] V.M.Abazov *et al.*, Phys. Lett. B **675**, 289 (2009).
[8] T.Aaltonen *et al.*, www-cdf.fnal.gov/physics/exotic/r2a/20090319.stop_dilepton
[9] V.M.Abazov *et al.*, Phys. Lett. B **680**, 34 (2009).
[10] V.M.Abazov *et al.*, Phys. Rev. Lett. **103**, 081802 (2009)

TESTING THE STANDARD MODEL WITH TOP QUARKS

Erich W. Varnes [a]

Department of Physics, University of Arizona, Tucson 85721, USA

Abstract. The top quark, by far the most massive known fermion, provides a unique laboratory in which to study phyiscs at the electroweak scale. I report on recent top quark measurements from the CDF and DØ experiments at the Fermilab Tevatron $p\bar{p}$ collider, including the first observation of single top quark production, measurement of the top quark mass, the $t\bar{t}$ production rate, and several searches for new physics in the properties of the top quark, and in its production and decay.

1 Introduction

The top quark is both familiar, in that in the standard model it occupies the isospin $+1/2$ location in the third quark family, and also unique, in that it is far more massive than the other fermions. The large mass of the top quark means that only the CDF and DØ experiments at the Fermilab Tevatron $p\bar{p}$ collider have collected top quark samples. In the context of the standard model (SM), the large mass also means that the top quark is a rather uninteresting object: it decays before it can form a color-singlet bound state, meaning that the sets of hadrons associated with the other quarks is absent for the top quark. Further, given the known values of the CKM matrix elements, the top quark must decay nearly always to Wb, so that the only variety in top quark signatures is provided by the various decay modes of the W boson. On the other hand, the top quark is the only fermion with a Higgs Yukawa coupling ≈ 1, which may indicate that it plays a special role in electroweak symmetry breaking, making close examination of the top quark's production and properties a critical test of the SM. The current status of this examination is presented here.

2 Single Top Quark Production

Electroweak single top quark production is of interest because, among other things, its cross section is proportional to the strength of the tWb coupling, and therefore to the CKM matrix element $|V_{tb}|$. This mode is far more difficult to observe than $t\bar{t}$ production due to its lower cross section (≈ 3 pb at the Tevatron in the SM, compared to 7.5 pb for $t\bar{t}$), less distinctive signature, and larger W+jet background. To extract a signal one must employ a multivariate discriminant that estimates the probability that a given event arose from single top production rather than background. The number of events with low signal probability normalizes the background production rate, and allows the observation of single top production as an excess of events at high signal probability. Both CDF and DØ use this method, with either a neural network,

[a] e-mail: varnes@physics.arizona.edu

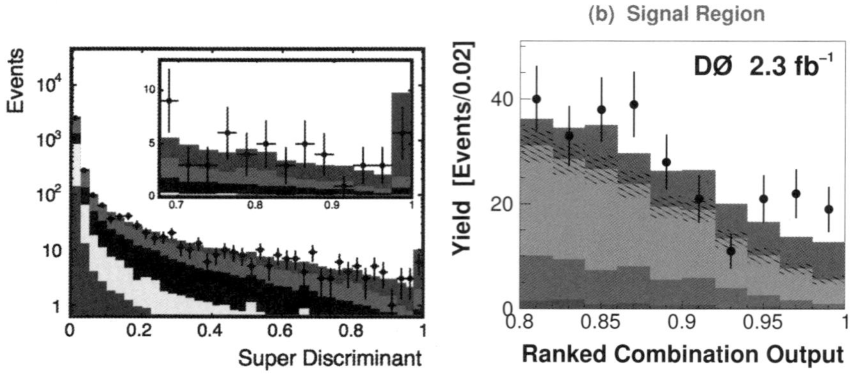

Figure 1: Single top quark production signatures observed at CDF (left) and DØ (right). In each experiment a multivariate discriminant is used to constrain the background level and to search for single top quark production, which would appear as an excess of events at large discriminant values. Both experiments observe such an excess with a significance of greater than 5σ.

boosted decision tree, matrix element calculation or classical likelihood taken as the discriminant, with the results shown in Fig. 1. In both experiments an excess is in fact observed in the signal region with a statistical significance of 5 standard deviations, allowing both to claim observation of the single top production process. The cross section measured by CDF is $\sigma = 2.3^{+0.6}_{-0.5}$ pb, which corresponds to a constraint that $|V_{tb}| > 0.71$ at 95% confidence level (CL) [1]. DØ measures $\sigma = 3.84 \pm 0.88$ pb, corresponding to $|V_{tb}| > 0.71$ at 95% CL [2]. Combination of the CDF and DØ results yields $\sigma = 2.76^{+0.58}_{-0.47}$ pb [3].

3 Top quark pair production

Most of the information we have about the top quark comes from events in which $t\bar{t}$ pairs are produced by gg fusion or $q\bar{q}$ annihilation. The most accessible decay mode is the $\ell+$ jets channel $t\bar{t} \to W^+bW^-\bar{b} \to \ell^+\nu b q\bar{q}'\bar{b}$, but the dilepton $t\bar{t} \to W^+bW^-\bar{b} \to \ell^+\nu\ell^-\bar{\nu}\bar{b}$ and all-hadronic $t\bar{t} \to W^+bW^-\bar{b} \to q\bar{q}'bq\bar{q}'\bar{b}$ modes are also used in top quark measurements.

Measurements of the top quark production rate provide excellent tests of perturbative QCD and also probe for new physics in either the production or decay of top quarks. Both CDF and DØ have produced many such measurements in a variety of $t\bar{t}$ final states, and find consistent results across all of those final states. In addition, the combined measurements $\sigma = 7.50 \pm 0.31$ (stat.) $\pm$ 0.34 (syst.)$\pm$0.15 (lumi.) pb from CDF [4] and $\sigma = 7.8 \pm 0.5$ (stat.)$\pm$0.6 (syst.)$\pm$ 0.5 (lumi.) pb from DØ [5] are consistent with the SM expectation at the Tevatron. CDF has extended the test of perturbative QCD by measuring the cross section for at least one jet to be produced in association with a $t\bar{t}$ pair to be $\sigma = 1.6 \pm 0.2$ (stat.) $\pm$ 0.5 (syst.) $\pm$ 0.5 (lumi.) pb [6], which is also consistent with the expectation from QCD.

4 Measurement of the top quark mass

The fact that the top quark mass m_t is large compared to the QCD scale allows its mass to be measured to a fractional precision better than that of any other quark. Since tb loops provide the dominant correction to the tree-level W boson mass, with Higgs boson loops providing another correction, precise measurement of both the top and W masses constraints the Higgs boson mass in the context of the SM.

The best measurements of m_t come from the ℓ+jets channel, using a technique known as the Matrix Element Method in which the set of observed decay products in the detector are compared directly to the mass-dependent physics of $t\bar{t}$ production, integrating over the detector response. This method provides superior statistical precision, and allows an *in situ* measurement of the jet energy by using information from the $W \rightarrow q\bar{q}\prime$ decays. Results are typically presented as a function of both jet energy scale and m_t as shown for both CDF and DØ in Fig. 2. CDF measures $m_t = 172.6 \pm 0.9$ (stat.) $\pm$ 1.3 (syst.) GeV [7] while DØ finds $m_t = 173.7 \pm 0.8$ (stat.) $\pm$ 1.6 (syst.) GeV [8].

DØ has also measured the t and $\bar{t}$ masses separately using the matrix element method, providing a test of the requirement from CPT conservation that these masses be equal, and finds $m_t - m_{\bar{t}} = 3.8 \pm 3.4$ (stat.) $\pm$ 1.2 (syst.) GeV [9].

While the matrix element method currently offers the most precise measurement of m_t, other techniques that are less sensitive to systematic variations may prove better in the long run (particularly at the LHC). CDF has pioneered to the use of techniques that are independent of the jet response, with one result being a measurement based on the lepton p_T distributions, for which they find $m_t = 178.2 \pm 7.2$ (stat.) $\pm$ 2.3 (syst.) GeV when the ℓ+jets and dilepton channels are combined [10].

Both CDF and DØ have also measured m_t by applying the matrix element method to dilepton events, with the results $m_t = 171.2 \pm 2.7$ (stat.)$\pm$2.9 (syst.) GeV from CDF [11] and $m_t = 174.7 \pm 2.9$ (stat.)$\pm$2.4 (syst.) GeV from DØ [12]. In addition, CDF has performed a measurement using events from the all-hadronic final state, finding $m_t = 174.8 \pm 2.4$ (stat.)$^{+1.2}_{-1.0}$ (syst.) GeV [13].

Measurements of m_t from all channels from both CDF and DØ are combined to yield a world average $m_t = 173.1 \pm 0.6$ (stat.) $\pm$ 1.1 (syst.) GeV [14]. The fractional precision of 0.8% on m_t makes the mass of the top quark more precisely known than the mass of any other quark. Using this measurement along with the world average W boson mass and other electroweak measurements, the mass of the Higgs boson is constrained to be < 157 GeV at 95% CL (< 186

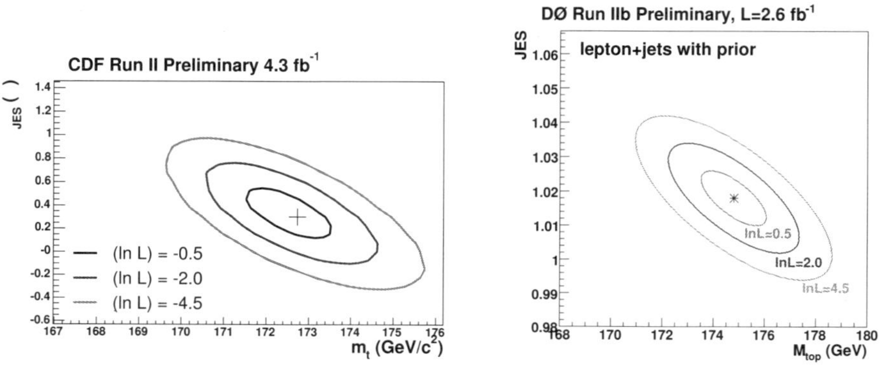

Figure 2: Result of the two-dimensional fit of top quark mass and data/MC jet response ratio in CDF (left) and DØ (right).

GeV if the direct search exclusion is taken as a prior probability) [15].

5 Top quark properties

Beyond m_t, several other production and decay properties of the top quark have been measured. One of these is the forward-backward production asymmetry A_{FB}, which is expected to be $\approx 5\%$ but could be enhanced by new physics at the production vertex. CDF finds $A_{FB} = 19 \pm 7 \pm 2\%$ [16] while DØ measures $A_{FB} = 12 \pm 8 \pm 1\%$ [17]. These numbers are not directly comparable since CDF corrects for acceptance effects while DØ does not.

Both experiments also measure the correlation between the spins of the two top quarks in each event. Since the top quark decays so rapidly, the correlation present when the pair is created is expected to be largely preserved in the decay, yielding a correlation coefficient $C \approx 0.8$. Using dilepton events, CDF finds $C = 0.320^{+0.545}_{-0.775}$ [18] while DØ finds $C = -0.17^{+0.64}_{-0.53}$ [19] (the experiments use slightly different spin quantization axes).

6 Search for new particles related to the top quark

Since the three-generation structure of the SM is unexplained, it is natural to search for a fourth generation. CDF has searched for the isospin 1/2 quark of the fourth generation, denoted as t' by looking for an excess of events with large reconstructed top mass or large jet p_Ts. They observe no significant excess, and set a limit that $m_{t'} > 311$ GeV at 95% CL [20].

CDF and DØ have searched for a charged Higgs boson under the assumption that the H^+ mass is low enough to allow the decay $t \to H^+ b$. The experiment

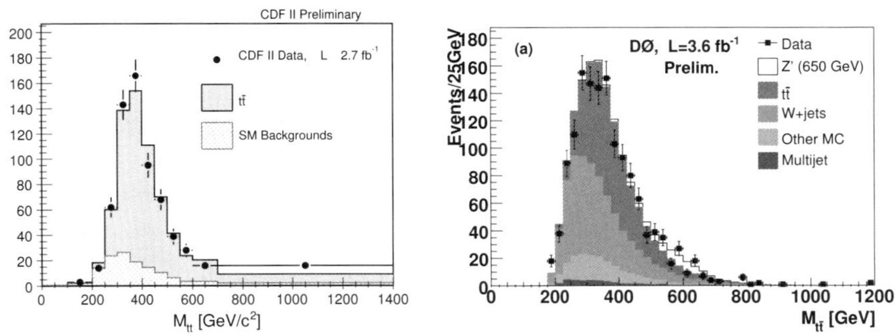

Figure 3: Distribution of $m_{t\bar{t}}$ observed at CDF (left) and DØ (right).

use different techniques, as CDF searches for a peak in the invariant mass of the $c\bar{s}$ from top quark decay [21], while DØ compares the measured cross section for several final states (including those with τ leptons, to which the H^+ may preferentially decay) [5]. Neither experiment observes an H^+ signature.

The possibility of resonant $t\bar{t}$ production has also been explored by both CDF and DØ by comparing the invariant mass of the $t\bar{t}$ system $m_{t\bar{t}}$ to the expectation from the SM. The $m_{t\bar{t}}$ distributions observed in both experiments in the ℓ+jets channel are shown in Fig. 3, and both are consistent with the SM. CDF interprets its result as a limit that $\chi/M_{\rm pl} > 0.16$ at 95% C.L. for a 600 GeV graviton [22], while DØ sets a limit on a leptophobic Z' bosons mass of $m_{Z'} > 820$ GeV at 95% C.L. [23]. CDF also explores the $m_{t\bar{t}}$ distribution in the all-hadronic final state, and finds agreement with the SM, leading to a limit $m_{Z'} > 805$ GeV at 95% C.L. [24].

7 Summary

The large data sample collected by the Tevatron in recent has allowed the top quark to be studied with unprecedented precision. Highlights of recent measurements include the observation of electroweak single top quark production, and the measurement of m_t with a precision of 0.8%. Many other measurements of the top quark's production and decay properties have also been done, with results consistent with SM expectations. As the Tevatron continues to collect data and the Large Hadron Collider comes on line, these tests will be repeated with far greater precision in the years to come.

References

[1] T. Aaltonen *et al.*, Phys. Rev. Lett. **103**, 092002 (2009).

[2] V.M. Abazov *et al.*, Phys. Rev. Lett. **103**, 092001 (2009).
[3] Tevatron Electroweak Working Group, arXiv:0908.2171 [hep-ex] (2009).
[4] CDF Collarboration, CDF note 9913 (2009). `http://www-cdf.fnal.gov/physics/new/top/confNotes/cdf9913_ttbarxs4invfb.ps`
[5] V.M. Abazov *et al.*, Phys. Rev. D **80**, 071102 (2009).
[6] CDF Collaboration, CDF note 9850 (2009). `http://www-cdf.fnal.gov/physics/new/top/confNotes/cdf9913_ttbarxs4invfb.ps`
[7] CDF Collaboration, CDF/PHYS/TOP/PUBLIC/9880 (2009). `http://www-cdf.fnal.gov/physics/new/top/confNotes/cdf9850_schwarzt_ttj.pdf`
[8] DØ Collaboration, D note 5877-CONF (2009). `http://www-d0.fnal.gov/Run2Physics/WWW/results/prelim/TOP/T78/T78.pdf`
[9] V.M. Abazov it et al., Phys. Rev. Lett. **103**, 132001 (2009).
[10] CDF Collaboration, CDF/PHYS/TOP/PUBLIC/9881 (2009). `http://www-cdf.fnal.gov/physics/new/top/confNotes/cdf9881_Combined_Mtop_lepPt.pdf`
[11] T. Aaltonen *et al.*, Phys. Rev. Lett. **102**, 152001 (2009)
[12] DØ Collaboration, DNote 5897-CONF (2009). `http://www-d0.fnal.gov/Run2Physics/WWW/results/prelim/TOP/T80/T80.pdf`
[13] T. Aaltonen *et al.*, arXiv: 1002.0365 [hep-ex] (2009).
[14] Tevatron Electroweak Working Group, arXiv:0903.2503 [hep-ex] (2009).
[15] LEP Electroweak Working Group, CERN-PH-EP/2009-023 (2009).
[16] CDF Collaboration, CDF/ANAL/TOP/PUBLIC/9724 (2009). `http://www-cdf.fnal.gov/physics/new/top/2009/tprop/Afb/cdfnote_9724_public_v01.pdf`
[17] V.M. Abazov *et al*, Phys. Rev. Lett. **100**, 062004 (2008).
[18] CDF Collaboration, CDF note 9824 (2009). `http://www-cdf.fnal.gov/physics/new/top/confNotes/cdf9824_spincorr2.8fb-1.pdf`
[19] DØ Collaboration, DØ note 5950-CONF (2009). `http://www-d0.fnal.gov/Run2Physics/WWW/results/prelim/TOP/T84/T84.pdf`
[20] CDF Collaboration, CDF/PUB/TOP/PUBLIC/9446 (2008). `http://www-cdf.fnal.gov/physics/new/top/confNotes/cdf9446_tprime_public_2.8.pdf`
[21] T. Aaltonen *et al*, Phys. Rev. Lett. **103**, 101803 (2009).
[22] T. Aaltonen *et al*, Phys. Rev. Lett. **102**, 222003 (2009).
[23] DØ Collaboration, DØ Note 5882-CONF (2009). `http://www-d0.fnal.gov/Run2Physics/WWW/results/prelim/TOP/T83/T83.pdf`
[24] CDF Collaboration, CDF note 9844 (2009). `http://www-cdf.fnal.gov/physics/new/top/confNotes/cdf9844_AllHadMtt_pub.pdf`

SEARCH FOR PHYSICS BEYOND THE STANDARD MODEL WITH THE ATLAS EXPERIMENT AT THE LHC

Eduardo Ros [a]

IFIC, Univ.Valencia-CSIC, apartado 22085, 46071-Valencia, Spain

Abstract. A brief summary of searches for new phenomena using the ATLAS detector at the LHC is presented.

1 Introduction

The aim of this talk is to give an overview of the potential of the ATLAS experiment at the LHC to discover beyond the standard model (BSM) physics. In this short summary, supersymmetry (SUSY) is excluded since it is already covered by another talk in these proceedings. Even after excluding SUSY, the field of BSM physics is extremely large, and there is no intention to cover it in any detail. Rather than that, we focus on a small number of simple signatures, namely 2-body decay signatures, and finally, as typical examples of more complex signatures, we focus on models like the 'left-right symmetric model', the 'little higgs model' and the 'twin higgs model'. It is by now well known that the first data at the LHC will be collected at a center-of-mass (cms) energy of $\sqrt{s} = 7$ TeV, rather than at the nominal cms energy of 14 TeV. The question of what can be achieved with this reduced cms energy is also covered by another talk in these proceedings and is not discussed here. In the following, we assume therefore $\sqrt{s} = 14$ TeV and luminosities typically in the range of 1 to 100 fb^{-1}. These luminosities require of course several years of data taking. For more detailed reviews, we refer to reference [1], that describes a large amount of potential BSM searches, using a fast simulation of ATLAS, and also to reference [2], with a reduced selection of searches, but with a detailed simulation of the ATLAS detector.

Even if no details can be given here, it is worth mentioning that ATLAS is a general purpose detector, capable of detecting SM and BSM signatures at the LHC. Its weight is 7000 tons and overall dimensions 46×22 m^2. It has a central cylindrical solenoid with a 2 T magnetic field and external toroidal magnets to cover barrel and end-cap muon chambers. The central tracker yields a momentum resolution for charged particles of $/p_T = 5 \cdot 10^{-4} p_T \oplus 0.01$ (p_T in GeV). The calorimeters measure energy with electromagnetic resolution of $/E = 10\%/\sqrt{E} \oplus 0.01$ and hadronic resolution of $/E = 50\%/\sqrt{E} \oplus 0.03$, where E is the energy in GeV. Finally a muon momentum resolution of 2% and 10% can be achieved for muons of 50 GeV and 1 TeV, respectively.

2 Selection of channels: 2-body decays

The simplest searches consist of 2-body decays of a resonance R, typically a Z' but any narrow resonance is covered by the searches described here. These searches in-

[a] e-mail: Eduardo.Ros@ific.uv.es

clude the following final states $R \to e^+e^-$, $^+$ $^-$, $^+$ $^-$, $b\bar{b}$, $t\bar{t}$. Leptonic decays of a resonance are well known signatures for new physics at hadron colliders. Examples of important discoveries using leptons are the $J/$ at Brookhaven in 1973, the at Fermilab in 1977, and finally the Z at CERN in 1984. ATLAS has of course the capability to discover a new narrow resonance decaying into e^+e^- or $^+$ $^-$, up to masses of several TeV. Figure 2 shows, as an example, the signal that might be observed for a 3.5 TeV resonance decaying into electrons and muons. We observe that the signal is much more narrow for electrons than for muons. The widths are 20 GeV and 400 GeV, respectively. This is expected, since the colorimeter energy resolution improves with energy as $1/\sqrt{E}$, whereas the muon momentum resolution is proportional to p_T. This result by no means implies that the muon channel is to be neglected, since on one side a new resonance R might decay preferably into muons, and on the other side the muon background is expected to be smaller.

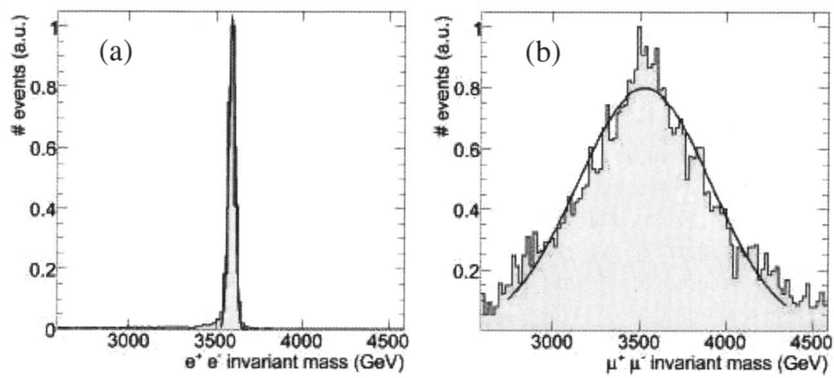

Figure 1: Simulation of the signal observed for a 3.5 TeV resonance decaying into (a) e^+e^- and (b) $^+$ $^-$. Only the signal peak is shown.

As a novelty compared to past colliders, ATLAS is able to observe as well the decay of a resonance into pairs, b-quarks and t-quarks. As for the channel $R \to$ $^+$ $^-$, a significance of 8 can be achieved if the mass of R is 600 GeV, and the luminosty at least 1 fb^{-1}. The mass resolution of the decays $R \to b\bar{b}$ and $R \to t\bar{t}$, where R could be an excited gluon g^* or a Kaluza-Klein state g_{KK}, has been found to be about 150 GeV, for a resonance mass of 1 TeV. These resolutions are dominated by detector resolution and fragmentation effects, since the natural width of a resonance R decaying hadronically does not exceed 10% of its mass in most models. In both b and t channels, it is possible to reach masses up to 2 TeV for 10 fb^{-1} luminosities. The b channel is somewhat more difficult due to a larger irreducible background.

3 Selection of channels: complex signatures

As examples of complex signatures, we study in the following the LR symmetric model, the Little Higgs model and the Twin Higgs model. All them are characterized by the appearance of several new particles, with correlated masses, widths and cross-sections. All 3 models imply however the presence of a heavy Z' decaying into leptons, in the way discussed in the previous section. Furthermore, all 3 require that the Z' is accompanied by a heavy W'. The signature of a heavy W' is also well-known. It consists of a hight p_T lepton, in correlation with large transverse missing energy. It is not possible in this case to reconstruct the invariant mass of the resonance, but rather the so-called 'transverse mass', with a jacobean peak at the position of the mass.

A very simple extension of the SM is the LR symmetric model, requiring Z_R, W_R, and a heavy neutrino N, possibly of Majorana type. The signature for this model would be the decay $W_R \to lN_l$, followed by $N_l \to lq\bar{q}$, so the final state consist of 2 leptons (sign-like for the Majorana case) and 2 jets, without missing energy. Examples of this search are heavy W bosons with masses of 1.5 and 1.8 TeV, and heavy neutrino masses of 500 and 300 GeV, respectively. In both cases a significant signal can be observed provided the luminosity is at least 100 pb^{-1}. Two somewhat more complex extensions of the SM are the Little Higgs and the Twin Higgs model. Both require the presence of many new particles, in particular a heavy top state T and a new heavy scalar sector ^{0}, $^+$ and $^{++}$, in addition to a heavy Z and W. The Twin Higgs model has additional scalar particles h_1^0, h_2^0 and h^+, that could explain dark matter. There are typical decays of these models that can be used as characteristic signatures. For example the decay $Z_H \to Zh$, where h is the SM Higgs boson, is typical of the Little Higgs model. The branching ratio (BR) of this decay could be as large as 10%. Another typical decay is $T \to Zt$, with a BR of 25%. The decay $Z_H \to Zh$ has a characteristic final state of 2 leptons and 2 b-jets. Assuming a Z_H mass of 1 TeV, a significant signal can be observed for a luminosity above 10 fb^{-1} (see figure 3a). The decay $T \to Zt$ can also be easily identified by its final state consisting in 2 leptons, a W and a b-jet. For a T mass of 1 TeV, a significant signal can be can be observed for a luminosity above 100 fb^{-1} (see figure 3b).

In all models discussed before there are free parameters, normally coupling constants, that could enhance or decrease the predicted signals. An interesting decay to look at, in case a heavy boson W_H is detected, is the hadronic decay $W_H \to tb$. The BR for this decay is 25% for most models, but only 3% in the Twin Higgs model. The detection of this decay is therefore challenging but, if measured, could be a clear hint in favor of this model. In this case, assuming a mass of 1 TeV for W_H, the maximum expected luminosity of 300 fb^{-1} should be used to detect this decay. We mention finally another characteristic decay of W_H in the Twin Higgs model, namely the cascade decay $W_H \to Tb \to {}^+bb \to tbbb \to Wbbbb$ that yields a final state with a W and 4 b-jets. This decay is extremely challenging to obverse, and is one out of the many examples of interesting signatures that will be very difficult to observe at the LHC.

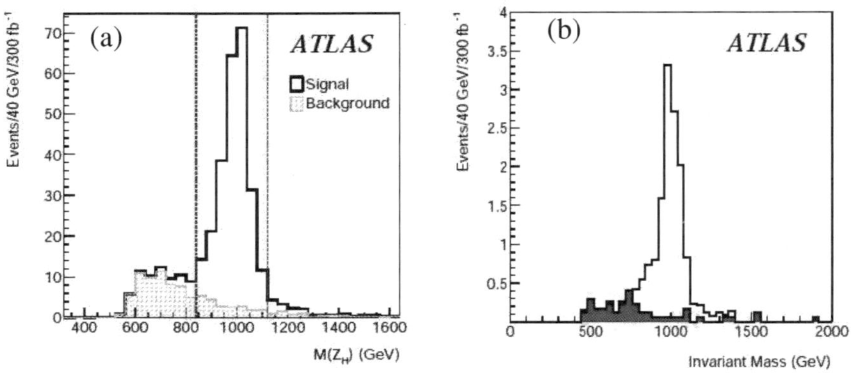

Figure 2: Reconstructed mass for the decays (a) $Z_H \to Zh \to e^+e^-bb$ and (b) $T \to Zt \to e^+e^-bW$, that are predicted in the little higgs model.

4 Summary and outlook

The ATLAS detector at the LHC has the capability to discover resonances decaying into leptons, but also into b or t quarks. These resonances are typically heavy Z or W states, that are predicted in many BSM models. The detection of more complex decays is possible, depending on the value of masses, widths and cross-sections. A beautiful example of a complex decay that might be observed at the LHC consists of a heavy W decaying into 2 leptons and 2 jets, without missing energy. This decay is expected for a heavy Majorana neutrino.

Acknowledgments

I would like to thank my ATLAS colleagues for providing material for this talk, and the organization of the conference for providing a pleasant stay at Moscow.

References

[1] ATLAS coll., ATLAS detector and physics performance, Technical Design Report (2 vol.), CERN/LHCC/99-14,99-15 (1999).

[2] ATLAS coll., The ATLAS experiment at the LHC collider, *JINST* 2 (2008) s088003.

Neutrino Physics

NEUTRINO PHYSICS WITH OPERA

Alessandro Bertolin [a]
on behalf of the OPERA Collaboration
Istituto Nazionale di Fisica Nucleare, Sezione di Padova,
Via Marzolo 8, 35131 Padova, Italy

Abstract. Neutrino physics with the OPERA experiment will be discussed in this paper. First the OPERA physic goal will be presented. A description of the neutrino beam and of the detector will follow. The analysis of the beam induced neutrino interactions will then be presented.

1 The OPERA physics goal

Studies of the atmospheric neutrino interactions by the Super–Kamiokande experiment in 1998 showed that the so called "atmospheric neutrino anomaly" can be explained in terms of neutrino flavor oscillations. The CHOOZ results excluded the ν_μ to ν_e oscillation as the dominant process responsible for the atmospheric ν_μ disappearance. The atmospheric ν_μ disappearance signal was further confirmed by $K2K$ and $MINOS$. In this context the ν_μ to ν_τ oscillation seems an interesting possibility.
As a consequence direct observation of ν_τ appearance from ν_μ oscillations is the primary goal of the OPERA experiment. In this case the decay products of a τ lepton are identified in the final state of an incoming ν_μ interaction.

2 The CNGS beam

The CERN Neutrino To Gran Sasso [1], CNGS, primary source is a 400 GeV proton beam from the SPS hitting a graphite target. The ν_μ beam emerges beyond a decay tunnel with a magnetic system to focus charged particles of the appropriate charge. μ detectors are used to quantify the intensity of the ν_μ beam. The ν_e plus $\bar{\nu}_e$ and $\bar{\nu}_\mu$ contaminations are 0.87 % and 4 % of the ν_μ flux. The prompt ν_τ component is negligible. The average energy of the ν_μ beam is 17.9 GeV. The expected intensity is 4.5 10^{19} p.o.t./year, assuming 200 days of beam operation per year. With a target mass of 1.3 k tons this leads to $\sim$ 22000 changed or neutral current ν_μ events and to $\sim$ 70 – 150 ν_τ charged current events (for $\Delta m^2 = 2 - 3\ 10^{-3}$ eV^2 and $\sin^2 2\theta_{23} = 1$). In each SPS super-cycle three cycles are devoted to the CNGS beam, the CNGS cycles. Each CNGS cycle consists of two ν_μ shots, 50 ms apart, each shot lasting 10.5 μs. It is thus possible to establish if the interaction observed in the OPERA detector is in time with the CNGS beam, while the observed events "not in time" are surely due to cosmic rays.

[a] e-mail: bertolin@pd.infn.it

3 The OPERA detector

The OPERA detector is located $\sim$ 730 km away from the neutrino source in the Laboratori Nazionali del Gran Sasso, $\sim$ 150 km eastbound from Roma. More precisely the detector is located underneath the Gran Sasso mountain, 2914 m height. The mountain thus provides a powerful "natural" filter against cosmic rays. A special branch of the highway tunnel from and to Rome is used to access the experimental area.

The detector is bimodular. Each module consists of an instrumented target, with a lead–photographic emulsion sandwich and orthogonal scintillator bars, and of an instrumented iron spectrometer with a dipolar magnetic field. The field is vertical with a strength of 1.55 T. The spectrometer is instrumented with 22 planes of resistive plate chambers, RPC, and 6 planes of drift tubes, called High Precision Trackers, HPT. The drift tubes perform a measurement in the bending plane only and hence are lined up vertically. Each HPT plane has three layers of drift tubes. The RPCs have readout in both the vertical and horizontal planes by means of an orthogonal set of copper strips. Combining the RPC and the HPT data it is possible to identify and measure the momentum of penetrating particles. Upstream with respect to the first module, two planes of glass resistive plate chambers are used to tag incoming charged tracks produced by the ν_μ beam in the rock surrounding the detector.

The passive material of each target module consist of $\sim$ 75000 "bricks" arranged in 31 walls. Each brick is made by a lead–photographic emulsion sandwich. Moreover 31 walls of horizontal and vertical scintillator bars, the active material, are used for a rough identification of the neutrino interactions.

Combining the informations on penetrating particles with the patterns of energy deposits in the scintillator bars it is first possible to determine with little uncertainty if the primary vertex of the event occurred inside the target. With a more refined analysis it is then possible to locate the brick that with highest probability contains the primary vertex. Each brick is made of 56 lead plates and 57 photographic emulsion plates. Each emulsion plate consists of $\sim$ 44 μm thick emulsion layers with a $\sim$ 212 μm thick plastic base in between. The primary vertex brick is taken out of the detector and the emulsion layers developed. As a final step, a primary vertex search is performed in the emulsion layers.

Therefore OPERA is a hybrid detector: the electronic detectors are used to tag in almost real time the brick containing the primary vertex. High resolution tracking, O(few μm) resolution, is performed later on in the emulsion layers of the tagged brick.

4 Real time analysis in the electronic detectors

The first neutrino interaction induced events have been recorded with the electronic detectors in 2006. At that time the OPERA target was empty. The effective running time was $\sim$ 10 days with only 7.6 10^{17} integrated p.o.t.; less that 2 % of the nominal expected amout of p.o.t. per year. These data have however been useful for the electronic detectors [2]. Fig. 1 shows the distribution of the vertical angle of the reconstructed three dimensional tracks. The data are shown by the black bullets. The vertical error bars are statistical only. The expected cosmic ray contribution is shown by the yellow histogram. The cosmic rays were normalized using the effective running time. Fig. 1 clearly shows a peak at about 3.4^o not described by the cosmic rays contribution. This peak is due to the beam induced muon tracks which due to the Earth curvature are expected to be seen in OPERA with an angle of 3.3^o. The measured value was in agreement with the geodetic expectations.
Requiring a time coincidence with the neutrino beam only the events of the insert are left. All the cosmic ray background is thus removed.

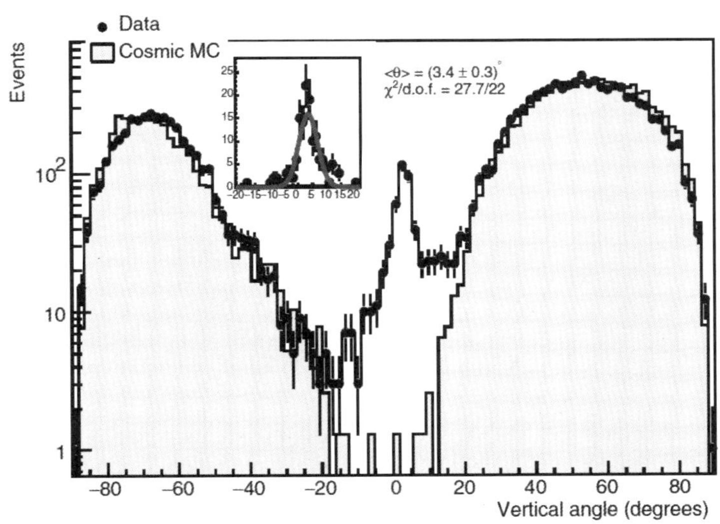

Figure 1: Distribution of the vertical angle of the reconstructed three dimensional tracks using the 2006 data sample. More details are given in the text.

Data taking was resumed in September 2007, still at rather low beam intensity and with a target about half filled. Unfortunately due to a fault of the CNGS beam the physics run lasted only a few days. About 0.08 10^{19} p.o.t. were accumulated corresponding to $\sim$ 3.6 days of running at the nominal intensity. 465 events on time with the beam have been recorded out of which 35 occurred in the region of the target filled with bricks [3].

The 2008 run started in June and lasted till November. The target was almost fully filled. About 1.8 10^{19} p.o.t. were recorded, $\sim$ 40 % of the nominal value. 10100 events were in coincidence with the neutrino beam out of which 1663 were automatically classified as occurring in the target volume [3].
The analysis of the vertical angle of the beam induced three dimensional tracks was repeated with significantly higher statistics. The result is given in the left part of Fig. 2: the data are shown by the black bullets, the vertical error bars are statistical only. The MC prediction is given by the yellow histogram. From this comparison the vertical beam tilt is consistent with expectations. The momentum times charge distribution is shown in the right part of Fig. 2. As expected most of the measured tracks are negative muons from ν_μ changed current interactions. A positive muon component is also seen. It is due to the $\bar{\nu}_\mu$ contamination of the beam. From the agreement observed between data and MC for positive tracks the $\bar{\nu}_\mu$ contamination is fairly consistent with expectations.

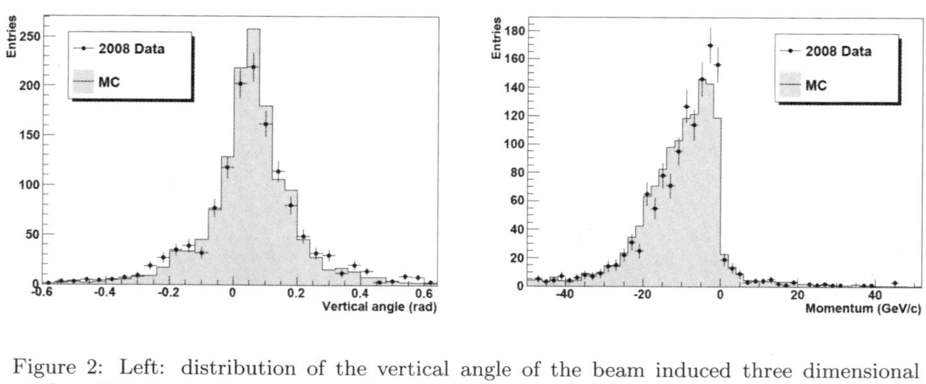

Figure 2: Left: distribution of the vertical angle of the beam induced three dimensional tracks. Right: momentum times charge distribution for the same tracks sample. More details are given in the text.

The 2009 run started on June the 1st. The run start was delayed by an earthquake that occurred in the region of the Laboratori Nazionali del Gran Sasso. The earthquake had a large and dramatic impact on the local population, 306 killed and 1600 injured, but the effect in the underground experimental area was dumped by a factor $\sim$ 5 thus no damage occurred to the detector.

At the time these proceedings are being written the run period ended with $\sim 3.5\ 10^{19}$ recorded p.o.t. Likely without the delay due to the earthquake the nominal luminosity may have been reached.

5 Analysis in the emulsions

As mentioned in Sec. 3 the electronic detectors data allows the automatic selection of the events with the primary vertex inside the target volume and the selection of the brick that with highest probability contains the primary vertex.
At this point the analysis in the emulsion starts with the extraction from the detector of the interesting brick and the development of the corresponding emulsion films.
The developed films are being analyzed in Europe with customized commercial optics and mechanics, and in Japan with custom made hardware. The tracks search is based on the tomographic acquisition of the emulsion layers. The scanning speed is $\sim 20 - 75\ \mathrm{cm}^2$ / hour according to the hardware being used.

The full analysis of a typical charged current events is shown in Fig. 3. The upper left and right plots show the event as it appears in the electronic detectors. A penetrating muon track stopping in the iron of the first spectrometer is apparent. The track emerges from a significant energy deposit in the scintillator bars. The incoming ν_μ direction is shown to guide the eye.
Using the emulsion data the lower left, middle and right plots show the event primary vertex and the track leaving it in the X vs Z, Y vs Z and Y vs X projections, respectively. The thicker segments along the tracks correspond to the particle path inside the emulsion layers. The Y vs X projection, front view, is particularly interesting. The isolated continuous track on the lower left corner is the muon one. The cloud of dots also in the lower left corner is due to an electromagnetic shower. In the opposite hemisphere with respect to the muon track a large number of charged particles are generated from the primary vertex. One of these tracks has an obvious kink (upper side of the primary vertex almost on the vertical with respect to it). The large kink is visible also in the other projections. A more detailed kinematic analysis shows that this event is consistent with being a charm candidate.

6 Conclusions

The first neutrino interactions have been observed in the OPERA detector in 2006 in parallel to the commissioning of the CNGS beam. Since then and at the time these proceedings are being written OPERA integrated $5.28\ 10^{19}$ p.o.t. The electronic detectors are fully operational and are doing very well. All steps of the emulsion analysis have been successfully validated. Events are

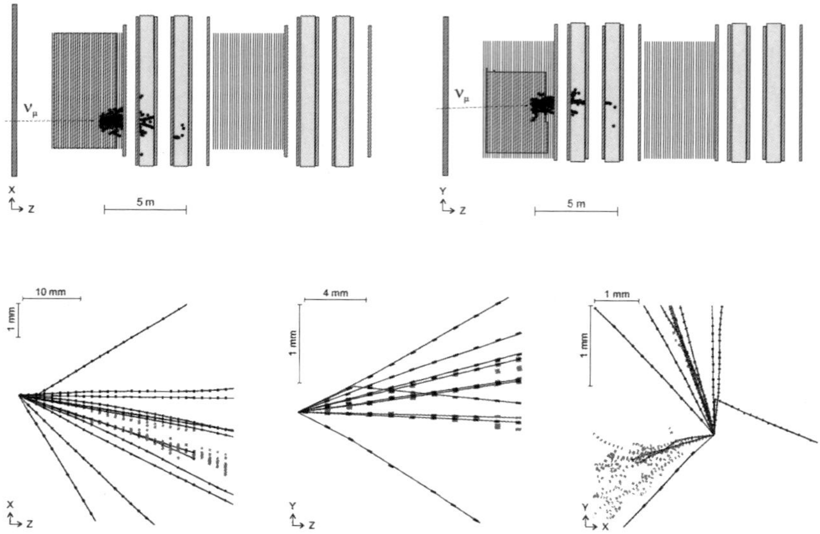

Figure 3: Event display in both the electronic detectors, upper part, an in the emulsion layers, lower part, of a charm candidate event. More details are given in the text.

being regularly analyzed up to the primary vertex. Charm events have been identified. As statistic is a key point in order to achieve the direct observation of ν_μ to ν_τ oscillation, OPERA relies on the CERN support to have soon the CNGS running at nominal intensity, 4.5 10^{19} p.o.t./year.

Acknowledgments

It is a pleasure to thanks Prof. Alexander I. Studenikin and all the Lomonosov conference team for their interest to the OPERA physics case and for their support during the conference.

References

[1] CNGS project: on line at *http://proj-cngs.web.cern.ch/proj-cngs/*
[2] R. Acquafredda et al., *New Journal of Physics* **8**, (2006), 303.
[3] N. Agafonova et al., *2009 JINST*, 4, P06020.

NEW RESULTS FROM THE FNAL SCIBOONE NEUTRINO EXPERIMENT (FNAL E954)

Tsuyoshi NAKAYA [a] for the SciBooNE collaboration
Faculty of Science, Kyoto University, Sakyo-ku, Kyoto 606-8502, JAPAN

Abstract. We report new results from Fermilab E954, a neutrino scattering experiment SciBooNE. SciBooNE is motivated to measure the neutrino-nucleus cross sections precisely in the energy region of ~ 1 GeV. The measurements are important to understand the interactions of neutrinos with the nucleus and information is essential for future accelerator neutrino oscillation experiments, such as T2K. We report the measurements of the charged-current quasi-elastic scattering, the measurements of pion productions by neutrinos, and the status of the short baseline neutrino disappearance search together with the MiniBooNE experiment.

1 Introduction

The neutrino-nucleus cross sections are recently drawn much attention to by particle and nuclear physics communities. Data of the cross sections are not precise enough to reveal the underlying mechanism of nuclear physics in neutrino interactions. In addition, precise knowledge of neutrino cross sections on nuclei is desired by the next generation accelerator neutrino oscillation experiments such as T2K [1], of which ultimate goals are the complete studies of neutrino oscillations including the CP violation.

The SciBooNE experiment is designed to measure the neutrino cross sections on carbon in the energy region of ~ 1 GeV. In SciBooNE, we study quasi-elastic scattering, resonant single π production, coherent π production and deep-inelastic scattering of neutrinos and anti-neutrinos in both charged-current (CC) and neutral-current (NC) reactions. In addition, since the SciBooNE detector is located on the same neutrino beam line of the MiniBooNE experiment, the measurement of the neutrino flux at SciBooNE could be contributed to the better prediction of the neutrino flux at MiniBooNE which is useful to improve the studies of neutrino oscillations at MiniBooNE.

2 Experimental Setup

SciBooNE shares the Fermilab Booster Neutrino Beam (BNB) with the MiniBooNE experiment. The 8 GeV protons from the Booster are extracted to BNB and hit a 71.1 cm long beryllium target. The hadorons, mainly pions, are focused by an electromagnetic horn. The neutrinos are produced from the decay of hadrons, and the mainly muon neutrinos from $\pi \to \mu\nu$ decays are the dominant component in the beam with the peak energy of 0.7 GeV. The neutrino fluxes at SciBooNE predicted by the beam Monte Carlo Simulation [2]

[a] e-mail: t.nakaya@scphys.kyoto-u.ac.jp

are shown in Figure 1. By switching the polarity of the horn current, the BNB provides both the neutrino and the anti-neutrino beams.

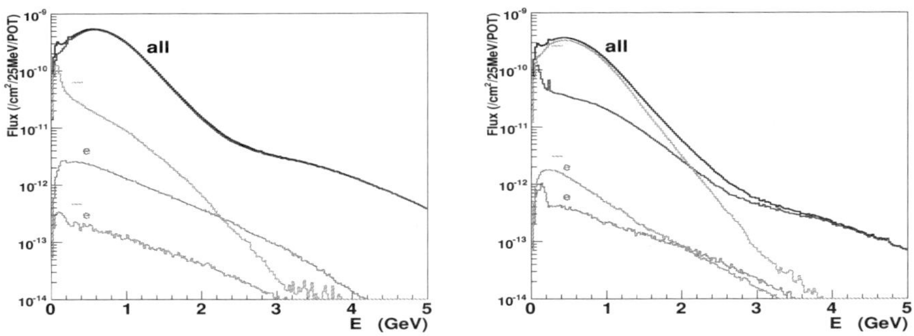

Figure 1: Neutrino flux predictions at SciBooNE as a function of neutrino energy. The neutrino beam mode (left) and the anti-neutrino beam mode (right) are shown.

The SciBooNE detector located 100 m downstream from the target consists of three detectors: a fully-active fine-segmented tracking detector (SciBar) [3] as a neutrino target and a vertex detector, an electro-magnetic calorimeter (EC) [4] and a muon range detector (MRD). The SciBar consists of 14,336 extruded plastic scintillator strips, and each strip is $1.3 \times 2.5 \times 300$ cm^3. The scintillators are arranged vertically and horizontally to construct a $3 \times 3 \times 1.7$ m^3 volume with a total mass of 15 tons. Based on the hits information with energy, we reconstruct the tracks of charged particles and identify a proton track from others by the dE/dx measurement. The EC is located just downstream of SciBar to measure the electron neutrino component in the beam and to detect the photons from π^0 decay. The EC is made of 32 vertical and 32 horizontal modules with the cross-section of 262×256 cm^2. Each module with the dimension of $262 \times 8 \times 4$ cm^3 is a spaghetti-type calorimeter comprised of 1 mm diameter scintillating fibers embedded in lead foil. The MRD is installed downstream of the EC to measure the momentum of a muon. Twelve 5 cm thick iron plates are sandwiched between 6 mm thick scintillator hodoscopes. The 13 alternating horizontal and vertical scintillator hodoscopes are mounted. The cross section of the iron plate is 274×305 cm^2 and the total mass of absorber material is 48 tons. The MRD measures the momentum of the muon up to 1.2 GeV/c using the range of the muon.

The SciBooNE experiments took data from June 2007 to August 2008 with 2.64×10^{20} protons on target (POT) in total. The available data for analysis are 0.99×10^{20} POT with the neutrino beam and 1.53×10^{20} POT with the anti-neutrino beam.

3 Physics Results

In the analysis, we use the NEUT program library [5, 6] to simulate the interactions of neutrinos with nuclei. In the neutrino beam running, 53,000 Charged-Current (CC) quasi-elastic scattering (QE) (41 %), 30,000 CC single π via resonances (23 %), 23,000 Neutral-Current (NC) elastic (18 %), 6,900 NC single π^0 via resonances (5.4 %), 6,100 CC DIS (4.7 %), 4,700 NC single meson but π^0 (3.7 %), 1,800 CC coherent π^+ (1.4 %), 1,800 NC DIS (1.4 %), 1,100 NC coherent π^0 (0.9 %), and 800 CC single meson but π (0.7 %) interactions are estimated by NEUT in the 10.6 ton fiducial volume of SciBar[b].

3.1 Charge-Current Quasi-Elastic Scattering

Charged-Current quasi-elastic (CC-QE) scattering is a dominant process in the neutrino interactions with nuclei below 1 GeV energy. CC-QE is a useful reaction in which the parent neutrino energy could be reconstructed only by the muon momentum with information of the neutrino beam direction. We study CC-QE events by requiring one muon track reaching the MRD[c]. Then, we count the number of tracks from the common vertex and identify another track as a proton or a pion based on the dE/dx information. In Table 1, we summarize three samples: the 1-track, the 2-track (μ + p) and the 2-track (μ + π) samples. The 1-track and 2-track (μ + p) samples contain the CC-QE events with high purity. The 2-track (μ + π) sample contains the other type of events, by which the amount of the background events on CC-QE is studied in data.

Sample	# Events	Efficiency (%)	Purity (%)
1 track	13586	52.9	65.2
2 track (μ + p)	2915	11.1	68.5
2 track (μ + π)	1628	2.7	32.3

Table 1: Summary of three samples for the CC-QE cross section measurement. The efficiency and purity are for the CC-QE events estimated by NEUT.

The ν_μ-CCQE cross section per nucleon is measured in the neutrino energy range between 0.6 and 1.6 GeV. The result is shown in Figure 2 [7], which is consistent with the prediction by NEUT. A main systematic uncertainty is due to the uncertainty of the neutrino flux predictions at SciBooNE.

[b]The SciBar fiducial volume is defined as $-130 < x, y < 130$ cm and $2.62 < z < 157.2$ cm where the z axis is the beam direction.

[c]We also have another analysis not requiring the MRD matching track.

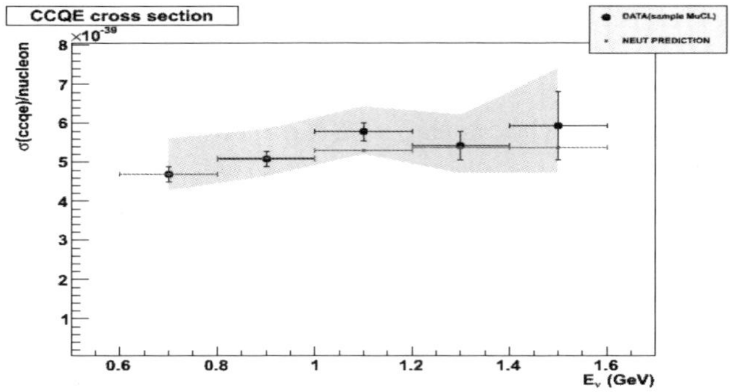

Figure 2: The ν_μ-CCQE cross section measured in SciBooNE for the SciBar-MRD analysis (circle dots) in comparison with the NEUT prediction (triangle dots). The statistics (bars) and the flux systematic errors (yellow [or light-gray]) band are shown.

3.2 Pion Production by Neutrinos

Pion production by neutrinos is one of the most interesting and popular subjects in the neutrino scattering physics. CC π production is a major background to CC-QE process which is used to reconstructed neutrino energy in the neutrino oscillation experiments. NC π^0 production is a major background events in search for $\nu_\mu \to \nu_e$ appearance signal which probes the unknown mixing angle θ_{13}. Recently, we published the result of no evidence of CC coherent π^+ [8]. We set an upper limit on the cross section ratio of CC coherent π^+ production to the total CC cross section at 0.67×10^{-2} at mean neutrino energy 1.1 GeV and 1.36×10^{-2} at mean neutrino energy 2.2 GeV in 90% confidence level. The result is consistent with and confirm the result by K2K [9].

After the publication, we further investigate the signal and find the excess of the events with the pion scattered more forward ($< 35°$) [10]. Those events have very low Q^2 and just a muon and a pion tracks observed. So the events have a features of the coherent π production although the kinematic distributions and the production cross section do not match with the model prediction by Rein-Sehgal [11,12]. So, further theoretical studies on coherent π are interesting and expected together with the SciBooNE results.

In the anti-neutrino data, the better signal discrimination for CC coherent π is expected because of less background events. We looks for the signal of CC coherent π in the anti-neutrino data, and find the excess of the signals [13] as shown in Figure 3. In the anti-neutrino data, we observed the excess of the signal over the background by 4 σ level (statistical uncertainty only). We plan to evaluate the systematical error including the estimate of the wrong sign background components which only appear in the analysis of the anti-neutrino beam data.

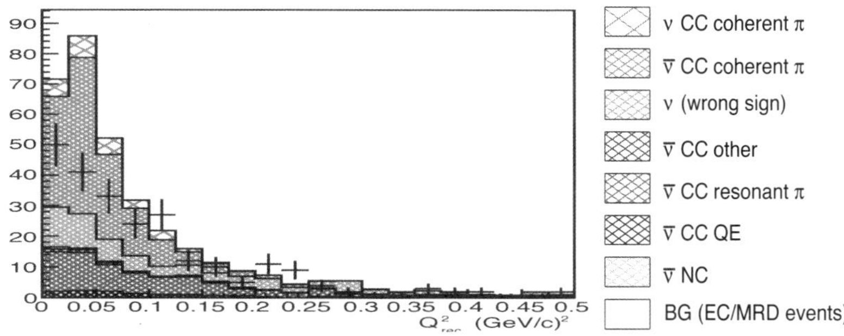

Figure 3: The reconstructed Q^2 distribution in the anti-neutrino charged current coherent pion sample. The expected signal level in MC is based on a model by Rein-Sehgal

The π^0 productions are intensively studied by several modern experiments [14–18]. We also measured the NC π^0 production cross section ratio to the total CC cross section to be $[7.7 \pm 0.5(\text{stat.}) \pm 0.5(\text{sys.})] \times 10^{-2}$, while the MC prediction is 6.8×10^{-2}. For the measurements, we select the events with two γ's converted in the SciBar fiducial volume. We rejects events with positively identified a proton track or a muon track. The invariant mass and the momentum distributions of the events are shown in Figure 4. We observed 657 NC π^0 candidate events with the efficiency of 5.3 % over 240 background events estimated by the MC simulation. The measurements are consistent with the Rein-Sehgal model implemented in the MC simulation.

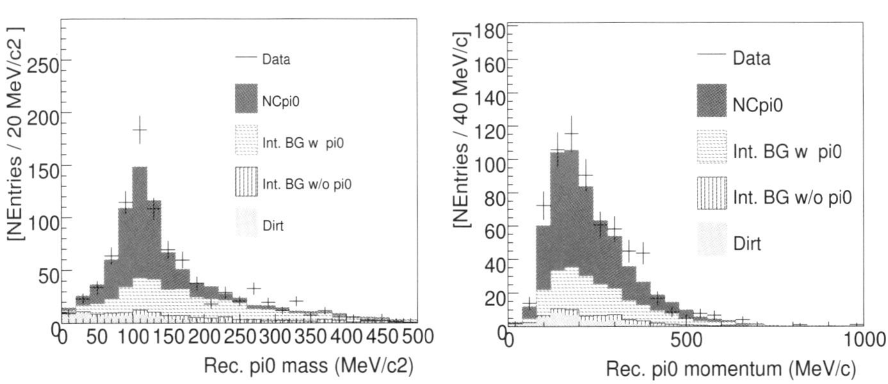

Figure 4: The invariant mass and the momentum distributions of the π^0 candidate events.

With the NC π^0 candidate events, we study the NC coherent π^0 production and the result is under preparation.

3.3 Short Baseline Neutrino Oscillation Search with the SciBooNE and MiniBooNE detectors

A short baseline neutrino oscillation search in the Δm^2 region of $0.5 - 40$ eV^2 is conducted by using the SciBooNE as a front detector and the MiniBooNE as a far detector. At SciBooNE, we measure the neutrino flux, constrain the uncertainties of neutrino interactions, and provide the better prediction of the neutrino events at MiniBooNE. The expected distributions of the reconstructed neutrino energy at MiniBooNE is shown in Figure 5. With the predicted neutrino events at MiniBooNE, muon neutrino disappearance by oscillation is studied and the expected sensitivity is shown in Figure 5. Now, we are doing the final check to publish the result of the neutrino oscillation.

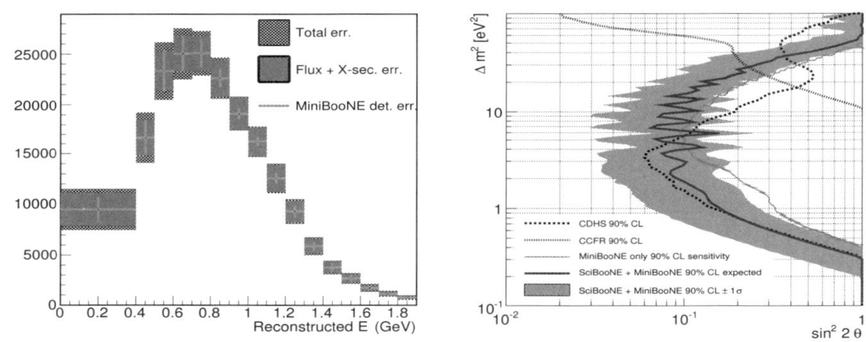

Figure 5: (Left) The reconstructed neutrino energy distribution predicted at MiniBooNE from SciBooNE. The MiniBooNE detector error, flux and cross-section uncertainties, and the total systematic uncertainty are separately shown. (Right) The expected sensitivity for ν_μ disappearance. The dotted curve shows the 90% CL limits from CDHS and CCFR experiments. The thin solid curve is the MiniBooNE-only 90% CL sensitivity. The thick solid curve and the filled region are the 90% CL sensitivity and $\pm 1\sigma$ band from SciBooNE-MiniBooNE joint analysis, respectively.

4 Summary and Future Prospect

In the paper, we report some highlight results from the SciBooNE experiment. We are in the middle of the active analysis stage and we expect more results. In the neutrino beam data, we more study the NC elastic scattering [19] and the components of electron neutrinos in our beam. In the anti-neutrino beam data, we also study the copious channels of the results presented here. We are providing high precision data of neutrino interactions with nuclei which could give a great contribution to the future of neutrino physics. We are very much looking forward to exploring the neutrino oscillations and the CP violation.

Acknowledgments

We acknowledge support from grants and contracts from DOE, NSF (U.S.), JSPS, MEXT (Japan), INFN (Italy) and Spanish Ministry of Education and Science. The work was supported by MEXT and JSPS with the Grant-in-Aid for Scientific Research A 19204026, Young Scientists S 20674004, Young Scientists B 18740145, Scientific Research on Priority Areas "New Developments of Flavor Physics", the global COE program "The Next Generation of Physics, Spun from Universality and Emergence", and the Japan/U.S. Cooperation Program in the field of High Energy Physics.

References

[1] Y. Itow *et al.* [The T2K Collaboration], arXiv:hep-ex/0106019.
[2] A. A. Aguilar-Arevalo *et al.* [MiniBooNE Collaboration], Phys. Rev. D **79**, 072002 (2009) [arXiv:0806.1449 [hep-ex]].
[3] K. Nitta *et al.*, Nucl. Instrum. Meth. A **535** (2004) 147 [arXiv:hep-ex/0406023].
[4] S. Buontempo *et al.*, Nucl. Instrum. Meth. A **349** (1994) 70.
[5] Y. Hayato, Nucl. Phys. Proc. Suppl. **112**, 171 (2002).
[6] G. Mitsuka, AIP Conf. Proc. **981**, 262 (2008).
[7] J. L. Alcaraz-Aunion and J. Walding [SciBooNE Collaboration], AIP Conf. Proc. **1189**, 145 (2009) [arXiv:0909.5647 [hep-ex]].
[8] K. Hiraide *et al.* [SciBooNE Collaboration] Phys. Rev. D **78**, 112004 (2008).
[9] M. Hasegawa *et al.* [K2K Collaboration], Phys. Rev. Lett. **95**, 252301 (2005) [arXiv:hep-ex/0506008].
[10] K. Hiraide, AIP Conf. Proc. **1189**, 249 (2009) [arXiv:0909.5127 [hep-ex]].
[11] D. Rein and L. M. Sehgal, Nucl. Phys. B **223**, 29 (1983).
[12] D. Rein and L. M. Sehgal, Phys. Lett. B **657**, 207 (2007) [arXiv:hep-ph/0606185].
[13] H. K. Tanaka, AIP Conf. Proc. **1189** (2009) 255 [arXiv:0910.4754 [hep-ex]].
[14] S. Nakayama *et al.* [K2K Collaboration], Phys. Lett. B **619**, 255 (2005) [arXiv:hep-ex/0408134].
[15] S. Mine *et al.* [K2K Collaboration], Phys. Rev. D **77**, 032003 (2008) [arXiv:0801.0182 [hep-ex]].
[16] A. A. Aguilar-Arevalo *et al.* [MiniBooNE Collaboration], Phys. Lett. B **664**, 41 (2008) [arXiv:0803.3423 [hep-ex]].
[17] Y. Kurimoto *et al.* [SciBooNE Collaboration] , arXiv:0910.5768 [hep-ex].
[18] A. A. Aguilar-Arevalo *et al.* [MiniBooNE Collaboration], arXiv:0911.2063 [hep-ex].
[19] H. Takei, AIP Conf. Proc. **1189** (2009) 181.

STATUS OF THE T2K EXPERIMENT

Masahiro Shibata [a], for the T2K collaboration
Institute of Particle and Nuclear Studies, High energy accelerator research organization (KEK), 1-1 Oho, Tsukuba, Ibaraki, Japan

Abstract. Tokai to Kamioka experiment (T2K) is an accelerator based long baseline neutrino oscillation experiment. Neutrino beams are produced at Japan proton accelerator research complex (J-PARC) and injected to Super-Kamiokande (SK). Our goals are discovery of $\nu_\mu \rightarrow \nu_e$ mode and precise measurement of ν_μ disappearance mode. The neutrino facility was completed in March 2009 and first beam commissioning was performed in April and May. All components worked as designed and neutrino production was confirmed with the muon monitor.

1 Introduction

Neutrino oscillation is a phenomenon that probability of measured lepton flavor of a neutrino oscillates with time. Neutrino oscillation can happen when weak interaction and mass eigenstate of neutrino is different and mass eigenvalues are different between eigenstates. Relation between weak interaction eigenstates $(\nu_e, \nu_\mu, \nu_\tau)$ and mass eigenstates (ν_1, ν_2, ν_3) is described with Maki-Nakagawa-Sakata (MKS) matrix [1].

$$\begin{pmatrix} \nu_e \\ \nu_\mu \\ \nu_\tau \end{pmatrix} = U \begin{pmatrix} \nu_1 \\ \nu_2 \\ \nu_3 \end{pmatrix} \tag{1}$$

$$U = \begin{pmatrix} 1 & 0 & 0 \\ 0 & C_{23} & S_{23} \\ 0 & -S_{23} & C_{23} \end{pmatrix} \begin{pmatrix} C_{13} & 0 & S_{13}e^{-i\delta} \\ 0 & 1 & 0 \\ -S_{13}e^{-i\delta} & 0 & C_{13} \end{pmatrix} \begin{pmatrix} C_{12} & S_{12} & 0 \\ -S_{12} & C_{12} & 0 \\ 0 & 0 & 1 \end{pmatrix} \tag{2}$$

Where S_{ij} (C_{ij}) stands for $\sin\theta_{ij}$ $(\cos\theta_{ij})$. This matrix has four parameters, namely three mixing angle $(\theta_{12}, \theta_{23}, \theta_{13})$ and phase of CP asymmetry (δ).

The T2K experiment has two goals. The first goal is discovery of $\nu_\mu \rightarrow \nu_e$ mode, which is the last undiscovered mode and determine θ_{13}. The oscillation probability is

$$P(\nu_\mu \rightarrow \nu_e) \approx \sin^2 2\theta_{13} \sin^2 2\theta_{23} \sin^2 \left(\frac{1.27\Delta m^2_{31} L}{E} \right). \tag{3}$$

Where $\Delta m^2_{31} = m^2_3 - m^2_1$ is difference of squares of mass for each eigenstate in eV^2, L is length of the baseline in km and E is neutrino energy in GeV. Measurement of θ_{13} is important because θ_{13} just has upper limit [2] and is related to possibility of CP violation measurement in neutrino. The sensitivity

[a] e-mail: masahiro.shibata@kek.jp

is $\sin^2 2\theta_{13} = 0.006$ at 90% C.L. with integrated proton beam power of 3.75 MW $\times$ 10^7 s.

Another goal is precise measurement of ν_μ disappearance mode. The probability of disappearance is

$$P(\nu_\mu \to \nu_x) \approx \sin^2 2\theta_{23} \sin^2 \left(\frac{1.27 \Delta m^2_{23} L}{E} \right). \tag{4}$$

We are aiming to determine $\sin^2 2\theta_{23}$ and $\Delta \mathrm{m}^2_{23}$ with precision 1% and 1×10^{-4} eV^2, respectively.

The T2K experiment has three features. First one is high intensity proton beam of J-PARC. The design intensity is 750 kW and highest in the world. It greatly helps to reduce necessary time to achieve our goals.

Second one is off axis beam method [3]. In the T2K experiment, the proton beam axis is 2.5 degrees tilted from the direction of the neutrino detector. With this method, the peak of neutrino energy spectrum coincides with the peak of oscillation probability and high energy neutrino is suppressed to reduce background (fig.1).

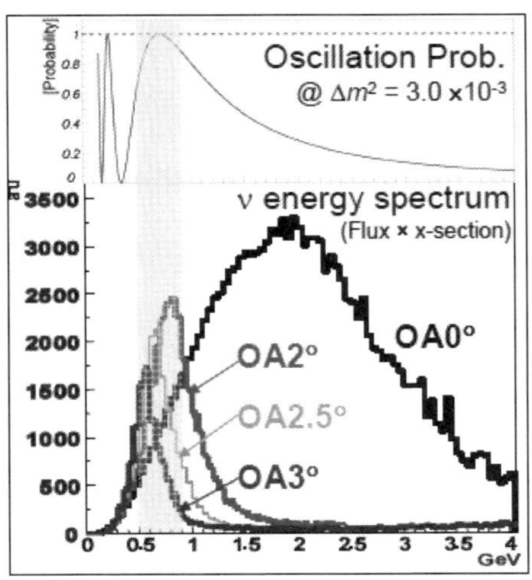

Figure 1: (top) Oscillation probability. (bottom) Energy spectrum of neutrino beam.

Third one is SK, the world largest 50 kt water Cherenkov detecor (fiducial volume 22.5 kt). SK has good performance of neutrino detection efficiency, momentum measurement and particle identification from several hundred MeV to 1 GeV, which is neutrino energy range of the T2K experiment.

2 Experimental setup

Fig.2 shows the experimental setup of the T2K experiment. 30 GeV proton beams are extracted from the main ring of J-PARC and transported to the graphite target. Superconducting conbined-function (bending + focusing) magnets are used to bend proton beams toward SK direction. Pions generated at the target are focused with toroidal magnetic fields generated by three electromagnetic horns. These pions decay in the decay volume and muon neutrinos and muons are generated. Arrays of Si PIN photodiodes and ion chambers are placed to monitor muons at downstream side of the beam dump which absorbs remaining hadrons. With this muon monitor neutrino beam intensity and direction are checked from muon intensity and distribution. At 280 m downstream from the target, a near neutrino detector is installed. It consists of on-axis detector and off-axis detector. The on-axis detector is an array of modules of scintillator planes sandwiched by iron targets and placed on the axis of the beamline. The on-axis detector measures neutrino beam direction and profile directly. The off-axis detector is a complex of scintillators, time projection chambers and electromagnetic calorimeters in the UA1 magnet which is donated from CERN. The off-axis detector is installed in the direction toward SK and measures neutrino flux and energy spectrum before oscillation. It also measures electron neutrino contamination and neutrino cross section.

Neutrino oscillation is searched by compareing neutrino flux and energy spectrum measured at SK with expected ones. They are estimated from proton numbers on the target, muon intensity and neutrino flux and energy spectrum measured at the near detector using neutrino cross section with the target material and the efficiency of each detector.

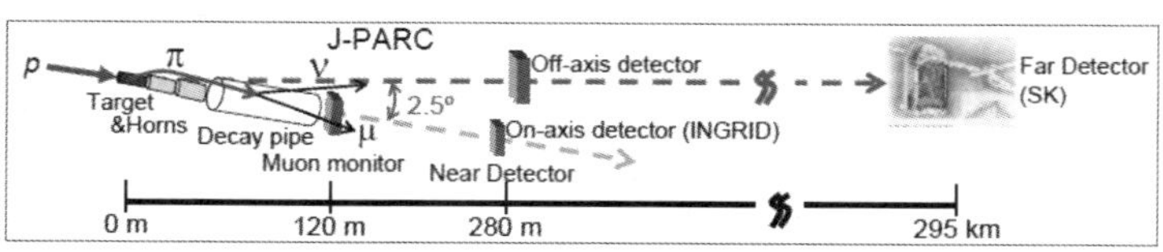

Figure 2: Experimental setup of the T2K experiment.

3 Neutrino facility commissioning

Commissioning of the neutrino facility was performed in April and May 2009 for nine days. Integrated proton number measured with the most down stream beam intensity monitor was 1.8×10^{14}. The goal of the commissioning was to comfirm fanctionality of the neutrino facility. The commissioning was performed with the first horn. And beam intensity was limited up to 4×10^{11}

protons per spill, because radio activation level around the target should have been kept small enough for installation of second and third horn during summer 2009.

After confirmation of functionality of the proton beam monitors (intensity, position, profile and loss) and magnets, extracted proton beam orbit was tuned within 3 mm from design orbit (fig.3). Beam position at the target was confirmed with a luminescence plate installed just in front of the target. Neutrino production was confirmed with the muon monitor by detecting muons produced with neutrinos (fig.4). Stability of beam intensity and position for 30 minutes continuous operation was better than 1% and 0.2 mm, respectively.

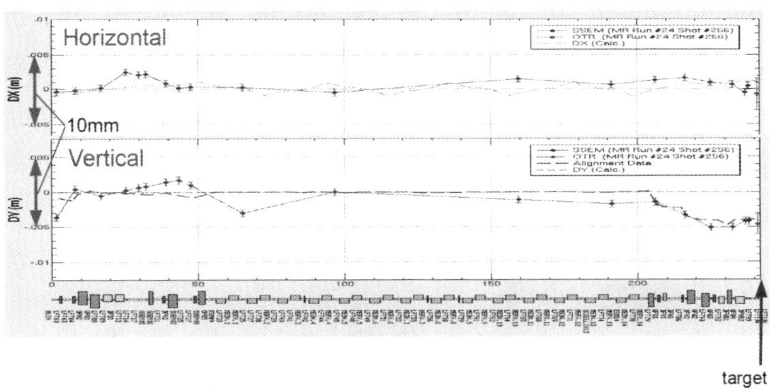

Figure 3: Proton beam orbit.

Focusing effect of the electromagnetic horn was confirmed. When the horn is operated with 273 kA, signal of the muon monitor becomes larger and 2D profile becomes sharper than those measured without horn operation (fig.5).

4 Future prospect

During 2009 summer and fall, the second and third horn and the near detector were installed. Commissioning of the facility was resumed from November 2009. Beam intensity was increased to 20 kW and neutrino events were observed in the near detector.

The physics run will start from early 2010. In 2010, we plan to search θ_{13} with better sensitivity than current CHOOZ limit with 100 kW $\times$ 10^7 s integrated beam power (fig.6). The goal is to discover θ_{13} with 3.75 MW $\times$ 10^7 s physics run.

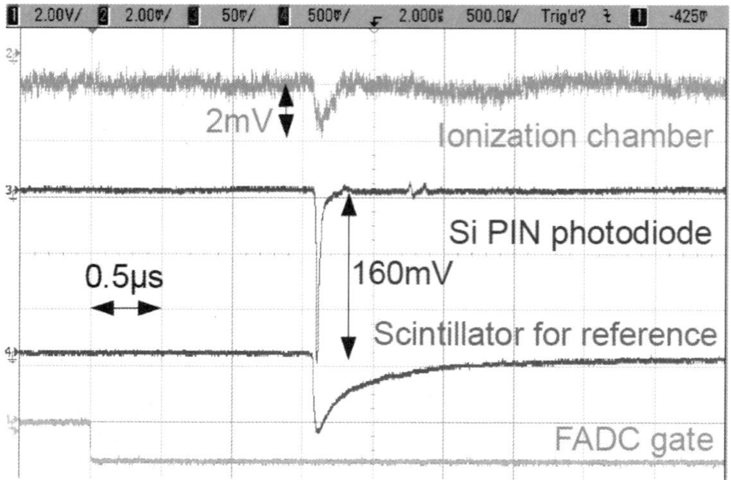

Figure 4: Signal observed with the muon monitor.

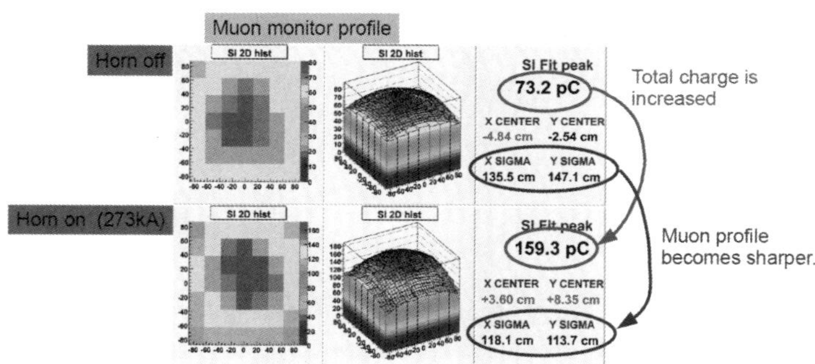

Figure 5: Horn focusing effect. 2D distribution of the muon monitor signal is shown for horn off (top) and 273 kA operation (bottom).

5 Summary

T2K is an accelerator based long base line neutrino oscillation experiment, aiming to discover θ_{13} and precisely measure θ_{23} and Δm^2_{23} with high intensity and narrow energy band neutrino beams and Super-Kamiokande. Accumulating data with 3.75 MW $\times$ 10^7 s integrated beam power, sensitivity for $\sin^2 2\theta_{13}$ is 0.006 at 90 % C.L., $\delta(\sin^2 2\theta_{23}) < 0.01$ and $\delta(\Delta m^2_{23}) < 1 \times 10^{-4}$ eV^2.

First beam commissioning was performed in April and May 2009. Neutrino production was confirmed with muon monitors. And basic performance of the

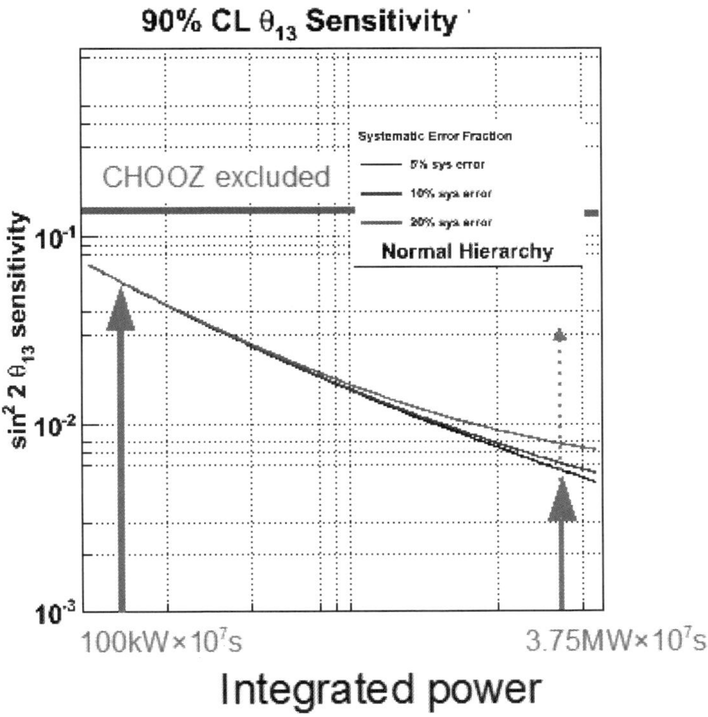

Figure 6: Sensitivity of θ_{13}.

neutrino facility was confirmed. We plan to start physics run from early 2010 and aim to search θ_{13} with sensitivity better than current CHOOZ limit in 2010.

References

[1] Z.Maki, N.Nakagawa and S.Sakata, *Prog.Theor.Phys.* 28, 870 (1962).

[2] M.Apollonio, et al. *Eur.Phys.J* **C** 27, 331 (2003), M.Apollonio, et al. *Phys.Lett.* **B** 466, 415 (1999).

[3] D.Beavis, A.Carroll, I.Chiang, et al. Proposal of BNL AGS E-889 (1995).

RECENT RESULTS OF THE BOREXINO EXPERIMENT

Evgeny Litvinovich [a]
(on behalf of the Borexino collaboration)
RRC "Kurchatov Institute", 123182, Kurchatov sq., 1, Moscow, Russia

Abstract. Borexino detector has performed the first real-time measurement of ^{7}Be solar neutrinos flux. Direct measurement of ^{8}B solar neutrinos flux with the lowest threshold ever achieved by real-time detectors was done as well. Simultaneous spectral measurements in vacuum-dominated (^{7}Be ν_e) and matter-enhanced (^{8}B ν_e) oscillation (LMA) regions were performed for the first time by single detector.

1 Borexino and the solar neutrinos

Borexino detector is continuously taking data since May of 2007 in the Gran Sasso underground laboratory (Italy). The main physical goal is solar neutrinos studies, primarily ^{7}Be neutrinos of the energy $E_\nu = 0.862$ MeV, which are emitted in the Sun's core in the process ^{7}Be(e^-,ν_e)^{7}Li. Detection mechanism is neutrino-electron elastic scattering in the large volume (278 tons) of the organic liquid scintillator. Detailed description of the detector can be found at [1].

1.1 ^{7}Be neutrinos

The sun shines by the energy produced in a thermonuclear reaction chain built up on the fusion of four protons to produce one ^{4}He nucleus. More than 99% of neutrinos from the sun have energies below 2 MeV. Beryllium ν_e contribute about 10% of the flux. According to Standart solar model (SSM), the prediction of ^{7}Be ν_e flux is 5.08×10^9 cm^{-2}s^{-1}. The uncertainty is currently 6% [2].

Current spectroscopic data on solar ν_e are confined to energies greater than 5 MeV (0.01% of the total solar neutrinos flux). The separation of the neutrino events in the Borexino detector is performed in the region of high background radioactivity of materials, so the success of the experiment depends directly on the degree of detector radiopurity. The possibility of purifying organic liquid scintillators to the level of $10^{-17} - 10^{-18}$ g/g of ^{238}U and ^{232}Th was successfully demonstrated in 1996 - 2004 with the prototype of the Borexino detector - Counting Test Facility (CTF). The radiopurity level achieved now in Borexino is summarized in Table 1. Extremely low radioactive contamination of the detector is a result of more than 15 years of work.

Borexino result on ^{7}Be solar neutrinos is based on 192 live days of data-taking [3]. Events are selected by means of the following cuts: (i) Events must have a unique time cluster of photomultipliers hits to reject pile-up of multiple events in the same acquisition window. (ii) Cosmic muons and all events within a time window of 2 ms after a muon are rejected. (iii) Decays

[a] e-mail: e.litvinovich@kiae.ru

due to radon daughters occurring before ^{214}Bi-^{214}Po delayed coincidences are vetoed. (iv) Events must be reconstructed within a spherical fiducial volume (FV) corresponding approximately to 1/3 of the scintillator volume in order to reject external γ background. Additionally, z coordinate of the event is required to be $|z|<1.7$ m in order to escape the background near the poles of the inner nylon vessel, where the Borexino scintillator is enclosed.

Table 1: Borexino radiopurity.

Source	Typical abundance	Level achieved
^{14}C/^{12}C	10^{-12} g/g (cosmogenic)	$(2.7 \pm 0.6) \times 10^{-18}$
^{238}U (by ^{214}Bi-^{214}Po)	2×10^{-5} g/g (dust)	$(1.6 \pm 0.1) \times 10^{-17}$
^{232}Th (by ^{212}Bi-^{212}Po)	2×10^{-5} g/g (dust)	$(6.8 \pm 1.5) \times 10^{-18}$
^{222}Rn (by ^{214}Bi-^{214}Po)	100 atoms/cm^3 (air)	$\sim 10^{-17}$ ($\sim$1 cpd [b] /100 tons)
^{210}Po	Surface contamination	May 2007: $\sim$70 cpd/ton, Sep 2008: $\sim$7 cpd/ton
^{40}K	2×10^{-6} (dust)	$< 3 \times 10^{-18}$ (90%) g/g
^{85}Kr	1 Bq/m^3 (air)	(28 ± 7) cpd/100 tons
^{39}Ar	17 mBq/m^3 (air)	$\ll^{85}$Kr

In order to determine the counting rate for ^{7}Be ν_e, two different approaches were used. In both cases consistent results have been obtained. The counting rate for beryllium solar neutrinos of energy 0.862 MeV was found to be $49 \pm 3_{stat} \pm 4_{syst}$ counts/(day · 100 tons). The expected signal for nonoscillated solar ν_e in the high metallicity solar model is 74 ± 4 counts/(day · 100 tons). Thus, Borexino measurements confirm the solar neutrinos deficit, observed by all radiochemical and water Ĉerenkov solar neutrino detectors. In the MSW-LMA scenario of solar neutrino oscillations, the expected ^{7}Be ν_e signal is 48 ± 4 counts/(day · 100 tons), in a good agreement with the Borexino result.

Solar ^{7}Be ν_e flux in the no-oscillation scenario is found to be $(2.74 \pm 0.27) \times 10^9$ cm^{-2}s^{-1}. In case of the MSW-LMA scenario of neutrino oscillations the flux is $(5.18 \pm 0.51) \times 10^9$ cm^{-2}s^{-1}. This value is in a good agreement with the SSM prediction.

Several sources, as summarized in Table 2, contribute to the systematic error. Analysis of the calibration data will allow to reduce the uncertainties related to fiducial mass definition and detector response function.

In the MSW-LMA scenario, neutrino oscillations are dominated by matter effect above 3 MeV and by vacuum regime below 0.5 MeV [4]. Borexino

[b] counts per day

Table 2: Estimated systematic uncertainties [%].

Total scintillator mass	0.2
Fiducial mass ratio	6.0
Live time	0.1
Detector response function	6.0
Efficiency of cuts	0.3
Total systematic error	8.5

measurements lead to survival probability for ^{7}Be electron neutrinos with the energy 0.862 MeV $P_{ee} = 0.56 \pm 0.10$. The no-oscillation hypothesis, $P_{ee} = 1$, is rejected at 4σ C.L.

1.2 ^{8}*B neutrinos*

Boron solar neutrinos are originated from the reaction $^8\mathrm{B} \rightarrow {}^8\mathrm{Be}^* + e^+ + \nu_e$. They have continuous energy spectrum ended at 14.06 MeV. Borexino result on ^{8}B solar neutrinos is based on 245.9 live days of data-taking with a target mass of 100 tons, defined by a FV cut of radius 3 m [5].

Table 3: Effect of the sequence of cuts on the observed counts.

Cut	Counts 2.8-16.3 MeV	Counts 5.0-16.3 MeV
All counts	20449	14304
Muon cut	3363	1135
Neutron cut	3280	1114
FV cut	567	372
Cosmogenic cut	71	26
^{10}C removal	65	26
^{214}Bi	62	26
Expected ^{208}Tl	14 ± 3	0
Measured ^{8}B ν	48 ± 8	26 ± 5
SSM BS07(GS98) ^{8}B ν	50 ± 5	25 ± 3

Prior to all cuts, the counting rate in the ^{8}B ν_e energy range between 2.8 and 16.3 MeV, the maximum recoiled electron energy from ^{8}B neutrino scattering, recorded in the entire active mass (278 tons) is about 1500 cpd. Starting from here, the analysis is based upon consecutive removing all the known classes of background events in the energy range of interest. Table 3 gives the effect of the sequence of cuts on the observed counts.

After all cuts, the counting rates in the energy region above 2.8 MeV (5.0 MeV) is $0.26 \pm 0.04_{stat} \pm 0.02_{syst}$ counts/(day $\cdot$ 100 tons) ($0.14 \pm 0.03_{stat} \pm 0.01_{syst}$ counts/(day $\cdot$ 100 tons)). The equivalent unoscillated ^{8}B ν_e flux is $(2.65 \pm 0.44_{stat} \pm 0.18_{syst}) \times 10^6$ cm^{-2}s^{-1} ($(2.75 \pm 0.54_{stat} \pm 0.17_{syst}) \times 10^6$ cm^{-2}s^{-1}).

The boron ν_e survival probability at the mean energy of 8.6 MeV for ^{8}B neutrinos is found to be $\overline{P}_{ee} = 0.35 \pm 0.10$. Taking into account the experimentally determined P_{ee} for ^{7}Be ν_e at the energy of 0.862 MeV (0.56 ± 0.10), one can conclude that the Borexino data confirm the presence of a transition between vacuum-driven and matter-enhanced solar neutrino oscillations. Borexino is thus the first experiment detecting at the same time neutrinos in the low energy, vacuum-dominated- and in the high energy, matter-enhanced region.

2 Borexino calibrations

Several calibration campaigns (on- and off-axis) were performed in the Borexino experiment within October 2008 - July 2009. The main goal is to reduce systematic uncertainties and to achieve at least 5% error in the determination of ^{7}Be solar neutrinos flux. Calibrations will also allow to study α, γ and proton quenching in the Borexino liquid scintillator.

The position of the radioactive source inside the Borexino detector is known within 2 cm accuracy. The following sources were deployed into different positions inside the inner vessel of the detector: several γ-sources in the wide energy region ($E_\gamma = 122 \div 1461$ keV), α-source ^{222}Rn, neutron source ^{241}Am-^{9}Be. Analysis of the calibration data is currently in progress.

Acknowledgments

Borexino experiment is funded by: INFN (Italy), NSF (USA), BMBF, DFG and MPG (Germany), ROSNAUKA (Russia), MNiSW (Poland).

References

[1] G. Alimonti *et al.* (Borexino collab.), *Nucl.Instrum.Methods* **A** 600, 568 (2009).

[2] J.N. Bahcall, A.M. Serenelli, and S. Basu, *Astrophys. J. Suppl. Ser.* **165**, 400 (2006); astro-ph/0909.2668v2.

[3] C. Arpesella *et al.* (Borexino collab.), *Phys.Rev.Lett.* **101**, 091302 (2008).

[4] A. Yu. Smirnov, in Proceedings of the 11th International Workshop on Neutrino Telescopes, Venice, 2003; hep-ph/0305106.

[5] G. Bellini *et al.* (Borexino collab.), astro-ph/0808.2868v1.

DOUBLE CHOOZ PROJECT: STATUS OF A REACTOR EXPERIMENT AIMED AT SEARCH FOR NEUTRINO OSCILLATIONS

Sergey Sukhotin [a]
(on behalf of the Double Chooz Collaboration)
Russian Research Centre "Kurchatov Institute"
123182 Kurchatov sq., 1,Moscow,Russia

Abstract. Double Chooz is an experiment that is devoted to searches for reactor-antineutrino oscillations at the CHOOZ nuclear power plant. This project is aimed at measuring the unknown mixing angle θ_{13}. It is assumed that the value of θ_{13} will be extracted from an analysis of the distortion of the antineutrino spectra obtained in relative measurements at two distances from the nuclear reactors by means of two identical detectors. The method makes it possible to minimize systematic errors of the experiment and to improve the sensitivity to the sought parameter. To date, the most stringent constraint on the parameter θ_{13} was obtained from the CHOOZ experiment in 1995 – 1997 ($\sin^2(2\theta_{13}) < 0.19$, with the difference of the squares of the neutrino masses being $\Delta m^2_{13} = 2.5 \times 10^{-3} eV^2$).

1 Introduction

Pontecorvo's hypothesis [1] of neutrino oscillations has been confirmed in experiments with natural and artificial sources. Experiments with solar neutrinos [2–5] and with reactor antineutrinos (KamLAND Collaboration [6]) resulted in determining the parameters of $\nu_e \to \nu_\mu$ oscillations, which explain the deficit of solar neutrinos:

$$\Delta m^2_{12} = (7.9 \pm 0.3) \times 10^{-5} eV^2$$
$$\sin^2(2\theta_{12}) = 0.80 \pm 0.04$$

The parameters of oscillations belonging to the $\nu_\mu \to \nu_\tau$ type were measured in experiments with atmospheric neutrinos [7], as well as in experiments at accelerators [8, 9]. The results are:

$$\Delta m^2_{23} = (2.5 \pm 0.2) \times 10^{-3} eV^2$$
$$\sin^2(2\theta_{23}) = 1.0 \pm 0.08$$

The set of experiments performed thus far determined Δm^2_{13}, thereby facilitating searches for the mixing angle θ_{13}, which is the last unknown parameter. Oscillations of the $\nu_e \to \nu_\tau$ type have not yet been observed. The most stringent constraint, $\sin^2(2\theta_{13}) < 0.19$ at the differences of the masses squared in the region around $\Delta m^2_{13} = 2.5 \times 10^{-3} eV^2$, was obtained in the CHOOZ experiment [10, 11], which was performed in 1995 – 1997. A reactor experiment that belongs to a new type and which would involve simultaneously measuring antineutrino spectra by two identical detectors positioned at different distances from the reactor was proposed in 2000 [12]. In 2004, a group of physicists

[a] e-mail: sukhotin@kiae.ru

from France, Germany and Russia (Kurchatov Institute) announced the formation of an international collaboration with the aim of implementing such an experiment. In 2005, the project was supported by several groups from the United States of America. In 2006, laboratories from the United Kingdom, Brazil, Spain, and Japan joined the project. The project was called Double Chooz [13].

2 ATOMIC POWER PLANT AND DETECTOR POSITIONS

The Double Chooz experiment will be performed at the French nuclear power plant situated at the Chooz town near the border with Belgium. The nuclear power plant has two nuclear reactors of total thermal power 8.4 GW. Thus, the distances to the reactors from the far detector are already known, 997.9 and 1114.6 m. The near detector will be positioned at a mean distance of about 390 m from the reactors. The farther laboratory is arranged in a tunnel at a depth of about 300 mwe. The nearer laboratory will be situated at a depth of about 100 mwe.

3 DETECTOR FOR ELECTRON ANTINEUTRINOS

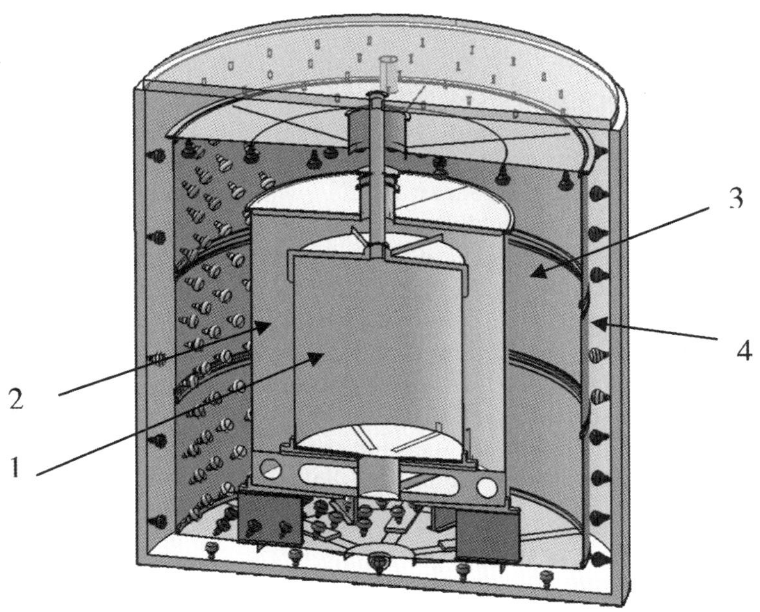

Figure 1: Design of the Double Chooz detectors: (1) Gadolinium scintillator target, (2) absorber of gamma rays, (3) mineral oil buffer, (4) inner veto

The design of the detector is illustrated in Fig. 1. The detector will be a four-zone assembly of concentric cylinders inserted into one another [13]. The first zone (target) is a cylinder 2.3 m in diameter and 2.46 m in height. This zone will be filled with a liquid scintillator specially developed for the Double Chooz project. The scintillator has a high content of hydrogen and features a small addition of gadolinium (1 g/l). The second zone (gamma-ray absorber) is a cylinder 3.4 m in diameter and 3.57 m in height. It will be filled with the scintillator free from gadolinium. The third zone (buffer) is a cylinder 5.5 m in diameter and 5.67 m in height. This zone, filled with pure mineral oil, will protect the detector target from the natural radioactivity of the materials of the outer zones. Photomultiplier tubes viewing the gamma-ray absorber and the target will be arranged at its walls. The fourth zone (active shield from cosmic rays - veto) is a cylinder 6.6 m in diameter and 6.64 m in height. This zone will be filled with a scintillator. It will ensure a reliable shield for the detector from cosmic-ray muons. The inner volume of this zone will be viewed by photomultiplier tubes recording muons.

4 DETECTION OF REACTOR ANTINEUTRINOS

The interactions of reactor antineutrinos in the detector will be recorded by the inverse beta-decay reaction occurring on hydrogen and resulting in the production of a positron and a neutron:

$$\widetilde{\nu}_e + p \rightarrow e^+ + n \qquad \text{Ethr} = 1.806 \text{ MeV}. \qquad (1)$$

The delayed coincidences of signals from a positron and neutron capture in the time gate equal approximately to the tripled neutron lifetime in the detector will be used to single out events of reaction (1). A neutron, after being moderated, is captured either by gadolinium nuclei (with a probability of about 80%) or by hydrogen (with a probability of about 20%), this being accompanied by the emission of gamma rays whose total energy is about 8 MeV in the case of gadolinium or 2.2 MeV in the case of hydrogen. Owing to the presence of gadolinium in the target scintillator, the neutron lifetime decreases from 200 μs (scintillator without gadolinium) to 30 μs. The neutron detection threshold can be set to about 6 MeV. This makes it possible to reduce the background of random coincidences substantially, since the natural radioactivity spectrum does not go beyond 4 MeV, the neutron detection efficiency remaining high (about 80%). The positron-detection threshold will be set to about 0.5 MeV.

5 SENSITIVITY OF THE EXPERIMENT TO THE MIXING ANGLE θ_{13}

In the CHOOZ experiment [10,11], the search for oscillations was performed by comparing the antineutrino spectrum measured by the detector with its calculated counterpart. No effect of neutrino oscillations was discovered. The ratio (R) of the expected number of reactor-antineutrino interactions to their measured number proved to be:

$$R = 1.01 \pm 2.8\%(stat.) \pm 2.7\%(syst.)$$

The values of the statistical and systematic errors determined the constraint on the mixing angle. In order to improve the sensitivity substantially, it is necessary to reduce the experimental errors greatly. The statistical error can easily be reduced to 0.5% by increasing the target volume and the time of data acquisition in relation to the previous CHOOZ experiment. The main systematic error in the CHOOZ experiment was associated with the determination of the flux and energy spectrum of antineutrinos corresponding to the current reactor power and fuel burnup. In the new project, it is proposed to perform simultaneous measurements with two identical detectors, whereby systematic errors related to the reactors are avoided almost completely, since the two detectors are similarly affected by any variations in the reactor power and in the composition of the reactor area. The systematic errors related to the properties of the detectors will also be removed to a considerable extent owing to the identity of the detectors. In particular, it is unnecessary to know, in the new experiment, the absolute number of hydrogen nuclei in the target, but this was a source of a significant systematic error in the first experiment. Our analysis revealed that the total systematic error related to the detectors would be about 0.6%.

6 STATUS OF THE DOUBLE CHOOZ PROJECT

The assembling of the far detector is completed in December 2009. The data taking will statrt in April 2010. Within 1.5 years of data accumulation with the far detector, an improvement of the sensitivity to the mixing angle 013 and the attainment of values in the range $\sin^2(2\theta_{13}) < 0.06$ are expected in the Double Chooz experiment. The construction of the nearer laboratory will have been completed at the end of 2010. The near detector will have been assembled in November 2011. Thus, it is planned to begin collecting data with two detectors at the end of 2011. Within three years after that, this would make it possible to set the constraint $\sin^2(2\theta_{13}) < 0.03$ on the mixing angle θ_{13} or to measure it if $\sin^2(2\theta_{13}) > 0.05$.

7 CONCLUSIONS

The Double Chooz project is aimed at measuring the mixing angle θ_{13}. This measurement is of importance not only for obtaining deeper insight into the nature of neutrino oscillations but also for studying phenomena that could play a significant role for developing a more fundamental theory of the subatomic world, a theory that would describe the evolution of the Universe, and so on.

Acknowledgments

This work was supported by the Russian Foundation for Basic Research (project 05-02-19649) and also by the state contract 02.513.12.3090.

References

[1] B. M. Pontekorvo, *Sov. Phys. JETP* 6, 429 (1957)51.
[2] R. Davis, D. S. Harmer, K. C. Hoffman, *Phys. Rev. Lett* 20, 1205(1968).
[3] J. N. Abdurashitov, T. J. Bowles, C. Cattadori, et al. *Astropart. Phys.* 25, 349 (2006).
[4] M. Altmann, M. Balata, P. Belli, et al. *Phys. Lett.* **B** 616, 174(2005).
[5] SNO Collaboration., *Phys. Rev.* **C** 72, 055502 (2005).
[6] K. Eguchi, S. Enomoto, K. Furuno, et al. *Phys. Rev. Lett.* 92, 071301 (2004).
[7] Super-Kamiokande Collaboration., *Phys. Rev. Lett.* 97, 171801 (2006).
[8] MINOS Collaboration., *Phys. Rev. Lett.* 97, 191801 (2006).
[9] K2K Collaboration., *Phys. Rev.* **D** 74, 072003 (2006).
[10] M. Apollonio, A. Baldini, C. Bemporad, et al., *Phys. Lett.* **B** 466, 415 (1999).
[11] M. Appolonio et al., *Eur. Phys. J.* **C** 27, 331 (2003).
[12] L. A. Mikaelyan and V. V. Sinev, *Eur. Phys. J.* 63, 1002 (2000).
[13] Proposal Double Chooz, hep-ex/0606025.

THE ANTARES EXPERIMENT IN THE MEDITERRANEAN SEA: OVERVIEW AND FIRST RESULTS

D. Zaborov [a] on behalf of the ANTARES collaboration
Institute for Theoretical and Experimental Physics, 117218 Moscow, Russia
*Abstract.*A general overview on the ANTARES neutrino telescope is given. The status of the experiment is briefly reviewed. First results of atmospheric muon and neutrino analyses are presented.

1 The ANTARES neutrino telescope

ANTARES is a large water Cherenkov detector installed in the Mediterranean Sea, 40 km offshore from Toulon (France) at a depth of 2475 m [1]. The elements of the detector are arranged in twelve flexible structures (lines) attached to the seabed with an anchor and kept taut by a buoy (see Fig. 1). A detector line comprises 25 storeys distributed along the length of the 450 m long electromechanical cable, starting 100 m above the seabed. The storeys are separated by 14.5 m. A storey contains three transparent glass spheres (17-inch in diameter, arranged in 120° distance to each other), each housing a large-area hemispherical 10-inch photo-multiplier tube (PMT) [2,3]. The PMTs are oriented downwards at an angle of 45° from the vertical. The storey also contains an electronics module for the readout of the PMTs [4]. Some storeys are also equipped with additional devices aimed for calibration purposes (LED Beacons) and acoustic positioning (hydrophones). In addition, each storey has a digital compass and a tiltmeter. The detector lines are installed on the seabed with a separation of about 70 m from each other and connected to the Junction Box, also installed on site, which is in turn connected to shore via a long-distance electro-optical cable. In total the detector contains nearly 900 PMTs.

The experiment is optimized for the detection of upward-going muons produced by neutrino interactions inside or below the instrumented volume. Arrival times of the Cherenkov photons emitted by the muons are used to reconstruct the muon direction of origin, which is at high energies well connected with the neutrino direction. With advanced reconstruction techniques, an angular resolution up to 0.2^o can be achieved at TeV energies. An estimate of the muon energy is made using the reconstructed track length and the produced amount of light. Hadronic showers from neutral current interactions of all neutrino types, as well as charged current interaction of ν_e and ν_τ, can also be detected and reconstructed with dedicated algorithms.

Scientific objectives of the experiment comprise the search for high-energy neutrino sources outside the solar system, indirect Dark Mater searches (e.g. neutralino annihilation in the Sun), diffuse neutrino flux, exotic searches (magnetic monopoles, nuclearites), and other related studies.

[a] e-mail: dmitry.zaborov@itep.ru

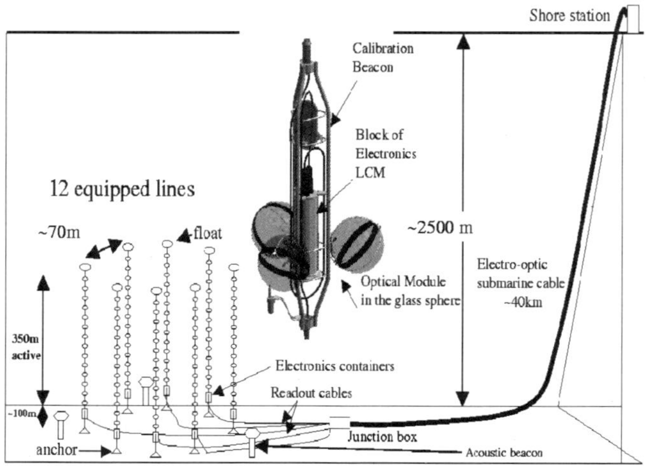

Figure 1: Schematic view of the ANTARES detector (only 8 lines out of 12 are shown).

2 Detector status and in situ calibration

The detector has been operated in partial configurations since March 2006 and was completed in May 2008. Currently (January 2010) 10 lines out of 12 are working, one more is awaiting an interlink cable connection by a submarine, and another one is to be recovered for maintenance works. In addition, a short instrumentation line (IL07) is operating. IL07 hosts a number of instruments dedicated to the measurements of water properties (temperature, salinity, sound velocity, transmission of light) and water currents, an oxygen probe, a seismometer, and 2 sensitive cameras with an infrared flash light. Three of the 6 storeys of IL07 and 3 storeys of Line 12 are equipped with hydrophones for acoustic noise and acoustic detection R&D studies. Line 12 also has a robotized Oxygen probe (IODA) capable of taking and analysing water samples in automatic mode.

Before deployment, each detector line is calibrated in the laboratory. Later on the calibration parameters are regularly verified for possible changes in situ. A central role in the time calibration is played by LED Beacons [5]. The ensemble of the detector storeys is synchronized by a dedicated optical fiber system, which uses round-trip delay measurements. The time offsets between PMTs located on the same storey can also be verified using the coincidence signal from Potassium-40 decays in sea water. The ^{40}K technique is also used to monitor the efficiencies of the PMTs [6], which are related to the PMT gains and electronics thresholds. The PMT gains are monitored independently by measuring the position of the single photoelectron peak in the recorded data. The elec-

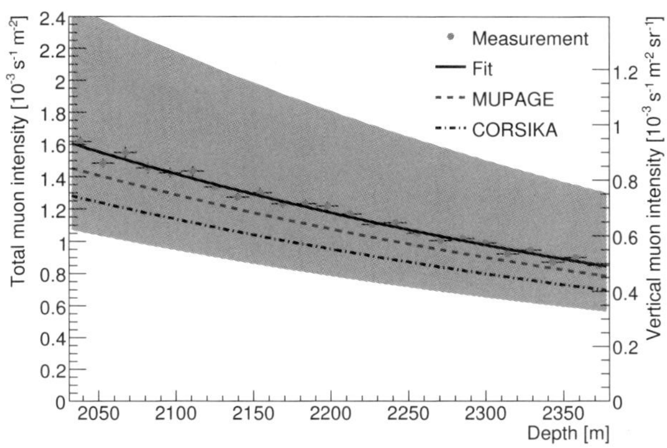

Figure 2: The measured muon flux as a function of depth, interpreted in terms of total muon intensity (left axis) and vertical muon intensity (right axis). The grey band shows the normalization uncertainty of the data. The predictions of the Monte Carlo simulations based on MUPAGE and CORSIKA are shown by dashed and dash-dotted lines, respectively.

tronics thresholds are also monitored in situ. In case of a significant drift, the change is compensated by adjusting the corresponding threshold parameter or the PMT high voltage. The performance of most of the optical modules after about 2 years of operation remains good. The precision of the time calibration meets the design goal of $\sigma \approx 0.5$ ns.

3 First results

A measurement of the muon flux and its dependence on the depth in ANTARES has been recently published in [6]. A novel approach was used, which is based on the observation of coincidence signals in adjacent storeys of the detector. This yields a low energy threshold of 4 GeV. The depth intensity relation was obtained directly from the coincidence rates measured in different detector storeys (see Fig. 2). The results are found compatible with a Monte Carlo simulation using MUPAGE [7], as well as a CORSIKA [8] simulation using the NSU model of the primary cosmic ray spectrum [9]. Another measurement of the muon flux was performed using 5-line detector data taken in June and July 2007. Like in an earlier study [10], the depth intensity relation was obtained from the zenith angle distribution of reconstructed muons. The result is compatible with the analyses [6] and [10].

The data sample acquired so far by ANTARES includes about 10^8 downward-going muon events, and over 2000 neutrino candidates. A zenith angle distribution for a combined data set is shown in Fig. 3. As one can see, the data agree reasonably well with expectations. Further analysis is ongoing.

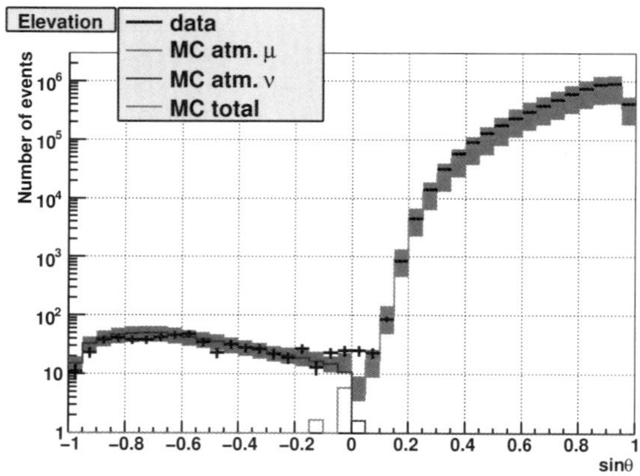

Figure 3: Elevation angle of muons recorded by ANTARES in 2008. The Monte Carlo curve for atmospheric muons was obtained using CORSIKA. The simulation of atmospheric neutrinos includes oscillations with $\sin^2 2\theta = 1$ and $\Delta m^2 = 2.4\ 10^{-3}\ \mathrm{eV}^2$.

4 Outlook

The ANTARES neutrino telescope is currently operating in a nearly nominal configuration. The performance of the optical modules, calibration and positioning systems conforms to the design specification. The flux of atmospheric muons has been measured using two different techniques and found to be in agreement with Monte Carlo simulations and earlier measurements. The angular distribution of upward-going neutrino-induced events has been also measured and agrees with simulations. Further physics analyses are ongoing.

References

[1] J.Carr for ANTARES collab., *J. Phys. Conf. Ser.* 136 (2008) 022047.
[2] P.Amram et al., *Nucl. Instrum. Meth.* **A** 484, 369 (2002).
[3] J.A.Aguilar et al., *Nucl. Instrum. Meth.* **A** 555, 132 (2005).
[4] J.A.Aguilar et al., *Nucl. Instrum. Meth.* **A** 570, 107 (2007).
[5] M.Ageron et al., *Nucl. Instrum. Meth.* **A** 578, 498 (2007).
[6] J.A.Aguilar et al., doi:10.1016/j.astropartphys.2009.12.002.
[7] G.Carminati, A.Margiotta and M.Spurio, *Comput. Phys. Commun.* 179, 915 (2008).
[8] D.Heck, G.Schatz, T.Thouw, J.Knapp and J.N.Capdevielle, Forschungszentrum Karlsruhe Report FZKA-6019, 1998.
[9] S.I.Nikolsky, I.N.Stamenov and S.Z.Ushev, *Sov. Phys. JETP* 60, 10 (1984); *Zh. Eksp. Teor. Fiz.* 87, 18 (1984).
[10] M.Ageron et al., *Astropart. Phys.* 31, 277 (2009).

HIGH-ENERGY NEUTRINOS FROM GALACTIC SOURCES

Alexander Kappes [a]

Friedrich-Alexander-Universität Erlangen-Nürnberg, Erlangen Centre for Astroparticle Physics, Erwin-Rommel-Str. 1, 91058 Erlangen, Germany

Abstract. Even 100 years after the discovery of cosmic rays their origin remains a mystery. In recent years, TeV gamma-ray detectors have discovered and investigated many Galactic sources where particles are accelerated up to energies of 100 TeV. However, it has not been possible up to now to identify these sites unambiguously as sources of hadronic acceleration. The observation of cosmic high-energy neutrinos from these or other sources will be a smoking-gun evidence for the sites of the acceleration of cosmic rays.

1 Introduction

Up to now, information on objects in our galaxy and beyond has nearly exclusively been obtained using electromagnetic waves as cosmic messengers. Modern astronomy started in 1610 with optical photons, when Galileo Galilei took one of the first telescopes and pointed into the sky. Only in the last century, people started to extended the observations to lower and higher energies of the electromagnetic spectrum, where today's instruments span an enormous range of 20 orders of magnitude from radio waves to TeV gamma-rays.

In addition, we know from measurements of the cosmic-ray spectrum that there exist sources in the universe which accelerate protons or heavier nuclei up to energies of $\sim 10^{20}\,\mathrm{eV}$, 10^7 times higher than the most energetic man-made accelerator, the LHC at CERN. These highest energies are believed to be reached in extra-Galactic sources like gamma-ray bursts or active galactic nuclei whereas Galactic sources like supernova remnants (SNRs) or micro-quasars are thought to accelerate particles at least up to energies of $3 \times 10^{15}\,\mathrm{eV}$, also called the "knee" region of the cosmic-ray spectrum. Though these phenomena are quite different, the basic acceleration mechanism is believed to be very similar. Particles are injected into shock fronts which develop when fast moving matter collides with other matter. In these shock fronts, the injected particles are accelerated in a repeating process (Fermi acceleration) where the energy gain per cycle is only small. Other phenomena where a "one-shot" acceleration of cosmic rays to high energies might take place are objects with strong magnetic fields (up to $10^{15}\,\mathrm{G}$) like pulsars and magnetars. However, despite the detailed measurements of the cosmic-ray spectrum and 100 years after their discovery by Victor Hess, we still do not know what the sources of the cosmic rays are as they are deflected in the Galactic and extra-Galactic magnetic fields and hence have lost all information about their origin when reaching Earth. Only at the highest energies beyond $\sim 10^{19.6}\,\mathrm{GeV}$ cosmic rays might retain enough directional information to locate their sources.

[a] e-mail: kappes@physik.uni-erlangen.de

Alternative messengers for locating the sources of the cosmic rays must have two distinct properties: they have to be electrically neutral and essentially stable. Only two of the known elementary particles meet these requirements: photons and neutrinos. Both particles are inevitably produced when the accelerated protons or nuclei collide with matter or photons inside or near the source. In these reactions neutral and charged pions are produced which then decay into high-energy photons and neutrinos, respectively. The well known ratio between the production of neutral and charged pions yields a direct link between the photon and neutrino flux from proton-proton and proton-gamma interactions, respectively.

The number of sources detected in TeV gamma-rays has increased dramatically during the last decade which was made possible by significant advances in the technique of air Cherenkov telescopes. Today, over 100 sources are known, both Galactic (e.g., SNRs, micro-quasars, pulsars) and extra-Galactic (e.g., active galactic nuclei, starburst galaxies). However, it has not been possible up to now to unambiguously identify these sources as sites of cosmic-ray acceleration as the observed TeV photons could also have been generated in the up-scattering of photons in reactions with accelerated electrons (inverse-Compton scattering). High-energy neutrinos, on the other hand, are only produced in reactions of accelerated hadrons and hence are a smoking-gun evidence for the sources of cosmic rays.

2 Potential Galactic neutrino sources

From the discussion above it follows naturally that TeV gamma-ray sources are prime candidates for high-energy neutrino emission. Today, the list of Galactic objects with observed TeV gamma-ray emission contains a.o. SNRs, pulsar wind nebulae (PWNe), binary systems and molecular clouds. In the following, we will discuss the prospects for neutrino emission from these objects and the expected fluxes. For an overview and more details about the sources see [1] and references therein.

Supernova remnants: SNRs are the prime candidate for the acceleration of the Galactic cosmic rays. This has several reasons: There exists a plausible model for the acceleration of hadrons in the shock fronts generated when the ejecta of the supernova interact with the surrounding medium. Furthermore, energetic considerations show [2] that the observed supernova rate of about 3 per century and a conversion of 10% of the 10^{51} erg released per nova into cosmic-ray acceleration matches quite well the energy required to keep the energy in cosmic rays constant over time. Up to now, 11 SNRs have been observed in TeV gamma-rays. Under the assumption that all gamma-ray photons origin from the interaction of accelerated hadrons and utilizing the connection between gamma-rays and neutrinos one can make rather precise predictions

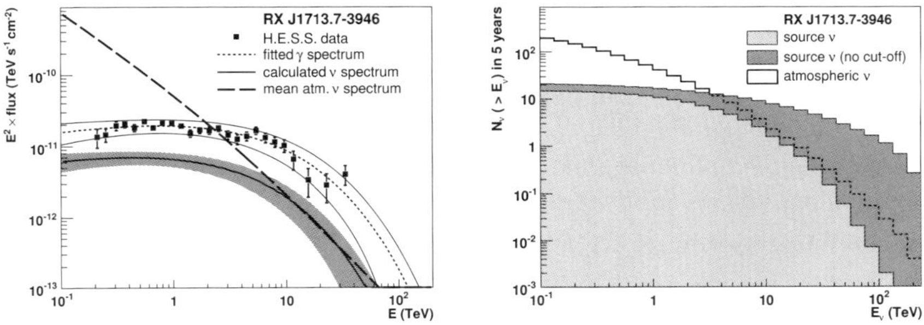

Figure 1: Left: Measured photon and expected neutrino flux from RX J1713.7−3946. The area formed by the solid lines and the shaded area represent the uncertainty (including systematics) on the respective spectrum. Right: Expected event rates from RX J1713.7−3946 in a cubic-kilometer sized neutrino telescope with and without a high-energy cut-off. Also shown is the rate of atmospheric neutrinos. Taken from [3].

for the expected neutrino flux from these sources [3]. An example of such a calculation is shown in the left plot of Fig. 1. Using the effective muon neutrino area of a neutrino telescope one can calculate the integrated number of observed events as a function of the threshold energy. This is displayed in the right plot of Fig. 1 for a water Cherenkov neutrino telescope with an instrumented volume of 1 km^3. The importance of the cut-off at high energies for the detectability of the source becomes apparent when comparing the light and dark shaded areas with the expected background from atmospheric neutrinos. The strong contribution of high-energy neutrinos to the number of detected events origins from the strong rise of the effective area with the neutrino energy which is mainly caused by the increasing neutrino-nucleon cross section and the extended range of the muons produced. On the other hand, the number of background events scales with the size of the emission region which in the case of RX J1713.7−3946 is quite large with 1.3° diameter. These two features, the low energy cut-off and the large emission region, make the detection of this and similar sources like RX J0852.0−4622 with neutrino telescopes challenging. In addition, there exits a population of SNRs where no cut-off could be determined from the measured gamma-ray spectrum due to the large error bars at high energies. However, these spectra are rather steep and only a small number of events comparable to those from atmospheric neutrinos is expected [3].

Binary systems: The binary systems observed in gamma-rays consist of a compact object (either a neutron star or a black hole) and a companion star. The compact object accretes material from the companion star producing jets where particles are accelerated. One of these objects is LS 5039 discovered with the H.E.S.S. telescope. The system has a period of 3.9 days and the TeV gamma-ray emission region is point-like for neutrino telescopes. A calculation of the expected neutrino event rate from the measured gamma-ray flux yields less than one event per year for a cubic-kilometer neutrino telescope [3]. However, due to the dense photon fields originating from the companion star the measured gamma-ray flux may be significantly suppressed by up to a factor 100 [4]. If this is indeed true then this and similar sources would be one of the most promising targets for neutrino telescopes.

Pulsar wind nebulae: A PWN is powered by the wind from a pulsar which streams into the ambient medium creating shock fronts. In these shock fronts particles are accelerated. In general, PWNe are believed to accelerate mainly electrons. In [5] however, the authors argue that there also might be a significant fraction of nuclei in the pulsar wind. In this case PWNe like Vela X pose interesting targets for neutrino telescopes.

Molecular clouds: Gamma-ray emission from molecular clouds origins from the interaction of cosmic rays from a nearby source, e.g., a SNR, with the nuclei in the cloud. Hence, molecular clouds are a "guaranteed" source of neutrino emission. Such an emission region was located by H.E.S.S. near the Galactic center. Unfortunately, the expected neutrino fluxes derived from the measured gamma rays are rather low and the emission region is quite large making the detection of this and other molecular clouds in neutrinos very challenging.

3 The missing Pevatrons

The existence of the "knee" in the cosmic ray spectrum tells us that there must exist Galactic cosmic-ray sources which accelerate protons up to energies of several PeV. These "Pevatrons" will produce pionic gamma rays whose spectrum extends to several hundred TeV without cut-off in interactions with the interstellar medium. However, none of the observed gamma-ray sources shows such a spectrum. One reason for this observation could be the fact that the highest energies in SNRs are only reached during the first few hundred years after the supernova explosion (a SNR has a typical lifetime of the order of 10,000 years). In this case, it could simply be possible that currently there does not exist an observable SNR in this early phase. Nevertheless, the detection of these cosmic rays could still be possible by observing the gamma rays produced in their interaction with interstellar medium, in particular, with dense molecular clouds as shown in [6]. In [7] the authors argue that some of the gamma-ray sources discovered by Milagro [8] might origin from such secondary interactions of cosmic rays from Pevatrons. Future measurements with air Cherenkov and neutrino telescopes have to show whether this is true. On the other hand, one of the suggested sources, MGRO 1908+06, has now been plausibly associated with a pulsar [9], disfavoring it as a region of proton acceleration.

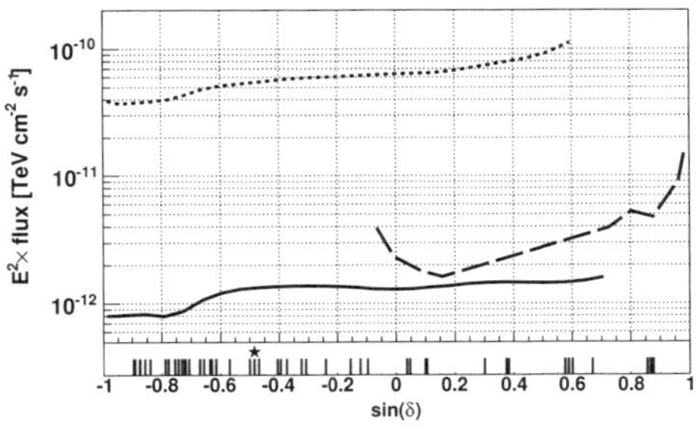

Figure 2: Upper part: Sensitivity at 90% CL of neutrino telescopes to an E^{-2} neutrino flux as a function of the source declination: ANTARES (dotted, predicted) [10], IceCube (dashed, predicted) [12], KM3NeT (solid, not final), [13]. Lower part: declination of Galactic objects with observed TeV gamma-ray emission. The position of the Galactic Center is marked with a star.

4 Sensitivity of current and future neutrino telescopes

All current event rate calculations indicate that neutrino telescopes with instrumented volumes of cubic kilometer size are necessary to identify the first cosmic source of high-energy neutrino emission. Due to the large background of downgoing muons produced in interactions of cosmic rays with the Earth's atmosphere, neutrino telescopes have their highest sensitivity when looking through the Earth. Hence, in order to cover the full sky with high sensitivity both a telescope in the Northern and in the Southern hemisphere is necessary.

Currently, the most sensitive neutrino telescope in the Northern hemisphere is the ANTARES detector [10], installed at a depth of 2500 m off the coast of south France in the Mediterranean Sea. The detector instruments a volume of about 0.01 km^3 and therefore is probably too small for cosmic neutrino detection. In the Southern hemisphere, the currently largest and hence most sensitive neutrino telescope, IceCube, is close to completion planned for next year. With an instrumented volume of 1 km^3 it is the first cubic-kilometer-class detector. The sensitivities of both experiments to an E^{-2} neutrino flux are shown in Fig. 2. It is apparent, that in order to achieve a high-sensitivity coverage over the full sky a cubic kilometer class detector in the Northern hemisphere is needed which will cover most of the Galactic plane including the Galactic center. Such a detector with an instrumented volume of about 5 km^3, KM3NeT [11], is currently in its planning phase. Data taking is planned to start around 2014.

5 Summary and outlook

The observation of cosmic high-energy neutrinos will open a new window to the universe and help to solve long-standing mysteries like the question of the origin of the cosmic rays. With the recent advances in gamma-ray astronomy we were able to identify many good Galactic candidate sources of neutrino emission. However, due to the expected low fluxes and the early high-energy cut-offs of the predicted neutrino spectra the detection of these sources is challenging.

IceCube will be the first detector to advance into the "discovery region" but will likely only scratch it. Therefore, a factor five to ten more sensitive detector is needed in the Northern hemisphere which will also cover the inner Galactic Plane and Center. This detector, KM3NeT, is supposed to start data taking around 2014. With IceCube and KM3NeT combined, neutrino astronomy will hopefully soon become reality.

Acknowledgments

Supported by the BMBF under project 05A08WEA. The author also acknowledges the support by the EU Marie Curie OIF program.

References

[1] F. Aharonian, J. Buckley, T. Kifune, and G. Sinnis. *Rep. Prog. Phys.*, 71:096901, 2008.

[2] F. Halzen. *J. Phys.: Conf. Series*, 171:012014, 2009.

[3] A. Kappes, J. Hinton, C. Stegmann, and F. A. Aharonian. *ApJ*, 656:870, 2007.

[4] F. A. Aharonian, L. A. Anchordoqui, D. Khangulyan, and T. Montaruli. *J. Phys.: Conf. Series*, 39:408, 2006.

[5] D. Horns, F. Aharonian, A. Santangelo, A. I. D. Hoffmann, and C. Masterson. *A&A*, 451:L51, 2006.

[6] S. Gabici and F. A. Aharonian. *ApJ*, 665:L131, 2007.

[7] F. Halzen, A. Kappes, and A. Ó Murchadha. *Phys. Rev.*, D78:063004, 2008.

[8] A. A. Abdo, (Milagro Collaboration), et al. *ApJ*, 664:L91, 2007.

[9] A. A. Abdo, (Fermi-LAT Collaboration), et al. *ApJ*, 711:64, 2010.

[10] P. Coyle, (ANTARES Collaboration), et al., Preprint arXiv:1002.0754, 2010.

[11] P. Bagley, (KM3NeT Consortium), et al. KM3NeT, conceptual design for a deep-sea research infrastructure incorporating a very large volume neutrino telscope in the Mediterranean Sea, 2008. ISBN 978-90-6488-031-5.

[12] J. Dumm, (IceCube Collaboration), et al. All-sky point-source search with 40 strings of IceCube. to be published in Proc. of ICRC'09.

[13] U. F. Katz, (KM3NeT Consortium), et al. The KM3NeT project. to be published in Proc. of the VLVnT'09 Workshop.

THE GSI TIME ANOMALY: FACTS AND FICTION

Carlo Giunti [a]
INFN, Sezione di Torino, Via P. Giuria 1, I–10125 Torino, Italy
Abstract. The claims that the GSI time anomaly is due to the mixing of neutrinos in the final state of the observed electron-capture decays of hydrogen-like heavy ions are refuted with the help of an analogy with a double-slit experiment. It is a consequence of causality. It is shown that the GSI time anomaly may be caused by quantum beats due to the existence of two coherent energy levels of the decaying ion with an extremely small energy splitting (about 6×10^{-16} eV) and relative probabilities having a ratio of about 1/99.

A GSI experiment [1] observed an anomalous oscillatory time modulation of the electron-capture decays

$$^{140}\mathrm{Pr}^{58+} \to {}^{140}\mathrm{Ce}^{58+} + \nu_e \,, \qquad ^{142}\mathrm{Pm}^{60+} \to {}^{142}\,\mathrm{Nd}^{60+} + \nu_e \,. \tag{1}$$

The hydrogen-like ions $^{140}\mathrm{Pr}^{58+}$ and $^{142}\mathrm{Pm}^{60+}$ were produced by fragmentation of a beam of $^{152}\mathrm{Sm}$ with 500-600 MeV energy per nucleon on a $^{9}\mathrm{Be}$ target and stored in the ESR cooler-storage ring where they circulated with a frequency of about 2 MHz and were monitored by Schottky Mass Spectrometry. The electron capture data are fitted by an oscillatory decay rate with a period $T \simeq 7\,\mathrm{s}$ and an amplitude $A \simeq 0.2$ [1].

It has been proposed [1–4] that the GSI anomaly is due to the interference of the massive neutrinos which compose the final electron neutrino state,

$$|\nu_e\rangle = \cos\vartheta|\nu_1\rangle + \sin\vartheta|\nu_2\rangle \,, \tag{2}$$

where ϑ is the solar mixing angle (see Ref. [5]).

In order to assess the viability of this explanation of the GSI anomaly, it is necessary to understand that interference is the result of the superposition of two or more waves [6]. If the waves come from the same source, interference can occur if the waves evolve different phases by propagating through different paths. Therefore, interference occurs after wave propagation, not at the wave source. In the case of the GSI experiment, there cannot be any interference effect of ν_1 and ν_2 in the electron-capture decays (1), which are the sources of ν_1 and ν_2.

Let us illustrate these remarks through an analogy with the well-known double-slit interference experiment with classical or quantum waves depicted in Fig. 1a. In a double slit experiment an incoming plane-wave packet hits a barrier with two tiny holes, generating two outgoing spherical wave packets which propagate on the other side of the barrier. The two outgoing waves are coherent, since they are created with the same initial phase in the two holes. Hence, the intensity after the barrier, which is proportional to the squared modulus of

[a] e-mail: giunti@to.infn.it

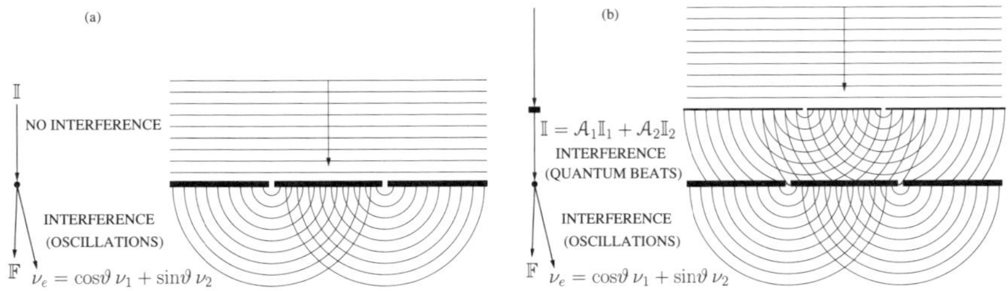

Figure 1: Analogy between the electron-capture decay process (3) and a double-slit interference experiment with one source (a) and two coherent sources (b).

the sum of the two outgoing waves, exhibits interference effects. The interference depends on the different path lengths of the two outgoing spherical waves after the barrier.

For the analogy with the double-slit experiment, let us write schematically an electron-capture decay process of the type in Eq. (1) as

$$\mathbb{I} \to \mathbb{F} + \nu_e \,. \tag{3}$$

Taking into account the neutrino mixing in Eq. (2), we have two different decay channels: $\mathbb{I} \to \mathbb{F}+\nu_1$ and $\mathbb{I} \to \mathbb{F}+\nu_2$. The initial state in the two decay channels is the same. In our analogy with the double-slit experiment, the initial state $\mathbb{I}$ is analogous to the incoming wave packet. The two final states $\mathbb{F} + \nu_1$ and $\mathbb{F} + \nu_2$ are analogous to the two outgoing wave packets. Different weights of ν_1 and ν_2 production due to $\vartheta \neq \pi/4$ correspond to different sizes of the two holes in the barrier.

In the analogy, the decay rate of $\mathbb{I}$ corresponds to the fraction of intensity of the incoming wave which crosses the barrier, which depends only on the sizes of the holes. It does not depend on the interference effect which occurs after the wave has passed through the barrier. In a similar way, the decay rate of $\mathbb{I}$ cannot depend on the interference of ν_1 and ν_2 which occurs after the decay has happened[b].

The above argument is a simple consequence of causality: the interference of ν_1 and ν_2 occurring after the decay cannot affect the decay rate.

Causality is explicitly violated in Ref. [2], where the decaying ion is described by a wave packet, but it is claimed that there is a selection of the momenta of the ion caused by a final neutrino momentum splitting due to the mass

[b]Of course, flavor neutrino oscillations caused by the interference of ν_1 and ν_2 can occur after the decay of $\mathbb{I}$, in analogy with the occurrence of interference of the outgoing waves in the double-slit experiment, regardless of the fact that the decay rate is the incoherent sum of the rates of production of ν_1 and ν_2 and the fraction of intensity of the incoming wave which crosses the barrier is the incoherent sum of the fractions of intensity of the incoming wave which pass trough the two holes.

difference of ν_1 and ν_2. This selection violates causality. In the double-slit analogy, the properties of the outgoing wave packets are determined by the properties of the incoming wave packet, not vice versa. In a correct treatment, all the momentum distribution of the wave packet of the ion contributes to the decay, generating appropriate neutrino wave packets.

The authors of Refs. [3,4] use a different approach: they calculate the decay rate with the final neutrino state

$$|\nu\rangle = \sum_{k=1}^{3} |\nu_k\rangle \, . \tag{4}$$

This state is different from the standard electron neutrino state, which is given by

$$|\nu_e\rangle = \sum_{k=1}^{3} U_{ek}^* \, |\nu_k\rangle \, , \tag{5}$$

where U is the mixing matrix (in the two-neutrino mixing approximation of Eq. (2), $U_{e1} = \cos\vartheta$, $U_{e2} = \sin\vartheta$, and $U_{e3} = 0$). It is not even properly normalized to describe one particle ($\langle\nu|\nu\rangle = 3$). Moreover, it leads to a decay rate which is different from the standard decay rate, given by the incoherent sum of the rates of decay into the different massive neutrinos final states weighted by the corresponding element of the mixing matrix (see Ref. [5]). The decay rate is given by the integral over the phase space of the decay probability

$$P_{\mathbb{I}\to\mathbb{F}+\nu} = |\langle\nu, \mathbb{F}|\mathsf{S}|\mathbb{I}\rangle|^2 = \left|\sum_{k=1}^{3} \langle\nu_k, \mathbb{F}|\mathsf{S}|\mathbb{I}\rangle\right|^2 , \tag{6}$$

where S is the S-matrix operator. Considering the S-matrix operator at first order in perturbation theory, $\mathsf{S} = 1 - i\int \mathrm{d}^4x\, \mathcal{H}_W(x)$, with the effective four-fermion interaction Hamiltonian

$$\begin{aligned} \mathcal{H}_W(x) &= \frac{G_F}{\sqrt{2}} \cos\theta_{\mathrm{C}}\, \overline{\nu_e}(x)\gamma_\rho(1-\gamma^5)e(x)\, \overline{n}(x)\gamma^\rho(1-g_A\gamma^5)p(x) \\ &= \frac{G_F}{\sqrt{2}} \cos\theta_{\mathrm{C}} \sum_{k=1}^{3} U_{ek}^* \overline{\nu_k}(x)\gamma_\rho(1-\gamma^5)e(x)\, \overline{n}(x)\gamma^\rho(1-g_A\gamma^5)p(x) \, , \end{aligned} \tag{7}$$

where θ_{C} is the Cabibbo angle, one can write the matrix elements in Eq. (6) as $\langle\nu_k, \mathbb{F}|\mathsf{S}|\mathbb{I}\rangle = U_{ek}^* \mathcal{M}_k$, with

$$\mathcal{M}_k = \frac{G_F}{\sqrt{2}} \cos\theta_{\mathrm{C}} \, \langle\nu_k, \mathbb{F}|\overline{\nu_k}(x)\gamma_\rho(1-\gamma^5)e(x)\, \overline{n}(x)\gamma^\rho(1-g_A\gamma^5)p(x)|\mathbb{I}\rangle \, . \tag{8}$$

Therefore, the decay probability is given by $P_{\mathbb{I}\to\mathbb{F}+\nu} = \left|\sum_{k=1}^{3} U_{ek}^* \mathcal{M}_k\right|^2$. It is different from the standard one (see Ref. [5]), which is obtained by summing incoherently over the probabilities of decay into the different massive neutrinos final states weighted by the corresponding element of the mixing matrix:

$$P = \sum_{k=1}^{3} |U_{ek}|^2 \, |\mathcal{M}_k|^2 \, . \tag{9}$$

The analogy with the double-slit experiment and the causality argument discussed above support the correctness of the standard decay probability P. Another argument against the decay probability $P_{\mathbb{I}\to\mathbb{F}+\nu}$ is that in the limit of massless neutrinos it does not reduce to the decay probability in the Standard Model,

$$P_{\text{SM}} = |\mathcal{M}_{\text{SM}}|^2 \, , \tag{10}$$

with

$$\mathcal{M}_{\text{SM}} = \frac{G_F}{\sqrt{2}} \cos\theta_{\text{C}} \, \langle \mathbb{F}, \nu_e^{\text{SM}} | \overline{\nu_e^{\text{SM}}}(x) \gamma_\rho (1-\gamma^5) e(x) \, \overline{n}(x) \gamma^\rho (1 - g_A \gamma^5) p(x) | \mathbb{I} \rangle \, , \tag{11}$$

where ν_e^{SM} is the SM massless electron neutrino. Indeed, for $m_k \to 0$ we have

$$\mathcal{M}_k \to \mathcal{M}_{\text{SM}} \, , \tag{12}$$

leading to $P_{\mathbb{I}\to\mathbb{F}+\nu} \to |\mathcal{M}_{\text{SM}}|^2 \left| \sum_{k=1}^{3} U_{ek}^* \right|^2$. This is different from the SM decay probability in Eq. (10). Notice that the contribution of the elements of the mixing matrix should disappear automatically[c] in the limit $m_k \to 0$.

We conclude that the state in Eq. (4) does not describe the neutrino emitted in an electron-capture decay process of the type in Eq. (3) and Refs. [3, 4] are flawed.

The correct normalized state ($\langle \nu_e | \nu_e \rangle = 1$) which describes the electron neutrino emitted in an electron-capture decay processes of the type in Eq. (3) is [5, 7]

$$\begin{aligned} |\nu_e\rangle &= \left(\sum_j |\langle \nu_j, \mathbb{F} | \mathsf{S} | \mathbb{I} \rangle|^2 \right)^{-1/2} \sum_{k=1}^{3} |\nu_k\rangle \, \langle \nu_k, \mathbb{F} | \mathsf{S} | \mathbb{I} \rangle \\ &= \left(\sum_j |U_{ej}|^2 |\mathcal{M}_j|^2 \right)^{-1/2} \sum_{k=1}^{3} U_{ek}^* \mathcal{M}_k \, |\nu_k\rangle \, . \end{aligned} \tag{13}$$

In experiments which are not sensitive to the differences of the neutrino masses, as neutrino oscillation experiments, we can approximate $\mathcal{M}_k \simeq \overline{\mathcal{M}}$ and the

[c] In fact, even in the SM one can define the three massless flavors neutrinos ν_e, ν_μ, ν_τ as arbitrary unitary linear combinations of three massless neutrinos ν_1, ν_2, ν_3. However, all physical quantities are independent of such an arbitrary transformation.

state (13) reduces to the standard electron neutrino state in Eq. (5) (apart for an irrelevant phase $\overline{\mathcal{M}}/|\overline{\mathcal{M}}|$).

With the electron neutrino state in Eq. (13), the decay probability is given by

$$P_{\mathbb{I}\to\mathbb{F}+\nu_e} = |\langle\nu_e, \mathbb{F}|\mathsf{S}|\mathbb{I}\rangle|^2 = \sum_{k=1}^{3} |\langle\nu_k, \mathbb{F}|\mathsf{S}|\mathbb{I}\rangle|^2 = \sum_{k=1}^{3} |U_{ek}|^2 \, |\mathcal{M}_k|^2 \,. \tag{14}$$

This is the correct standard result in Eq. (9): the decay probability is given by the incoherent sum over the probabilities of decay into different massive neutrinos weighted by the corresponding element of the mixing matrix.

Using Eq. (12) and the unitarity of the mixing matrix, one can also easily check that $P_{\mathbb{I}\to\mathbb{F}+\nu_e}$ reduces to P_{SM} in Eq. (10) in the massless neutrino limit.

Although the GSI time anomaly cannot be due to the interference of the two massive neutrinos in the final state of the electron-capture process, it can be due to the interference of two coherent energy states of the decaying ion which produces quantum beats. Also in this case we can draw an analogy with a double-slit experiment. However, we must change the setup, considering the double-slit experiment with two coherent sources of waves depicted in Fig. 1b. The two coherent sources are produced by an incoming plane-wave packet hitting a first barrier with two tiny holes. The two coherent outgoing waves interfere in the space between the first and the second barrier. The interference at the holes in the second barrier induces a modulation of the intensity which crosses the barrier. The role of causality is clear: the interference effect is due to the different phases of the two coherent incoming waves at the holes in the second barrier, which have developed during the propagation of the two waves along different path lengths between the two barriers. Analogously, quantum beats in the GSI experiment can be due to interference of two coherent energy states of the decaying ion which develop different phases before the decay. The two coherent energy states could be produced in the creation process of the ion, which in GSI occurs through fragmentation of a beam of heavier ions on a target [1], as illustrated in Fig. 1b. Alternatively, the two coherent energy states could be due to interactions of the decaying ion in the storage ring.

The quantum mechanical description of quantum beats is rather simple. If the two energy states of the decaying ion $\mathbb{I}_1$ and $\mathbb{I}_2$ are produced at the time $t = 0$ with amplitudes $\mathcal{A}_1$ and $\mathcal{A}_2$ (with $|\mathcal{A}_1|^2 + |\mathcal{A}_2|^2 = 1$), we have $|\mathbb{I}(t=0)\rangle = \mathcal{A}_1 |\mathbb{I}_1\rangle + \mathcal{A}_2 |\mathbb{I}_2\rangle$. Assuming, for simplicity, that the two states with energies E_1 and E_2 have the same decay rate Γ, at the time t we have [d]

$$|\mathbb{I}(t)\rangle = \left(\mathcal{A}_1 \, e^{-iE_1 t} |\mathbb{I}_1\rangle + \mathcal{A}_2 \, e^{-iE_2 t} |\mathbb{I}_2\rangle\right) e^{-\Gamma t/2} \,. \tag{15}$$

[d] Since the measuring apparatus monitors the ions through elastic electromagnetic interactions with a frequency of the order of the revolution frequency in the ESR storage ring, about 2 MHz, the coherence can be preserved only if the interaction with the measuring apparatus does not distinguish between the two states. In this case the interaction is coherent,

The probability of electron capture at the time t is given by

$$P_{\text{EC}}(t) = |\langle \nu_e, \mathbb{F}|\mathsf{S}|\mathbb{I}(t)\rangle|^2 = [1 + A\,\cos(\Delta E t + \varphi)]\,\overline{P}_{\text{EC}}\,e^{-\Gamma t}\,. \qquad (16)$$

where S is the S-matrix operator, $A \equiv 2|\mathcal{A}_1||\mathcal{A}_2|$, $\Delta E \equiv E_2 - E_1$, $\overline{P}_{\text{EC}} = |\langle \nu_e, \mathbb{F}|\mathsf{S}|\mathbb{I}_1\rangle|^2 = |\langle \nu_e, \mathbb{F}|\mathsf{S}|\mathbb{I}_2\rangle|^2$, and φ is a constant phase which takes into account possible phase differences of $\mathcal{A}_1$ and $\mathcal{A}_2$ and of $\langle \nu_e, \mathbb{F}|\mathsf{S}|\mathbb{I}_1\rangle$ and $\langle \nu_e, \mathbb{F}|\mathsf{S}|\mathbb{I}_2\rangle$.

The fit of GSI data presented in Ref. [1] gave $\Delta E \simeq 6\times 10^{-16}$ eV and $A \simeq 0.2$. Therefore, the energy splitting is extremely small and the oscillation amplitude A is significantly smaller than one.

The authors of Ref. [1] noted that the splitting of the two hyperfine $1s$ energy levels of the electron is many order of magnitude too large. It is difficult to find a mechanism which produces a smaller energy splitting. Furthermore, since the amplitude $A \simeq 0.2$ of the interference is rather small, it is necessary to find a mechanism which generates coherently the states $\mathbb{I}_1$ and $\mathbb{I}_2$ with probabilities $|\mathcal{A}_1|^2$ and $|\mathcal{A}_2|^2$ having a ratio of about 1/99!

In conclusion, the GSI time anomaly cannot be due to neutrino mixing. It may be due to quantum beats due to the existence of two coherent energy levels of the decaying ion.

[1] Yu. A. Litvinov et al. Observation of Non-Exponential Orbital Electron Capture Decays of Hydrogen-Like ^{140}Pr and ^{142}Pm Ions. *Phys. Lett.*, B664:162–168, 2008.

[2] Harry J. Lipkin. The GSI method for studying neutrino mass differences - For Pedestrians. 2008.

[3] A. N. Ivanov, R. Reda, and P. Kienle. On the time–modulation of the K–shell electron capture decay of H-like $^{140}\text{Pr}^{58+}$ ions produced by neutrino–flavour mixing. 2008.

[4] Manfried Faber. Kinematics and Quantum Field Theory of the Neutrino Oscillations Observed in the Time-modulated Orbital Electron Capture Decay in an Ion Storage Ring. 2008.

[5] C. Giunti and C. W. Kim. *Fundamentals of Neutrino Physics and Astrophysics.* Oxford University Press, Oxford, UK, 2007.

[6] Carlo Giunti. Rates of Processes with Coherent Production of Different Particles and the GSI Time Anomaly. *Phys. Lett.*, B665:92–94, 2008.

[7] Carlo Giunti. Neutrino Flavor States and the Quantum Theory of Neutrino Oscillations. *J. Phys. G: Nucl. Part. Phys.*, 34:R93–R109, 2007.

---

i.e. the two states suffer the same phase shift. Since the energy splitting ΔE is extremely small, coherence is maintained for a long time if $\mathbb{I}_1$ and $\mathbb{I}_2$ have the same electromagnetic properties.

UPDATED CONSTRAINTS ON NONSTANDARD NEUTRINO INTERACTIONS

O. G. Miranda [a]
Departamento de Física, Centro de Investigación y de Estudios Avanzados del IPN, Apartado Postal 14-740 México D F 07000, México

Abstract. The current status of the constraints for non-standard interactions (NSI) is discussed. This type of interactions arise naturally in models that consider physics beyond the Standard Model, especially when one tries to explain the origin of neutrino masses. Future perspectives for improving these constraints will also be briefly discussed.

The great progress of neutrino physics in the last decade has demonstrated that neutrinos change flavor when they propagate either in vacuum or in media [1]. The explanation for this phenomena is given by the neutrino oscillation mechanism and implies a nonzero neutrino mass, which is a hypothesis that appears in many extensions of the Standard Model of Elementary Particles (SM).

It is natural to wonder if neutrino experiments could test the nature of the underlying mechanism that give rise to neutrino masses. In particular, for different models beyond the SM it is common to have corrections to the $V - A$ structure of the theory. These corrections can be parametrized by using an effective Lagrangian that describes, in a model independent way, the neutrino nonstandard interactions (NSI). Such NSI constitute an unavoidable characteristic of gauge models of neutrino mass, for example, models of the generic seesaw type [2].

We will focus on nonstandard interactions that can be parametrized with the effective low–energy four–fermion operator for neutral currents:

$$\mathcal{L}_{NSI} = -\varepsilon_{\alpha\beta}^{fP} 2\sqrt{2} G_F \left(\bar{\nu}_\alpha \gamma_\mu L \nu_\beta\right) \left(\bar{f}\gamma^\mu P f\right), \tag{1}$$

where P = L, R and f is a first generation fermion: e, u, d. The coefficients $\varepsilon_{\alpha\beta}^{fP}$ denote the strength of the NSI between the neutrinos of flavors α and β and the P–handed component of the fermion f. We will study the case when the fermion f is either a d type quark or an electron.

For the case of interactions of neutrinos with a d-type quark it is possible to obtain constraints for the neutrino NSI either from the solar neutrino data or from accelerator experiments. For the solar neutrino case, a global analysis can be developed if one includes the results from the radiochemical experiments Homestake [3], SAGE [4] and GALLEX/GNO [5], the zenith-spectra data from Super-Kamiokande I [6], the data reported from the three different phases of the SNO experiment [7] and the measurement of the ^{7}Be solar neutrino rate

[a] e-mail: omr@fis.cinvestav.mx

Table 1: Constraints on the NSI couplings at 90% C L from the combined analysis of solar and accelerator experiments [11] .

$-0.5 < \varepsilon_{ee}^{dV} < 1.2$	$-1.8 < \varepsilon_{\tau\tau}^{dV} < 4.4$
$-0.4 < \varepsilon_{ee}^{dA} < 1.4$	$-1.5 < \varepsilon_{\tau\tau}^{dA} < 0.7$

performed by the Borexino collaboration [8]. Besides, in this analysis it is also necessary to include the data from the KamLAND reactor experiment [9], in order to obtain better constraints for the NSI parameters.

The results reported by the CHARM collaboration, an accelerator experiment measuring the ratio of the neutral current to the charge current cross section for electron (anti)neutrinos off quarks [10], could also be very useful for obtaining constraints on the NSI of neutrinos with a d-type quark.

Constraints to the NSI parameters coming from an analysis that uses these experiments [11] are shown in Table 1. It is possible to see from these results that there is still a lot of room for new physics effects, especially in the case of the tau neutrino.

For the case of neutrino interactions with electrons, it is possible to focus again on the solar neutrino data [12] or to consider laboratory experiments [13].

In the case of laboratory constraints one can study the cross section $e^+e^- \to \nu\bar{\nu}\gamma$ [14] that in the presence of NSI will change its expected value. Besides, the $\nu_e e$ and $\bar{\nu}_e e$ scattering processes can also give constraints to the NSI. For the case of the interaction $\nu_e + e^- \to \nu_e + e^-$, there are results reported by the Liquid Scintillator Neutrino Detector (LSND) [15], while for the antineutrino case, the $\bar{\nu}_e e$ process has been measured by reactor experiments in Irvine [16], Rovno [17] and also in the most recent MUNU experiment [18].

For the case of muon type neutrinos, the CHARM II collaboration results give very restrictive constraints for the NSI parameters. The collaboration measured the $\nu_\mu + e \to \nu_\mu + e$ and $\bar{\nu}_\mu + e \to \bar{\nu}_\mu + e$ scattering processes. The neutrinos were produced by a 450 GeV proton beam accelerated in the Super Proton Synchrotron (SPS) for 2.5×10^{19} protons on target [19].

Finally, for the case of the solar neutrino data, a similar analysis to the one discussed above for the NSI of neutrinos with d-type quark can be developed. The constraints obtained from using either the solar neutrino data [12] or laboratory experiments [13] are summarized in Table 2. We can see that in this case it is possible to obtain stronger constraints than in the case of the interaction with d-type quarks.

Besides the current constraint on the NSI parameters, there are also different experimental proposals that could improve the bounds on the parameters shown

Table 2: Constraints on the NSI couplings at 90% C L. We show the most stringent constraints coming either from the solar analysis [12] or from a combined laboratory analysis [13].

$-0.036 < \varepsilon_{ee}^{eL} < 0.063$	$-0.03 < \varepsilon_{ee}^{eR} < 0.18$
$-0.033 < \varepsilon_{\mu\mu}^{eL} < 0.055$	$-0.040 < \varepsilon_{\mu\mu}^{eR} < 0.053$
$-0.16 < \varepsilon_{\tau\tau}^{eL} < 0.11$	$-0.4 < \varepsilon_{\tau\tau}^{eR} < 0.6$

here. For the case of the interactions with quarks, different experimental setups, such as long baseline neutrino experiments and neutrino factories, have been considered [20], as well as the case of coherent neutrino nucleus scattering [21]. For the case of NSI with electrons there are also new experimental proposals for measuring the neutrino scattering off electrons, such as Texono [22] and NuSong [23].

Non standard interactions of neutrinos is a promising subject that could shed light on the physics beyond the Standard Model. In this work I have discussed the current status for the constraints on the NSI parameters.

Acknowledgments

I would like to thank to all my collaborators: Juan Barranco, Azucena Bolaños, Francisco Escrihuela, Celio Moura, Timur Rashba, Mariam Tórtola and José Valle, with whom the works summarized in this proceedings have been done. This work has been supported by CONACyT.

References

[1] T. Schwetz, M. A. Tortola and J. W. F. Valle, New J. Phys. **10**, 113011 (2008) [arXiv:0808.2016 [hep-ph]].

[2] P. Minkowski, Phys. Lett. **B67**, 421 (1977); R. N. Mohapatra and G. Senjanovic, Phys. Rev. Lett. **44**, 912 (1980); J. Schechter and J. W. F. Valle, Phys. Rev. **D22**, 2227 (1980).

[3] B. T. Cleveland *et al.*, Astrophys. J. **496**, 505 (1998).

[4] J. N. Abdurashitov *et al.* [SAGE Collaboration], Phys. Rev. C **80**, 015807 (1990)

[5] GNO collaboration, M. Altmann *et al.*, Phys. Lett. **B616**, 174 (2005), [hep-ex/0504037];

[6] Super-Kamiokande collaboration, S. Fukuda *et al.*, Phys. Lett. **B539**, 179 (2002), J. Hosaka *et al.* [Super-Kamkiokande Collaboration], Phys. Rev. D **73**, 112001 (2006).

[7] SNO collaboration, Q. R. Ahmad *et al.*, Phys. Rev. Lett. **89**, 011301 (2002), SNO collaboration, Q. R. Ahmad *et al.*, Phys. Rev. Lett. **89**,

011302 (2002), SNO collaboration, S. N. Ahmed *et al.*, Phys. Rev. Lett. **92**, 181301 (2004), SNO collaboration, B. Aharmim *et al.*, Phys. Rev. **C72**, 055502 (2005), SNO collaboration, B. Aharmim *et al.*, Phys. Rev. Lett. **101**, 111301 (2008),

[8] Borexino collaboration, C. Arpesella *et al.*, Phys. Rev. Lett. **101**, 091302 (2008), C. Galbiati *et al.*, J. Phys. Conf. Ser. **136**, 022001 (2008).

[9] S. Abe *et al.* [KamLAND Collaboration], Phys. Rev. Lett. **100**, 221803 (2008)

[10] J. Dorenbosch *et al.* [CHARM Collaboration], Phys. Lett. B **180**, 303 (1986).

[11] F. J. Escrihuela, O. G. Miranda, M. A. Tortola and J. W. F. Valle, Phys. Rev. D **80**, 105009 (2009) [Erratum-ibid. D **80**, 129908 (2009)] [arXiv:0907.2630 [hep-ph]].

[12] A. Bolanos et. al., Phys. Rev. D **79**, 113012 (2009)

[13] J. Barranco, O. G. Miranda, C. A. Moura and J. W. F. Valle, Phys. Rev. D **77**, 093014 (2008)

[14] ALEPH collaboration, R. Barate *et al.*, Phys. Lett. **B420**, 127 (1998), ALEPH collaboration, R. Barate *et al.*, Phys. Lett. **B429**, 201 (1998); ALEPH collaboration, A. Heister *et al.*, Eur. Phys. J. **C28**, 1 (2003); DELPHI collaboration, P. Abreu *et al.*, Eur. Phys. J. **C17**, 53 (2000), L3 collaboration, M. Acciarri *et al.*, Phys. Lett. **B415**, 299 (1997); L3 collaboration, M. Acciarri *et al.*, Phys. Lett. **B444**, 503 (1998); L3 collaboration, M. Acciarri *et al.*, Phys. Lett. **B470**, 268 (1999), OPAL collaboration, K. Ackerstaff *et al.*, Eur. Phys. J. **C2**, 607 (1998),OPAL collaboration, G. Abbiendi *et al.*, Eur. Phys. J. **C8**, 23 (1999), OPAL collaboration, G. Abbiendi *et al.*, Eur. Phys. J. **C18**, 253 (2000),

[15] LSND collaboration, L. B. Auerbach *et al.*, Phys. Rev. **D63**, 112001 (2001), [hep-ex/0101039].

[16] F. Reines, H. S. Gurr and H. W. Sobel, Phys. Rev. Lett. **37**, 315 (1976).

[17] A. I. Derbin *et al.*, JETP Lett. **57**, 768 (1993).

[18] MUNU collaboration, Z. Daraktchieva *et al.*, Phys. Lett. **B564**, 190 (2003), [hep-ex/0304011].

[19] CHARM-II collaboration, P. Vilain *et al.*, Phys. Lett. **B335**, 246 (1994).

[20] N. Cipriano Ribeiro et. al. JHEP 0712:002 (2007); Kopp, Ota and Winter, Phys. Rev. **D78** 053007 (2008); S. Antusch, J. P. Baumann and E. Fernandez-Martinez, Nucl. Phys. B **810**, 369 (2009).

[21] J. Barranco, O. G. Miranda and T. I. Rashba, JHEP 0512:021 (2005).

[22] M. Deniz and H. T. Wong [TEXONO Collaboration], arXiv:0810.0809

[23] T. Adams *et al.* [NuSOnG Collaboration], Int. J. Mod. Phys. A **24**, 671 (2009)

KATRIN EXPERIMENT IN 2009 - SYSTEMATIC IMPROVEMENTS

N. Titov (for the KATRIN Collaboration) [a]
Institute for Nuclear Research, Russian Academy of Sciences, 117312 Moscow, Russia

Abstract. The Karlsruhe Tritium Neutrino experiment (KATRIN) is designed to measure the mass of electron antineutrinos from tritium β-decay with an unprecedented sensitivity of 0.2 eV/c^2. As the experimentally measured parameter is the square of the neutrino mass, such a sensitivity can be reached only by improving both the statistical and systematic errors by two orders of magnitude relative to previous experiments. Special measures will be taken to improve the source thickness measurements, to prevent tritium penetration to the spectrometers, to improve the high voltage measurement, to control space charge effects in the gaseous source and the tritium adsorption on the rear wall of the source, etc.

1 Introduction

KATRIN is an international experiment currently being assembled at Karlsruhe to measure the absolute value of the electron antineutrino mass at the 0.2 eV/c^2 level. It will study the shape of the tritium β-decay spectrum near the end-point with an electrostatic spectrometer with adiabatic magnetic collimation. In order to accumulate high precision data and to reduce the systematic uncertainties, a windowless gaseous tritium source (WGTS) will be used to produce an unprecedented number of decay electrons [1]. Systematic uncertainties as well as statistic errors should be reduced by two orders in magnitude (for m_ν^2) in comparison with previous experiments.

The discovery of neutrino oscillations by experiments with atmospheric and solar neutrinos proves that neutrinos mix and that they have non-zero masses in contrast to their previous description in the Standard Model of particle physics. Unfortunately, these oscillation experiments are sensitive only to the differences of squared neutrino mass states $\delta m_{ij}^2 = m^2(i) - m^2(j)$, but not directly to the neutrino masses $m(i)$ themselves. With defined mixing parameters, the neutrino mass spectrum can be obtained if one neutrino mass is measured absolutely.

Different Standard Model extensions lead to both "quasi-degenerate" and "hierarchical" mass schemes, see *e.g.* [2]. The sensitivity of the KATRIN experiment will allow to confirm or practically exclude the "quasi - degenerate" scenario (see fig.1).

Free streaming relic neutrinos with the masses > 0.3 eV/c^2 scale usually are considered to be in contradiction with the observed large scale structures in the Universe. It was found recently that the relic neutrino mass is highly correlated with the dark energy equation of state $w = P/\rho c^2$, [3, 4]. A laboratory

[a] e-mail: titov@inr.ru

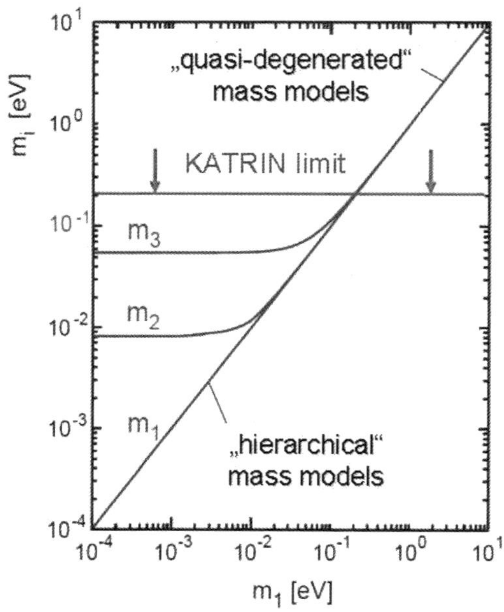

Figure 1: Neutrino mass eigenstates masses as a function of the lightest one (normal hierarchy). KATRIN is probing "quasi-degenerate" regime when all masses are close to each other. "Quasi-degenerate" regime could be confirmed or rejected by KATRIN results.

measurement of $m_\nu > 0.2eV/c^2$ could thus imply $w < -1$ (quintessence).

2 History

First neutrino mass evaluations were performed 70 years ago by E.M. Lyman [5] and by A. I. Alichanian, A. I. Alichanov and B. S. Dzhelepov [6] with the use of the RaE (^{85}At) β-spectrum. The spectra were found to be consistent with $m_\nu = (0.3 \ldots 0.8)\ m_e$. It was recognized soon that the tritium β - spectrum is much better suited for the neutrino mass measurement due to the following reasons:

1. superallowed transition: the matrix element M is not energy dependent;
2. low endpoint energy: the decay fraction at the endpoint sensitive to the neutrino mass is comparatively high;
3. short half life: only a low amount of source material is required, only a small fraction of the electrons are subject to inelastic scattering inside the source;
4. As a hydrogen isotope it has a simple atomic shell: final states are precisely calculable.

First two arguments from the list above were given by S.C. Curran, J. Angus, and A.L. Cockcroft in their publication in the 1948 [7]. This year we are celebrating the 60-th anniversary of the first experiments of the neutrino mass measurement with tritium. This pioneer experiment was performed with a gaseous proportional counter filled with counting gas mixed with tritium. An observed signal amplitude spectrum contradicted both the zero and non-zero neutrino mass hypothesis and gave a wrong endpoint position [8]. This experiment revealed once more a rule that each improvement in the statistical sensitivity should be accompanied by the improved study of the increasing list of the systematic effects. The same year G.C. Hanna and B. Pontecorvo published results of a similar experiment but they achieved a better determination of the counter response function [9]. The crucial influence came from the reduced signal amplification at the ends of the counter, where signal wire holders have a larger diameter than the wire itself [10]. As a result, the measured spectrum was in agreement with a zero neutrino mass and a limit $m_\nu < 1\ keV$ was obtained and measured endpoint value within it error coincides with cotemporal one. Thus, switching to tritium as a β -source immediately improved the sensitivity to the neutrino mass by two orders of magnitude.

A review of the further progress of the neutrino mass measurements in the tritium β-decay up to the middle-80'es can be found in [11]. One has to mention the invention of the spectrometer with toroidal magnetic field [12] and an elaborate systematic analysis first performed in [13]. The experimental sensitivity at that time reached a limit posed by the trade between source luminosity and spectrometer resolution. A further breakthrough was achieved by invention of the electrostatic spectrometer with adiabatic magnetic collimation [14]. The

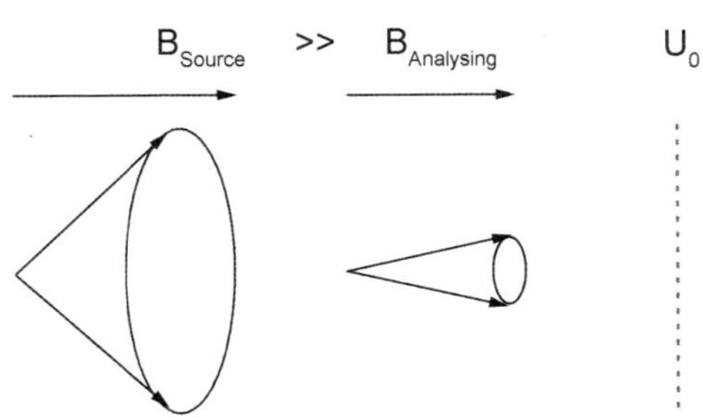

Figure 2: Electrostatic spectrometer with adiabatic magnetic collimation principle. Electron born in the strong magnetic field and travel to the week one. Due to the adiabaticity of his motion transversal electron momentum is almost totally transferred into longitudinal one. In the week magnetic field electron energy is analyzed by the electrostatic spectrometer.

operation principle of the new spectrometer spectrometer is based on the conservation of the adiabatic invariant $\mu = \frac{E_\perp}{B}$. Once an electron is born in the strong magnetic field of the source it will have to travel to the weak field at the spectrometer. Due to the adiabaticity of the electron motion, transversal electron momentum is almost totally transferred into longitudinal one. In the weak magnetic field, the electron energy is analyzed by the electrostatic spectrometer that serves as a high-pass energy filter with a threshold eU_0, were U_0 is a spectrometer stopping potential. The spectrometer resolution (full width) is $eU_0 \frac{B_{analysis}}{B_{max}}$ for any electron trajectory if an adiabaticity motion criteria $\lambda = \frac{\omega_c R}{V} << 1$ (were ω_c - electron cyclotron frequency, R - trajectory curvature radius, V -electron velocity) is fulfilled. Thus, *the spectrometer resolution became decoupled from the source luminosity* and the subsequently increased statistical accuracy has made it possible to improve the neutrino mass upper limit by the order of magnitude [15].

3 KATRIN experiment - probing subelectronvolt neutrino masses

The KATRIN experiment [1] is based on the recent advances in modern technology and combines the best features of its ancestors [15,16] with new solutions that make possible a reduction of both the statistic and systematic errors by one more order of magnitude. One has to have in mind that actually *two* orders of magnitude improvements are required, as the measured parameter is the neutrino mass square m_ν^2. The Windowless Gaseous Tritium Source (WGTS) of KATRIN is a 90 mm diameter, 10 m long pipe with a tritium density at its center of $10^{15} mol/cm^3$. The KATRIN spectrometer vessel is a 10m diameter 24 m long cylinder, with residual pressure $10^{-11} mbar$ (that is formed by the H_2 outgassing from the walls). The transport system between WGTS and spectrometer includes both turbo-molecular and cryogenic pumping stations that will reduce the tritium flow toward the spectrometer by a factor 10^{14} and together with spectrometer pumping will achieve tritium pressure in the spectrometer at the $10^{-20} mbar$ level.

Both the basic spectrometer and source parameters should be known with high accuracy: the stopping potential $\delta U_0 \pm 30 mV$, the spectrometer resolution $\delta(\frac{B_{analysis}}{B_{max}}) / \frac{B_{analysis}}{B_{max}} < 3 \cdot 10^{-2}$, the source thickness $\delta(\rho d)/\rho d < 1 \cdot 10^{-3}$, the source thickness spirality correction $\delta(\frac{B_{source}}{B_{max}}) / \frac{B_{source}}{B_{max}} < 4 \cdot 10^{-3}$. The stopping potential, magnetic fields have radial and azimuthal variation and should be carefully mapped. It is proposed to use for this purpose a test electron beam generated by the electron gun of a new type. This electron gun makes use of the Ultra Violet Light Emitting Diode (UV LED) in pulsed mode in order to produce short (up to 40 ns) signals for the time - of - flight measurements [20]. Further improvement is an electron gun that emits electrons within a narrow angle with respect to the external magnetic field direction.

The spectrometer resolution could be perturbed by an external magnetic field in the analyzing plane. Special precautions were undertaken that paid back.

Particularly, the building concrete near the analyzing plane of the spectrometer was armed by the stainless steel . This year the external magnetic field variation was measured to be less than $1.7\mu T$ compared with $100\mu T$ specification [17].

In order to reduce T_2 source thickness variation, the WGTS temperature will be maintained constant within 30 mK. A temperature stabilization system based on 2-phase boiling Ne was developed [18]. Tests of the system will take place in 2010 with WGTS full scale model ("Demonstrator").

Contributions to the T_2 source thickness from the admixture of other hydrogen isotopes will be controlled on-line with the use of the Raman spectroscopy. First results were published last year [19] and continuing efforts promise that the required sensitivity will be reached.

A precision 35 kV high voltage divider was developed for monitoring the spectrometer high voltage in the collaboration with the Physikalisch - Technische Bundesanstalt (PTB), German National Metrology Institute that provides scientific and technical services . As a result of careful design, use of the new type of high stability resistors, etc the total measurement uncertainty with the new divider is proved to be better than $1 \cdot 10^{-6}$ [21]. After successful tests of the first divider a second one with more improvements is being manufactured. Comparison of the two divider readings will allow to monitor HV drifts beyond their quite conservative specifications.

A complementary spectrometer potential monitoring method is a position measurement of the conversion lines of ^{83m}Kr by an independent monitoring spectrometer connected to the same HV supply. It is shown that two different approaches to the ^{83m}Kr source preparation lead to the reproducible results within KATRIN specifications $\delta E \pm 30\ meV$ [22, 23].

The tritium decay rate inside the WGTS is planned at the $1.2 \cdot 10^{11} sec^{-1}$ level. Together with the processes of secondary ionization, thermalization and charge transport this leads to an ion/electron pair density inside WGTS at the level of $10^7 \ldots 10^8 cm^{-3}$. At the operation temperature 30 K these charges will behave as a plasma [24]. One has to be confident that corresponding space charge potential do not exceeds 200 meV. Our first estimations showed that this could be achieved due to the high conductance between plasma and WGTS rear wall with well defined potential. Ions leaving the WGTS follow the magnetic field lines and special equipment is developed in order to prevent them from entering the spectrometer. Ion beam toward spectrometer will be analyzed in the 3-pole Penning trap by the Fourier Transform Ion Cyclotron Resonance (FT-ICR) method [25]. The plasma effects will be modeled in the Differential Pumping Section (DPS2F) that is a part of the transport channel under commissioning in Karlsruhe now. Use of the artificial ion- and plasma sources is planned during experiments in the middle of 2010.

The beginning of T_2 measurements by KATRIN is scheduled at 2012.

Acknowledgments

The author would like to acknowledge numerous valuable discussions with colleagues about their contribution to KATRIN systematic improvements. This report was prepared under support from the Karlsruhe Institute of Technology (KIT), Karlsruhe, Germany by a Fellowship Stipend.

References

[1] J.Angrik *et al*, (KATRIN Design Report 2004), FZK Scientific Report 7090, *http://www-ik.fzk.de/katrin.*
[2] P.D.Bari, *these issue.*
[3] S. Hannestad,*arXiv: 0710.1952v1 [hep-ph].*
[4] G. La Vacca, J.R.Kristiansen, L.P.L.Colombo, R.Mainini, S. A. Bonometto, *arXiv: 0906.3369v1 [astro-ph.CO]* .
[5] E.M.Lyman, *Phys. Rev.* 51, 1 (1937).
[6] A.I.Alichanian, A.I.Alichanov, B.S.Dzhelepov, *Phys. Rev.* 53, 766 (1938).
[7] S.C.Curran, J. Angus, A.L.Cockcroft, *Nature (London, United Kingdom)* 162, 302 (1948).
[8] S.C.Curran, J. Angus, A.L.Cockroft, *Phys. Rev.* 76, 853 (1949).
[9] G.C.Hanna, B.Pontecorvo, *Phys. Rev.* 75, 983 (1949).
[10] G.C.Hanna, D.H.W.Kirkwood, B.Pontecorvo, *Phys. Rev.* 75, 985 (1949).
[11] R.G.H.Robertson, D.A.Knapp *Ann. Rev. Nucl. Sci.* 38 185 (1988).
[12] E. F.Tretyakov et al. *Izv. Akad. Nauk SSSR, Ser. Fiz.* 40(10), 1 (1976).
[13] K.-E.Bergkvist *Nuclear Physics* **B** 39, 317 (1972).
[14] V.M.Lobashev, P.E.Spivak *Nucl. Inst. and Meth.* **A** 240, 305 (1985).
[15] Ch.Kraus et. al.*Eur. Phys. J.* **C** 40, 447 (2005).
[16] A.I. Belesev et al. *Phys. Lett.* **B** 350, 263 (1995).
[17] J.Reich, *Diploma Thesis*, (University of Karlsruhe, Karlsruhe) 2009.
[18] S.Grohmann, *Cryogenics* 49, 413 (2009).
[19] R.J.Lewis et al., *Laser Physics Letter* 5, 522 (2008).
[20] K.Valerius et al.,*New J.Phys.* 11, 063018 (2009).
[21] Th.Thuemmler, R.Marx, Ch.Weinheimer, *New J.Phys.* 11, 103007 (2009).
[22] D.Venos et al., *arXiv:0902.0291 [physics.ins-det]*
[23] B.Ostrick, *PhD Thesis*, (University of Muenster, Muenster) 2009.
[24] A.F.Nastoyashchii et. al., *Fusion Science and Technology* 48, 743 (2005).
[25] M. Ubieto-Daz et al, *International Journal of Mass Spectrometry* 1, 288 (2009).

SEARCH FOR NEUTRINOLESS DOUBLE BETA DECAY OF ^{76}Ge WITH THE GERmanium Detector Array "GERDA"

Alberto Garfagnini [a]
on behalf of the GERDA Collaboration
Physics Department, Padova University, I-35131 Padova, Italy

Abstract. The study of neutrinoless double beta decay (DBD) is the only presently known approach to the fundamental question if the neutrino is a Majorana particle, i.e. its own anti-particle. The observation of neutrinoless DBD would not only establish the Majorana nature of the neutrino but also represent a determination of its effective mass if the nuclear matrix element is given. So far, the most sensitive results have been obtained with ^{76}Ge, and the group of Klapdor-Kleingrothaus has made a claim of discovery. Future experiments have to reduce radioactive backgrounds to increase the sensitivity. GERDA is a new DBD experiment which is currently under construction in the INFN Gran Sasso National Laboratory, Italy. It is implementing a new shielding concept by operating bare Ge diodes - enriched in ^{76}Ge - in high purity liquid argon supplemented by a water shield. The aim of GERDA is to scrutinize the recent claim of discovery, and, in a second phase, to achieve a two orders of magnitude lower background index than recent experiments, increasing the sensitive mass and reaching an exposure of 100 kg yr. The paper will discuss design, physics reach, and status of construction of GERDA.

1 Introduction

Since their discovery neutrinos have been an object of extensive experimental study and the knowledge about their properties has advanced our understanding of weak interactions significantly. Still unanswered, however, is the very fundamental question whether the neutrino is a Majorana particle like most extensions of the Standard Model assume. The study of DBD is the only one presently known approach capable to answer this question; if the decay occurs without the emission of neutrinos then their Majorana nature is proven. In the following, a brief summary of the theory and the present experimental status is given; finally the GERDA experiment, its goals, the detector and the present status are described.

2 Theory and present experimental programs

Neutrinoless DBD, $\beta\beta(0\nu)$, is a very slow lepton-number violating nuclear transition that happens if neutrinos have mass and are their own antiparticles. An initial nucleus (Z, A) decays to $(Z+2, A)$, emitting two electrons. A related transition, called two-neutrino DBD, $\beta\beta(2\nu)$, results in the emission of two electron antineutrinos in addition to the electrons. The latter is a very rare process but it does not violate any rule: in fact its has been observed in a number of experiments. If the $\beta\beta(0\nu)$ decay is mediated by the exchange of a

[a]e-mail: alberto.garfagnini@pd.infn.it

light Majorana neutrino, the half-life is

$$[T^{0\nu}_{1/2}(0^+ \to 0^+]^{-1} = G^{0\nu}(Q_{\beta\beta}, Z)|M^{0\nu}|^2 \langle m_{\beta\beta}\rangle^2 \tag{1}$$

where $G^{0\nu}$ is the exactly calculable phase space integral, $M^{0\nu}$ is the nuclear matrix element and $\langle m_{\beta\beta}\rangle$ is the effective Majorana mass of the electron neutrino:

$$\langle m_{\beta\beta}\rangle \equiv |\sum_k m_k U^2_{ek}| \tag{2}$$

Here the m'_ks are the masses of the three light neutrinos and U is the matrix that transforms states with well-defined mass into states with well-defined flavour. So the observation of neutrinoless DBD would not only establish the Majorana nature of the neutrino but also provide a measurement of its effective mass, $m_{\beta\beta}$. Moreover, through Eq. (2), the absolute mass scale for the neutrino can be investigated. In Eq. (1) the nuclear matrix element $M^{0\nu}$ is not well known: presently the most accurate calculations give the same result for a given matrix element to within a factor of 2-3.

At present there is only one claim for a positive $\beta\beta(0\nu)$ observation by Klapdor *et al.* [1] as part of the Heidelberg-Moscow Collaboration: $T^{0\nu}_{1/2}(^{76}\mathrm{Ge}) = 2.23^{+0.44}_{-0.31} \times 10^{25}$ yr at 1σ. The other most sensitive limits are from the IGEX experiment [2]: $T^{0\nu}_{1/2}(^{76}\mathrm{Ge}) \geq 1.6 \times 10^{25}$ yr (90% C.L.), and the CUORICINO experiment [3]: $T^{0\nu}_{1/2}(^{130}\mathrm{Te}) \geq 3.0 \times 10^{24}$ yr (90% C.L.). These experiments contained 5-10 kg of the parent isotope and ran for several years. Hence in order to increase the half-life sensitivity by a factor of about 100, the goal of the next generation of experiments will require hundreds of kg of parent isotope and a significant decrease in background beyond the present state of the art (~ 0.1 counts/(keV kg yr)).

3 The GERDA experiment

The GERDA experiment [4] located in Hall A of the Laboratori Nazionali del Gran Sasso (LNGS) has the aim to search for neutrinoless DBD using enriched ^{76}Ge detectors. The experiment will proceed in several phases. In phase I, existing enriched ^{76}Ge of Heidelberg-Moscow and IGEX experiments for a total mass of 18 kg will be used. The goal for the background index is 10^{-2} counts/(keV kg yr). In these conditions, after an exposure of 15 kg yr, a limit of $T^{0\nu}_{1/2} > 2.2 \times 10^{25}$ yr will be reached, scrutinizing the Klapdor claim. In phase II, with additional 20 kg of enriched Ge detectors and a background of 10^{-3} counts/(keV kg yr) and 100 kg yr exposure, GERDA will reach the sensitivity of $T^{0\nu}_{1/2} > 1.4 - 2.0 \times 10^{26}$ yr [b] at 90% C.L., corresponding to about 130 meV

[b]Depending on the number of background events

(with the Rodin et al. [5] matrix elements). Depending on the results obtained in phase I and II, a new phase III is foreseen with the aim of reaching the 10 meV scale: this requires $\mathcal{O}(1\text{t})$ of enriched germanium and represents a big challenge, which can be afforded only in the context of a world-wide collaboration. GERDA, shown as an artist view in Fig. 1, will use naked Ge crystals immersed in liquid Ar. The shielding of external environmental backgrounds is realized using an onion-like structure consisting of, from outside to inside: hyper-pure water in a 10 m diameter tank shown open in Fig. 1, a stainless steal cryostat with internal lining of hyperpure copper and ultra-pure liquid Ar. The Ge diodes will be assembled into an array suspended in the centre of

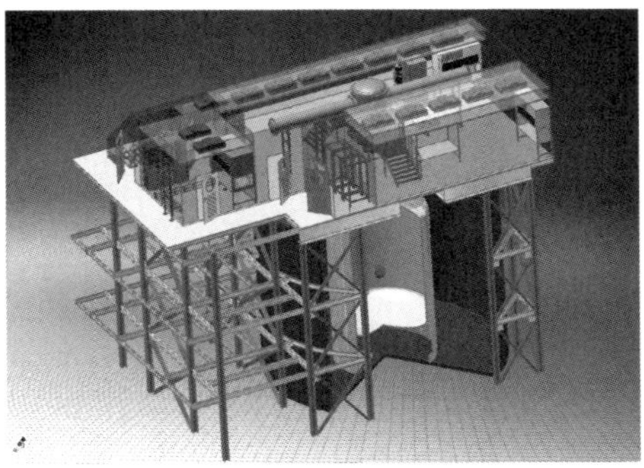

Figure 1: Artistic view of the GERDA experiment.

the liquid Ar volume. In Phase I the existing detectors are expected to contribute with about 10^{-2} counts/(keV kg yr), due to their intrinsic cosmogenic ^{60}Co contamination. This will dominate the background budget. However this will be enough to confirm or refute the Klapdor-Kleingrothaus et al. claim in an exposure of one year. Notice that this test needs to be performed on the same isotope, due to the uncertainty on the nuclear matrix elements. To reach the background index 10^{-3} counts/(keV kg yr), no individual component of the background budget should be larger than 10^{-4} counts/(keV kg yr). The method followed by GERDA is the following.

- The quantitative evaluation of all background sources in every component of the apparatus by analytical methods or Monte Carlo simulations. These calculations provide the allowed specific activity for each component, depending on its location relative to the detectors.

- Shielding: γ's from the external environment (like ^{208}Tl) with the H_2O (3 m thickness) and the LAr (2 m thickness) shields; from the ^{228}Th in the stainless steel cryostat with a Cu shield and LAr; μ-induced prompt signals with two vetoes. a plastic scintillator roof and the water used as Cherenkov medium.

- Mechanical design of minimum mass suspension and contact system for the Ge detectors. Notice also that in GERDA the detectors are surrounded by low Z materials, a fact that minimizes the muon induced neutron background and makes the experiment feasible at the depth of Gran Sasso. Low mass ASIC and front-end electronics.
- Geometrical localization and pulse shape analysis to discriminate the backgrounds intrinsic to the Ge diodes. The two electrons of the signal deposit their energy within a few millimeters, whilst the gammas, which constitute the principal background, with a few Compton scatters separated by a few centimeters. Consequently, the signal events are "single site" (SS); background events are "multiple site" (MS). SS can be distinguished from MS with pulse shape analysis and anti-coincidence between diodes.

The detectors for phase II are presently under development. Two different types are considered: detectors [6] with segmented electrodes [c], and point-contact diodes [7] developed as a commercial product by the company Canberra. Both detectors give similar suppression factors for the MS events and similar efficiency for SS events. The choice will be taken in 2010.

4 Outlook

Presently (February 2010) all existing detectors made from enriched material have been tested to work properly and are ready to be installed. The water tank, the cryostat (filled with LAr), the clean room have been already installed at LNGS. The final tests for the FE electronics and DAQ are well advanced. The commissioning is in an advanced phase and the data-taking will follow in spring 2010.

[1] H.V. Klapdor-Kleingrothaus et al., *Phys. Lett.* **B 586** 198 (2004); H.V. Klapdor-Kleingrothaus et al., *Mod. Phys. Lett.* **A 21** 1547 (2006).
[2] C.E. Aalseth et al., *Phys. Rev.* **D 65** 092007 (2002); C.E. Aalseth et al., *Phys. Rev.* **D 70** 078302 (2004).
[3] C. Arnaboldi et al., *Phys. Rev.* **C 78** 035502 (2008).
[4] I. Abt et al. *GERDA The GERmanium Detector Array for the search of neutrinoless $\beta\beta$ decays ^{76}Ge at LNGS* LNGS P38/04 Sept 2004.
[5] V.A. Rodin et al. *Nucl. Phys.* **A 766** 107 (2006); V.A. Rodin et al. *Nucl. Phys.* **A 793** 213 (2007).
[6] I. Abt et al., *N.I.M.* **A 577** 574 (2007).
[7] D. Dudjáš et al., *IEEE Conference 2008*, 19-25 Oct 08, Dresden, Germany.

[c] The segmentation will be 6-fold in azimuth and three-fold in height.

CUORE: NEUTRINOLESS DOUBLE BETA DECAY WITH BOLOMETERS

Sergio Di Domizio [a] on behalf of the CUORE Collaboration
Dipartimento di Fisica dell'Università di Genova, Via Dodecaneso 33, 16146 Genova (Italy)

Abstract. Cuore is a 1 ton-scale bolometric experiment devoted to the search of neutrinoless double beta decay of ^{130}Te. The experiment will start data taking in 2012 at the Laboratori Nazionali del Gran Sasso, Italy. Cuore will have a sensitivity for the half life of the searched process of the order of $10^{26}y$, and will be able to touch the inverted neutrino mass-hierarchy region. The success of the experiment will rely on the results that will be achieved in the reduction of the background.

1 Introduction

The search for neutrinoless double beta ($0\nu\beta\beta$) decay [1] is the most promising approach to investigate the nature of neutrino mass: besides proving that neutrinos are Majorana particles, its observation could give informations on the absolute scale of neutrino masses and on the so called neutrino mass-hierarchy.

The experimental signature of $0\nu\beta\beta$ decay is a sharp peak in the energy spectrum whose position is given by the mass difference between the parent and the daughter nucleus, minus twice the electron mass (Q-Value). In order to search for this rare process, $0\nu\beta\beta$ experiments require big source mass, low background, and good energy resolution.

The Cuore collaboration uses bolometric detectors to search for $0\nu\beta\beta$ decay in ^{130}Te (Q-Value = 2527.5 keV). Bolometers are cryogenic calorimeters capable of measuring the energy released by a particle by detecting the thermal pulse it produces in an absorber crystal.

The Cuoricino experiment completed data taking in 2008. It set the best limit for $0\nu\beta\beta$ decay of ^{130}Te and demonstrated the feasibility of Cuore, a 1 ton scale experiment that will begin data taking in 2012.

2 Cuoricino detector

Cuoricino [2] was installed underground at the Laboratori Nazionali del Gran Sasso (LNGS), under a 3200 m w.e. rock shield against cosmic rays. The detector consisted in an array of TeO_2 bolometers with a total mass of 40.2 kg (11.8 kg in ^{130}Te), maintained at ~8 mK by a ^{3}He/^{4}He dilution refrigerator. The bolometers were arranged in a tower with 13 floors. Eleven floors were composed by four crystals with size of 5×5×5 cm^3 while the remaining two floors were composed by nine crystals with size of 3×3×6 cm^3. Four of the small crystals were enriched, two to 82.3% in ^{128}Te and two to 75% in ^{130}Te,

[a] e-mail: sergio.didomizio@ge.infn.it

while all other crystals had natural abundance of ^{130}Te (34%). The crystals were held by a structure made of PTFE and OFHC copper which served both as mechanical support and as thermal bath. Thermal signals were read using Neutron Transmutation Doped germanium thermistors with a doping concentration corresponding to the Variable Range Hopping regime. For 1 MeV of released energy the typical thermal pulse had an amplitude of about 0.1 mK and a signal of about 100 μV was obtained at the sensor output.

The detectors were protected from environmental radioactivity by two lead shields. An inner 1 cm thick layer, thermalized at 600 mK, protected the detectors from radioactive contaminations in the cryostat materials, while another 20 cm thick layer was located at room temperature outside the cryostat. The external lead shield was surrounded by a 10 cm layer of borated polyethylene to reduce background from neutrons, and a continuous overpressure of nitrogen was maintained around the cryostat to reduce background from radon in the air. The experimental apparatus was enclosed in a Faraday cage to reduce electromagnetic disturbances.

All the materials used for the construction of the detector underwent strict radioactivity controls and were treated with dedicated cleaning procedures, and particular attention was paid to the parts that are very close to the crystals, where shielding is not possible. All the assembly procedures were performed in nitrogen atmosphere.

3 Cuoricino Performances and Results

Cuoricino took data for five years between 2003 and 2008. During this period a statistics of 18 kg·y in ^{130}Te was accumulated. The energy calibration was performed by inserting for three days per month a Th source between the cryostat and the external lead shield. The energy resolution measured on the 2615 keV line from ^{208}Tl visible during calibration measurement was on average of 8 keV (FWHM) with the best detectors reaching a resolution of 4 keV. No evidence for a peak at the Q-Value of the decay was found in the energy spectrum summed over all the detectors, and the measured background rate in the region of interest was of 0.18 counts/(keV·kg·y). The lower limit on the half life, obtained by a maximum likelihood fit where the big, small and enriched detectors were considered separately, was evaluated on the anti-coincidence sum spectra, resulting in

$$T^{0\nu}_{1/2} > 2.94 \times 10^{24} y \ (90\% C.L.)$$

Using nuclear matrix elements from several authors this value translates into an upper limit on the effective Majorana mass of $m_{\beta\beta} < 0.2 \div 0.7$ eV, where the interval ranges from the most favorable to the less favorable calculation.

4 Cuore

The Cuore experiment [3], currently under construction at the LNGS, will be a tightly packed array of 988 TeO_2 bolometers with natural abundance of ^{130}Te. It will have a total mass of 750 kg (204 kg in ^{130}Te). The bolometers will be arranged in 19 towers, each tower being very similar to Cuoricino. The Cuore collaboration aims at reaching a background of 0.01 counts/(keV·kg·y) in the region of the $0\nu\beta\beta$ peak. In five years of data taking this would result in a sensitivity on the $0\nu\beta\beta$ decay half life of 2.1×10^{26} y, which translates into to $m_{\beta\beta} < 20 \div 100$ meV.

The background reduction around the Q-Value of ^{130}Te is the main challenge to be faced by the Cuore collaboration. The Cuoricino background was dominated in this region by three contributions: multi-Compton scattering from the ^{208}Tl γ line (40%) and α surface contaminations in the crystals (10%) and in the copper structure (50%). A full Monte Carlo simulation of the Cuore detector shows that the background from the ^{208}Tl line, originating from thorium contaminations in the cryostat radiation shields, can be reduced below the target background level by proper design of the detector shieldings. In contrast background from surface contaminations in the crystals and in the copper structure cannot be shielded and must be removed by proper material cleaning. For this purpose several R&D measurements were performed in the cryogenic test facility in the Hall C at LNGS, in which new cleaning techniques were employed. Compared to Cuoricino, the Hall C tests showed improvements in material cleanliness for both crystal an copper surfaces. However, while an adequate background reduction was obtained for the crystals, only a partial improvement was obtained for copper, for which a background between 0.02 counts/(keV·kg·y) and 0.04 counts/(keV·kg·y) was measured. While the baseline for the assembly of the Cuore detector is to employ the cleaning techniques whose effectiveness was demonstrated in the Hall C measurements, further R&D is still ongoing with the purpose of reaching the target background of 0.01 counts/(keV·kg·y).

References

[1] S. R. Elliott, P. Vogel, Ann. Rev. Nucl. Part. Sci. **52** (2002) 115
[2] C. Arnaboldi *et al.*, Phys. Rev. C **78** (2008) 035502
[3] R. Ardito *et al.*, arXiv:hep-ex/0501010.

THE ABSOLUTE MASS OF NEUTRINO AND THE FIRST UNIQUE FORBIDDEN β-DECAY OF ^{187}Re

Rastislav Dvornický[1], Fedor Šimkovic[1,2], Kazuo Muto[3]
[1]*Department of Nuclear Physics, Comenius University, Bratislava, Slovakia*
[2]*Laboratory of Theoretical Physics, JINR Dubna, Russia*
[3]*Department of Physics, Tokyo Institute of Technology, Tokyo, Japan*

Abstract. The planned rhenium β-decay experiment MARE might probe the absolute mass scale of neutrinos with the same sensitivity as the tritium β-decay experiment KATRIN, which will start data taking in 2011 and will proceed for five years. We present the energy distribution of emitted electrons for the first unique forbidden β-decay of ^{187}Re. It is found that the p-wave emission of electron dominates over the s-wave. By assuming mixing of three neutrinos the Kurie function for the rhenium β-decay is derived. It is shown that the Kurie plot near the endpoint is within a good accuracy linear in the limit of massless neutrinos like the Kurie plot of the superallowed β-decay of ^3H.

1 Introduction

The recent atmospheric, solar, reactor and accelerator neutrino experiments have convinced us that neutrinos are massive particles. However, the problem of absolute values of their masses is still waiting for a solution. Neutrino oscillations depend on the differences of neutrino masses, not on their absolute values. Apparently three kinds of neutrino experiments have a chance to determine the light neutrino masses. These are: i) cosmological measurements, ii) the tritium and rhenium single β-decay experiments, iii) neutrinoless double β-decay experiments.

The measurement of the electron spectrum in β-decays provides a robust direct determination of the values of neutrino masses. In practice, the most sensitive experiments use tritium β-decay, because it is a super-allowed transition with a low Q-value. The effect of neutrino masses m_k (k=1,2,3) can be observed near the end point of the electron spectrum, where $Q - T \sim m_k$. T is the electron kinetic energy. A low Q-value is important, because the relative number of events occurring in an interval of energy ΔT near the endpoint is proportional to $(\Delta T/Q)^3$.

The current best upper bound on the effective neutrino mass m_β given by,

$$m_\beta^2 = \sum_{k=1}^{3} |U_{ek}|^2 m_k^2, \tag{1}$$

have been obtained in the Mainz and Troitsk experiments: $m_\beta < 2.2\ eV$ [1]. U_{ek} is the element of neutrino mixing matrix. In near future, the KATRIN experiment will reach a sensitivity of about $0.2\ eV$ [1]. In these experiments the β-decay of tritium is investigated with electrostatic spectrometers.

Calorimetric measurements of the β-decay of rhenium where all electron energy released in the decay is recorded, appear complementary to those carried out with spectrometers. The unique first forbidden β-decay,

$$^{187}\mathrm{Re} \rightarrow {}^{187}\mathrm{Os} + e^- + \overline{\nu}_e, \tag{2}$$

is particularly promising due to its low transition energy of $\sim 2.47\ keV$ and the large isotopic abundance of ^{187}Re (62.8%), which allows the use of absorbers made with natural Rhenium. Measurements of the spectra of ^{187}Re have been reported by the Genova and the Milano/Como groups (MIBETA and MANU experiments). The achieved sensitivity of $m_\beta < 15\ eV$ was limited by statistics [1]. The success of rhenium experiments has encouraged the micro-calorimeter community to proceed with a competitive precision search for a neutrino mass. The ambitious project is planned in two steps, MARE I and MARE II. MARE I is to meet the existing upper limit of 2 eV around 2011, MARE II is to challenge the KATRIN goal of 0.2 eV [2].

The aim of this contribution is to derive the form for the endpoint spectrum of emitted electrons for the β-decay of ^{187}Re, which is needed for extracting the effective neutrino mass m_β or limit on this quantity from future MARE I and II experiments.

2 First unique forbidden β-decay of ^{187}Re

The ground-state spin-parity is $5/2^+$ for ^{187}Re and $1/2^-$ for the daughter nucleus ^{187}Os, and the decay is associated with $\Delta J^\pi = 2^-$ ($\Delta L = 1$, $\Delta S = 1$) of the nucleus, *i.e.*, classified as unique first forbidden β-decay. The emitted electron and neutrino are expected to be, respectively, in $p_{3/2}$ and $s_{1/2}$ states or vice versa. The emission of higher partial waves is strongly suppressed due to a small energy release in this nuclear reaction.

The differential decay rate is a sum of two contributions associated with emission of the $s_{1/2}$ and the $p_{3/2}$ state electrons. By considering the finite nuclear size effect the theoretical spectral shape of the β-decay of ^{187}Re is

$$\begin{aligned} N(E_e) &= \frac{d\Gamma}{dE_e} = \sum_{k=1}^{3} |U_{ek}|^2 \frac{G_F^2 V_{ud}^2}{2\pi^3} B R^2 p_e E_e (E_0 - E_e)\sqrt{(E_0 - E_e)^2 - m_k^2} \\ &\times \frac{1}{3}\left[F_1(Z,E_e)p_e^2 + F_0(Z,E_e)((E_0 - E_e)^2 - m_k^2)\right]\theta(E_0 - E_e - m_k) \end{aligned} \tag{3}$$

with

$$B = \frac{g_A^2}{6} | < 0^-_{1/2} || \sqrt{\frac{4\pi}{3}} \sum_n \tau_n^+ \frac{r_n}{R} \{\sigma_1(n) \otimes Y_1(n)\}_2 || 0^+_{5/2} > |^2. \tag{4}$$

G_F is the Fermi constant and V_{ud} is the element of the Cabbibo-Kobayashi-Maskawa (CKM) matrix. p_e, E_e and E_0 are the momentum, energy, and maximal endpoint energy (in the case of zero neutrino mass) of the electron, respectively. R is the nuclear radius. The Fermi functions $F_0(Z, E)$ and $F_1(Z, E)$ in (3) are due to a distortion of the $s_{1/2}$- and the $p_{3/2}$- electron wave states in the Coulomb field of final nucleus, respectively. The value of nuclear matrix element in (4) can be determined from the measured half-life of the β-decay of ^{187}Re.

Experimentally, it was found that $p_{3/2}$-state electrons are predominantly emitted in the β-decay of ^{187}Re [3]. By performing numerical analysis of partial decay rates associated with emission of the $p_{3/2}$ and the $s_{1/2}$ electrons (terms with $F_1(Z, E_e)$ and $F_0(Z, E_e)$ in Eq. (3), respectively) we conclude that about 10^4 more $p_{3/2}$-state electrons are emitted when compared with emission of $s_{1/2}$-state electrons. The reasons of it are as follows: i) $F_1(Z, E_e) \gg F_0(Z, E_e)$ for $E_e - m_e < Q$ (see Fig. 1), ii) the maximal momentum of electron (~ 49.3 keV) is much larger when compared to maximal momentum of neutrino (~ 2.5 keV). Henceforth, we shall neglect a small contribution to the differential decay rate given by an emission of the $s_{1/2}$-state electrons.

For a normal hierarchy (NH) of neutrino masses with $m_3 > m_2 > m_1$ the Kurie function of the β-decay of ^{187}Re is given by

$$\begin{aligned} K(y) &= \mathcal{B}_{Re}\sqrt{y+m_1}\Big[|U_{e1}|^2\sqrt{y(y+2m_1)} \\ &+|U_{e2}|^2\sqrt{(y+m_1-m_2)(y+m_1+m_2)}\theta(y+m_1-m_2) \\ &+|U_{e3}|^2\sqrt{(y+m_1-m_3)(y+m_1+m_3)}\theta(y+m_1-m_3)\Big]^{1/2}, \end{aligned} \tag{5}$$

with

$$\mathcal{B}_{Re} = \frac{G_F V_{ud}\sqrt{B}}{\sqrt{2\pi^3}}\sqrt{\frac{R^2}{3}\frac{p_e^2 F_1(Z,E_e)}{F_0(Z,E_e)}} \tag{6}$$

and $y = (E_0 - E_e - m_1) \geq 0$. The ratio $\left(p_e^2\, F_1(Z, E_e)\right)/F_0(Z, E_e)$ depends only weakly on the electron momentum in the case of the β-decay of rhenium. With a good accuracy the factor $\mathcal{B}_{Re}$ can be considered to be a constant.

So far the rhenium β-decay experiments will not see any effect due to neutrino masses, it is possible to approximate $m_k \ll Q - T$ and obtain

$$K(y) \simeq \mathcal{B}_{Re}\left((y+m_\beta)\sqrt{y(y+2m_\beta)}\right)^{1/2}, \tag{7}$$

where $y = (E_0 - E_e - m_\beta)$. In Fig. 2 we show the Kurie plot for β-decay of ^{187}Re versus y near the endpoint. We see that the Kurie plot is linear near the endpoint for $m_\beta = 0$. However, the linearity of the Kurie plot is lost if m_β is not equal to zero.

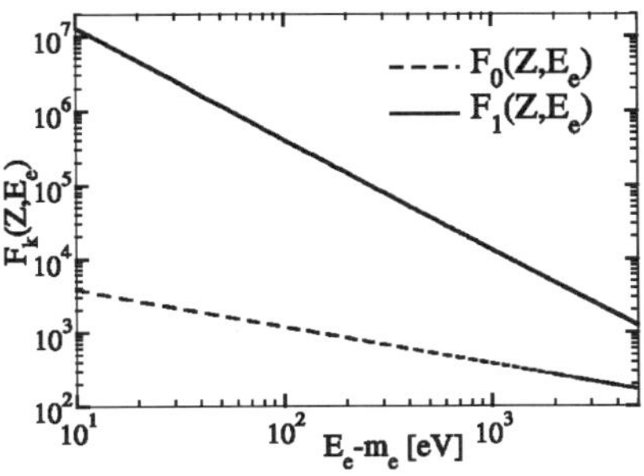

Figure 1: Fermi functions for $k = 0$ and $k = 1$ as functions of electron energy E_e in energy interval relevant for the β-decay of ^{187}Re.

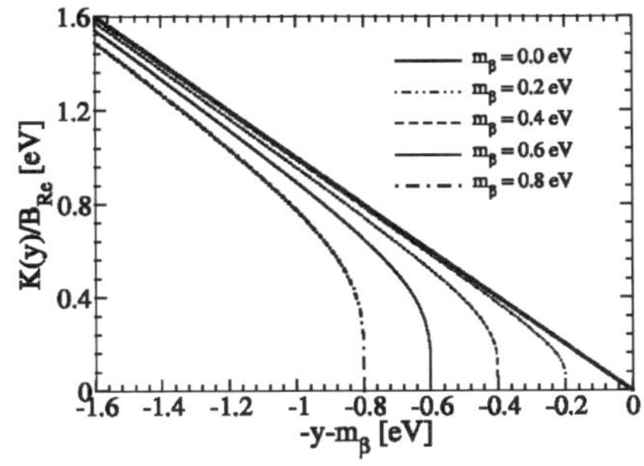

Figure 2: Endpoints of the Kurie plot of the rhenium β-decay for various values of the effective neutrino mass: $m_\beta =$ 0, 0.2, 0.4, 0.6 and 0.8 eV.

3 Conclusions

For the first unique forbidden β-decay of ^{187}Re to ground state of ^{187}Os we derived the theoretical spectral shape. It is found that electrons are emitted dominantly in the $p_{3/2}$-wave state in agreement with available data analysis [3]. The Kurie function of the rhenium β-decay close to the endpoint coincides up to a factor to that for the superallowed β-decay of tritium [4]. These foundings are important for the planned MARE experiment, which will be sensitive to neutrino mass in sub eV region.

Acknowledgments

One of the authors (R.D.) acknowledges the financial support from the JINR grant for collaboration with Slovakia.

References

[1] E.W. Otten, C. Weinheimer, Rept. Prog. Phys. **71**, 086201 (2008).

[2] MARE Collaboration, E. Andreotti et al., Nucl. Instrum. Meth. A **572**, 208 (2007).

[3] C. Arnaboldi et al., Phys. Rev. Lett. **96**, 042503 (2006).

[4] F. Šimkovic, R. Dvornický and A. Faessler, Phys. Rev. C **77**, 055502 (2008).

NEUTRINOLESS DOUBLE BETA DECAY: SEARCHING FOR NEW PHYSICS WITH COMPARISON OF DIFFERENT NUCLEI

A. Ali [a]
Deutsches Electronen-Synchrotron, DESY, 22607 Hamburg, Germany
A. V. Borisov [b]
Faculty of Physics, Moscow State University, 119991 Moscow, Russia
D. V. Zhuridov [c]
Department of Physics, National Tsing Hua University, 30013 Hsinchu, Taiwan

Abstract. The neutrinoless double beta decay is analyzed using a general Lorentz invariant effective Lagrangian for various decaying nuclei of current experimental interest: ^{76}Ge, ^{82}Se, ^{100}Mo, ^{130}Te, and ^{136}Xe. We work out the half-lives and angular correlation coefficients of the outgoing electrons in several scenarios for new physics: the left-right symmetric models, the R-parity-violating SUSY and models with leptoquarks. The theoretical uncertainty in the nuclear matrix elements is discussed.

1. The Majorana nature of neutrino masses is anticipated by most of the theories created to explain the observable lightness of neutrinos, in particular, seesaw mechanism and models with radiative neutrino mass generation (see [1, 2] and references therein). Experimental evidence for the neutrinoless double beta decay ($0\nu 2\beta$) would deliver a conclusive confirmation of the Majorana nature of neutrinos, in contrast to the Dirac nature of all other known fermions. There is the overriding interest in carrying out these experiments and in the related phenomenology [3]. We recall that $0\nu 2\beta$ decays are forbidden in the Standard Model (SM) by lepton number (LN) conservation. However, an extended version of the SM could contain terms that violate LN and allow the $0\nu 2\beta$ decay. Probable mechanisms of LN violation may include exchanges by: Majorana neutrinos (ν_Ms), SUSY Majorana particles, scalar bilinears, e.g. doubly charged Higgs, leptoquarks (LQs), right-handed W_R bosons etc. [1]. These various contributions will have to be disentangled to extract information from the $0\nu 2\beta$ decay on the characteristics of the sources of LN violation, in particular, on the neutrino masses and mixing. Measurements of the $0\nu 2\beta$ decay in different nuclei will help to determine the underlying physics mechanism [4–6]. In Ref. [7] the $0\nu 2\beta$ decay angular correlation for the ^{76}Ge nucleus was investigated in order to discriminate among the various possible mechanisms contributing to this decay. However much more new physics one can extract using the experimental data for various decaying nuclei. In this report, we generalize the analysis of Ref. [7] for the case of the following set of nuclei: ^{76}Ge, ^{82}Se, ^{100}Mo, ^{130}Te, and ^{136}Xe.

2. Following Ref. [7], we use the general effective Lagrangian for the $0\nu 2\beta$ decay

$$\mathcal{L} = \frac{G_{\rm F} V_{ud}}{\sqrt{2}} [(U_{ei} + \epsilon^{V-A}_{V-A,i}) j^{\mu i}_{V-A} J^{+}_{V-A,\mu} + \sum_{\alpha,\beta}{}' \epsilon^{\beta}_{\alpha i} j^{i}_{\beta} J^{+}_{\alpha} + \text{H.c.}] , \quad (1)$$

where the hadronic and leptonic currents are defined as: $J^{+}_{\alpha} = \bar{u} O_{\alpha} d$ and $j^{i}_{\beta} = \bar{e} O_{\beta} \nu_i$; the leptonic currents contain neutrino mass eigenstates and the

[a] e-mail: ahmed.ali@desy.de
[b] e-mail: borisov@phys.msu.ru
[c] e-mail: zhuridov@phys.nthu.edu.tw

index i runs over the light eigenstates; a summation over the repeated indices is assumed; $\alpha, \beta = V \mp A, S \mp P, T_{L,R}$ ($O_{T_\rho} = 2\sigma^{\mu\nu} P_\rho$, $\sigma^{\mu\nu} = \frac{i}{2}[\gamma^\mu, \gamma^\nu]$, $P_\rho = (1 \mp \gamma_5)/2$ is the projector, $\rho = L, R$); the prime indicates the summation over all the Lorentz invariant contributions, except for $\alpha = \beta = V - A$, U_{ei} is the PMNS mixing matrix [8] and V_{ud} is the CKM matrix element [9]. The coefficients $\epsilon^{\beta}_{\alpha i}$ encode new physics, parametrizing deviations of the Lagrangian from the standard $V - A$ current-current form and mixing of the non-SM neutrinos. Eq. (1) describes the so-called long range mechanism of the $0\nu 2\beta$ decay mediated by light Majorana neutrinos.

The differential width for the $0^+(A, Z) \to 0^+(A, Z+2)e^- e^-$ transitions is [7]

$$d\Gamma/d\cos\theta = (\ln 2/2)|M_{\rm GT}|^2 \mathcal{A}(1 - K\cos\theta), \quad K = \mathcal{B}/\mathcal{A}\ , \ -1 < K < 1, \qquad (2)$$

where θ is the angle between the electron momenta in the rest frame of the parent nucleus, $M_{\rm GT}$ is the Gamow–Teller nuclear matrix element, and K is the angular correlation coefficient. Eq. (2) is derived taking into account the leading contribution of the parameters $\epsilon^{\beta}_{\alpha} = U_{ei}\epsilon^{\beta}_{\alpha i}$. The expressions for $\mathcal{A}$ and $\mathcal{B}$ for different choices of $\epsilon^{\beta}_{\alpha}$, with only one nonzero coefficient considered at a time, are given in Ref. [7].

Using the data on various decaying nuclei we have considered the two particular cases for the parameter space: A) $\epsilon^{\beta}_{\alpha} = 0$, $|\langle m \rangle| \neq 0$ (SM plus ν_Ms), B) $\epsilon^{\beta}_{\alpha} \neq 0$, $|\langle m \rangle| = 0$ (vanishing effective Majorana mass). Only the terms with $\epsilon^{V\mp A}_{V\mp A}$ (see Ref. [7] for their expressions through the parameters of the nonstandard models with W_Rs, LQs, and RPV SUSY) are taken into account as the corresponding nuclear matrix elements have been worked out in the literature [10]. The results for more general cases C) $\epsilon^{\beta}_{\alpha} \neq 0$, $|\langle m \rangle| \neq 0$, $\cos\psi_i = 0$ and D) $\epsilon^{\beta}_{\alpha} \neq 0$, $|\langle m \rangle| \neq 0$, $\cos\psi_i \neq 0$, where ψ_i are the relative phases for the non-SM contributions, will be published elsewhere.

The differences in the half-lives and angular coefficients for various nuclei are described by the ratios $\mathcal{R}^{\beta}_{\alpha}(^{A}{\rm X}) = T_{1/2}(\epsilon^{\beta}_{\alpha}, {}^{A}{\rm X})/T_{1/2}(\epsilon^{\beta}_{\alpha}, {}^{76}{\rm Ge})$ and $\mathcal{K}^{\beta}_{\alpha}(^{A}{\rm X}) = K(\epsilon^{\beta}_{\alpha}, {}^{A}{\rm X})/K(\epsilon^{\beta}_{\alpha}, {}^{76}{\rm Ge})$, for the choice of only one nonzero coefficient $\epsilon^{\beta}_{\alpha}$ which characterize specific alternative new physics contributions (we make a comparison with ^{76}Ge as it is the best tested isotope to date). The numerical values of $\mathcal{R}$, $\mathcal{K}$ and $\mathcal{R}^{\beta}_{\alpha}$, $\mathcal{K}^{\beta}_{\alpha}$ corresponding to cases A) and B), respectively, are given in Tables 1 and 2 for two different nuclear models: QRPA without and with p-n pairing [10].

Table 1: The ratios of the half-lives $\mathcal{R}$ and $\mathcal{R}^{\beta}_{\alpha}$ for various nuclei in QRPA without (with) p-n pairing [10].

Nucleus	$\mathcal{R}=\mathcal{R}^{V-A}_{V-A}$	$\mathcal{R}^{V-A}_{V+A}$	$\mathcal{R}^{V+A}_{V-A}$	$\mathcal{R}^{V+A}_{V+A}$
^{82}Se	0.42 (0.15)	0.37 (2.76)	2.10 (3.07)	0.24 (0.03)
^{100}Mo	1.08 (195.18)	52.87 (0.59)	1.11 (0.49)	1.06 (0.79)
^{130}Te	0.24 (0.11)	0.21 (0.16)	0.20 (0.12)	0.15 (0.03)
^{136}Xe	0.53 (0.15)	0.40 (0.37)	0.41 (0.22)	0.34 (0.06)

The entries for the ratios of the half-lifes are well separated, besides the $\mathcal{R}^{V-A}_{V-A}$, which is equal to $\mathcal{R}$. However, they are dominated by the uncertainties of the nuclear model. On the other hand, the angular coefficients $\mathcal{K}$ and

Table 2: The ratios of the angular coefficients $\mathcal{K}$ and $\mathcal{K}_{\alpha}^{\beta}$ for various nuclei in QRPA without (with) p-n pairing [10].

Nucleus	$\mathcal{K}=\mathcal{K}_{V\pm A}^{V-A}$	$\mathcal{K}_{V-A}^{V+A}$	$\mathcal{K}_{V+A}^{V+A}$
^{82}Se	1.08	1.11 (1.11)	1.13 (0.95)
^{100}Mo	1.08	1.14 (1.14)	1.13 (0.84)
^{130}Te	1.04	1.07 (1.07)	1.01 (0.90)
^{136}Xe	1.03	1.06 (1.06)	0.98 (0.91)

$\mathcal{K}_{V\pm A}^{V-A}$ do not depend on the nuclear matrix elements, and coefficients $\mathcal{K}_{V-A}^{V+A}$ essentially do not depend on the uncertainties of the nuclear model. Moreover, the ratios of the angular correlations are not discriminating among the various underlying theories as within the anticipated experimental uncertainty they are all consistent with unity. The most sensitive to the listed ratios is ^{100}Mo, except for the ratio $\mathcal{R}_{V-A}^{V+A}$ to which the most sensitive is ^{82}Se. From the measurements of the half-lives, the most sensitive to the effects of $\epsilon_{V+A}^{V\pm A}$ and ϵ_{V-A}^{V+A} are the pairs ^{100}Mo – ^{130}Te and ^{82}Se – ^{130}Te, correspondingly. From the measurements of the angular coefficients, the most sensitive to the effects of $\epsilon_{V\pm A}^{V+A}$ is the pair ^{76}Ge – ^{100}Mo.

3. In conclusion, the comparison of the half-lives and the electron angular correlations for the selected decaying nuclei would help to minimize the theoretical uncertainties in the nuclear matrix elements and identify the dominant mechanism underlying these decays. The experimental facilities that can measure not only the half-lives but also the angular correlations in the $0\nu2\beta$ decays are the running experiment NEMO3 [11] (it has already measured the electron angular distributions for the two neutrino double beta decays of ^{100}Mo and ^{82}Se) and the proposed ones SuperNEMO [12], MOON [13], and EXO [14].

References

[1] R. N. Mohapatra, *"Unification and Supersymmetry: The Frontiers of Quark-Lepton Physics"* (Springer-Verlag, New York, 2003).
[2] C. S. Chen, C. Q. Geng, D. V. Zhuridov, *Phys. Lett.* B **666**, 340 (2008); arXiv:0803.1556 [hep-ph]; 0806.2698 [hep-ph].
[3] P. Vogel, arXiv:hep-ph/0611243; 0807.1559 [hep-ph].
[4] F. Deppisch, H. Päs, *Phys. Rev. Lett.* **98**, 232501 (2007).
[5] V. M. Gehman, S. R. Elliott, *J. Phys.* G **34**, 667 (2007).
[6] G. L. Fogli, E. Lisi, A. M. Rotunno, *Phys. Rev.* D **80**, 015024 (2009).
[7] A. Ali, A. V. Borisov, D. V. Zhuridov, *Phys. Rev.* D **76**, 093009 (2007).
[8] B. Pontecorvo, *Sov. Phys. JETP* **6**, 429 (1958); Z. Maki, M. Nakagawa, S. Sakata, *Prog. Theor. Phys.* **28**, 870 (1962).
[9] Particle Data Group: C. Amsler et al., *Phys. Lett.* B **667**, 1 (2008).
[10] G. Pantis, F. Šimkovic, J. D. Vergados, A. Faessler, *Phys. Rev.* C **53**, 695 (1996).
[11] A. S. Barabash [NEMO Collaboration], arXiv:0807.2336 [nucl-ex].
[12] Yu. Shitov [SuperNEMO Collaboration], arXiv:0807.3078 [nucl-ex].
[13] M. Nomachi et al., *Nucl. Phys. B (Proc. Suppl.)* **138**, 221 (2005).
[14] D. Akimov et al., *Nucl. Phys. B (Proc. Suppl.)* **138**, 224 (2005).

STUDY OF QUASI-ELASTIC $\nu_\mu(\bar{\nu}_\mu)$ SCATTERING IN THE NOMAD EXPERIMENT

Jean-Michel Levy [a]

LPNHE, T43 rdc, 4 place Jussieu, 75252 Paris Cedex 05, France

*Abstract.*The legacy of the NOMAD neutrino oscillation experiment is a unique data set with a large physics studies potential. We present here the results obtained for the quasi-elastic $\nu_\mu n \to \mu^- p$ and $\bar{\nu}_\mu \mathrm{p} \to \mu^+ \mathrm{n}$ reactions.

1 Introduction

The NOMAD (CERN WA96) experiment purpose was to look for $\nu_\mu \to \nu_\tau$ oscillations through the topological and kinematical identification of the decay products of the CC generated τ lepton, requiring both abundant and precise data. This was obtained through the use of a thinly segmented active target (drift chambers made of low Z material to minimize multiple scattering) complemented by various detectors for identification and calorimetry. The resulting data set is both abundant and kinematically accurate and its exploitation has yielded many important results beyond limits on oscillation parameters. In this talk, we presented the measurement of the quasi-elastic (anti)neutrino nucleon cross-section and of the axial mass. Full details as well as further references on the experiment and its results can be found in [1].

2 Method

2.1 Detector and data sample

The NOMAD detector is shown in Fig.1. The drift chambers allow for momentum reconstruction with an uncertainty approximately given by :
$\frac{\sigma_p}{p} \approx \frac{0.05}{\sqrt{L}} \oplus \frac{0.008p}{\sqrt{L^5}}$ (momentum p is in GeV/c and track length L in m). They are used to determine the event topology, the vertex(tices) position(s) and type(s), and the track parameters at each vertex. Together with the muon chambers and the veto and trigger planes, these DC's are the most important elements for the present analysis. The most upstream chamber is used as an additional off line veto against through going muons.

The data sample used here consists of about 751000 ν_μ and 23000 $\bar{\nu}_\mu$ charged-current (CC) interactions in a reduced detector fiducial volume. The average energy of the incoming ν_μ is 25.9 GeV (17.6 GeV for $\bar{\nu}_\mu$) Candidate neutrino events must comprise a vertex in the fiducial volume and an unvetoed negative

[a] e-mail: jmlevy@in2p3.fr

track identified as a muon, with or without a positive track compatible with being the proton (1T and 2T events in what follows). Unvetoed single positives identified as antimuons are taken as anti-neutrino event candidates.

2.2 Modeling neutrino interactions

Since target nuclei are mostly carbon, Fermi momentum and final state interactions forbid identification of quasi-elastic events by kinematical fits. Therefore, the method used relies on a detailed description of the neutrino primary interaction and nuclear effects; particularly important are Fermi motion and rescattering of the produced hadrons in the nucleus, since they modify observable distributions and therefore bear on the definitions of the cuts used to reduce background in the above described data samples. A modified Fermi gas model is used for the nucleon momentum distibution. The parameters of the formation zone model used for rescattering are tuned by demanding equality of the cross-sections derived from the 1T and 2T samples, ensuring consistency of the corresponding efficiencies. It is further found that results (cross sections and axial mass) are least sensitive to these parameters when calculated with the combination of both samples. Finally, comparison of the 1T and 2T extracted cross sections allows to estimate the systematic errors due to this modeling.

2.3 Procedure

The event generator and detector Monte Carlo are used to compute efficiencies for signal (QEL events) and background which may arise from single pion resonant production (RES), deep inelastic scattering (DIS) and coherent pion production on the target nucleus (COH). RES and DIS event numbers are predicted using analytical formulae, corrected with the MC computed efficiencies and subtracted from the observed numbers of QEL event candidates. The resulting numbers are further corrected for efficiencies in view of cross-section and axial mass calculations. The expected number of coherent π^+ is estimated at 2700 but the demands imposed on the recoil proton candidate reduce the contamination from this channel to less than .4% of the QEL sample.
In these calculations, DIS charged current events above 40 GeV in the same fiducial volume are used to give the overall normalization, which is subjected to a further and independant check performed by determining the number of $\nu_\mu e^- \rightarrow \mu^- \nu_e$ events through a one-parameter fit of the $P_T(\mu^-)$ distribution.

3 Results

With the background subtracted and efficiency corrected QEL event numbers as a function of energy and the overall normalization provided by DIS, cross sections are calculated for ν and $\bar{\nu}$. The axial mass parameter is found by demanding that the flux averaged QEL cross section calculated within the Smith-Moniz [2] model be equal to our measured one. The same parameter is also extracted from the fit of the shape and the shape plus normalization of the Q^2 distribution for events with a reconstructed proton track.

The flux averaged neutrino cross-section over the energy interval 3-100 GeV and the axial mass inferred from it are as follow:

$$\begin{aligned} \langle\sigma_{qel}\rangle_{\nu_\mu} &= (0.92 \pm 0.02(stat) \pm 0.06(syst)) \times 10^{-38}\ \mathrm{cm}^2 \\ M_A &= 1.05 \pm 0.02(stat) \pm 0.06(syst)\ \mathrm{GeV} \end{aligned}$$

This result for M_A is in agreement with both the standard (shape) fit of the Q^2 distribution and the fit of the Q^2 shape plus normalization:

$$M_A = 1.07 \pm 0.06(stat) \pm 0.07(syst)\ \mathrm{GeV}$$

$$M_A = 1.06 \pm 0.02(stat) \pm 0.06(syst)\ \mathrm{GeV}$$

Finally, the results for the antineutrino case are:

$$\begin{aligned} \langle\sigma_{qel}\rangle_{\bar{\nu}_\mu} &= (0.81 \pm 0.05(stat) \pm 0.08(syst)) \times 10^{-38}\ \mathrm{cm}^2 \\ M_A &= 1.06 \pm 0.07(stat) \pm 0.10(syst)\ \mathrm{GeV} \end{aligned}$$

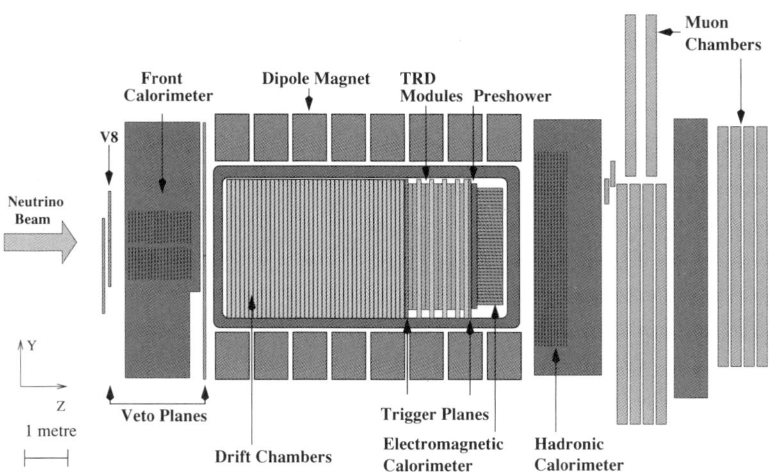

References

[1] V. Lyubushkin et al Eur. Phys. J. C Vol. 63, pp 355-381 (Oct. 2009)
[2] R.A. Smith, E.J. Moniz, Nucl. Phys. B43, p 605 (1972)

PRECISION MEASUREMENTS BY NOMAD EXPERIMENT

Dmitry V.Naumov[a]

Joint Institute for Nuclear Research, 141980, Dubna, Russia

*Abstract.*Recently completed precision measurements and other ongoing analyses by NOMAD experiment are reviewed.

1 NOMAD experiment

The Neutrino Oscillation MAgnetic Detector (NOMAD) experiment at CERN used a neutrino beam with E_ν integrated relative composition of $\nu_\mu, \bar{\nu}_\mu, \nu_e, \bar{\nu}_e$ neutrino species 1.00 : 0.025 : 0.015 : 0.0015. Thus, 95% of ν-events are due to ν_μ-interactions. The NOMAD detector consisted of several subdetectors crucial for an accurate measurement of interaction verteces and particle momenta. The active target of 2.7 tons in mass comprised 132 planes of 3×3 m^2 drift chambers (DC) with an average density similar to that of liquid hydrogen (0.1 g/cm^3). Downstream of the DC, there were nine modules of transition radiation detectors (TRD), followed by a preshower (PRS) and a lead-glass electromagnetic calorimeter (ECAL). The ensemble of DC, TRD, and PRS/ECAL was placed within a dipole magnet providing a 0.4 T magnetic field orthogonal to the neutrino beam line. The charged tracks in the DC were measured with an approximate momentum (p) resolution of $\frac{\sigma_\mathbf{p}}{\mathbf{p}} \approx \frac{\mathbf{0.05}}{\sqrt{\mathbf{L}}} \oplus \frac{\mathbf{0.008 \times p}}{\sqrt{\mathbf{L^5}}}$ (p in GeV/c and L in meters). Full event kinematics was reconstructed as well. The details of the NOMAD detector can be found in [1]. The experiment collected the following numbers of ν-interactions on various targets: over 1.7 millions in Drift Chambers, over 2 millions in Al-coil, and over 20 millions in Fe-scintillator calorimeter. A good accuracy in kinematics reconstruction supplimented by world record statistics of neutrino interactions collected allows us to work on a number of precision measurements. Let me briefly recall some of these below.

2 NOMAD precision measurements

2.1 Inclusive and QEL cross-sections

We measured the muon neutrino-nucleon inclusive charged current cross-section, off an isoscalar target, in the neutrino energy range $2.5 \leq E_\nu \leq 40$ GeV [2]. The signicance of this measurement is its precision, ±4% in $2.5 \leq E_\nu \leq 10$ GeV, and ±2.6% in $10 \leq E_\nu \leq 40$ GeV regions, where signicant uncertainties in previous experiments still exist. In fig. 1 it is shown the measured inclusive $\nu_\mu N$ cross-section as a function of neutrino energy.

[a]e-mail: dnaumov@jinr.ru

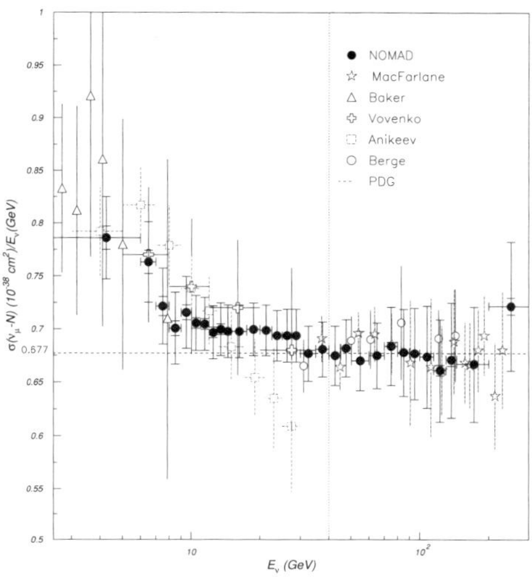

Figure 1: Inclusive $\nu_\mu N$ cross-section as a function of neutrino energy.

The systematics error of the measured inclusive cross-section is dominated by uncertainties in the flux prediction.

Also NOMAD measured with the best precision the quasi-elastic cross-section of $\nu_\mu n \to \mu^- p$ reaction and with a somewhat worse resolution (due to a lack of statistics primarily) the cross-section of $\bar{\nu}_\mu p \to \mu^+ n$ [3] both discussed at this conference in details by J. M. Levy. The results for the flux averaged QEL cross-sections in the (anti)neutrino energy interval 3-100 GeV are $\sigma_{\nu_\mu}^{qel} = (0.92 \pm 0.02(stat) \pm 0.06(syst)) \times 10^{-38} cm^2$ and $\sigma_{\bar{\nu}_\mu}^{qel} = (0.81 \pm 0.05(stat) \pm 0.08(syst)) \times 10^{-38} cm^2$ for neutrino and antineutrino, respectively. The axial mass parameter M_A extracted from the measured quasi-elastic neutrino cross-section is $M_A = 1.05 \pm 0.02(stat) \pm 0.06(syst)$ GeV.

2.2 Coherent π^0 production

Another interesting analysis deals with an exclusive neutral pion production in neutrino-nucleus neutral current interactions. Such analysis is of great importance for long base line (LBL) neutrino experiments with massive detectors because such π^0 if collinear to neutrino beam will make an electromagnetic shower resulting in about 60% of the expected backgrounds for ν_e appearance. We measure with best precision the coherent π^0 production cross-section to be $(72.6 \pm 10.6) \cdot 10^{-40} cm^2$ per nucleon [4].

2.3 Ongoing analyses

There are several ongoing precision analyses in NOMAD. In one of these we study $\pi^\pm$ multiplicities as functions of various kinematics variables. In fig. 2 we show a subsample of NOMAD data comprizing only about $\frac{1}{4}$ of the total statistics in DC compared to the HERMES measurements as a function of relative energy of the hadronic jet carried out by a pion. One can observe a factor 30 smaller value of the left edge of the distribution measured by NOMAD. We plan to extract also fragmentation and fracture functions using these data. Another analysis near to be completed measures the dimuons cross-section of the process $\nu_\mu N \to \mu^+ \mu^- X$ which is dominated by $\nu_\mu s(d) \to \mu^- c, c \to$

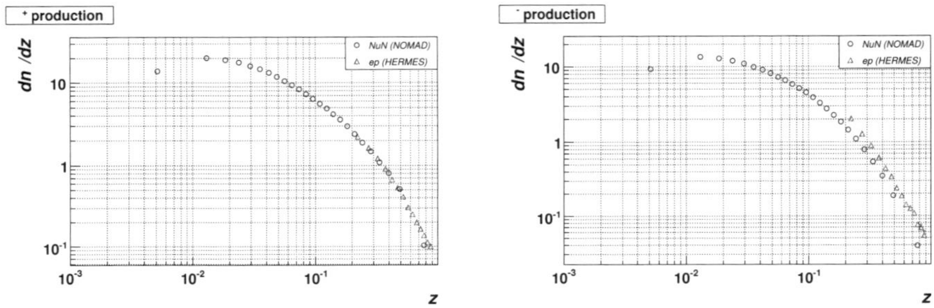

Figure 2: π^+ (right) and π^- (left) multiplicities measured by NOMAD and HERMES as a function of z.

$\mu^+ s(d)$ subprocesses. A precise measurement of this cross-section improves our knowledge of strange content of the nucleon and about the charm quark mass. The accuracy obtained ranges from less than 7% at low E_ν to about 3% at large E_ν yielding the most precise measurements of these quantitities up to date.

At last but not least let me mention another ongoing analysis - a measurement of the ratio of neutral to charged current cross-sections as a function of hadronic energy. This quantity provides a sensitive constrain on partonic density functions, admixture of (quasi)elastic, resonance and deep-inelastic events and it is expected to be helpful for future LBL experiments designs and analyses.

Acknowledgments

I would like to thank L. Camillieri, L. DiLella, S. Mishra, R. Petti and O. Samoylov for useful discussions about the details of the reported analyses.

References

[1] J. Altegoer *et al.*, [NOMAD Collaboration], Nucl. Instr. and Meth. **A404** (1998) 96

[2] Q. Wu *et al.* [NOMAD Collaboration], Phys. Lett. B **660** (2008) 19 [arXiv:0711.1183 [hep-ex]].

[3] V. Lyubushkin *et al.* [NOMAD Collaboration], Eur. Phys. J. C **63** (2009) 355 [arXiv:0812.4543 [hep-ex]].

[4] C. T. Kullenberg *et al.* [NOMAD Collaboration], Phys. Lett. B **682** (2009) 177 [arXiv:0910.0062 [hep-ex]].

SOME REMARKS ON NEUTRINO DETECTION FROM SN1987A

V.L. Dadykin, O.G. Ryazhskaya [a]

Institute for Nuclear Research RAS, 117312 Moscow, Russia

*Abstract.*In the paper the authors comment on the results concerning the detection of neutrino radiation from SN 1987A presented in complete bulk and published from 1987 to 1992 and also suggest their considerations on possible interpretation of the experiment.

1 Introduction

On February,23, 1987 astronomers observed a supernova in a nearby galaxy - the Large Magellanic Cloud. That day at about 2:52 UT LSD detector (Italy-USSR, Mont Blanc) and at about 7:35 UT BUST (USSR, Baksan), KII (Japan-USA, Kamioka) and IMB (USA, Cleveland) detectors measured neutrino radiation from collapsing star for the first time in the history of science. The theory predicted and the experiment confirmed. Only the fact of the neutrino detection from supernova itself is an outstanding achievement. But the details are also important. One group of the results corresponds to theoretical predictions. The other part of results was comprehended only after many years. However, there are also other results which are not success to interprete upto now in the frame of the modern models of stellar evolution. Our paper is about all this.

The experiment on searching for neutrino radiation from collapsing stars was based on the idea about spherically symmetric, nonrotating, nonmagnetic star [1]. In this case 1) the advantage for neutrino detection has the reaction $\overline{\nu_e} p \rightarrow e^+ n$ due to its large cross section and 2) the duration of neutrino burst is very short: from 5 to 20 s.

Basically, the experiment is simple. The detector, a scintillation or Cherenkov one, is filled with a hydrogenous material. A neutrino burst is identified by a series of scintillations in the range of amplitudes 5 - 50 MeV in a time less than 20 s. The number of pulses in the series is proportional to the hydrogen mass in the detector. Fluctuations in the frequency of background pulses can imitate the effect. To reduce the background, the detectors are placed deep underground and anticoincidence systems are used. For success of a searching experiment the detectors should have a target mass of hundreds or thousands of tons. A concurrent operation of several detectors placed at various points of the globe is very important for the reliability of the results.

[a] e-mail: ryazhskaya@lvd.ras.ru

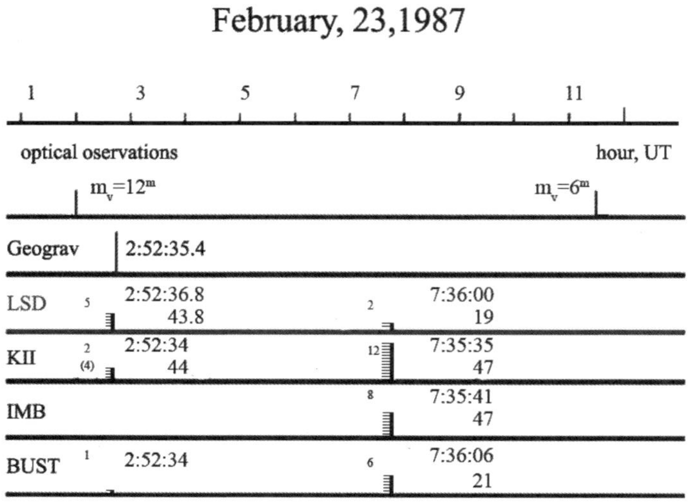

Figure 1: Time sequence of events observed with various detectors on February 23, 1987, on the scale of Universal Time (UT). For each neutrino detector, the number of pulses in the packet is shown along the vertical axis; the arrival times of the first and last pulses are given nearby. Geograv is the gravitational wave antenna in Rome.

2 On Experimental Results

Two scintillation detectors, LSD (90 t of C_nH_{2n+2} + 200 t of Fe) [2] and BUST (200 t of C_nH_{2n+2} + 160 t of Fe) [3], and two Cherenkov detectors, KII (2000 t of H_2O) [4] and IMB (5000 t of H_2O) [5], were operational on February 23, 1987. The timing diagram of effects in the detectors is shown in the figure 1. We see that the effects in the detectors are small and have poor statistics. This leads to quite understandable scepticism with regard to the significance of the effects themselves and the schemes for their interpretation. However, this experiment was unique up to now and therefore all the information obtained at that time requires very careful treatment. The effects in all detectors at about 7:35 UT were undoubtedly related to the Supernova explosion. They coincided between themselves in time to within a minute and with the optical observations of the Supernova within several hours. This assertion concerns mainly the KII and IMB detectors, where the effects are noticeable and the probability of imitations by the background is negligible. Given the difference in hydrogen masses, the effects in LSD and BUST at this time correspond, within the error limits, to those in KII and IMB.

The fact could testify on the success of the model of standard collapse under assumption that the pulses in all detectors at about 7:35 UT were caused by $\overline{\nu_e}p$-

reaction. If an angular distribution of particles in Cherenkov counters would be isotropic, it could prove that just this reaction took place. However, 15 of 20 events (the sum of KII and IMB pulses) fall into the forward hemisphere, and 3 of 5 rest ones are near boundary of forward and backward hemisphere. Therefore the reliable identification of neutrino type was not successful. The main intrigue of the experiment is connected with LSD. At about 2:52 UT this detector observed a burst of 5 pulses during 7 s. Clearly these were not caused by $\overline{\nu_e}p$-reaction. If it would be so, KII should measure about 50 pulses at the same time. The fact that at 2:52 UT any of other three detectors had not significant effect cannot compromise LSD one. This means only that LSD observed something that other detectors could not measure. 200 t of iron in the LSD structure is the reason of this. Thick scintillation counters like LSD provided a sufficiently high detection efficiency of the products of electron or muon neutrino interaction with iron. The details are discussed in [6, 7]. They show that there is no reason to believe the effect at 2:52 UT to be a freak of statistics or noise. The effect is of high quality and is undoubtedly related to the SN1987A explosion and should be taken into account on a par with the effects at 7:35 UT when the experiment is analysed. Our interpretation of the signal in LSD is entirely consistent with a popular scheme of neutrino oscillations.

It is clear that the model of standard collapse can not explain the value of the LSD effect at 2:52 UT and the existence of two neutrino bursts separated by a time interval of about 5 h. Only later on when V.S. Imshennik suggested the theory of a rotating mechanism of supernova explosion for massive main-sequence stars [8], it was done reasonable explanation of the neutrino observation from SN 1987A taking into account both the effects at 2:52 and 7:35 UT [9]. In this so called model of rotating collapsar the rotational energy of the presupernova core at the neutronization stage leads to a strong deformation and, subsequently, fragmentation of the core with the formation of a close neutron star binary. As a result of this system evolution the gravitational collapse has two stages. The short intense pulse of electron neutrinos with energies 30 - 55 MeV is produced during the first stage (binary formation). It was detected at 2:52 UT with LSD thanks to the reaction ($\nu_e^{56}Fe$). The second neutrino burst is generated after about 5 hrs and contains all types of neutrino with average energies of about 15 - 20 MeV probably according to the standard collapse model. It was detected at about 7:35 UT.

Apart from the repeatedly discussed results concerning the signals in the four neutrino detectors at 2:52 and 7:35 UT on February, 23, 1987, there are also other, less-known, results obtained at the same time that day. These include the papers in which the coincidencies between the pulses in the different neutrino detectors (LSD, BUST, KII) in a very narrow time window of about 1 s in width [10–13] as well as the coincidences between the signals in these detectors and signals from the Maryland and Rome gravitational antennas were studied [14]. Finally, note the papers of the LSD collaboration about studies on the double pulses in LSD that day [15]. The consideration of the significance and the place of the papers in the experiment on searching for neutrino radiation from SN 1987A are discussed in our paper [16].

In the normal mode of operation the neutrino detectors record noise pulses with a rate about $10^{-2}s^{-1}$. The expected rate of chance coincidences between pulses of two detectors is about 1 per hour. This value for the pairs of detectors LSD-KII and LSD-BUST was measured for the long time interval of about one month around February, 23, 1987, and was exactly the same except of the time interval when the supernova explosion was observed. The result is shown in figure 2. It also shows the rare series of double pulses detected with LSD. It consists of nine pairs of pulses arrived in the interval from 5:42:48 to 10:13:04 UT. We selected the cases where the time between two pulses was no more than 2 s; on average it was 1.5 s. It was required that the amplitude of one of the pulses, the first or the second, was larger than 25 MeV. The amplitude of the other pulse should exceed 5 MeV. The mean rate of such background double pulses calculated from February 10, to March 7, 1987 is 0.275 h^{-1}. Hence, such event (nine pairs) could coincide by chance with the day of supernova explosion once in 3×10^3 years.

The figure 2 shows that the excess of coincidences concentrates near 2:52 UT, while the excess of double pulses concentrates near 7:35 UT. Such effect could be simulated by chance extremely rarely [10–13]. This is convincing evidence for the genetic relationship of the effects considered to the supernova explosion. Our figure shows only the results of three works [10–12]. The results of all the remaining works mentioned above would look similar to the figure 2.

3 On The Interpretation of Some Results

In our opinion, the excess of coincidences near 2:52 UT may be indicative of a very-low-intensity pulsed neutrino source operating in the regime of a pulsar at the presupernova stage. This pulsar has an emission time of about 1 s, a pulse repetition period much longer than 1 s, and a lifetime of about 2 h. It has such an intensity that each pulsar triggering gives, on average, much fewer than one neutrino interaction in a LSD-type detector. As regards the understanding of the double pulses note that a mechanism of this phenomenon in the spirit of the ideas of the rotating collapsar model was proposed in [6, 17]. These pulses can be the decay products of the pions and muons generated in the collisions of iron nuclei orbiting a heavy neutron star with the surrounding iron gas. This is the star that is involved at the final stage of collapse and is responsible for the neutrino signal at 7:35 UT.

The study of the coincidences between the signals from gravitational an-

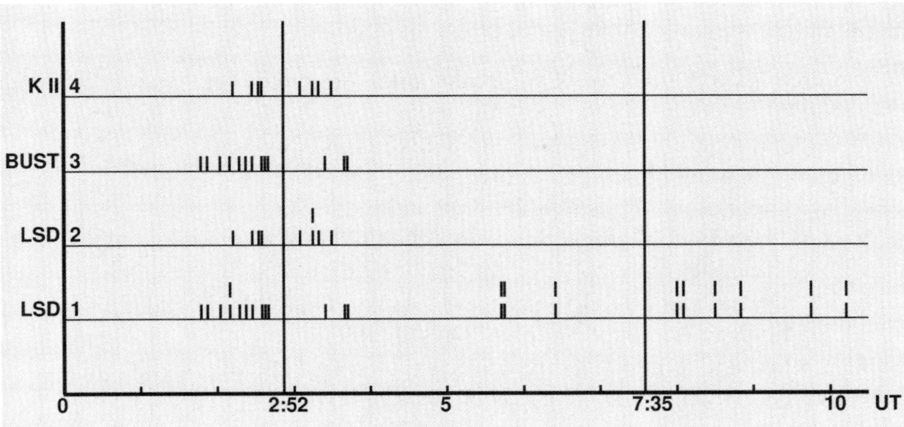

Figure 2: Timing diagram of the BUST pulses coincident with the LSD pulses within 1 s and similar coincidences for the K2 and LSD detectors as well as double pulses in LSD over the period from 0:00 to 10:00 UT on February 23, 1987. 1,2,3,4 - duplicate scales for LSD, BUST and KII. The rate of coincidences over 2 hour time interval was 8 (scales 2,4) for LSD-KII and 13 (scales 1,3) for LSD-BUST, of which 2 and 3 coincidences, respectively, could be chance ones. No background pulses for the coincidences outside the interval 1:45 - 3:45 UT and no background double pulses outside the interval 5:42 - 10:13 UT are shown in the figure. These data are absent in the original publications.

tennas and the pulses in neutrino detectors shows an excess of coincidences near 2:52 UT. Of course, the sensitivities of the antennas are too low to detect gravitational effects at a distance of 50 kpc. The situation could be improved significantly by assuming that the gravitational radiataion from our star is anisotropic. In any case, the logic of the experiment, a periodically repeated coincidence of the signal in the gravitational wave detector with the signal in a neutrino detector within narrow time windows, points to the possible existence of gravitational pulsar operating synchronously with a neutrino one.

Acknowledgments

We wish to thank V. V. Boyarkin for help in preparing the manuscript for publication. This work was supported by the Russian Foundation for Basic Research (project nos. 09-02-300-a and 08-02- 91005CERN_a) and the Scientific School of Academician G.T. Zatsepin (grant no. SSchool-959.2008.2 from the President of Russia).

References

[1] D.K. Nadyozhin and V.S. Imshennik, *Intern.J.Mod.Phys.A* **20** 6597 (2005).
[2] M.Aglieta et al., *Europhys.Lett.* **3** 1315 (1987).
[3] E.N. Alekseev et al., *JETP.Lett.* **45** 589 (1987).
[4] K. Hirata et al., *Phys.Rev.Lett.* **58** 1490 (1987).
[5] R.M. Bionta et al.,*Phys.Rev.Lett.* **58** 1497 (1987).
[6] O.G.Ryazhskaya, *Phys.Usp.* **49** 1017 (2006).
[7] V.L.Dadykin and O.G.Ryazhskaya, *Astron.Lett.* **34** 581 (2008).
[8] V.S.Imshennik, *Space Sci.Rev.* **74** 325 (1995).
[9] V.S.Imshennik and O.G. Ryazhskaya, *Astron.Lett.* **30** 14 (2004).
[10] M.Aglietta et al.,*Ann.New York Acad.Sci.* **571** 584 (1989)
[11] M.Aglietta et al., *Nuovo Cim.* **C12** 75 (1989)
[12] A.E. Chudakov, *Ann. New York Acad. Sci.* **571** 577 (1989)
[13] M.Aglietta et al., *Nuovo Cim.* **C14** 171 (1991)
[14] E. Amaldi et al.,*Ann. New York Acad. Sci.* **571** 561 (1989)
[15] V.L.Dadykin et al., *JETP Lett.* **56** 426 (1992)
[16] V.L.Dadykin and O.G. Ryazhskaya, *Astron.Lett.* **35** 384 (2009).
[17] V.S.Imshennik and K.V.Manukovskii,*Astron.Lett.* **33** 468 (2007).

FLAVOR OSCILLATIONS OF LOW ENERGY NEUTRINOS IN THE ROTATING NEUTRON STAR

Maxim Dvornikov [a]
Departamento de Física y Centro de Estudios Subatómicos, Universidad Técnica Federico Santa María, Casilla 110-V, Valparaíso, Chile and IZMIRAN, 142190, Troitsk, Moscow region, Russia

*Abstract.*We study flavor oscillations of low energy neutrinos propagating in dense matter of a rotating neutron star. On the basis of the exact solutions of the wave equations for neutrinos mass eigenstates we derive the transition probability for neutrinos having big initial angular momentum. It is found that flavor oscillations of neutrinos with energies of several electron-Volts can be resonancely enhanced.

It is known that neutrinos play a significant role at the last stages of the evolution of massive stars. For example, almost 99% of the gravitational energy of a protoneutron star is carried away during the supernova explosion. The remaining dense, compact object, a neutron star, can have extreme properties: central density $\sim 10^{14}\,\mathrm{g/cc}$, magnetic field $\sim 10^{15}\,\mathrm{G}$ and angular velocity $\sim 10^{3}\,\mathrm{s}^{-1}$. In this short note we examine the influence of the neutron star rotation on flavor oscillations of neutrinos. Note that the propagation of neutrinos in rotating matter was also studied in Refs. [1, 2].

Let us study the evolution of the two flavor neutrinos system (ν_α, ν_β) interacting with the background matter by means of the electroweak forces. The Lagrangian for this system has the form,

$$\mathcal{L} = \sum_{\lambda=\alpha,\beta} \bar{\nu}_\lambda (\mathrm{i}\gamma^\mu \partial_\mu - f_\lambda^\mu \gamma_\mu^{\mathrm{L}}) \nu_\lambda - \sum_{\lambda\lambda'=\alpha,\beta} m_{\lambda\lambda'} \bar{\nu}_\lambda \nu_{\lambda'}, \qquad (1)$$

where $\gamma_\mu^{\mathrm{L}} = \gamma_\mu(1-\gamma^5)/2$. Supposing that matter is electroneutral and all the background fermions rotate as a rigid body we can express the external fields f_λ^μ for $\alpha = \mu$ or τ and $\beta = e$ oscillations channel as,

$$f_\alpha^\mu = -\frac{G_{\mathrm{F}}}{\sqrt{2}} j_n^\mu, \quad f_\beta^\mu = \frac{G_{\mathrm{F}}}{\sqrt{2}}(2j_e^\mu - j_n^\mu), \quad j_{e,n}^\mu = (n_{e,n}, n_{e,n}\mathbf{v}), \qquad (2)$$

where G_{F} is the Fermi constant, $n_{e,n}$ is the number density of electrons and neutrons and $\mathbf{v} = (\boldsymbol{\Omega} \times \mathbf{r})$ is the velocity of the background matter.

To study the evolution of the system (1) we should introduce the neutrino mass eigenstates ψ_a to diagonalize the mass matrix $(m_{\lambda\lambda'})$, $\nu_\lambda = (\exp[-\mathrm{i}\sigma_2\theta])_{\lambda a} \times \psi_a$, where θ is the vacuum mixing angle. We suggest that the mass eigenstates are Dirac particles. In the basis of the mass eigenstates ψ_a neutrinos have definite masses m_a.

In the limit of small neutrino masses the wave equations for the upper ξ_a and lower η_a chiral components of the spinor $\psi_a^{\mathrm{T}} = (\xi_a, \eta_a)$ decouple. Therefore

[a] e-mail: maxim.dvornikov@usm.cl

using cylindrical coordinates (r, ϕ, z) with $\mathbf{\Omega} = \Omega \mathbf{e}_z$ we can write the general expression for the two component wave function η_a in the form (see Ref. [2]),

$$\eta_a(r,\phi,t) = \sum_{\mathrm{n,s}=0}^{\infty} \Big(a_{\mathrm{n}s}^{(a)}(t)\ u_{a,\mathrm{n}s}^{+}(r,\phi) \exp[-\mathrm{i}E_{\mathrm{n}}^{(a)+}t] \\ + b_{\mathrm{n}s}^{(a)}(t)\ u_{a,\mathrm{n}s}^{-}(r,\phi) \exp[-\mathrm{i}E_{\mathrm{n}}^{(a)-}t] \Big), \tag{3}$$

where the energy levels

$$E_{\mathrm{n}}^{(a)\pm} = -V_a \pm \sqrt{4V_a\Omega\mathrm{n} + m_a^2}, \quad \mathrm{n} = 0, 1, 2, \ldots \tag{4}$$

have the discrete values and the basis spinors

$$u_{a,\mathrm{ns}}^{(\pm)}(r,\phi) = \sqrt{\frac{V_a\Omega}{2\pi}} \begin{pmatrix} I_{\mathrm{n}-1,s}(\rho_a)e^{\mathrm{i}(l-1)\phi} \\ \mp\mathrm{i}I_{\mathrm{n},s}(\rho_a)e^{\mathrm{i}l\phi} \end{pmatrix}, \quad l = \mathrm{n} - s, \tag{5}$$

are expressed in terms of the Laguerre functions $I_{\mathrm{n},s}(\rho_a)$ of the dimensionless argument $\rho_a = V_a\Omega r^2$, $V_1 = G_{\mathrm{F}}(n_n - 2n_e \sin^2\theta)/\sqrt{2}$ and $V_2 = G_{\mathrm{F}}(n_n - 2n_e\cos^2\theta)/\sqrt{2}$ are the potentials of the interaction of mass eigenstates with background matter. Note that in Eqs. (3)-(5) we study neutrinos propagating in the equatorial plane with $z = 0$.

In Ref. [2] we obtained the general differential equations for the coefficients $a_{\mathrm{n}s}^{(a)}(t)$ and discussed the situation of the small initial angular momentum: $l \ll s$. Now we study neutrino flavor oscillations for $l \gg s$, i.e. particles with big initial angular momentum. Using the results of our work [2] we get that in this situation the differential equations for different l and s decouple and we can describe the evolution of the system with help of the single Schrödinger equation,

$$\mathrm{i}\frac{\mathrm{d}}{\mathrm{d}t}\begin{pmatrix} \tilde{a}_l^{(1)} \\ \tilde{a}_l^{(2)} \end{pmatrix} = \begin{pmatrix} \omega/2 & \Delta \\ \Delta & -\omega/2 \end{pmatrix}\begin{pmatrix} \tilde{a}_l^{(1)} \\ \tilde{a}_l^{(2)} \end{pmatrix}, \tag{6}$$

where the components $\tilde{a}_l^{(a)}$ of the "wave function", which now can be enumerated with the single quantum number "l" are related to the coefficients in Eq. (3) by the formula, $\tilde{a}_l^{(a)} = (\exp[-\mathrm{i}\sigma_3\omega/2])_b^a a_l^{(b)}$. The parameters of the effective Hamiltonian in Eq. (6) have the form,

$$\Delta = \left(\frac{G_{\mathrm{F}}}{\sqrt{2}} - \frac{k}{2n_n}\right) n_e \sin 2\theta, \quad \omega = \frac{\delta m^2}{2k}, \quad k = \sqrt{4V\Omega l}, \tag{7}$$

where k is the effective momentum of neutrinos and $V = G_{\mathrm{F}}n_n/\sqrt{2}$.

Using Eqs. (6) and (7) we can obtain the transition probability in the form,

$$P_{\beta\to\alpha}(x) = \frac{(\Delta\cos 2\theta + \omega \sin 2\theta/2)^2}{\Delta^2 + (\omega/2)^2} \sin^2\left(\sqrt{\Delta^2 + (\omega/2)^2}x\right). \tag{8}$$

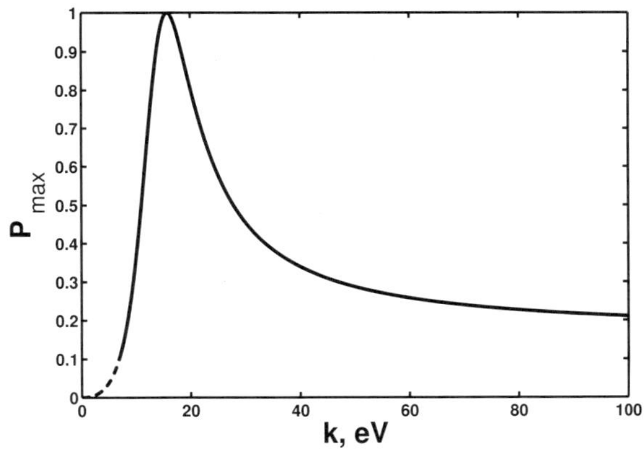

Figure 1: The dependence of the maximal transition probability on the neutrino energy for the case $l \gg s$. This plot corresponds to $\nu_e \to \nu_\mu$ oscillations, with $\delta m^2 \approx 8.1 \times 10^{-5}\,\mathrm{eV}^2$ and $\sin^2\theta \approx 0.3$ and matter with $n_n = 10^{38}\,\mathrm{cm}^{-3}$ and $Y_e = n_e/n_n = 3 \times 10^{-3}$. The very low energy part of the curve, which cannot be treated in frames of the quantum mechanical approach, is shown by the dashed line.

Let us discuss the oscillation scheme $\nu_e \to \nu_\mu$. In Fig. 1 we present the maximal transition probability as a function of the neutrino energy, built on the basis of Eq. (8). Note that for the nuclear matter in β equilibrium the number density of electrons has the following value: $n_e \approx 3\pi^2 n_n^2/(2m_n)^3$. As one can see from this picture, the transition probability has a resonance behaviour. The maximal transition probability is reached at $\sim 16\,\mathrm{eV}$. At very large energies the transition probability approaches to the limit $P_{\max} \to \cos^2(2\theta) \approx 0.16$, a result that can also be inferred from Eq. (8). We should notice that the solution presented in Eq. (8) is not valid for very large neutrino energies, because in that case the condition $l \gg s$ is violated.

Acknowledgments

The work has been supported by Conicyt (Chile), Programa Bicentenario PSD-91-2006. The author is thankful to C. O. Dib and for helpful discussions and the organizers of 14th Lomonosov conference for the invitation.

References

[1] A. V. Grigoriev, *et al.*, *Russ. Phys. J.* 50, 845 (2007);
A. I. Studenikin, *J. Phys.* **A** 41, 164047 (2008).

[2] M. Dvornikov and C. O. Dib, 0907.1445 [astro-ph]; M. Dvornikov, to be published in *Appl. Comput. Math.*, 1001.2516 [hep-ph].

DIRAC NEUTRINO MAGNETIC MOMENT AND THE DYNAMICS OF A SUPERNOVA EXPLOSION

Alexander V. Kuznetsov [a], Nickolay V. Mikheev [b], Alexander A. Okrugin [c]
Division of Theoretical Physics, Department of Physics, Yaroslavl State University, Sovietskaya 14, 150000 Yaroslavl, Russian Federation

Abstract. The double conversion of the neutrino helicity $\nu_L \to \nu_R \to \nu_L$ has been analyzed for supernova conditions, where the first stage is due to the interaction of the neutrino magnetic moment with plasma electrons and protons in the supernova core, and the second stage, due to the resonance spin flip of the neutrino in the magnetic field of the supernova envelope. It is shown that, in the presence of the neutrino magnetic moment in the range $10^{-13}\,\mu_B < \mu_\nu < 10^{-12}\,\mu_B$ and a magnetic field of $\sim 10^{13}$ G between the neutrinosphere and the shock-stagnation region, an additional energy of about 10^{51} erg, which is sufficient for a supernova explosion, can be injected into this region during a typical shock-stagnation time.

Two main obstacles are encountered in numerical simulations of a supernova explosion [1–5]. First, a mechanism stimulating the damping shock, which is likely necessary for the explosion, has not yet been well developed. Recall that shock damping is mainly due to energy losses on the dissociation of nuclei. Second, the energy release of the "theoretical" supernova explosion is much lower than the observed kinetic energy $\sim 10^{51}$ erg of an envelope. This is called the FOE (ten to the Fifty One Ergs) problem. It is believed that a self-consistent description of the explosion dynamics requires an energy of $\sim 10^{51}$ erg to be transferred via some mechanism from the neutrino flux emitted from the supernova central region to the envelope.

Dar [6] proposed a possible way for solving the above-mentioned problems. His mechanism is based on the assumption that the neutrino has a magnetic moment which is not very small. Left-handed electron neutrinos ν_e intensely generated in the collapsing supernova core are partially converted into right-handed neutrinos due to the interaction of the neutrino magnetic moment with plasma electrons and protons. In turn, the right-handed neutrinos sterile with respect to weak interactions freely leave the central part of the supernova if the neutrino magnetic moment is not too large, $\mu_\nu < 10^{-11}\,\mu_B$, where μ_B is the Bohr magneton. Some of these neutrinos can be inversely converted to left-handed neutrinos due to the interaction of the neutrino magnetic moment with the magnetic field in the supernova envelope. According to contemporary views, the magnetic field strength in this region can be of the order of the critical magnetic field $B_e = m_e^2/e \simeq 4.41 \times 10^{13}$ G [d] or even higher [7–9]. The born again left-handed neutrinos can transfer additional energy to the envelope

[a] e-mail: avkuzn@uniyar.ac.ru
[b] e-mail: mikheev@uniyar.ac.ru
[c] e-mail: okrugin@uniyar.ac.ru

[d] Hereafter, we use the natural system of units in which $c = \hbar = 1$, and $e > 0$ is the elementary charge.

by virtue of the beta-type absorption $\nu_e n \to e^- p$.

In our opinion, the mechanism of the double conversion of neutrino helicity should be analyzed more carefully. It was shown in our recent paper [10] that the flux and luminosity of right-handed neutrinos from the central region of the supernova were significantly underestimated in previous works. Here, we reconsider the process $\nu_L \to \nu_R \to \nu_L$ under supernova conditions and analyze the possibilities for stimulating the damping shock.

A consistent analysis of the processes

$$\nu_L \to \nu_R + \gamma^*, \quad \nu_L + \gamma^* \to \nu_R \tag{1}$$

with neutrino-helicity conversion due to the interaction with both plasma electrons and protons via a virtual plasmon and with the inclusion of polarization effects of the plasma on the photon propagator was given in [10]. In particular, according to the numerical analysis, the contribution of the proton component of the plasma is not only significant, but even dominant. As a result, using the data on supernova $SN1987A$, a new astrophysical limit was imposed on the electron-neutrino magnetic moment:

$$\mu_\nu < (0.7 - 1.5) \times 10^{-12}\, \mu_{\rm B}\,, \tag{2}$$

This is a factor of two better than previous constraints.

The effect of the magnetic field on neutrinos with nonzero magnetic moments can be illustrated most conveniently using the equation for neutrino-helicity evolution in the uniform external magnetic field [11–17]. The additional energy gained by left-handed electron neutrinos in the matter

$$C_L = \frac{3\,G_{\rm F}}{\sqrt{2}}\,\frac{\rho}{m_N}\left(Y_e + \frac{4}{3}\,Y_{\nu_e} - \frac{1}{3}\right) \tag{3}$$

deserves a special analysis. Here, the ratio $\rho/m_N = n_B$ is the nucleon number density and $Y_e = n_e/n_B = n_p/n_B$, $Y_{\nu_e} = n_{\nu_e}/n_B$, n_{e,p,ν_e} are the number densities of electrons, protons, and neutrinos, respectively.

It is important that this quantity can vanish in the considered region of the supernova envelope. In turn, this is a criterion for the resonance transition $\nu_R \to \nu_L$. Since the neutrino number density in the supernova envelope is fairly low, this yields the resonance condition in the form $Y_e = 1/3$. It is remarkable that only one such point with $\mathrm{dY_e/dr} > 0$ exists (see [2, 4]). Note that the so-called adiabatic condition should also be satisfied.

Our analysis [18] shows that the Dar mechanism of the double conversion of neutrino helicity $\nu_L \to \nu_R \to \nu_L$ exists under the following not very severe conditions: the Dirac-neutrino magnetic moment should be in the range $10^{-13}\,\mu_{\rm B} < \mu_\nu < 10^{-12}\,\mu_{\rm B}$ and a magnetic field of $\sim 10^{13}$ G should exist in the region between the neutrinosphere and the shock-stagnation region. In this case, an additional energy of about

$$\Delta E \simeq L_{\nu_R}\, \Delta t \sim 10^{51}\, \text{erg}\,, \tag{4}$$

is injected into this region during the shock-stagnation time $\Delta t \sim 0.2$–0.4 s. This energy is sufficient to solve the problem.

Acknowledgments

We are grateful to M.I. Vysotsky for helpful discussion. This work was performed in the framework of realization of the Federal Target Program "Scientific and Pedagogic Personnel of the Innovation Russia" for 2009 - 2013 (project no. NK-410P-69) and was supported in part by the Ministry of Education and Science of the Russian Federation under the Program "Development of the Scientific Potential of the Higher Education" (project no. 2.1.1/510).

References

[1] V. S. Imshennik, D. K. Nadezhin, *Usp. Fiz. Nauk* 156, 561 (1988).
[2] H. A. Bethe, *Rev. Mod. Phys.* 62, 801 (1990).
[3] G. G. Raffelt, *"Stars as Laboratories for Fundamental Physics"* (Univ. of Chicago Press, Chicago), 1996.
[4] R. Buras et al., *Astron. Astrophys.* 447, 1049 (2006).
[5] H.-Th. Janka et al., *Phys. Rep.* 442, 38 (2007).
[6] A. Dar, Inst. Advanced Study Preprint-87-0178, Princeton, 1987.
[7] G. S. Bisnovatyi-Kogan, *Astron. Zh.* 47, 813 (1970).
[8] G. S. Bisnovatyi-Kogan, *"Stellar Physics 2: Stellar Evolution and Stability"* (Nauka, Moscow), 1989; (Springer, New York), 2001.
[9] N. V. Ardeljan, G. S. Bisnovatyi-Kogan, S. G. Moiseenko, *Mon. Not. Roy. Astron. Soc.* 359, 333 (2005).
[10] A. V. Kuznetsov, N. V. Mikheev, *J. Cosmol. Astropart. Phys.* 11, 031 (2007).
[11] M. B. Voloshin, M. I. Vysotsky, L. B. Okun, *Yad. Fiz.* 44, 677 (1986).
[12] M. B. Voloshin, M. I. Vysotsky, *Yad. Fiz.* 44, 845 (1986).
[13] L. B. Okun, *Yad. Fiz.* 44, 847 (1986).
[14] M. B. Voloshin, M. I. Vysotsky, L. B. Okun, *Zh. Eksp. Teor. Fiz.* 91, 754 (1986); *Zh. Eksp. Teor. Fiz.* 92, 368 (1987).
[15] L. B. Okun, *Yad. Fiz.* 48, 1519 (1988).
[16] M. B. Voloshin, *Phys. Lett.* **B** 209, 360 (1988).
[17] S. I. Blinnikov, L. B. Okun, *Pisma Astron. Zh.* 14, 867 (1988).
[18] A. V. Kuznetsov, N. V. Mikheev, A. A. Okrugin, *Pis'ma v Zh. Eksp. Teor. Fiz.* 89, 97 (2009).

REEXAMINATION OF A BOUND ON THE DIRAC NEUTRINO MAGNETIC MOMENT FROM THE SUPERNOVA NEUTRINO LUMINOSITY

Alexander V. Kuznetsov [a], Nickolay V. Mikheev [b], Alexander A. Okrugin [c]
Division of Theoretical Physics, Department of Physics, Yaroslavl State University, Sovietskaya 14, 150000 Yaroslavl, Russian Federation

Abstract. The neutrino helicity-flip process under the conditions of the supernova core is reinvestigated. Instead of the uniform ball model for the SN core used in previous analyses, realistic models for radial distributions and time evolution of physical parameters in the SN core are considered. A new upper bound on the Dirac neutrino magnetic moment is obtained from the limit on the supernova core luminosity for ν_R emission.

The neutrino helicity flip $\nu_L \to \nu_R$ under physical conditions corresponding to the central region of a supernova has been studied in a number of works. All previous analyses [1–4] were based on a very simplified model of the supernova core as the uniform ball with some averaged values of physical parameters. Moreover, the parameter values look, in modern views, rather too high than typical.

We reinvestigate the right-handed neutrino luminosity caused by the neutrino helicity-flip process under the conditions of the supernova core and make the estimation of the upper bound on the Dirac neutrino magnetic moment from the limit on the supernova core luminosity for ν_R emission by a more reliable way, with taking account of radial distributions and time evolution of physical parameters from some realistic models of the supernova core.

The very recent model was developed by H.-Th. Janka with collaborators who presented us the results of their simulations [5] of the O-Ne-Mg core collapse supernovae which were a continuation of model simulations of Refs. [6, 7]. The successful explosion results for this case have recently been independently confirmed by the Arizona/Princeton SN modelling group, [8, 9] which found very similar results. So we were provided with a model whose explosion behavior was comparatively well understood and generally accepted.

The upper bounds are obtained on the combination of the effective magnetic moments of the electron, muon and tau neutrinos from the condition of not-spoiling the subsequent Kelvin—Helmholtz stage of the supernova explosion by emission of right-handed neutrinos during a few seconds after the collapse. In a general case the combined limit on the effective magnetic moments of the electron, muon and tau neutrinos is

$$\left[\mu^2_{\nu_e} + 0.71 \left(\mu^2_{\nu_\mu} + \mu^2_{\nu_\tau}\right)\right]^{1/2} < 3.7 \times 10^{-12} \, \mu_{\rm B} \,. \quad (1)$$

[a] e-mail: avkuzn@uniyar.ac.ru
[b] e-mail: mikheev@uniyar.ac.ru
[c] e-mail: okrugin@uniyar.ac.ru

This limit is less stringent than the bound obtained in the frame of the uniform ball model for the SN core, but it is surely more reliable. The similar procedure of evaluation was performed with using of the results of Ref. [10–12].

The upper bound on the flavor- and time-averaged neutrino magnetic moment at the Kelvin-Helmholtz phase of the supernova explosion occurs to be

$$\bar{\mu}_\nu < (1.1 - 2.7) \times 10^{-12} \, \mu_{\rm B} \,, \tag{2}$$

depending on the explosion model.

Acknowledgments

We are grateful to Hans-Thomas Janka and Bernhard Müller for providing us with detailed data on radial distributions and time evolution of physical parameters in the supernova core, obtained in their model of the SN explosion. We thank Oleg Lychkovskiy for useful discussion.

This work was performed in the framework of realization of the Federal Target Program "Scientific and Pedagogic Personnel of the Innovation Russia" for 2009 - 2013 (project no. NK-410P-69) and was supported in part by the Ministry of Education and Science of the Russian Federation under the Program "Development of the Scientific Potential of the Higher Education" (project no. 2.1.1/510).

References

[1] R. Barbieri, R. N. Mohapatra, *Phys. Rev. Lett.* 61, 27 (1988).

[2] A. Ayala, J. C. D'Olivo, M. Torres, *Phys. Rev.* **D** 59, 111901 (1999).

[3] A. Ayala, J. C. D'Olivo, M. Torres, *Nucl. Phys.* **B** 564, 204 (2000).

[4] A. V. Kuznetsov, N. V. Mikheev, *Journ. Cosmol. Astropart. Phys.* 11, 031 (2007).

[5] B. Müller, H.-Th. Janka, unpublished results (private communication).

[6] F. S. Kitaura, H.-Th. Janka, W. Hillebrandt, *Astron. Astrophys.* 450, 345 (2006).

[7] H.-Th. Janka, B. Müller, F. S. Kitaura, R. Buras, *Astron. Astrophys.* 485, 199 (2008).

[8] L. Dessart, A. Burrows, C. D. Ott, E. Livne, S.-C. Yoon, N. Langer, *Astrophys. J.* 644, 1063 (2006).

[9] A. Burrows, L. Dessart, C. D. Ott, E. Livne, *Phys. Rep.* 442, 23 (2007).

[10] R. Buras, M. Rampp, H.-Th. Janka, K. Kifonidis, *Astron. Astrophys.* 447, 1049 (2006).

[11] J. A. Pons, S. Reddy, M. Prakash, J. M. Lattimer, J. A. Miralles, *Astrophys. J.* 513, 780 (1999).

[12] W. Keil, H.-Th. Janka, *Astron. Astrophys.* 296, 145 (1995).

Astroparticle Physics and Cosmology

COSMIC ANTIMATTER: MODELS AND PHENOMENOLOGY

A.D. Dolgov [a],
ITEP, 117218, Moscow, Russia
INFN, Ferrara 40100, Italy
University of Ferrara, Ferrara 40100, Italy

Abstract. The possibility of creation of cosmologically significant antimatter are analyzed in different scenarios of baryogenesis. It is argued that there may exist plenty of antimatter even in our Galaxy. Possible forms of antimatter objects and their observational signatures are discussed.

Prediction of antimatter made by P.A.M. Dirac [1] and the discovery of anti-electron, i.e. positron [2] was surely the most impressive result of quantum field theory of the XX century. The whole new world of elementary particles was predicted and Dirac naturally suggested that " If we accept the view of complete symmetry between positive and negative electric charge so far as concerns the fundamental laws of Nature, we must regard it rather as an accident that the Earth (and presumably the whole solar system), contains a preponderance of negative electrons and positive protons. It is quite possible that for some of the stars it is the other way about, these stars being built up mainly of positrons and negative protons. In fact, there may be half of stars of each kind."

It is striking that 30 years before Dirac and one year after discovery of electron (J.J. Thomson, 1897) A. Schuster [3] (another British physicist) conjectured that there might be other sign electricity, which he called *antimatter*, and supposed that there might be entire solar systems, made of antimatter and indistinguishable from ours. He even made an ingenious (or wild) guess that matter and antimatter are capable to annihilate and produce vast energy.

Today it is established beyond any doubt that for each particle there exists an antiparticle (except for a few absolutely neutral ones, as e.g. photons) but the expectations that half of the universe may consist of antimatter are more dead than alive. To a large extent this attitude is stimulated by the Sakharov theory of baryogenesis [4] which successfully explained the observed cosmological predominance of matter over antimatter.

Though observations strongly restrict possible existence of antimatter domains and objects, as is discussed below, it is still not excluded that antimatter may be abundant in the universe and even in the Galaxy, not too far from us. This remaining freedom will be either eliminated by the existing (BESS [5], AMS [6], PAMELA [7]) and planned (AMS-2 [8], PEBS [9], and GAPS [10]) missions for search of cosmic antimatter or anti-stars or total anti-solar systems, as was envisaged by Schuster and Dirac, will be discovered.

Possible existence of anti-worlds depends upon the mechanism of breaking of symmetry between particles and antiparticles, i.e. on physics of C and CP

[a] e-mail: dolgov@fe.infn.it

violation. From the cosmological point of view there are three types of such violation:
1. Explicit, by complex parameters in Lagrangian, as is usually assumed in particle physics.
2. Spontaneous, by a vacuum expectation value (v.e.v.) of a real pseudoscalar field [11] or a complex scalar or pseudoscalar field. Such mechanism is locally indistinguishable from the explicit one but globally leads to charge symmetric universe, consisting of equal amount of matter and antimatter. This mechanism suffers from a serious domain wall problem which jeopardizes homogeneity and isotropy of the observed universe [12].
3. Dynamical or stochastic CP-breaking [13] enforced by a (pseudo)scalar field shifted from the equilibrium position and not yet relaxed to equilibrium point at baryogenesis. It could operate only in the early universe and disappeared without trace today and thus it is free of the domain wall problem. Of course an additional source of CP-violation effective only in cosmology is at odds with the Occam razor, but nevertheless it must work in the early universe if there exists a scalar field with mass smaller than the Hubble parameter during inflation, $m < H_{inf}$.

Scenarios of baryogenesis with explicit C(CP)-violation lead to creation of the universe devoid of noticeable antimatter, at least in the simplest versions of the models.

Spontaneous C(CP)-violation evidently predicts alternating matter and antimatter domains but, according to the results of ref. [14] the domain size should be very large so that the nearest domain may be at least at the distance comparable to the cosmological horizon, $l_B \sim$ Gpc. However, in some versions of the scenarios of the domain wall formation and elimination this bound may be relaxed (work in progress).

An interesting possibility is the combined action of explicit plus spontaneous C(CP)-violation [15] but probably it is strongly restricted, if not excluded, because it leads to too large angular fluctuations of CMB and distortion of big bang nucleosynthesis.

The third mechanism of stochastic C(CP) breaking is probably most suitable for creation of cosmologically interesting antimatter, moreover such mechanism naturally works in popular spontaneous [16] and Affleck and Dine [17] scenarios of baryogenesis. However, the versions of the models of cosmic antimatter creation suggested ref. [18] seem to suffer from excessive density perturbations at large scales which are excluded by observations.

A promising scenario of significant antimatter creation in our neighborhood was proposed in ref. [19], which I will discuss in some details below, as well as its cosmological manifestations [20]. The model predicts that antimatter is mostly concentrated in compact stellar like objects (probably dead stars) or dense clouds of antimatter and usual matter. These dense matter and antimatter objects have equal share in the total cosmological mass density and may contribute significant fraction of dark matter in our galaxy and in the universe, possibly even 100%.

On the other hand, there are quite strong observational bounds on possible antimatter in the universe. The nearest antigalaxy in our supercluster cannot be closer than at ~10 Mpc [21]. Otherwise the annihilation of intergalactic gas from our supercluster inside such an anti-galaxy would create too large flux of gamma quanta with energy around 100 MeV. Analogous arguments exclude fraction of anti-matter inside two colliding galaxies (Bullet Cluster) larger than 3×10^{-6} [22]. These bounds together with cosmic ray data, that the fraction of $\bar{p}$ in cosmic rays is less than 10^{-4} and the fraction of antihelium is smaller than 3×10^{-8} [5] seem to exclude large amount of antimatter in our Galaxy. However these bounds are valid if antimatter makes exactly the same type objects as the OBSERVED matter, as is usually assumed. For example, compact objects made of antimatter may be abundant, live in the Galaxy, but still escape observations, as wee in what follows.

To create compact high density baryonic and antibaryonic objects we rely on the Affleck-Dine mechanism of baryogenesis [17]. According to this mechanism a scalar baryon, χ, condenses along flat directions of its potential and accumulates very big baryonic charge, B, later released in the decays of χ into quarks. If χ acquires high B value homogeneously in all space, it would end up with the universe with constant and large baryon or antibaryon asymmetry. However, if the window to the flat direction is open only during a short period, cosmologically small but possibly astronomically large bubbles with high B could be created, occupying a small fraction of the cosmological volume, while the rest of the universe would have the normal baryon asymmetry:

$$\beta = N_B/N_\gamma \approx 6 \times 10^{-10}, \tag{1}$$

while inside the high B-bubbles the asymmetry may be of order unity or even larger.

Such rather unusual inhomogeneous modification of the Affleck-Dine mechanism can be realized by a very simple modification of the potential of the Affleck-Dine field χ. We assume that χ has the usual Coleman-Weinberg potential [23] plus general renormalizable coupling to inflaton field Φ:

$$U(\chi,\Phi) = \lambda_1|\chi|^2(\Phi-\Phi_1)^2 + \lambda_2|\chi|^4\ \ln(|\chi|^2/\sigma^2) + m_0^2|\chi^2| + (m_1^2\chi^2 + h.c.), \tag{2}$$

where Φ_1 is some constant value which Φ passes in the course of inflation but not too long before the end of inflation, otherwise the high B-regions would be too large. The mass parameter m_1 may be complex but CP would be still conserved, because one can "phase rotate" χ to eliminate complex parameters in the Lagrangian. It is essential that the last term is not invariant with respect to $U(1)$ transformation, $\chi \to \exp(i\Theta)\,\chi$, and thus it breaks B-conservation.

Potential (2) has one minimum at $\chi = 0$ for large and small Φ and has a deeper minimum at non-zero χ when Φ is close to Φ_1. At that time the chances for χ to reach a high value at the other minimum are non-negligible.

The probability for χ to reach the deeper minimum is determined by the quantum diffusion at inflationary stage. It is governed by the diffusion equation [24]:

$$\frac{\partial \mathcal{P}}{\partial t} = \frac{H^3}{8\pi^2} \sum_{k=1,2} \frac{\partial^2 \mathcal{P}}{\partial \chi_k^2} + \frac{1}{3H} \sum_{k=1,2} \frac{\partial}{\partial \chi_k} \left[\mathcal{P} \frac{\partial U}{\partial \chi_k} \right] \tag{3}$$

where $\chi = \chi_1 + i\chi_2$. The inflation may be not exact and H_I may depend upon time but this does not significantly influence the spectrum of the produced bubbles with high baryonic density.

It can be shown [19] that the mass spectrum of the high B bubbles has log normal form:

$$\frac{dn}{dM} = C_M \exp\left[-\gamma \ln^2 (M/M_0)\right] \tag{4}$$

where C_M, γ, and M_0 are some constant model dependent parameters.

According to this scenario the universe would look as a huge piece of swiss cheese with high B bubbles instead of holes and homogeneous background with constant baryonic density. The mass of those high B objects is model dependent and can be of the order of stellar mass or even larger or much smaller. Despite their small size the mass fraction of the bubbles could be comparable or even larger than the observed baryonic mass fraction.

Initially the density contrast between the bubbles with high values of χ and the bulk with $\chi \sim 0$ was small. Later, when the mass of χ came into play, the matter inside the bubbles with a large amplitude of χ became nonrelativistic and the density contrast started to rise. The rise continued till χ decayed into light quarks and/or leptons and the matter inside became relativistic as in the bulk of the universe. The second period of the rising perturbations took place after the QCD phase transition at $T = T_{QCD} \sim 100$ MeV, when relativistic quarks confined to make nonrelativistic nucleons.

The final state of the high B-bubble depends upon their mass and radius. If the density contrast was large, $\delta\rho/\rho \geq 1$ at horizon crossing, the bubbles would form primordial black holes (PBH). The mass inside horizon at cosmological time t is equal to:

$$M_{hor} \approx m_{Pl}^2 t \approx 10^{38}\text{g}\,(\text{t/sec}) \approx 10^5 \text{M}_\odot (\text{t/sec}), \tag{5}$$

where $M_\odot$ is the Solar mass. At the moment of QCD phase transition the mass inside horizon is about $M_\odot$, while during big bang nucleosynthesis (BBN) the mass inside horizon varies from $10^5 M_\odot$ to to $10^7 M_\odot$. So as a by-product the considered scenario allows for formation of superheavy black holes whose origin is not well explained by conventional mechanisms.

One should keep in mind that not only black holes but some other compact objects could be formed too. If PBH had not been formed, the subsequent evolution of the B-bubble depends upon the relation between their mass and the Jeans mass. Bubbles with $\delta\rho/\rho < 1$ but with $M_B > M_{Jeans}$ at horizon would decouple from cosmological expansion and form compact stellar type objects or lower density clouds. Such anti-objects could survive against an early annihilation. Moreover, the annihilation products would be practically unnoticeable. According to the estimates of ref. [20], the fraction of extra energy produced by the annihilation on the surface of the compact stellar-like anti-objects would not exceed about 10^{-15} of CMB.

The bubbles with high baryonic number density, after they decoupled from the cosmological expansion, would have higher temperature than the background. The luminosity due to thermal emission into external space would be $L \approx 10^{39}$ erg/sec for the bubbles with solar mass. Even if these high-B bubbles make all cosmological dark matter, i.e. $\Omega_{BB} = 0.25$, the thermal keV photons would have the energy density not more than $10^{-4} - 10^{-5}$ of the energy density of CMBR, red-shifted today to the background light [20].

Some of those high density baryonic and antibaryonic objects might be similar to red giants core. The dominant internal energy source of these objects would be the nuclear helium burning through the reaction $3He^4 \rightarrow C^{12}$, however with larger T by factor ~ 2.5 in comparison with red giants. Since $L \sim T^{40}$, the nuclear power exhaust would be very fast, with life-time shorter than 10^9 s. The total energy outflux would be below 10^{-4} of CMBR. It is unclear it the nuclear reactions could lead to the B-ball explosion and creation of solar mass anti-cloud. Astrophysics of such early formed high density objects is not yet studied.

Thus sufficiently large compact anti-objects could survive in the early universe, especially if they are PBHs. A kind of early dense stars might be formed with initial pressure outside larger than that inside, which introduced an additional stabilizing factor for their survival. Such "stars" would evolve quickly and, in particular, could make early supernovae which would enrich the universe with heavy (anti)nuclei and re-ionize the universe. Energy release from stellar like objects in the early universe is small compared to the energy density of CMB. The objects with high (anti)baryonic number are not dangerous for the standard outcome of big bang nucleosynthesis since the volume of such B-bubbles is small. However, all above are the results of some simplified estimates and more rigorous calculations are necessary.

In the present day universe the antimatter bubbles created by the discussed above mechanism may form all kinds of antimatter objects populating the Galaxy or its halo. They could make dense clouds of antimatter, isolated antistars, which are most probably evolved and not shining, anti-black holes, which are indistinguishable from the black holes made of the usual matter, except for possible anti-atmosphere of non-collapsed primordial antimatter.

The observational signatures of such antimatter objects are almost evident. They include the standard list of 100 MeV gamma background, excessive antiprotons, and low energy positrons. For all these observations an alternative explanations cannot be a priori excluded, but an observation of cosmic antinuclei, starting from anti-helium and heavier ones would be unambiguous proof of existence of primordial antimatter.

One can also expect to see compact sources of γ radiation, which are not associated with shining stars. The chemical content of those objects maybe be different from the normal stars, because they were created from matter processed through BBN with unusually high baryon-to-photon ratio. So an observation of compact stellar-like objects in the sky or clouds with unusual chemistry would be a strong indication that such objects are made of antimatter and search for annihilation around them has chance to be successful.

There could also be some catastrophic phenomena generated by the star-antistar collisions, processes similar to hypernova explosions induced by transfer of material in double star-antistar system. A capture of compact antistar by a red giant would lead to a noticeable change of the red giant luminosity over a time interval of the order of a month.

To summarize both theory and observations allow for abundant cosmic antimatter in the Galaxy and in its halo and the search of cosmic antinuclei, excessive antiprotons and positrons, gamma-quanta from the antiproton and positron annihilation, and striking astrophysical phenomena due to star-antistar interactions have a non-negligible chance to be discovered.

References

[1] P.A.M. Dirac, *Proceedings of the Royal Society of London*, A117, 610 (1928).
[2] C.D. Anderson, *Phys. Rev.* 43, 491 (1933).
[3] A. Schuster, *Nature*, 58, 367 (1898).
[4] A.D. Sakharov, *Pisma Zh. Eksp. Teor. Fiz.* 5, 32 (1967); *JETP Lett.* 5, 24 (1967).
[5] M. Sasaki, et al, BESS Collaboration, *Advances in Space Research*, 42, Issue 3, 450 (2008).
[6] AMS Collaboration, arXiv:hep-ex/0002048v1
[7] M. Boezio et al, PAMELA Collaboration, *Journal of Physics: Conference Series*, 110, 062002 (2008). P. Picozza, A. Morselli, *Journal of Physics: Conference Series*, 120, 042004 (2008).
[8] AMS-2 home page: http://ams-02project.jsc.nasa.gov .

[9] H. Gast, P. von Doetinchem, T. Kirn, G.R. Yearwood, S. Schael, *Proceedings of the 30th International Cosmic Ray Conference*, R. Caballero, J.C.D. Olivo, G. Medina-Tanco, L. Nellen, F.A. Sánchez, J.F. Valdés-Galicia (eds.), Mexico City, Mexico, 2008 Vol. 2 (OG part 1), page 293.

[10] See e.g.: http://www.cita.utoronto.ca/ kaya/gaps.html

[11] T.D. Lee, *Phys. Rept.* 9, 43 (1974).

[12] Ya.B. Zeldovich, I.Yu. Kobzarev, L.B. Okun, *Zh. Eksp. Teor. Fiz.* 67, 3 (1974); *Sov. Phys. JETP* 40, 1 (1974).

[13] A.D. Dolgov, *Phys. Repts.* 222, 309 (1992);
A.D. Dolgov, "CP violation in cosmology" in Proc. of International School of Physics "Enrico Fermi": CP Violation: From Quarks to Leptons, Varenna, Italy, 19-29 Jul 2005. Published in "Varenna 2005, CP violation", 407-438; e-Print: hep-ph/0511213 and references therin.

[14] A.G. Cohen, A. De Rujula, S.L. Glashow, *Astrophys. J.*, 495, 539 (1998).

[15] V.A. Kuzmin, M.E. Shaposhnikov, I.I. Tkachev, *Phys. Lett.* B105, 167 (1981); V.A. Kuzmin, I.I. Tkachev, M.E. Shaposhnikov, *Pisma Zh. Eksp. Teor. Fiz.* 33, 557 (1981).

[16] A.G. Cohen, D.B. Kaplan, *Nucl. Phys.*, B308, 913 (1988).

[17] I. Affleck, M. Dine, *Nucl. Phys.*, B249, 361 (1985).

[18] M.Yu. Khlopov, S.G. Rubin, A.S. Sakharov, *Phys.Rev.* **D** 62, 083505 (2000);
M.Yu. Khlopov, S.G. Rubin, A.S. Sakharov, Talk given at 14th Rencontres de Blois: Matter - Anti-matter Asymmetry, Chateau de Blois, France, 17-22 Jun 2002, e-Print: hep-ph/0210012.

[19] A. Dolgov, J. Silk, *Phys. Rev.* **D** 47, 4244 (1993);
A.D. Dolgov, M. Kawasaki, N. Kevlishvili, *Nucl. Phys.* **B** 807, 229 (2009),.

[20] C. Bambi, A.D. Dolgov, *Nucl. Phys.* **B** 784, 132 (2007).

[21] G. Steigman, *Ann. Rev. Astron. Astrophys.* 14, 339 (1976).

[22] G. Steigman, arXiv:0808.1122.

[23] S.R. Coleman, E. Weinberg, *Phys. Rev.* **D** 7, 1888 (1973).

[24] A.A. Starobinsky, *Phys. Lett.* bf B 117, 175 (1982).

COSMIC RAY STUDIES WITH PAMELA EXPERIMENT

P. Picozza [1 a], R. Sparvoli[1], O. Adriani[2], G. Barbarino[3], G. A. Bazilevskaya[4], R. Bellotti[5], M. Boezio[6], E. A. Bogomolov[7], L. Bonechi[2], M. Bongi[2], V. Bonvicini[6], S. Borisov[1], S. Bottai[2], A. Bruno[5], F. Cafagna[5], D. Campana[3], R. Carbone[3], P. Carlson[8], M. Casolino[1], G. Castellini[9], L. Consiglio[3], M. P. De Pascale[1], C. De Santis[1], N. De Simone[1], V. Di Felice[1], A. M. Galper[10], W. Gillard[8], L. Grishantseva[10], G. Jerse[6], A. Karelin[10], S. V. Koldashov[10], S. Y. Krutkov[7], A. N. Kvashnin[4], A. Leonov[10], V. Malvezzi[1], L. Marcelli[1], W. Menn[11], V. V. Mikhailov[10], E. Mocchiutti[6], A. Monaco[5], N. Mori[2], N. Nikonov[1], G. Osteria[3], P. Papini[2], M. Pearce[8], C. Pizzolotto[6], M. Ricci[12], S. B. Ricciarini[2], L. Rossetto[8], M. Simon[11], P. Spillantini[2], Y. I. Stozhkov[4], A. Vacchi[6], E. Vannuccini[2], G. Vasilyev[10], S. A. Voronov[10], J. Wu[8], Y. T. Yurkin[10], G. Zampa[6], N. Zampa[6], V. G. Zverev[10]

[1] *INFN and Physics Department of University of Rome "Tor Vergata", I-00133 Rome, Italy*

[2] *INFN and Physics Department of University of Florence, I-50019 Sesto Fiorentino, Florence, Italy*

[3] *INFN and Physics Department of University of Naples "Federico II", I-80126 Naples, Italy*

[4] *Lebedev Physical Institute, Leninsky Prospekt 53, RU-119991 Moscow, Russia*

[5] *INFN and Physics Department of University of Bari, I-70126 Bari, Italy*

[6] *INFN, Sezione di Trieste, I-34012 Trieste, Italy*

[7] *Ioffe Physical Technical Institute, RU-194021 St. Petersburg, Russia*

[8] *Physics Department of the Royal Institute of Technology (KTH), SE-10691 Stockholm, Sweden*

[9] *IFAC, I-50019 Sesto Fiorentino, Florence, Italy*

[10] *Moscow Engineering and Physics Institute, RU-11540 Moscow, Russia*

[11] *Physics Department of Universitaet Siegen, D-57068 Siegen, Germany*

[12] *INFN, Laboratori Nazionali di Frascati, I-00044 Frascati, Italy*

*Abstract.*The instrument PAMELA, in orbit since June 15th, 2006 on board of the Russian satellite Resurs DK1, is daily delivering to ground 16 Gigabytes of data. The apparatus is designed to study charged particles in the cosmic radiation, with a particular focus on antiparticles for searching antimatter and signals of dark matter annihilation. A combination of a magnetic spectrometer and different detectors allows antiparticles to be reliably identified from a large background of other charged particles. New results on the antiproton-to-proton and positron-to-all electron ratios over a wide energy range (1-100 GeV) have been obtained from the PAMELA mission. These data are mainly interpreted in terms of dark matter annihilation or pulsar contribution.

1 Introduction

One hundred years ago Victor Hess discovered cosmic rays and an impressive experimental study was performed from then on. Cosmic rays are associated

[a]e-mail: piergiorgio.picozza@roma2.infn.it

with the most energetic events and active objects in the Universe and the energies of the observed particles far exceed the energies even of the most powerful accelerators. Their chemical composition and energy spectrum give extensive information for probing their origin, acceleration and propagation mechanisms. In this spectrum there are, quite hidden, the answers to the main questions in cosmic rays research: where are particles coming from? Are they galactic or also extragalactic? How and where are they accelerated? How do they propagate through the interstellar medium and what kind of interaction do they encounter? What role do they play in the energy budget of the interstellar medium? Do we find hints of the existing of exotic particles as relic from the early Universe, as antimatter and dark matter?

Twenty-one orders of magnitude in energy have been explored up to now, by direct methods - balloon borne and satellite experiments - up to 10^{14} eV and by indirect methods - ground large size apparata - at the highest energies. At medium energy the study is focused on the search of antimatter and antiparticle content in cosmic rays as a unique tool to investigate several physics and astrophysics phenomena. The search of antimatter is instead strictly connected with the baryon antibaryon asymmetry in the Universe and the detection of antimatter of primary origin in cosmic rays would be a discovery of fundamental significance. Other possible contributions could come from evaporation of primordial mini black holes by the Hawking process and from exotic particles annihilation.

Several observations show that the Universe is prominently composed of dark matter and dark energy. Among the most plausible candidates for dark matter there are weakly interacting massive particles (WIMP), with the supersimmetric neutralino as a favourite candidate. The neutralino arises naturally in supersymmetric extensions of the standard model, and has the attractive feature of giving a relic density adequate to explain cosmological dark matter in a large region of the parameter space. Neutralinos are Majorana fermions and can annihilate with each other in the halo, resulting in the symmetric production of particles and antiparticles, the latter providing an observable signature. Other models privilege lightest Kaluza-Klein particles in the Universal Extra Dimension scenario. These ideas had a great improvement from the discovery of antiprotons on the top of the atmosphere made from Robert Golden [1] and Edward Bogomolov [2] in 1979 by balloon borne experiments. They measured a rate of antiprotons much higher than expected from interactions of cosmic rays with the interstellar matter. Many other experiments followed these pioneer ones, performed mainly from the WiZard, HEAT and BESS collaborations on board balloons and from AMS-01 on board the Shuttle, using novel techniques developed for accelerator physics. Although the first historical results were not confirmed later, the way for a wide research for primary antimatter and dark matter signals in the cosmic rays was open. However, possible contributions from dark matter annihilation or other exotic sources are mixed with a huge

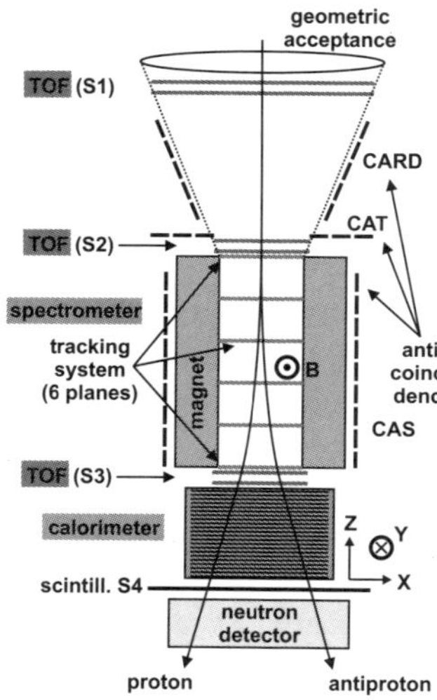

Figure 1: Schematic overview of the PAMELA apparatus.

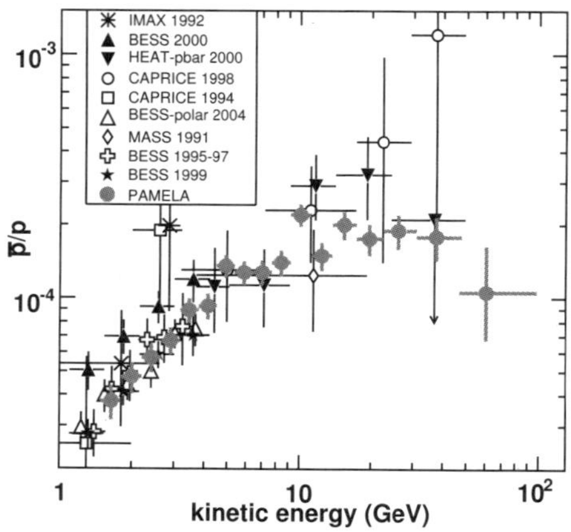

Figure 2: PAMELA antiproton-to-proton flux ratio compared with previous measurements.

background produced in the interactions of cosmic rays with the ISM, so that they appear as a distortion of the antiproton, positron energy spectra. Then, a better knowledge of the standard mechanisms of production, acceleration and transport of cosmic rays is required.

New satellite experiments have been devised with the task to measure at the same time antiprotons and positrons, but also experimental parameters included in the background calculation. In June 2006 the first of these satellite, PAMELA, was launched in orbit by a Soyuz-U rocket from the Bajkonur cosmodrome in Kazakhstan.

2 The PAMELA Instrument

The PAMELA experiment, part of the RIM (Russian-Italian Missions) program, is performed by an international collaboration composed of the Italian Universities and INFN divisions of Bari, Florence, Naples, Rome Tor Vergata and Trieste, the INFN National Laboratory of Frascati, the Russian Institutes MEPhI and FIAN Lebedev in Moscow and Joffe in S. Peterbourg, the University of Siegen in Germany and the Royal Technical Institute of Stockolm, Sweeden. Conceived mainly for searching primordial antimatter, signals from dark matter annihilation, exotic matter as strangelets, PAMELA achieves also other important tasks as the study of the mechanisms of acceleration and propagation of cosmic rays in the Galaxy, the cosmic ray solar modulation, the detection of solar flares. Studies of the interaction of particles with the terrestrial magnetosphere complete the PAMELA research program.

An overview of the PAMELA apparatus is shown in fig. 1. The core of the instrument is a magnetic spectrometer, made of a permanent magnet and a silicon tracking system for a maximum detectable rigidity of 1 TV. A time-of-flight system consisting of three double layers of segmented plastic scintillator provides timing and dE/dx measurements and defines the primary PAMELA trigger. The separation between hadronic and leptonic components is made by an imaging silicon-tungsten detector and a neutron counter. An imaging silicon calorimeter and a neutron detector assures a rejection of protons, compared to positrons, at the order of 10^5. The calorimeter permits also measurements of the electron energy up to 300 GeV, with a resolution of few per cent. A thick scintillator placed between the calorimeter and the neutron counters and an anticoindence system complete the instrument. PAMELA can measure electrons, positrons, antiprotons, protons and light nuclei in an energy range from tens MeV up to hundreds GeV. More details can be found in [3].

PAMELA has been inserted in a pressurized vessel and installed on board of the Russian satellite DK-1 dedicated to Earth observation. It was launched on June 15th 2006 in an elliptical orbit, ranging between 350 and 610 Km and with an inclination of 70 degrees. Since July 2006 PAMELA is daily delivering 16 Gigabytes of data to the Ground Segment in Moscow.

3 Data analysis and results

The results presented here correspond to the data-set collected between July 2006 and February 2008. More than 10^9 triggers were accumulated during a total acquisition time of approximately 500 days.

3.1 Antimatter and Dark Matter

Antiprotons identification was based on the spectrometer and the properties of the energy deposit and interaction topology in the calorimeter. The energy interval explored is limited by the "spillover", protons recognized as negative particles due to the intrinsic deflection uncertainty in spectrometer measurements at the highest energies and to a possible multiple scattering of the particles in the tracking system. Calorimeter was used to reject electrons.

The antiproton to-proton flux ratio [4] measured by the PAMELA experiment is shown in fig. 2, compared with theoretical calculations assuming pure secondary production of antiprotons by cosmic rays in the galaxy (see [5] and references therein). The ratio increases smoothly from about 2×10^{-5} at a kinetic energy of about 1 GeV and levels off at about 10^{-4} for energies above 10 GeV. The data do not present the features or structures expected from

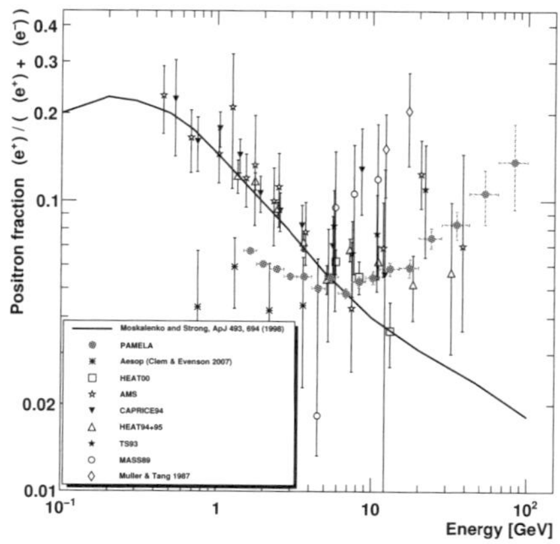

Figure 3: The positron fraction measured by the PAMELA experiment, compared with other recent experimental data.

exotic sources, so they place strong limits to dark matter annihilation models. Moreover, they set tight constraints on parameters relevant for secondary production calculations, e.g. the normalization and the index of the diffusion coefficient, the Alfven speed, and contribution of a hypothetical "fresh" local cosmic-ray component. It is worth to cite the results obtained by Kane et al. [6]; their calculation of the secondary antiproton production still leaves room for a contribution from an exotic source.

Positrons and electrons data need a very careful analysis, because of the possibility of misidentification of protons as positrons. Particle identification was based on the matching between the momentum measured by the magnetic spectrometer and the total energy measured by the calorimeter, the shower starting point, its lateral and longitudinal profiles and the neutron detector response. The positron to all electron ratio measured by the PAMELA experiment [7] is given in fig. 3, compared with other recent experimental results (see [5]). The data, covering the energy range 1.5–100 GeV, show two clear features. At low energies, below 5 GeV, the PAMELA results are systematically lower than data collected during the 1990's, but they are in agreement with data obtained from the AESOP balloon borne experiment [8] that flew from Scandinavian to Canada in June 2006. At high energies, above 10 GeV, PAMELA data show a positron fraction increasing significantly with energy.

Standard calculations of positrons secondary production predict a continuous decrease of the positron fraction, then the observed excess of positrons in the range 10-100 GeV has led to many theoretical models explaining its origin as due to annihilation or decaying of dark matter. The lack of particular features in the antiproton/proton ratio favors an explanation of the positron excess in terms of a direct leptonic annihilation channel for the WIMP's with a large boost factor. Another explanation claims for a contribution from nearby and young pulsars, objects well known as particle accelerators, or for some inhomogeneity in the SNR density in our Galaxy or for a production of secondary positrons in the same site where protons are accelerated. It is worth to remind that a reliable calculation of antiprotons and positrons secondary production is necessary to extract possible signals from exotic sources inside the huge astrophysics background. PAMELA is measuring with good precision and high statistics protons, ^{4}He, Carbon and Oxygen (primaries cosmic ray sources) together with ^{3}He, Li, Be, B (secondaries). These measurements will improve the propagation models describing the galactic structure and the various mechanisms involved. Data analysis for protons and helium fluxes and the B/C ratio measured by PAMELA is in progress.

3.2 Solar physics

Continuous monitoring of the solar activity and the detection of solar energetic particle events are other important issues addressed from PAMELA. It is well known that the low energy part of the cosmic ray energy spectra up to about 5–10 GeV is affected by solar modulation in a way depending on the particle electric charge sign. Moreover, this effect is different if the magnetic dipole projection on the solar rotational axis and the same axis is parallel (phase A^+) or anti-parallel (pase A^-). This is due to a systematic deviation from the reflection symmetry of the interplanetary magnetic field. The Parker field has opposite magnetic polarity above and below the equator, but the spiral field lines themselves are mirror images of each other. This anti-symmetry produces drift velocity fields that for positive particles converge on the heliospheric equator in the A^+ state or diverge from it in A^- state. PAMELA data have been collected during a A^- phase when the positrons are modulated more than electrons, and this explains the difference at low energy with the results obtained by previous experiments that were performed in A^+ phase. In fig. 4 phenomenological calculations for the positron/electron and antiproton/proton ratios for several solar phases A^+ and A^- and compared with data at 1.25 GV momentum of different experiments are shown. PAMELA data are very important also taking into account the long duration of its permanence in orbit in the recovery phase going towards solar maximum at cycle 24. It is interesting to stress that without a complete modeling of the solar modulation it becomes almost impossible to disentangle exotic signals at low energy.

PAMELA also detected the solar impulsive event of December 13th 2006. The observation of solar energetic particles (SEP) events with a magnetic spectrometer allows for several aspects of solar and heliospheric cosmic ray physics to be addressed for the first time.

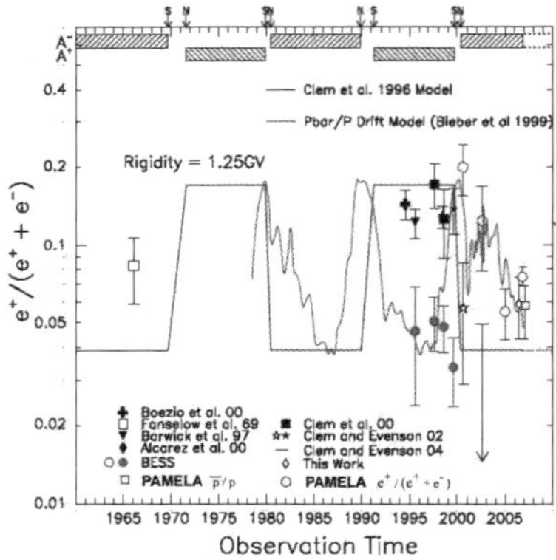

Figure 4: Phenomenological calculations for the positron-electron and antiproton-proton ratios along with the PAMELA and other experimental results.

4 Conclusion

PAMELA is a general purpose charged particle detector system exploring the antiparticle components of the cosmic radiation over a wide energy range. It has been in orbit since June 2006 and it is daily transmitting to ground 16 GB of data.

The main results obtained by PAMELA concern the antiproton-to-proton and the positron-to-electron ratios. The first appears to be in agreement with the standard secondary production in the collision between high energy cosmic rays and interstellar matter, while the positron fraction shows, instead, an increase in the ratio at energy above 10 GeV compared to expectations. This result is generally interpreted in terms of positron primary sources, such as dark matter annihilation or nearby pulsars contributions, or by non standard processes in secondary positrons production.

References

[1] Golden R. et al., Phys. Rev. Letter 43, 1196 (1979).
[2] Bogomolov E. et al., Proc. 16th ICRC 1, 330 (1979).
[3] Picozza P. et al., Astropart. Phys. 27, 296 (2007).
[4] Adriani O. et al., Phys. Rev. Lett. 102 , 051101 (2009).
[5] Boezio M. et al., New Journal of Physics 11 , 105023 (2009).
[6] Kane G., Lu R., and Watson S., Phys. Lett. B 681, 151 (2009).
[7] Adriani O. et al., Nature 458, 607 (2009).
[8] Clem J. and Evenson P., Proc. 30th Int. Conf. on Cosmic Rays (2007).

DAMA/LIBRA RESULTS

R. Bernabei [a], P. Belli, F. Montecchia, F. Nozzoli
Dip. di Fisica, Univ. di Roma "Tor Vergata" and
INFN sez. Roma "Tor Vergata", I-00133 Rome, Italy
F. Cappella, A. d'Angelo, A. Incicchitti, D. Prosperi
Dip. di Fisica, Univ. di Roma "La Sapienza" and
INFN, sez. Roma, I-00185 Rome, Italy
R. Cerulli
Laboratori Nazionali del Gran Sasso, I.N.F.N., Assergi, Italy
C.J. Dai, H.L. He, H.H. Kuang, X.H. Ma, X.D. Sheng, R.G. Wang, Z.P. Ye
Chinese Academy, P.O. Box 918/3, Beijing 100039, China

Abstract.
The DAMA/LIBRA set-up (about 250 kg highly radiopure NaI(Tl) sensitive mass) is running at the Gran Sasso National Laboratory of the I.N.F.N.. The first DAMA/LIBRA results published in 2008 have confirmed the model independent evidence for the presence of Dark Matter particles in the galactic halo, previously pointed out by the former DAMA/NaI experiment. After this conference, at beginning of 2010, the data of other two annual cycles have been released. The presence of Dark Matter particles in the galactic halo is now supported at 8.9 σ C.L. by an exposure of 1.17 ton $\times$ yr collected over 13 annual cycles by the former DAMA/NaI and the present DAMA/LIBRA. Related arguments are summarized.

DAMA is an observatory for rare processes located deep underground at the Gran Sasso National Laboratory of the I.N.F.N.. It is based on the development and use of low background scintillators. The main experimental set-ups are: i) DAMA/NaI ($\simeq$ 100 kg of highly radiopure NaI(Tl)) which completed its data taking on July 2002 [1–12]; ii) DAMA/LXe ($\simeq$ 6.5 kg liquid Kr-free Xenon enriched either in ^{129}Xe or in ^{136}Xe) [13]; iii) DAMA/R&D, a facility dedicated to tests on prototypes and to experiments for investigations on rare events [14]; iv) DAMA/Ge, dedicated to sample measurements and to specific measurements on rare events [15]; v) the second generation DAMA/LIBRA set-up ($\simeq$ 250 kg highly radiopure NaI(Tl)) [16–18]. Profiting of the low background features of these set-ups, many rare processes are studied.

In particular, DAMA/LIBRA is investigating the presence of Dark Matter (DM) particles in the galactic halo by exploiting the model independent DM annual modulation signature [19]. In fact, as a consequence of its annual revolution around the Sun, which is moving in the Galaxy travelling with respect to the Local Standard of Rest towards the star Vega near the constellation of Hercules, the Earth should be crossed by a larger flux of Dark Matter particles around $\sim$ 2 June (when the Earth orbital velocity is summed to the one of the solar system with respect to the Galaxy) and by a smaller one around $\sim$ 2 December (when the two velocities are subtracted). It is worth noting that the DM annual modulation signature has a different origin and peculiarities than

[a] e-mail: rita.bernabei@roma2.infn.it

the seasons on the Earth and than effects correlated with seasons (consider the expected value of the phase as well as the other requirements listed below). This signature offers an efficient model independent signature, able to test a large number of DM candidates, a large interval of cross sections and of halo densities.

The DM annual modulation signature is very distinctive since the corresponding signal must simultaneously satisfy all the following requirements: the rate must contain a component modulated according to a cosine function (1) with one year period (2) and a phase that peaks roughly around $\simeq 2^{nd}$ June (3); this modulation must only be found in a well-defined low energy range, where DM particle induced events can be present (4); it must apply only to those events in which just one detector of many actually "fires" (*single-hit events*), since the DM particle multi-interaction probability is negligible (5); the modulation amplitude in the region of maximal sensitivity must be $\lesssim$7% for usually adopted halo distributions (6), but it can be larger in case of some possible scenarios such as e.g. those in refs. [20, 21]. Only systematic effects or side reactions able to fulfil these requirements and to account for the whole observed modulation amplitude could mimic this signature; thus, no other effect investigated so far in the field of rare processes offers a so stringent and unambiguous signature.

In the following, we will just briefly summarize the results on the Dark Matter particle investigation obtained by DAMA/LIBRA, exploiting over six annual cycles the model independent DM annual modulation signature (exposure: 0.87 ton×yr), that is including the two annual cycles released at beginning 2010 after this Conference [17, 22]. The results have also been combined together with the previous data collected over 7 annual cycles by DAMA/NaI (0.29 ton×yr). Thus, the whole available data correspond to 13 annual cycles for a total exposure of 1.17 ton×yr, which is orders of magnitude larger than the exposures typically collected in the field.

The DAMA/NaI set up and its performances are reported in ref. [1, 3–5], while the DAMA/LIBRA set-up and its performances are described in ref. [16]. Here we just summarized the main features: i) the sensitive part of the set-up is made of 25 highly radiopure NaI(Tl) crystal scintillators placed in a 5-rows by 5-columns matrix; ii) the detectors' responses range from 5.5 to 7.5 photoelectrons/keV; iii) the hardware threshold of each PMT is at single photoelectron (each detector is equipped with two low background photomultipliers working in coincidence); iv) energy calibration with X-rays/γ sources are regularly carried out down to few keV in the same conditions as the production runs; v) the software energy threshold of the experiment is 2 keV.

Several analyses on the model-independent DM annual modulation signature have been performed (see ref. [17, 22] and references therein); here just few arguments are mentioned. In particular, Fig. 1 shows the time behaviour of the

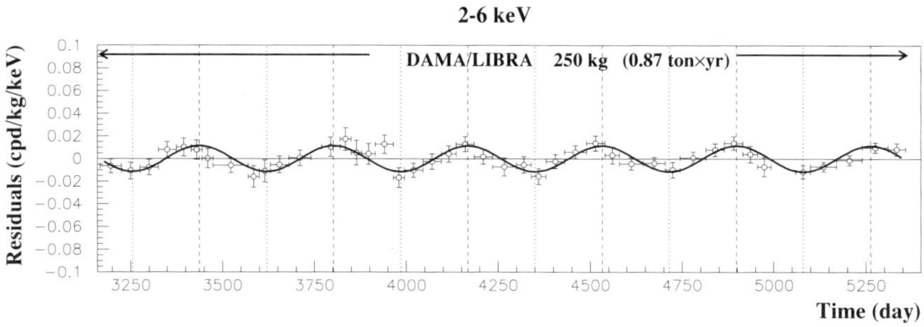

Figure 1: Experimental model-independent residual rate of the *single-hit* scintillation events, measured by DAMA/LIBRA,1,2,3,4,5,6 in the (2 – 6) keV energy intervals as a function of the time. The experimental points present the errors as vertical bars and the associated time bin width as horizontal bars. The superimposed curves are the cosinusoidal functions behaviors $A \cos \omega(t - t_0)$ with a period $T = \frac{2\pi}{\omega} = 1$ yr, with a phase $t_0 = 152.5$ day (June 2nd) and with modulation amplitudes, A, equal to the central values obtained by best fit over the whole data including also the exposure previously collected by the former DAMA/NaI experiment (1.17 ton × yr) The dashed vertical lines correspond to the maximum expected for the DM signal (June 2nd), while the dotted vertical lines correspond to the minimum [22].

experimental residual rates of the *single-hit* events collected by DAMA/LIBRA in the (2–6) keV energy interval [17, 22]. The superimposed curve is the cosinusoidal function: $A \cos \omega(t - t_0)$ with a period $T = \frac{2\pi}{\omega} = 1$ yr and with a phase $t_0 = 152.5$ day (June 2nd), and modulation amplitude, A, obtained by best fit over the seven cycles of DAMA/NaI [4,5] and the six of DAMA/LIBRA data. When the period and the phase parameters are also released in the fit, values well compatible with those expected for a DM particle induced effect are obtained for the cumulative exposure [22]: $A = (0.0116 \pm 0.0013)$ cpd/kg/keV, $T = (0.999 \pm 0.002)$ yr and $t_0 = (146 \pm 7)$ day in the cumulative (2–6) keV energy interval. Summarizing, the analysis of the *single-hit* residual rate favours the presence of a modulated cosine-like behaviour with proper features at 8.9 σ C.L. [22].

The same data of Fig.1 have also been investigated by a Fourier analysis, obtaining a clear peak corresponding to a period of 1 year [22]; this analysis in other energy region shows instead only aliasing peaks. Moreover, while in the (2–6) keV *single-hit* residuals a clear modulation is present, it is absent at energies just above [22]. In particular, in order to verify absence of annual modulation in other energy regions and, thus, to also verify the absence of any significant background modulation, the energy distribution measured during the data taking periods in energy regions not of interest for DM detection has also been investigated. In fact, the background in the lowest energy region is essentially due to "Compton" electrons, X-rays and/or Auger electrons, muon induced events, etc., which are strictly correlated with the events in the higher energy part of the spectrum; thus, if a modulation detected in the lowest energy region would be due to a modulation of the background (rather than to a signal), an equal or larger modulation in the higher energy regions should be present. The data analyses have allowed to exclude the presence of a background modulation in the whole energy spectrum at a level much lower than the effect found in the lowest energy region for the *single-hit* events [22].

A further relevant investigation has been done by applying the same hardware and software procedures, used to acquire and to analyse the *single-hit* residual rate, to the *multiple-hits* one. In fact, since the probability that a DM particle interacts in more than one detector is negligible, a DM signal can be present just in the *single-hit* residual rate. Thus, this allows the test of the background behaviour in the same energy interval of the observed positive effect. In particular, Fig. 2 shows the residual rates of the *single-hit* events measured over the six DAMA/LIBRA annual cycles, as collected in a single

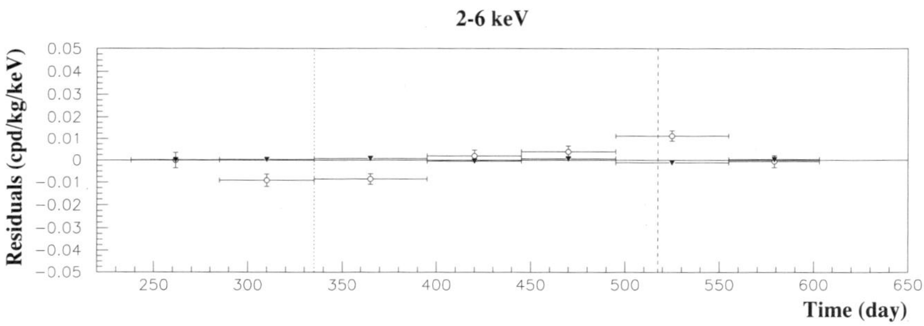

Figure 2: Experimental residual rates over the six DAMA/LIBRA annual cycles for *single-hit* events (open circles) (class of events to which DM events belong) and for *multiple-hits* events (filled triangles) (class of events to which DM events do not belong), in the energy interval (2 – 6) keV [22]. They have been obtained by considering for each class of events the data as collected in a single annual cycle and by using in both cases the same identical hardware and the same identical software procedures. The initial time of the scale is taken on August 7th. The experimental points present the errors as vertical bars and the associated time bin width as horizontal bars. Analogous results were obtained for the DAMA/NaI data [5].

annual cycle, together with the residual rates of the *multiple-hits* events, in the same energy interval. A clear modulation is present in the *single-hit* events, while the fitted modulation amplitude of the *multiple-hits* residual rate is well compatible with zero [22]. Similar results were previously obtained also for the DAMA/NaI case [5]. Thus, again evidence of annual modulation with proper features, as required by the DM annual modulation signature, is present in the *single-hit* residuals (events class to which the DM particle induced events belong), while it is absent in the *multiple-hits* residual rate (event class to which only background events belong). Since the same identical hardware and the same identical software procedures have been used to analyse the two classes of events, the obtained result offers an additional strong support for the presence of DM particles in the galactic halo further excluding any side effect either from hardware or from software procedures or from background.

The annual modulation present at low energy has also been analyzed by depicting the differential modulation amplitudes, $S_{m,k}$, as a function of the energy (the k index identifies the energy interval); the $S_{m,k}$ is the modulation amplitude of the modulated part of the signal obtained by maximum likelihood method over the data, considering $T = 1$ yr and $t_0 = 152.5$ day. The $S_{m,k}$ values are reported as function of the energy in ref. [22]. It has been also verified that the measured modulation amplitudes are statistically well distributed in all the crystals, in all the annual cycles and energy bins; these and other discussions can be found in ref. [22]. It is also interesting the results of the analysis performed by releasing the assumption of a phase $t_0 = 152.5$ day in the procedure of maximum likelihood to evaluate the modulation amplitudes from the data of the seven annual cycles of DAMA/NaI and the six annual cycles of DAMA/LIBRA. In this case alternatively the signal is written as: $S_{0,k} + S_{m,k} \cos\omega(t - t_0) + Z_{m,k} \sin\omega(t - t_0) = S_{0,k} + Y_{m,k} \cos\omega(t - t^*)$, where $S_{0,k}$ is the constant part of the signal in k-th energy interval. Obviously, for signals induced by DM particles one would expect: i) $Z_{m,k} \sim 0$ (because of the orthogonality between the cosine and the sine functions); ii) $S_{m,k} \simeq Y_{m,k}$; iii) $t^* \simeq t_0 = 152.5$ day. In fact, these conditions hold for most of the dark halo models; however, it is worth noting that slight differences in the phase can be expected in case of possible contributions from non-thermalized DM components, such as e.g. the SagDEG stream [7] and the caustics [23]. The 2σ contours in the plane (S_m, Z_m) for the (2–6) keV and (6–14) keV energy intervals and those in the plane (Y_m, t^*) are reported in ref. [22]. The best fit values for the (2–6) keV energy interval are (1σ errors): $S_m = (0.0111 \pm 0.0013)$ cpd/kg/keV; $Z_m = -(0.0004 \pm 0.0014)$ cpd/kg/keV; $Y_m = (0.0111 \pm 0.0013)$ cpd/kg/keV; $t^* = (150.5 \pm 7.0)$ day; while for the (6–14) keV energy interval (also shown there) are: $S_m = -(0.0001 \pm 0.0008)$ cpd/kg/keV; $Z_m = (0.0002 \pm 0.0005)$ cpd/kg/keV; $Y_m = -(0.0001 \pm 0.0008)$ cpd/kg/keV and t^* obviously not determined. These results confirm those achieved by other kinds of analyses. In particular, a modulation amplitude is present in the lower energy intervals for single hit events and the period and the phase agree with those expected for DM induced signals. For more detailed discussions see ref. [22]

Both the data of DAMA/LIBRA and of DAMA/NaI fulfil all the requirements of the DM annual modulation signature.

As previously done for DAMA/NaI [4, 5], careful investigations on absence of any significant systematics or side reaction effect in DAMA/LIBRA have been quantitatively carried out and reported in details in ref. [17] and some other arguments have also been addressed in ref [24, 25]. No systematics or side reactions able to mimic the signature (that is, able to account for the measured modulation amplitude and simultaneously satisfy all the requirements of the signature) has been found or suggested by anyone over more than a decade. For detailed quantitative discussions on all the related topics and for results see ref. [17, 22] and refs. therein.

In conclusion, DAMA/LIBRA has confirmed the presence of an annual modulation satisfying all the requirements of the DM annual modulation signature, as previously pointed out by DAMA/NaI; in particular, the evidence for the presence of DM particles in the galactic halo is cumulatively supported at 8.9 σ C.L..

It is worth recalling that no other experiment exists, whose result can be directly compared in a model-independent way with those by DAMA/NaI and DAMA/LIBRA, and that – more in general – results obtained with different target materials and/or different approaches cannot be directly compared among them in a model-independent way. This is in particular due to the existing experimental and theoretical uncertainties, not last e.g. how many kinds of dark matter particles can exist in the Universe[b], the nature, the interaction types, the different nuclear and/or atomic correlated aspects, the unknown right halo model, the right DM density, etc. as well as the uncertainties on the values of each one of the many involved experimental and theoretical parameter/assumption/approximation used in the calculations. Moreover, some experimental aspects of some techniques used in the field have also to be addressed [4, 25, 26]. Another relevant argument is the methodological robustness [27]. In particular, the general considerations on comparisons reported in Appendix A of ref. [17] still hold. Hence, claims for contradiction have no scientific basis. On the other hand, whatever possible "positive" result has to be interpreted and a large room of compatibility with DAMA annual modulation evidence is present.

Similar considerations can also be done for the indirect detection searches, since it does not exist a biunivocal correspondence between the observables in the direct and indirect experiments. However, if possible excesses in the positron to electron flux ratio and in the γ rays flux with respect to a modeling of the background contribution, which is expected from the considered sources, might be interpreted – under some assumptions – in terms of Dark Matter, this would also be not in conflict with the effect observed by DAMA experiments. It is worth noting that different possibilities either considering different background modeling or accounting for other kinds of sources can also explain the indirect observations [28].

After the first data release, a first upgrade of the DAMA/LIBRA set-up was performed in September 2008. One detector was recovered by replacing

[b]In fact, it is worth noting that, considering the richness in particles of the visible matter which is less than 1% of the Universe density, one could also expect that the particle part of the Dark Matter in the Universe may also be multicomponent.

a broken PMT and a new optimization of some PMTs and HVs was done. The transient digitizers were replaced with new ones, having better performances and a new DAQ with optical read-out was installed; since October 2008 DAMA/LIBRA was put again in operation.

Considering the relevance to lower the software energy threshold of the experiment, in order to improve the performance and the sensitivity of the experiment and to allow a highly precise determination of all the DM modulation parameters and deeper corollary information on the nature of the DM particle(s) and on the various related astrophysical, nuclear and particle Physics scenarios, the replacement of all the PMTs with new ones with higher quantum efficiency has been planned and work is in progress. DAMA/LIBRA will also study second order effects and several other rare processes as done by the former DAMA/NaI apparatus in the past [12] and by itself so far [18].

References

[1] R. Bernabei et al., *Il Nuovo Cim.* **A112**, 545–576 (1999).

[2] R. Bernabei et al., *Phys. Lett.* **B389**, 757–766 (1996); R. Bernabei et al., *Phys. Lett.* **B424**, 195–201 (1998); R. Bernabei et al., *Phys. Lett.* **B450**, 448–455 (1999); P. Belli et al., *Phys. Rev.* **D61**, 023512–9 (2000); R. Bernabei et al., *Phys. Lett.* **B480**, 23–31 (2000); R. Bernabei et al., *Phys. Lett.* **B509**, 197-203 (2001); R. Bernabei et al., *Eur. Phys. J.* **C23**, 61–64 (2002); P. Belli et al., *Phys. Rev.* **D66**, 043503–17 (2002).

[3] R. Bernabei et al., *Eur. Phys. J.* **C18**, 283–292 (2000).

[4] R. Bernabei el al., *La Rivista del Nuovo Cimento* **26** n.1, 1-73 (2003).

[5] R. Bernabei et al., *Int. J. Mod. Phys.* **D13**, 2127–2159 (2004).

[6] R. Bernabei et al., *Int. J. Mod. Phys.* **A21**, 1445–1469 (2006).

[7] R. Bernabei et al., *Eur. Phys. J.* **C47**, 263–271(2006).

[8] R. Bernabei et al., *Int. J. Mod. Phys.* **A22**, 3155–3168 (2007).

[9] R. Bernabei et al., *Eur. Phys. J.* **C53**, 205–213 (2008).

[10] R. Bernabei et al., *Phys. Rev.* **D77**, 023506–9 (2008).

[11] R. Bernabei et al., *Mod. Phys. Lett.* **A23**, 2125–2140 (2008).

[12] R. Bernabei et al., *Phys. Lett.* **B408**, 439–444 (1997); P. Belli et al., *Phys. Lett.* **B460**, 236–241 (1999); R. Bernabei et al., *Phys. Rev. Lett.* **83**, 4918–4921 (1999); P. Belli et al., *Phys. Rev.* **C60**, 065501–7 (1999); R. Bernabei et al., *Il Nuovo Cimento* **A112**, 1541–1552 (1999); R. Bernabei et al., *Phys. Lett.* **B515**, 6–12 (2001); F. Cappella et al., *Eur. Phys. J.-direct* **C14**, 1-6 (2002); R. Bernabei et al., *Eur. Phys. J.* **A23**, 7–10 (2005); R. Bernabei et al., *Eur. Phys. J.* **A24**, 51-56 (2005); R. Bernabei et al., *Astrop. Phys.* **4**, 45-54 (1995); R. Bernabei, in *The identification of Dark Matter*, World Sc. Pub., Singapore, 1997, pp. 574.

[13] P. Belli et al., *Astropart. Phys.* **5**, 217–219 (1996); P. Belli et al., *Nuovo*

Cim. **C19**, 537–544 (1996); P. Belli et al., *Phys. Lett.* **B387**, 222–226 (1996); *Phys. Lett.* **B389**, (1996) 783 (err.); R. Bernabei et al., *Phys. Lett.* **B436**, 379–388 (1998); P. Belli et al., *Phys. Lett.* **B465**, 315–322 (1999); P. Belli et al., *Phys. Rev.* **D61**, 117301–4 (2000); R. Bernabei et al., *New J. of Phys.* **2**, 15.1–15.7 (2000); R. Bernabei et al., *Phys. Lett.* **B493**, 12–18 (2000); R. Bernabei et al., *Nucl. Instr. & Meth* **A482**, 728–743 (2002); R. Bernabei et al., *Eur. Phys. J. direct* **C11**, 1–8 (2001); R. Bernabei et al., *Phys. Lett.* **B527**, 182–186 (2002); R. Bernabei et al., *Phys. Lett.* **B546**, 23–28 (2002). R. Bernabei et al., in the volume *Beyond the Desert 2003*, Springer, Berlin, 2003, pp. 365; R. Bernabei et al., *Eur. Phys. J.* **A27**, s01 35–41 (2006).

[14] R. Bernabei et al., *Astropart. Phys.* **7**, 73–76 (1997); R. Bernabei et al., *Nuovo Cim.* **A110**, 189–195 (1997); P. Belli et al., *Astropart. Phys.* **10**, 115–120 (1999); P. Belli et al., *Nucl. Phys.* **B563**, 97–106 (1999); R. Bernabei et al., *Nucl. Phys.* **A705**, 29–39 (2002); P. Belli et al., *Nucl. Instr. & Meth* **A498**, 352–361 (2003); R. Cerulli et al., *Nucl. Instr. & Meth* **A525**, 535–543 (2004); R. Bernabei et al., *Nucl. Instr. & Meth* **A555**, 270–281(2005); R. Bernabei et al., *Ukr. J. Phys.* **51** 1037–1043 (2006); P. Belli et al., *Nucl. Phys.* **A789**, 15–29 (2007); P. Belli et al., *Phys. Rev.* **C76**, 064603–10 (2007); P. Belli et al., *Phys. Lett.* **B658**, 193–197 (2008); P. Belli et al., *Eur. Phys. J.* **A36**, 167–170 (2008); P. Belli et al., *Nucl. Phys.* **A826**, 256–273 (2009).

[15] P. Belli et al., *Nucl. Instr. & Meth.* **A572**, 734–738 (2007); P. Belli et al., *Nucl. Phys.* **A806**, 388–397 (2008); P. Belli et al., *Nucl. Phys.* **A824**, 101–114 (2009); P. Belli et al., in the *Proceed. of the Int. Conf. NPAE 2008*, ed. INR-Kiev, Kiev, 2009, pp. 473.

[16] R. Bernabei et al., *Nucl. Instr. & Meth.* **A592**, 297–315 (2008).

[17] R. Bernabei et al., *Eur. Phys. J.* **C56**, 333–355 (2008).

[18] R. Bernabei et al., *Eur. Phys. J.* **C62**, 327–332 (2009).

[19] K. A. Drukier et al., Phys. Rev. **D33**, 3495–3508 (1986); K. Freese et al., Phys. Rev. **D37**, 3388–3405 (1988).

[20] D. Smith and N. Weiner, *Phys. Rev.* **D64**, 043502–11(2001); D. Tucker-Smith and N. Weiner, *Phys. Rev.* **D72**, 063509–8 (2005).

[21] K. Freese et al., Phys. Rev. **D71**, 043516–15 (2005); *Phys. Rev. Lett.* **92**, 111301–4 (2004).

[22] R. Bernabei et al., *Eur. Phys. J.* in publication, arXiv:1002.1028.

[23] F. S. Ling, P. Sikivie and S. Wick, *Phys. Rev.* **D70**, 123503–19 (2004).

[24] R. Bernabei et al., arXiv:0912.0660[astro-ph.GA], to appear in the Proceed. of Scineghe09, October 2009, Assisi (It).

[25] R. Bernabei et. al, *J. Phys.: Conf. Ser.* **203**, 012040 (2010) (arXiv:0912.4200); http://taup2009.lngs.infn.it/slides/jul3/nozzoli.pdf, talk given by F. Nozzoli.

[26] R. Bernabei et al., *Liquid Noble gases for Dark Matter searches: a synoptic survey*, Exorma Ed., Roma, ISBN 978-88-95688-12-1, 2009, pp. 1–53 (arXiv:0806.0011v2).
[27] R. Hudson, *Found. Phys.* **39**, 174–193 (2009).
[28] F. Donato et al., *Phys. Rev. Lett.* **102**, 071301 (2009).

SEARCHING FOR PROCESSES VIOLATING THE PAULI EXCLUSION PRINCIPLE IN Na AND I WITH DAMA/LIBRA

R. Bernabei, P. Belli, F. Montecchia, F. Nozzoli
Dip. di Fisica, Univ. di Roma "Tor Vergata" and INFN sez. Roma "Tor Vergata", I-00133 Rome, Italy
F. Cappella, A. d'Angelo, A. Incicchitti, D. Prosperi
Dip. di Fisica, Univ. di Roma "La Sapienza" and INFN, sez. Roma, I-00185 Rome, Italy
R. Cerulli [a]
Laboratori Nazionali del Gran Sasso, I.N.F.N., Assergi, Italy
C.J. Dai, H.L. He, H.H. Kuang, X.H. Ma, X.D. Sheng, Z.P. Ye
Chinese Academy, P.O. Box 918/3, Beijing 100039, China

Abstract.
A search for processes normally forbidden by the Pauli–Exclusion–Principle (PEP) has been carried out at LNGS by means of the DAMA/LIBRA set-up. The obtained upper limit for the spontaneous non-paulian emission rate of protons with energy $E_p \geq 10$ MeV in ^{23}Na and ^{127}I is 1.63×10^{-33} s^{-1} (90%C.L.). The corresponding limit on the relative strength (δ^2) for the transition is $\delta^2 \lesssim (3-4) \times 10^{-55}$ (90%C.L.). PEP violating electron transitions in Iodine atoms have also been investigated. Lifetimes shorter than 4.7×10^{30} s are excluded at 90%C.L. and a limit $\delta_e^2 < 1.28 \times 10^{-47}$ (90% C.L.) has been derived.

The exclusion principle was postulated by W. Pauli in 1925 to explain atomic spectra and regularities of the Periodic Table of the elements; it is related to the spin statistics and arises from the anti-commutation property of the fermion creation and destruction operators. Many theoretical attempts have been investigated to go beyond the Bose and Fermi statistics; an example is the case of the "q-commutation algebra" describing particles with a small probability (δ^2) of admixed symmetric component [1]; this model provides small PEP violations.

Many experimental tests of the PEP validity with improved sensitivities have been proposed and performed by using different approaches (see [2] for some discussions and references).

Recently, by analysing data collected with the help of the DAMA/LIBRA set-up [2, 3], two different possible Pauli–Exclusion–Principle violating processes have been investigated. The first process is the non-paulian emissions of protons with $E_p \geq 10$ MeV in ^{23}Na and in ^{127}I, while the second one is the possible PEP violating K-shell electron transitions in Iodine atoms.

Since the nucleus is a fermions' system, the nuclear structure is stable in case of an exact PEP. If small violations of PEP exist one of the nucleons in a higher energy shell may fall into a lower energy state, normally occupied, and another nucleon can acquire enough energy to reach the unbound region and to escape from the nucleus. The width of a single nucleon transition to an $i-th$

[a] e-mail: riccardo.cerulli@lngs.infn.it

occupied state, Γ_i, can be expressed according to [4,5] as: $\Gamma_i = \delta_i^2 \tilde{\Gamma}_i$, where δ_i^2 is the mixing probability of non-fermion statistics allowing the transition to the occupied state i, and $\tilde{\Gamma}_i$ is the width of the corresponding PEP-allowed transition whenever the final state i would be empty.

A search for non-paulian proton spontaneous emission with energy above 10 MeV has been performed by analysing the data collected in a devoted running period of 23.7 days, when the data taking was optimized for the very high energy region [2]. Considering all the 24 detectors in operation during the measurement (total exposure 1.9×10^{33} nuclei $\times$ s) 17 events above 10 MeV are observed by the ten top and bottom detectors in the matrix (11 of them in the four detectors placed in the corners of the matrix). In this very high energy region the expected background is due to the very high energy muons surviving the mountain. Generally they give rise to events, in which more than one detector fires, and thus they can be easily identified and are not competing background for the PEP violating processes. Only muons, impinging the sensitive part of the set-up with a direction that forbids them to hit more than one detector, can play a role. In order to have an accurate estimate of such a background, a suitable simulation has been realized on the basis of the set-up features, of the vertical muon intensity distribution and of the Gran Sasso rock overburden map of ref. [6]. The simulation has shown that the observed 17 events (11 in the four corner detectors and 6 in the remaining 6 detectors in the upper and in the lower rows in the matrix) are fully consistent with the expectations for this kind of background [2].

In the analysis, considering various detectors configurations, upper limits on the rates, λ, of non-paulian processes leading to emission of protons with $E_p \geq$ 10 MeV have been obtained. The final combined result is $\lambda \leq 1.63 \times 10^{-33}$ s^{-1}; it means $\Gamma = \Gamma(^{23}Na) + \Gamma(^{127}I) = \hbar\lambda \leq 1.1 \times 10^{-54} MeV$ (90% C.L.) [2]. This value represents an improvement of about a factor 3 with respect to the limits previously available. The corresponding δ^2 has been derived by considering two models for the momentum distribution functions of the nucleons in the bound state used to estimate $\tilde{\Gamma}$ [5]. Being the obtained values clearly model dependent, the cautious limit $\delta^2 \lesssim 3 - 4 \times 10^{-55}$ has been derived [2].

Finally, considering a $\tilde{\Gamma}$ energy behaviour as in ref. [5], one can roughly estimate a lower limit on the mean life for non-paulian proton emission, obtaining: $\tau_{Na} \gtrsim 2 \times 10^{25}$ yr and $\tau_I \gtrsim 2.5 \times 10^{25}$ yr [2]. These limits improve those previously available for ^{23}Na and ^{127}I [4,5].

Following a PEP violating electronic transition, X-rays and Auger electrons can be emitted due to both the transition itself and to the subsequent atomic shell rearrangements. The total energy release is expected to be at level of the ionization energy for the considered shell [2].

To investigate PEP violating electron transitions in atoms, the data collected by DAMA/LIBRA in the low energy range have been considered (exposure 0.53

ton × yr). In the analysis only the electron atomic transitions of Iodine atoms have been considered, being the energy of the transitions in Sodium atoms below the experimental energy threshold.

In the case of Iodine K-shell PEP violating process, a peak at 32 keV is expected [2]. No evidence for the presence of this peak has been observed and only a limit on the number of PEP violating K-shell transitions in Iodine have been obtained. In particular, lifetimes (τ_{PV}) shorter than 4.7×10^{30} s have been excluded at 90%C.L.. τ_{PV} is related to the lifetime (τ^0) of the PEP allowed transition of outer electrons into a free K-shell when an hole is present: $\tau^0 = \delta_e^2 \tau_{PV}$; δ_e^2 is the probability of the admixed symmetric component for the electron [7]. Since the normal electromagnetic dipole transition to Iodine K-shell is typically $\tau^0 \sim 6 \times 10^{-17}$ s [8], a limit $\delta_e^2 < 1.28 \times 10^{-47}$ (90% C.L.) has been obtained. It is worth noting that the achieved limit is one order of magnitude more stringent than the previously available ones [7,9] and that the sensitivity reached by VIP experiment [10] – by considering a different approach – for Cu electronic transition is of the order of 10^{-28}, with final goal at level of 10^{-31}.

In ref. [11] superficial violations of the PEP due to the possible substructure of electrons have been discussed in composite models of quarks and leptons where the possible finite size of the electron, r_0, is related to δ_e^2. In this framework an upper limit on the electron size: $r_0 < 5.7 \times 10^{-18}$ cm can be derived, corresponding to an energy scale of $E \gtrsim 3.5$ TeV.

References

[1] O. W. Greenberg, *Phys. Rev. Lett.* 64, (1990) 705; R. N. Mohapatra, *Phys. Lett.* **B** 242, (1990) 407; O. W. Greenberg and R. C. Hilborn, *Fund. Phys.* 29 (1999) 397.
[2] R. Bernabei et al., *Eur. Phys. J.* **C** 62, (2009) 327; R. Bernabei et al., *Found. of Phys.* DOI:10.1007/s10701-009-9362-1 (2009) in press.
[3] R. Bernabei et al., *Nucl. Instr. & Meth.* **A** 592, (2008) 297.
[4] H. Ejiri and H. Toki, *Phys. Lett.* **B** 306, (1993) 218.
[5] R. Bernabei et al., *Phys. Lett.* **B** 408, (1997) 439.
[6] M. Ambrosio et al, *Phys. Rev.* **D** 52, (1995) 3793.
[7] H. Ejiri et al., *Nucl. Phys.* **B** (Proc. Suppl.) 28A, (1992) 219.
[8] F. Reines and H. W. Sobel, *Phys. Rev. Lett.* 23, (1974) 954.
[9] P. Belli et al., *Phys. Lett.* **B** 460, (1999) 236.
[10] S. Bartalucci et al., *Phys. Lett.* **B** 641, (2006) 18; E. Milotti et al., *Int. J. Mod. Phys.* **A** 22, (2007) 242.
[11] K. Akama et al., *Phys. Rev. Lett.* 68, (1992) 1826.

ADVANCES IN THE EXPLORATION OF PARTICLE DARK MATTER SIGNALS

Nicolao Fornengo [a]

Department of Theoretical Physics, University of Torino and INFN/Torino, via P. Giuria 1, 1015 Torino, Italy

*Abstract.*The particle physics interpretation of the missing-mass, or dark-matter, problem of cosmological and astrophysical nature is going to be posed under deep scrutiny in the next years. From the particle physics side, accelerator physics will deeply test theoretical ideas of new physics beyond the Standard Model, where a particle physics candidate to dark matter is often naturally obtained. From the astrophysical side, many probes are already providing a great deal of independent information on the signals which can be produced by the galactic or extra-galactic dark matter. The ultimate hope is in fact to be able to disentangle a dark matter signal from the various sources of backgrounds and to extract a coherent picture of new physics from the accelerator physics, astrophysics and cosmology side. A very ambitious and far–reaching project, indeed!

1 Introduction

The presence of dark matter has been assessed on very different scales by a large number of experimental observations, ranging from dynamics of galaxy clusters, to the rotational curves of galaxies, weak lensing, the theory of structure formation and from the energy density budget of the Universe. Non–baryonic cold dark matter is needed, and this fact poses challenges to fundamental Physics since no viable Dark Matter (DM) candidate is present in the Standard Model. Extensions like Supersymmetry or theories of extra–dimensions typically accomodate succesfull DM candidates, like neutralinos or sneutrinos is Supersymmetric (SUSY) theories, or Kaluza–Klein excitations in theories of extra–dimensions.

2 Multichannel search of dark mater

Galactic DM may be sarched for in many ways: by looking at the recoil energy directly deposited in a low–background detector (direct detection) or by looking for annihilation products which are produced in the galactic environment (antimatter, gamma–rays) or in the Earth and Sun (neutrinos). In the following we will briefly report some recent results and give a comparative summary of the various searches.

3 Neutrinos as dark matter messengers

DM captured and accumulated inside bodies like the Earth and the Sun may annihilate and produce a neurtino flux which can escape the body. Recent

[a]e-mail: fornengo@to.infn.it – nicolao.fornengo@unito.it

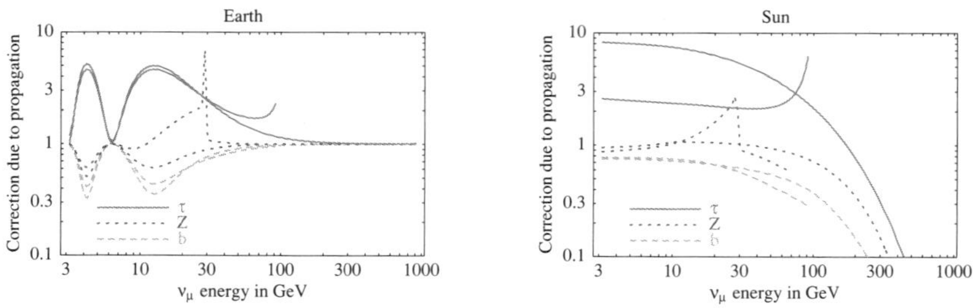

Figure 1: Modifications of neutrino fluxes from DM annihilation due to neutrino propagation [1]. The figures show the ratio of ν_μ fluxes 'with' over 'without' the effects of neutrino propagation (oscillations, absorptions, regeneration). The lines refer to neutrinos from DM annihilations into $\tau\bar{\tau}$ (continuous line), ZZ (dotted) and $b\bar{b}$ (dashed), for $m_{\rm DM} = 100, 1000\,{\rm GeV}$.

advances in the calculation of the theoretical fluxes have dealt with the effect induced by neutrino oscillation and by neutrino interactions with the medium (relevant for the Sun) [1, 2]. The relevance of the effect can be found in Ref. [1]. As a matter of example, for $m_{\rm DM} = 1000\,{\rm GeV}$, the rate is unaffected if annihilation into W^+W^- occurs in the Earth, while it gets reduced to 0.04 of its value if annihilations occur in the Sun. The largest enhancement of the rate due to oscillations occurs in the $\tau\bar{\tau}$ channel. Other channels cause a reduction. At large values of $m_{\rm DM}$ oscillations have a smaller impact.

4 Antiprotons

Annihilation in the galactic environment may produce antimatter, adding an exotic contribution to cosmic rays. The case of antiprotons is shown in Fig. 2, where predictions for the differential flux and for a scan of the SUSY parameter space of a low–energy realization of the Minimal Supersymmetric Standard Model (MSSM) where neutralino is the DM candidate are provided [3, 5]. Theoretical uncertainties of astrophysical origin are sizeable [3]. E.g., in the right panel of Fig.2 the scatter plot can be shifted upward or downward by about a factor of 6–10 [3], due to uncertainties in galactic propagation. Recently, PAMELA provided new reuslts on the $\bar{p}/p$ ratio [6]. Consequences have been derived in [7].

5 Antideuterons

Antideuterons as a DM indirect signal have been proposed in Ref. [8]. Recently a reanalysis has been developed, where also theoretical uncertainties have been quantified [9]. Some results are reported in Fig. 3, where it is shown that the low–energy spectrum offers a unique opportunity to desentangle a signal from

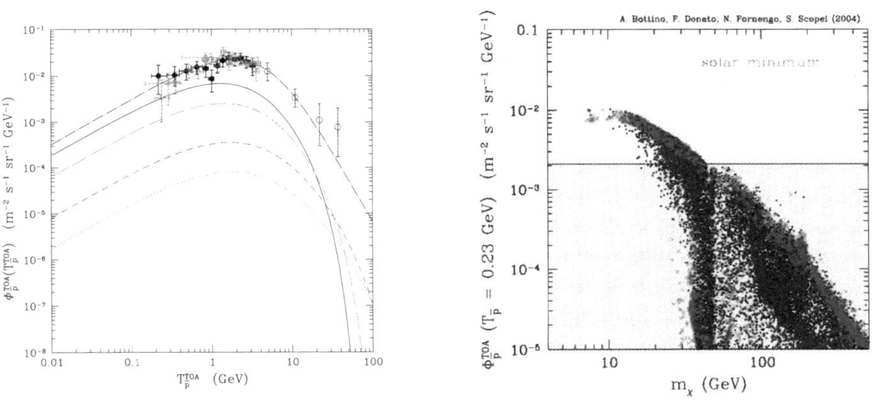

Figure 2: LEFT PANEL: Primary Top-Of-Atmosphere (TOA) antiproton fluxes as a function of the antiproton kinetic energy, for some representative spectra from neutralino annihilation [3]: the solid, long–dashed, short–dashed, dotted lines refer to $m_\chi = 60, 100, 300, 500$ GeV, respectively. The astrophysical parameters correspond to the median choice. Solar modulation is for minimal solar activity. The upper dot–dashed curve corresponds to the antiproton secondary flux [23,24]. Full circles, open squares, stars and empty circles show the data from BESS 1995-97 [25], BESS 1998 [26], AMS [27] and CAPRICE [28]. RIGHT PANEL: Antiproton flux at $T_{\bar{p}} = 0.23$ GeV vs. the neutralino mass, at solar minimum and for the best fit set for the astrophysical parameters [5]. A spherical isothermal DM density profile has been used. The scatter plots are derived by a full scan of the parameter space of non–universal gaugino models which predict low–mass neutralinos [29–31]. Crosses (red) and dots (blue) denote neutralino configurations with $0.095 \leq \Omega_\chi h^2 \leq 0.131$ and $\Omega_\chi h^2 < 0.095$, respectively. The shaded region denotes the amount of primary antiprotons which can be accommodated at $T_{\bar{p}} = 0.23$ GeV without entering in conflict with the experimental BESS data [25,26] and secondary antiproton calculations [32].

the background. The capability to probe the SUSY parameter space with a future experimental mission (GAPS) is shown in the rigth panel of the same figure. Neutralino consifurations with masses up to a few hundreds of GeV may be probed. and rates as large as 100 events are possibile.

6 Positrons

Positrons are currently the most interesting signal to look at, since recently PAMELA detector has released its first data on the positron fraction $e^+/(e^- + e^+)$ [10]. Novel theoretical analysis both for the signal component from DM annihilation in the Galaxy and for the astrophysical background have been recently derived [11, 12]. It has been shown that theoretical uncertainties are relevant also for the positron flux, and they are reported in Fig. 4. The importance of the electron flux in comparing theoretical predictions of the positron fraction with the data has been raised in Ref. [12]: this fact may have impact on the assessing of the presence of an excess in the PAMELA data and

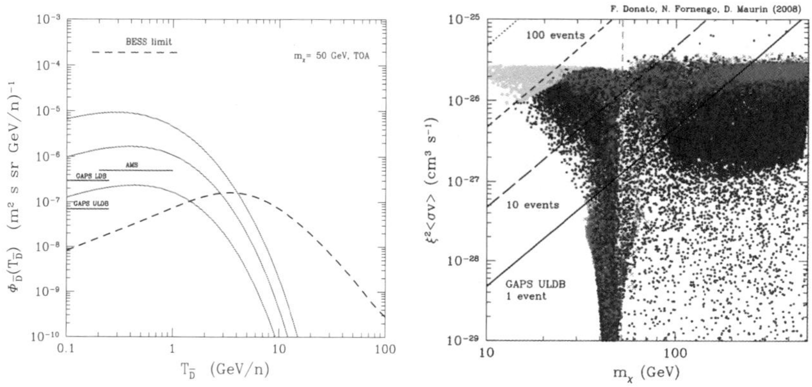

Figure 3: LEFT PANEL: TOA primary (solid lines) and secondary antideuteron fluxes, modulated at solar minimum, for a Weakly Interacting Massive Particle (WIMP) with $m_\chi = 50$ GeV and for the three propagation models which encompass astrophysical uncertainties [9]. The secondary flux (dashed line) is shown for the median propagation model. The upper dashed horizontal line shows the current BESS upper limit on the search for cosmic antideuterons. The three horizontal solid [blue] lines are the estimated sensitivities for (from top to bottom): AMS–02 [33], GAPS on a long (LDB) and ultra–long (ULDB) duration balloon flights [34–36]. RIGHT PANEL: GAPS ULDB reach compared to predictions for neutralino DM in low–energy supersymmetric models, shown in the plane effective annihilation cross section $\xi^2\langle\sigma_{\rm ann}v\rangle_0$ vs. neutralino mass m_χ [9]. The solid, long–dashed and short–dashed lines show the estimate of the capability of GAPS ULDB of measuring 1, 10 and 100 events, respectively, for the median propagation model. The scatter plot reports the quantity $\xi^2\langle\sigma_{\rm ann}v\rangle_0$ calculated in a low–energy MSSM (for masses above the vertical [green] dashed line) and in non–universal gaugino models which predict low–mass neutralinos [29–31]. [Red] Crosses refer to cosmologically dominant neutralinos, while [blue] dots stand for subdominant neutralinos. Grey point are excluded by antiproton searches.

an the determination of the size of such an effect.

7 Gamma rays

Gamma rays are another important tool in studying dark matter. The search for this signal will largely benefit from the FERMI/GLAST detector: a summary of its capabilities may be found in Ref. [13].

8 Summary

A brief comparative analysis of the various detection signals of particle DM may start from stating that antideuterons [8] are the signal which possesses the strongest feature, when compared to the expected background [9]: this occurs

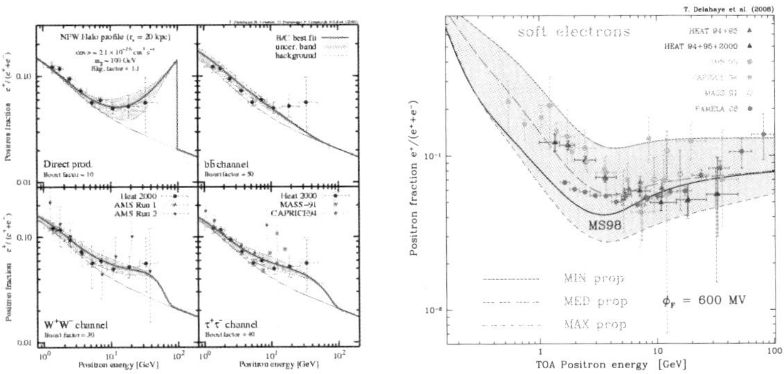

Figure 4: LEFT PANEL: Positron fraction $e^+/(e^- + e^+)$ versus the positron detection energy E for a DM particle with a mass of 100 GeV and for a Navarro-Frenk-White (NFW) profile [11]. The four cases refer to different annihilation final states : direct e^+e^- production (top left), $b\bar{b}$ (top right), W^+W^- (bottom left) and $\tau^+\tau^-$ (bottom right). In each panel, the thick solid [red] curve refers to the best–fit choice of the astrophysical parameters. The colored [yellow] area features the total uncertainty band arising from positron propagation. In each panel, the thin [brown] solid line stands for the background of Ref. [37,38]. Experimental data from HEAT [39], AMS01 [40,41], CAPRICE [42] and MASS [43] are also plotted. RIGHT PANEL: Positron fraction arising from pure cosmic rays interactions (background) as a function of the positron energy, for a soft (left panel) and hard (right panel) electron spectrum [12]. Data are taken from CAPRICE [42], HEAT [39], AMS [40,41], MASS [43] and PAMELA [10].

at low kinetic energies, which is therefore the place where experimental effort should concentrate. Antideuterons in fact appear to offer the best possibility to detect a signal, even in the absence of a boost factor. Foreseen experiment (GAPS, AMS) will have a unique chance to probe this signature directly in the next decade [9].

The antiproton signal at low–energies has a milder feature and when compared to the background the capability to clearly disentagle a signal from the background is hard, expecially when considering that astrophysical uncertainties will still be a major component in the theoretical determination of the signal [3]. In the case DM is heavy, the spectral feature could allow discrimination against the background, but this requires pretty strong boost factors, which appear to be disfavoured by recent studies [14]. Special annihilation mechanism, like the Sommerfeld enhancement [15], could prevent the necessity of large astrophysical boost factors. Current data from PAMELA on the $\bar{p}/p$ ratio nevertheless do not exhibit an excess ascribable to DM annihilation [6,7]. Antiprotons, instead, are suitable to set (potentially strong) bounds on an exotic component in the flux and therefore on the particle DM properties, once theoretical uncertainties are properly taken into account [3–5].

Positrons offer a very interesting possibility and have recently re–gained a lot of attention as a consenquence of the first release of the PAMELA data on the positron fraction [10]. The positron flux from DM annihilation may posses spectral features, depending on the final state of the particle DM annihilation [11]. Similary to the case of the other indirect detection signal, astrophysical uncertainties largely affect also the positron flux [11]. Large theoretical uncertainties affect also the background flux [12], and they have to be taken into consideration when comparison with data is attempted. Theoretical determinations agree with available experimental data [12], including the HEAT positron flux, when theoretical and experimental uncertainties are considered [12]. The most recent results are provided by the PAMELA experiment, but for the moment on the positron fraction, which requires, in the comparison with theoretical determination, to consider also the electron flux. Is has been shown that once astrophysical uncertainties are taken into account, the comparison between the predicted positron fraction and the PAMELA data is indicative of an excess in the case of a hard electron spectrum, while in the case of a soft electron spectrum the identifications of an excess is not conclusive [12] and requires a detailed study which properly takes into account the galactic propagation mechanisms. Typically, the positron signal requires sizeable boost factors in order to prevail over the background: theoretical uncertainties may actually be instrumental in reducing the amount of boost factors required to explain a possibile excess in the data [16] and therefore in making the PAMELA result fully compatible with the current understanding of the astrophysical properties of DM indirect detection signals.

Gamma rays are another important tool for studying DM annihilation in the Galaxy and to probe regions of the galactic environment which are partly different from those explored by charged cosmic rays. Spectral features of the gamma–ray signal are not typically very strong, execept for the case of direct annihilation into a gamma–line, which instead would be a striking signature of DM annihilation. The gamma line is typically strongly suppressed for suitable DM candidates, and therefore very hard to be probed. The gamma–ray signal typically requires (sizeable) boost factors in order to be observable on the top of the astrophysical gamma rays. FERMI will be in the next years a unique laboratory to study gamma rays and it will provide valuable insight also on the DM problem.

Finally, neutrinos from the Earth and the Sun, which can be studied at neutrino telescopes, are an important alternative which nicely complements the other indirect detection techniques. In this case, spectral and angular features may be exploited to desentangle the signal from the atmospheric neutrino background [1, 2]. Neutrino oscillation, and transport in the Sun, have been shown to be relevant effects, which cannot be neglected [1, 2]. The typical signature relies in the search for a muon neutrino flux, which induces upgoing muons in the neutrino telescope. On the other hand, since the DM annihilation and the

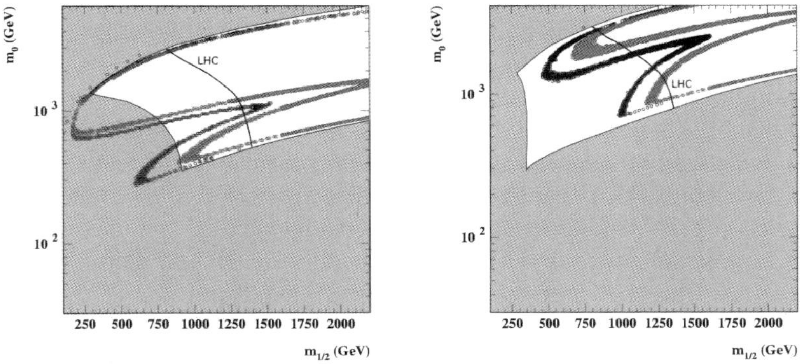

Figure 5: Minimal SUGRA parameter space m_0 (universal soft scalar mass) vs $m_1/2$ (gaugino mass) for $tan\beta = 45$ (left panel) and $tan\beta = 53$ (right panel) and common trilinear coupling $A_0 = 0$. The shaded areas are excluded by bounds on supersymmetry searches and supersymmetry contribution to rare processes. The dark (black) circles show the region of parameter space where the neutralino relic abundance matches the WMAP range for cold dark matter in standard cosmology. The light (red) points refer to the same situation in a scalartensor cosmology. From Ref. [45]. The solid line denotes the expected reach of LHC.

oscillation phenomenon produces also electron and tau neutrinos, additional signatures may be worthwhile to be explored [1].

Complementar to indirect detection is direct detection. A clear signature is offered by the annual modulation of the detection rate. The DAMA/NaI and DAMA/LIBRA detectors actually observe annual modulation in the low-energy single-hit events and this effect has now reached a statistical significance of 8.2σ [17]. In SUSY models, this effect is compatible with neutralinos in the MSSM or in gaugino non–universal schemes [18, 19] or sneutrinos in Left–Right (LR) models or models with see-saw neutrino masses [20]. Experiments which rely on the total counting rate and exploit rejection techniques (like e.g. CDMS [21] and XENON10 [22]) allow to set bounds on the scattering cross section of DM. These experiments currently probe a fraction of the MSSM parameter space for neutralino (see e.g. Ref. [19]) or sneutrino DM [20]. The actual extension of the probe depends on astrophysical (galactic halo properties) and nuclear physics (DM–nucleus interaction) assumptions [18]. Comparison with indirect searches may be found in Refs. [5, 18].

9 Accelerator physics and cosmology

Dark matter candidates are potentially present in almost any extension of the Standard Model of particle physics. In supersymmetric theories with R–parity conservation, both neutralinos and sneutrinos are successful cold dark matter candidates, although other possibilities are present, like e.g. gravitinos. In the next years LHC, and hopefully in the future the ILC, will probe these new physics models and a quite intriguing interconnection between high–energy physics studies, astophysics and cosmology will be posed under deep scrutiny. An example of this interplay is depicted in Fig. 5, where a section of the minimal SUGRA parameter space is shown, together with the expected reach of the LHC. A fraction of this parameter space is already excluded by LEP, Tevatron and studies of rare processes. In the allowed region, Fig. 5 shows the sector which is compatible with a relic neutralino able to explain the dark matter content of the Universe, a sector which is just a small fraction of the relevant parameter space. The same figure also shows the effect induced by the thermal history of the Universe: alternative cosmologies, different from the FRW cosmology, imply a modified decoupling epoch and an ensuing different relic abundance: therefore, the cosmologically relevant regions in parameter space are shifted. The example shown in Fig. 5 refers to scalar–tensor cosmologies. Reconstruction of the particle physics properties of dark matter and the underlying particle physics model represent a window also on the early Universe fisical properties [46].

Acknowledgments

Work supported by research grants funded by Ministero dell'Istruzione, della Università e della Ricerca (MIUR), Università di Torino, Istituto Nazionale di Fisica Nucleare (INFN), Italian Space Agency (ASI) under contract N° ASI-INAF I/088/06/0 and the spanish MICINNs Consolider–Ingenio 2010 Programme under grant MULTIDARK CSD2009-00064.

References

[1] M. Cirelli et al., Nucl. Phys. B727 (2005) 99, Erratum-ibid.B790 (2008) 338.
[2] M. Blennow, J. Edsjo, T. Ohlsson, JCAP 0801 (2008) 021.
[3] F. Donato et al., Phys. Rev. D69 (2004) 063501.
[4] A. Bottino, F. Donato, N. Fornengo, and P. Salati, Phys. Rev. D72 (2003) 083518.
[5] A. Bottino, F. Donato, N. Fornengo, and S. Scopel, Phys. Rev. D70 (2004) 015005.
[6] O. Adriani et al.(PAMELA Collaboration), arXiv:0810.4994.
[7] F. Donato, D. Maurin, P. Brun, T. Delahaye and P. Salati, arXiv:0810.5292.
[8] F. Donato, N. Fornengo, P. Salati, Phys. Rev. D62 (2000) 043003.

[9] F. Donato, N. Fornengo, D. Maurin, Phys. Rev. D 78 (2008) 043506.
[10] O. Adriani et al., arXiv:0810.4995.
[11] T. Delahaye et al., Phys. Rev. D 77 (2008) 063527.
[12] T. Delahaye et al., arXiv:0809.5268.
[13] M. Alvarez et al., arXiv:0712.1548.
[14] J. Lavalle, Q. Yuan, D. Maurin, X.J. Bi, arXiv:0709.3634.
[15] M. Lattanzi and J.I. Silk, arXiv:0812.0360 and references therein.
[16] N. Fornengo and R. Lineros, in preparation.
[17] R. Bernabei et al., Eur. Phys. J. C56 (2008) 333.
[18] A. Bottino, F. Donato, N. Fornengo, and S. Scopel, Phys. Rev. D77 (2008) 015002.
[19] A. Bottino, F. Donato, N. Fornengo, and P. Scopel, Phys. Rev. D78 (2008) 083520.
[20] C. Arina, N. Fornengo, JHEP 0711 (2007) 029
[21] Z. Ahmed et al., arXiv:0802.3530.
[22] J. Angle et al., Phys. Rev. Lett. 100 (2008) 021303.
[23] F. Donato et al., Ap. J. 563 (2001) 172.
[24] D. Maurin et al., astro-ph/0212111.
[25] S. Orito et al. (BESS Collaboration), Phys. Rev. Lett. 84 (2000) 1078.
[26] T. Maeno et al. (BESS Collaboration), Astropart. Phys. 16 (2001) 121.
[27] M Aguilar et al. (AMS Collaboration), Phys. Rep. 366 (2002) 331.
[28] M. Boezio et al. (CAPRICE Collaboration), Astrophys. J. 561 (2001) 787.
[29] A. Bottino, N. Fornengo, G. Polesello, and S. Scopel, Phys. Rev. D77 (2008) 115026.
[30] A. Bottino, F. Donato, N. Fornengo, and S. Scopel, Phys. Rev. D68 (2003) 043506.
[31] A. Bottino, N. Fornengo, and S. Scopel, Phys. Rev. D67 (2003) 063519.
[32] F. Donato et al., Ap. J. 563 (2001) 172.
[33] V. Choutko and F. Giovacchini (AMS Collab.), Proc. ICRC 2007, Vol. 4, p. 765 (2008).
[34] C.J. Hailey et al., Nucl. Inst. and Meth. in Phys. Res. B 214 (2004) 122.
[35] C.J. Hailey et al., JCAP 1 (2006) 7.
[36] J.E. Koglin, Proc. of TAUP 2007, JOP Conf. Series 120 (2008).
[37] E.A. Baltz and J. Edsjo, Phys. Rev. D59 (1999) 023511.
[38] I.V. Moskalenko and A.W. Strong, Ap. J. 493 (1998) 694.
[39] S. W. Barwick et al., Ap. J. 482 (1997) L191.
[40] M. Aguilar et al., Phys. Lett. B646 (2007) 145.
[41] J. Alcaraz et al., Phys. Lett. B 484 (2000) 10.
[42] M. Boezio, et al., Ap. J. 532 (2000) 653.
[43] C. Grimani, et al., A.&A. 392 (2002) 287.

[44] A. Bottino, F. Donato, N. Fornengo, and P. Scopel, Phys. Rev. D72 (2003) 083521.

[45] R. Catena, N. Fornengo, A. Masiero, M. Pietroni and M. Schelke, JHEP **0810** (2008) 003

[46] R. Catena, N. Fornengo, M. Pato, L. Pieri and A. Masiero, arXiv:0912.4421 [astro-ph.CO]; F. Donato, N. Fornengo and M. Schelke, JCAP **0703** (2007) 021; M. Schelke, R. Catena, N. Fornengo, A. Masiero and M. Pietroni, Phys. Rev. D **74** (2006) 083505.

DARK MATTER ANNIHILATION IN THE GALAXY

Veniamin Berezinsky[1], Vyacheslav Dokuchaev[2 a], Yury Eroshenko[2 b]
[1] *INFN, Laboratori Nazionali del Gran Sasso, I–67010 Assergi (AQ), Italy*
[2] *Institute for Nuclear Research RAS, 117312, Moscow, Russia*

Abstract. Surviving of small dark matter clumps is considered in early hierarchical structure formation and in the Galaxy for the standard power-law spectrum of primordial perturbations. The dense remnants of clumps provide a suitable contribution to the amplification of dark matter annihilation signal in the Galaxy. The formation and evolution of superdense clumps are also studied in the two non-standard scenarios: spike in the spectrum of primordial perturbations and cosmic strings loops as seeds.

1 Introduction

Dark matter (DM) particles may annihilate most effectively in the dense regions of Galactic halo, such as Galaxy center [1] or DM substructures in the form of DM clumps (minihaloes). As a result the bright point gamma-sources arise or the background signal is amplificated by several times or even by several orders of magnitude. This makes the DM annihilation more easy to detect in the future observations. However some witnesses of annihilation were already obtained in spectrum of cosmic rays and in the observations of Galactic center, the convincing results are still awaited just as the direct detection and/or production of DM particles in accelerator experiments.

The main topic of our report is the small-scale DM clumps of substellar masses. We discuss different possible conditions for the clumps formation in the early universe, their subsequent evolution, and the annihilation of DM inside the clumps.

2 Clumps in standard cosmological scenario with an inflationary-produced primordial fluctuation spectrum

In the case of neutralino, considered as DM particles, the formation of clumps begins from the minimum masses $M_{\min} \sim 10^{-6} M_{\odot}$, determined by dumping processes [2]. For the Harrison-Zeldovich spectrum of primordial fluctuations with CMB normalization the first small-scale DM clumps are formed at redshift $z \sim 60$ (for 2σ fluctuations) with a mean density 7×10^{-22} g cm^{-3}, virial radius 6×10^{-3} pc and internal velocity dispersion 80 cm s^{-1}.

The formed clumps are subjected to the tidal destruction at early stages of structures formation. We calculate [3] the mass function of clumps (fraction of

[a]e-mail: dokuchaev@ms2.inr.ac.ru
[b]e-mail: eroshenko@ms2.inr.ac.ru

DM in the form of clumps) within a hierarchical clustering as

$$\xi_{\rm int}\frac{dM}{M} \simeq 0.02(n+3)\,\frac{dM}{M}, \tag{1}$$

where an effective power-law index n is $n = -3(1 + 2\partial \log \sigma_{\rm eq}(M)/\partial \log M)$ and depends very weakly on M, and $\sigma_{\rm eq}(M)$ is the mean fluctuation at the time of the matter-radiation equality. This mass function gives that only a small fraction of these clumps, $\sim 0.1 - 0.5\%$, in each logarithmic mass interval $\Delta \log M \sim 1$ survives the stage of hierarchical clustering. The mass function is in good agreement with the numerical simulations [4].

The clumps survived the early stages can be disrupted further in the galaxies by tidal interactions with stars. We calculate [3] the destruction of clumps by stars in the bulge and halo and by the Galactic disk. It was demonstrated that clump remnants may survive through the tidal destruction during the lifetime of the Galaxy if the radius of the clumps cores are rather small. Destruction of the outer part of the clump affects the annihilation rate relatively weakly, because the annihilation signal is dominated by the dense part of the clump. The survived dense remnants of clumps provide a large contribution to the annihilation signal in the Galaxy. The uncertain parameter of the scenario is a value of core radius there central density in clump stops to grow.

3 Superdense clumps and superheavy DM

Spiky features in the spectrum of adiabatic inflationary perturbations may produce the superdense clumps even at radiation dominated stage of universe evolution. The production of primordial black holes constrains the allowed parameters of this model. The superdense clumps evolve as isolated objects for a long time after formation, and therefore they are not destroyed by tidal forces inside the later larger-scale structure. The ordinary $\sim$ 100 GeV neutralinos as DM particles are strongly disfavored in this scenario by too large gamma radiation from particles annihilation.

We consider the superheavy $\sim 10^{11}$ GeV neutralinos as possible constituent DM particles [5]. The minimum masses of clump $M_{\rm min}$ is extremely small for the superheavy DM. The two-body gravitational relaxation in central part of clumps may initiates the "gravithermal catastrophe", when the initial density profile turns into isothermal one $\rho \propto r^{-2}$ down to a new, extremely small core in the center, limited by Fermi degeneracy or impetuous annihilation. Small core and large density in the clumps centers may lead to the observed annihilation signal even for very small annihilation cross-section of superheavy particles. In the case of a higgsino, the annihilation signal is additionally enhanced by the Sommerfeld effect.

Therefore, these superdense clumps can in principle be observed through the

γ-radiation from DM annihilations and tidally by gravitational wave detectors. These result will be published in near future [6].

4 Clumps seeded by cosmic string loops

Linear topological defect — cosmic strings can be formed in early cosmological phase transitions. Along with infinite strings there are possibility of closed loops formation in a network of curved cosmic strings due to their interconnections [7]. The annihilation of the dark matter in clumps originated around cosmic string loops is studied. These clumps form at the radiation dominated stage and may have very large densities. There are possibility to avoid adiabatic restriction on the clumps densities [8]. Only low velocity loops can produce the clumps. The probability of low velocity loop formation is very small, but even such tiny fraction of formed loops may produce the dense clump population and significant annihilation signal.

We present the new constraint on tension μ (mass per unit length of the string) which obtained from the dark matter particles annihilation in the dense clumps seeded by the loops in comparison with observed γ-background as upper limit.

We conclude that 100 GeV neutralino DM is incompatible with range of strings tensions $1 \times 10^{-9} < G\mu/c^2 < 5 \times 10^{-9}$, where $\mu_{-8} = G\mu/(10^{-8}c^2)$, because the γ-ray signal exceeds the EGRET limit in this case.

Acknowledgments

This work was supported by the Russian Federal Agency for Science and Innovation under state contract 02.740.11.5092 and by the grants of the Leading scientific school 959.2008.2 and 438.2008.2.

References

[1] V.S.Berezinsky, A.V.Gurevich, K.P.Zybin, *Phys.Lett.* **B** 294, 221 (1992).

[2] E.Bertschinger, *Phys.Rev.* **D** 74, 063509 (2006).

[3] V.Berezinsky, V.Dokuchaev, Y.Eroshenko, *Phys.Rev.* **D** 77, 083519 (2008).

[4] J.Diemand, B.Moore, J.Stadel, *Nature* 433, 389 (2005).

[5] V.Berezinsky, M.Kachelries, M.A.Solberg, *Phys.Rev.* **D** 78, 123535 (2008).

[6] V.Berezinsky, V.Dokuchaev, Yu.Eroshenko, M.Kachelriess, M.A.Solberg, (2010). (in preparation)

[7] K.D.Olum, A.Vilenkin, *Phys.Rev.* **D** 74, 063516 (2006).

[8] E.W.Kolb, I.I.Tkachev, *Phys.Rev.* **D** 50, 769 (1994).

LEPTOGENESIS

Pasquale Di Bari [a]
School of Physics and Astronomy, University of Southampton, SO17 1BJ
Department of Physics and Astronomy, University of Sussex, Brighton, BN1 9QH
*Abstract.*The status of leptogenesis is reviewed, focusing in particular on the neutrino mass bounds and on the possibility to test New Physics.

1 The cosmological puzzles

All modern cosmological observations are explained, in a minimal way, by the ΛCDM model. The parameter space is over constrained by the observations that yield precise measurements of the involved cosmological parameters. The robustness and the precision of the cosmological results make therefore even more striking the existing clash between the ΛCDM model and the Standard Model of Particle Physics (SM) due to four basic features of the ΛCDM model unexplained within the SM: i) the nature of cold Dark Matter; ii) the matter-antimatter asymmetry of the Universe; iii) the nature of Inflation and finally iv) why the Universe is accelerating at present. Therefore, solutions to the cosmological puzzles, seem to provide a strong motivation for new physics. In particular the Dark Matter conundrum strongly points to the existence of some new particle that is stable on cosmological scales. In this review I will focus on the matter-antimatter asymmetry puzzle discussing recent developments in Leptogenesis, a model of Baryogenesis where the matter-antimatter asymmetry of the Universe is closely related to the observed neutrino masses and mixing.

2 Vanilla leptogenesis

From the acoustic peaks in the power spectrum of the Cosmic Microwave Background (CMB) anisotropies, the net baryon to photon number ratio is measured to be $|\Delta\eta_B| \equiv |\eta_B - \eta_{\bar{B}}| = (6.2 \pm 0.15) \times 10^{-10}$, in agreement with the Big Bang Nucleosynthesis estimation. These measurements are only sensitive to the η_B absolute value but not to the sign of η_B. However, combining CMB anisotropies and spectrum with cosmic rays observations, it is possible to rule out a symmetric (observable) Universe with matter-antimatter domains concluding that Δ_B is everywhere positive and therefore that there is indeed an overall matter-antimatter asymmetry. Notice moreover that so far any search of primordial anti-matter has produced negative results. Therefore, it is fair to conclude that $\Delta_B = \eta_B$, barring astrophysical contaminations.

The need of an inflationary stage in the history of the early Universe implies that the asymmetry had to be generated either after or at the end of inflation by some dynamical process, called baryogenesis. A successful model of baryogenesis has to fulfill the three Sakharov conditions and remarkably this

[a]e-mail: pdb1d08@soton.ac.uk

happens within the SM to some extent. However, because of the current experimental lower bound on the Higgs boson mass and of the current constraints on *CP* violation in the quark sector, any attempt to explain the observed baryon asymmetry fails and for this reason, a solution to the matter-antimatter asymmetry problem requires new physics.

A particularly intriguing solution is offered by leptogenesis [1] because it relies on the same minimal extension of the SM that can also explain the results from neutrino oscillation experiments and the stringent upper bound on the absolute neutrino mass scale. This extension is the (type I) see-saw mechanism.

By adding N right-handed (RH) neutrinos with Yukawa couplings h and a Majorana mass term M to the SM lagrangian, after electro-weak symmetry breaking a usual Dirac mass term $m_D = h\,v$ is generated by the Higgs vev. In the see-saw limit, $M \gg m_D$, the spectrum of mass eigenstates splits into a set of light neutrino masses $m_1 \leq m_2 \leq m_3$ described by the see-saw formula and into a set of heavy neutrino masses $M_1 \leq M_2 \leq \ldots \leq M_N$ almost coinciding with the eigenvalues of the Majorana mass term. For definiteness, we will consider the case $N = 3$ that is also the most reasonable choice, though current neutrino masses and mixing can well described by a choice $N = 2$ as well. We will see later on how a particularly attractive scenario of leptogenesis, the N_2-dominated scenario, supports $N \geq 3$. It is also reasonable to assume $m_D \gtrsim M_{\rm ew} \sim 100\,{\rm GeV}$ implying $M_i \gg 100\,{\rm GeV}$. The heavy RH neutrinos N_i decay into leptons and anti-Higgs with a decay rate Γ or into anti-leptons and Higgs with a decay rate $\bar{\Gamma}$. The total decay rate is given by $\Gamma_D = \Gamma + \bar{\Gamma}$. By assuming that that the reheating temperature of the early Universe $T_R \gtrsim M_1 \gg 100\,{\rm eV}$ has two important consequences: first, the RH neutrinos can be thermally produced from the thermal bath without the need of an external production mechanism and second, the lepton number produced in the decays is partly converted into a baryon number by ($B - L$ conserving) sphaleron processes.

In a traditional minimal description of leptogenesis (*vanilla leptogenesis*), the final asymmetry is dominated by the contribution of the lightest RH neutrino decays and can be simply calculated as $N_{B-L}^{\rm f} = \varepsilon_1\,\kappa^{\rm f}$, where ε_1 is the N_1 total *CP* asymmetry and $\kappa_1^{\rm f}$ is the final value of the N_1 efficiency factor. This is basically given by the number of N_1's that decay out-of-equilibrium [2] and can be calculated using simple analytical solutions of the set of Boltzmann equations. Vanilla leptogenesis relies on the following set of assumptions and approximations [3]: i) the flavour composition of the leptons in the final states is neglected; ii) the heavy RH neutrino mass spectrum is assumed to be such that $M_2 \gtrsim 3\,M_1$ and $M_3 \simeq M_2$; iii) there is no interference between the heaviest RH neutrino and the next-to-lightest RH neutrino, i.e. $(m_D^\dagger\,m_D)_{23} = 0$. If one additionally assumes that there are no fine tuned cancelations among the terms giving the RH neutrino masses in the see-saw formula, then an upper bound on the total *CP* asymmetry, $\varepsilon_1 \leq \varepsilon_1^{\rm max} \simeq 10^{-6}\,(M_1/10^{10}\,{\rm GeV})$, holds as well and, imposing $\eta_B^{\rm max} \equiv \varepsilon_1^{\rm max}\,\kappa_1^{\rm f} > \eta_B^{CMB}$, one obtains an allowed region in the

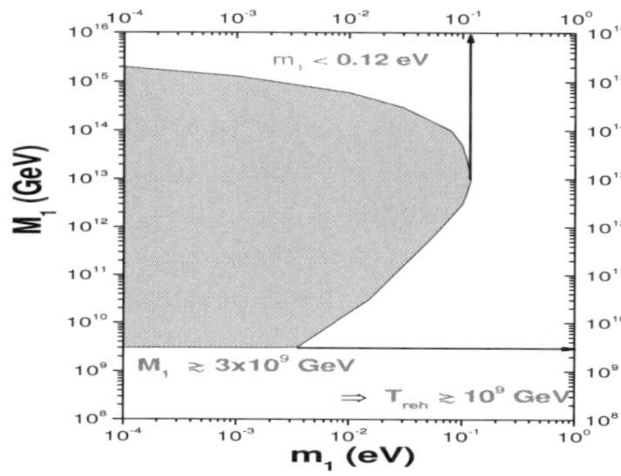

Figure 1: Neutrino mass bounds in the vanilla scenario.

plane (m_1, M_1) that is shown in Fig. 1. One can notice the lower bound on M_1, also implying a lower bound on the reheating temperature $T_{\rm reh}$, and the upper bound on m_1, incompatible with quasi-degenerate neutrino models.

3 Flavour effects

There are two kinds of flavour effects that the vanilla scenario neglects: heavy flavour effects, i.e. the influence of the heavier RH neutrinos on the final asymmetry, and light flavour effects, i.e. the influence of the flavour composition of the leptons produced in the decays on the final asymmetry. In the first two subsections we keep separate the discussion of the two effects but in the third one we discuss how actually their interplay can have very interesting applications.

3.1 Light flavour effects

Let us start assuming that the final asymmetry is influenced only by the decays of the lightest RH neutrinos neglecting heavy flavour effects. If $M_1 \gtrsim 5 \times 10^{11}\,{\rm GeV}$, then the flavour composition of the quantum states of the final leptons has no influence on the final asymmetry (unflavoured regime). This happens because nothing happens between the production of a lepton from a N_1 decay and the subsequent inverse decay with an Higgs and the lepton flavour composition does not play any role.

However, if $5 \times 10^{11}\,{\rm GeV} \gtrsim M_1 \gtrsim 10^9\,{\rm GeV}$, then between one decay and an inverse decay, on average, the produced leptons interact with tauons in a way that the lepton quantum state coherence breaks down. Therefore, at the inverse decays, the leptons are an incoherent mixture of a tauon component and of a (still coherent!) superposition of an electron and of a muon component that we will indicate with '$e + \mu$'. The important thing is that the fraction of asymmetry stored in each component and the fraction of wash-out strength acting on it, are not proportional in general. Therefore, the dynamics of the two flavour asymmetries, the tauon and the $e + \mu$ asymmetries, have to be calculated separately and in this way the final asymmetry can be considerably different from the result in the unflavoured regime. If $M_1 \lesssim 10^9\,\mathrm{GeV}$, then even the coherence of the electron and of the muon superposition breaks down between decays and inverse decays and a full three flavour regime is realized.

There are three kinds of consequences. First, the wash-out can be considerably reduced with respect to the unflavoured calculation. Second, the low energy phases play a direct role contributing to a second source of *CP* violation in the flavoured *CP* asymmetries and the most striking application is that the same source of *CP* violation that could have place in neutrino oscillations, can also be the unique source of *CP* violation for successful leptogenesis, though under some quite strict conditions on the RH neutrino mass spectrum. A third consequence is that the flavored *CP* asymmetries contain additional unbounded terms. In this way, by allowing some acceptable cancelations in the seesaw formula among the light neutrino mass terms and a mild RH neutrino mass hierarchy ($M_2/M_1 \sim 3$), the lower bound on the reheating temperature can be lowered by about one order of magnitude down to $10^8\,\mathrm{GeV}$.

3.2 Heavy flavour effects

There are basically two reasons why the influence of the heavier RH neutrinos on the final asymmetry can be neglected under the assumptions that define the vanilla scenario: either the asymmetry produced by the $N_{2,3}$'s is negligible because the *CP* asymmetries are suppressed in the hierarchical limit or because, even though a sizeable asymmetry is produced around $T \sim M_{2,3}$, this is later on washed out by the lightest RH neutrino inverse processes. However, there are different situations where heavy flavour effects have to be taken into account.

First of all one can always consider the quasi-degenerate limit when $(M_{2,3} - M_1)/M_1 \ll 1$. In this case the *CP* asymmetries are not suppressed and the wash-out from the lightest RH neutrinos is only partial. Second, even assuming a strong RH neutrino mass hierarchy, there is always a choice of the parameters where N_1 decouples and its wash-out vanishes. For the same choice of the parameters the N_2 *CP* asymmetry is unsuppressed if $M_3 \lesssim 10^{15}\,\mathrm{GeV}$. In this case a N_2-dominated scenario is realized [6]. Notice that the existence of a third heaviest RH neutrino species is crucial. Third, even assuming a strong mass hierarchy and an undecoupled N_1, if $M_1 \gtrsim 10^{12}\,\mathrm{GeV}$, then the asymmetry produced by the heavier RH neutrinos, in particular by the N_2's with unsuppressed total *CP* asymmetry, can be sizeable and in general is not completely washed-out by the lightest RH neutrino processes since there is an orthogonal component in the three light flavour space that escapes the N_1 wash-out. Notice that if the mass hierarchy is mild with $(M_3 - M_1)/M_1 \lesssim 10$, then, for the same reason, even the asymmetry produced by the N_3's can be sizeable and can avoid the N_1 and N_2 wash-out.

3.3 Flavoured N_2-dominated scenario

There is a fourth interesting case where the asymmetry from the heavier RH neutrinos dominates the final asymmetry due to the interplay of light and heavy flavour effects [7]. Indeed, even assuming a strong mass hierarchy, an undecoupled N_1 and $M_1 \lesssim 10^{12}\,\text{GeV}$, the wash-out from the lightest RH neutrinos can be avoided. If the production occurs in the two flavour regime, then the asymmetry produced by the heavier RH neutrinos splits into an incoherent mixture of light flavour components. The N_1 wash-out in one particular flavour can then be negligible in quite a wide region of the parameter space. This scenario extends the validity of the N_2-dominated scenario since it is not necessary that N_1 fully decouples but it is sufficient that it decouples only in some particular light flavor.

The applicability of the flavoured N_2-dominated scenario proves to be even larger when flavour coupling is considered [10]. Recently it has been realized that, accounting for the Higgs and for the quark asymmetries, the dynamics of the flavour asymmetries couple and the lightest RH neutrino wash-out can be circumvented for a larger region of the parameter space. Another new interesting effect is phantom leptogenesis. This is a pure quantum-mechanical effect that allows parts of the electron and of the muon asymmetries to be completely unwashed at the production when $T \sim M_2 \gg 10^9\,\text{GeV}$.

4 Grand unified theories and leptogenesis

Flavour effects are particularly relevant when testing theories beyond the SM with leptogenesis. A particularly interesting example is offered by scenarios that emerge quite naturally within grand-unified theories and in particular within $SO(10)$. In this case the neutrino Dirac mass matrix is somehow related to the quark mass matrices and in particular its eigenvalues typically exhibit the same strong hierarchical pattern. Moreover the matrix that acts on the lepton doublets when switching from the basis where the neutrino Dirac mass matrix is diagonal to that basis where the charged lepton mass matrix is diagonal, is related to the CKM matrix and is typically given by the combination of rotations with small mixing angles and in first approximation it can be assumed to be the identity matrix. Under these assumptions one has that $M_1 \ll 10^9\,\text{GeV}$ and within a vanilla scenario where, only the N_1 production is considered, the final asymmetry would not reproduce the observed one.

On the other hand $M_2 \gg 10^9\,\text{GeV}$ and it has been shown that when the N_2 production and light flavor effects are consistently taken into account, then there are different points in the parameter space where successful leptogenesis is possible [8]. Notice that within this framework all high energy see-saw parameters can be expressed in terms of the low energy parameters. It is therefore interesting that the allowed region in the parameters space for successful leptogenesis translate into precise regions in the low energy parameter space that can be tested in future low energy neutrino experiments.

5 Discrete flavour symmetries and leptogenesis

Pure heavy flavour effects seem to be relevant when leptogenesis is embedded within theories that try to explain the emerging tribimaximal mixing structure in the leptonic mixing matrix via flavour symmetries. It has been shown in particular that if the symmetry is unbroken then the *CP* asymmetries of the RH neutrinos would exactly vanish. On the other hand when the symmetry is broken, for the naturally expected values of the symmetry breaking parameters, then the observed matter-antimatter asymmetry can be successfully reproduced [9]. It is interesting that in this picture one has a RH neutrino mass spectrum with $10\,M_1 \gtrsim M_3 \gtrsim M_1 \gtrsim 10^{12}\,\text{GeV}$. One has therefore that all the asymmetry is produced in the unflavoured regime and the mass spectrum is only mildly hierarchical. At the same time the small symmetry breaking imposes a quasi-orthogonality of the three lepton quantum states produced in the RH neutrino decays. Under these conditions the wash-out of the asymmetry produced by one RH neutrino species from the inverse decays of a lighter RH neutrino species is essentially negligible. The final asymmetry then receives a non negligible contribution from the decays of all three RH neutrinos species.

In conclusion, a systematic study of flavor effects, light and heavy, seems to show that leptogenesis has entered a new interesting stage where new possibilities to test models beyond the SM seem to emerge.

Acknowledgments

I acknowledge financial support from the NExT Institute and SEPnet. I thank the organizers for the invitation to an enjoyable and stimulating conference.

References

[1] M. Fukugita, T. Yanagida, Phys. Lett. **B 174** (1986) 45.
[2] W. Buchmuller et al., Annals Phys. **315** (2005) 305.
[3] S. Blanchet and P. Di Bari, Nucl. Phys. B **807** (2009) 155.
[4] A. Abada et al. JCAP **0604**, 004 (2006); E. Nardi et al., JHEP **0601** (2006) 164.
[5] R. Barbieri et al., Nucl. Phys. B **575** (2000) 61.
[6] P. Di Bari, Nucl. Phys. B **727** (2005) 318.
[7] O. Vives, Phys. Rev. D **73** (2006) 073006.
[8] P. Di Bari and A. Riotto, Phys. Lett. B **671** (2009) 462.
[9] E. Bertuzzo et al., JHEP **0911** (2009) 036.
[10] S. Antusch, P. Di Bari, D. A. Jones and S. F. King, arXiv:1003.5132.

GRAVITATIONAL WAVES: FROM THE PAST TO THE FUTURE OF THE GLOBAL NETWORK OF DETECTORS

Fulvio Ricci [a]
Department of Physics, University of Rome Sapienza, I-00185 Roma, Italy
INFN Sezione di Roma, Piazza A. Moro 2, I-00185 Roma, Italy

Abstract. The global network of gravitational wav detectors is a reality: since May, 2007 the LIGO Scientific community and the Virgo collaboration started to analysis in common all the data produced by the gravitational wave interferometers running in Europe and in USA. Here we review some of the recent results obtained by the two collaboration and we discuss the impact of the present and future observation for fundamental physics and astrophysics.

1 Introduction

Since 1960 various laboratory experiments have been carried on to detect gravitational waves (GW). The detector technique evolved from the simpler room resonant detectors of J. Weber to the cryogenic ones and at present the field is dominated by the operation of the giant interferometric devices installed in Europe, Japan and USA. The device sensitivity and its bandwidth is increased progressively expanding the search domain on a larger variety of GW signals. However, since the Weber's epoch it was clear that this experimental challenge requires the coordinate effort of more than one detector installed in different Earth locations.

The first attempt to observe a gravitational wave signal using a world wide network was in 1986 the coincidence experiment of the three cryogenic resonant-mass detectors set up by the universities of the Louisiana state, Rome and Stanford. [1]. In the recent years the International Gravitational Event Collaboration , derived upper limit on the event rate of GW bursts using observational data of the resonant detectors in operation taken during almost three years of run [2].

Then, since May, 2007 at the start up of the first Virgo science run, the LIGO Scientific community and the Virgo collaboration agreed to analysis in common all the data produced by the interferometers of the GEO, LIGO and Virgo project. The data are produced by five interferometers: the LIGO ones are located in Hanford (H1 and H2, 4 and 2 km long respectively) and in Livingston (L1, 4 km long), while in Europe there is Virgo, a 3 km interferometer located in Cascina (Italy) and GEO, 0.6 km long installed in Hannover (Germany).

The scientific motivations for a global GW detector network are strong:

- The GW interferometers have broad sky coverage, but are not omni-directional: a globally distributed network allows for maximal sky coverage

[a] e-mail: fulvio.ricci@roma1.infn.it

- It permit the transient source location by triangulate (or N-angulate)
- It improves the detection confidence by the signal redundancy in multiple detectors
- It permits a more accurate estimates of amplitude and phase polarization, because the array of oriented detectors is sensitive to two polarizations
- The coherent combination of the data streams leads to better sensitivity and an optimal waveform and coordinate reconstruction

In conclusion, although the first detections might come from one single detector, the best use of the detectors consists in running all of them together in a VLBI kind of mode for insuring a strong impact on astrophysics. Moreover, the credibility of the first claims for detection would be infinitely higher if the event is seen by independent detectors and triggered by another kind of observation (optical, neutrinos or γ rays).

In the following sections we recall some of the recent results obtained by these interferometers looking for both transient and continuous GW signals. inally we discuss shortly the case of the GW stochastic background detection.

2 The search of GW transient signals.

With interferometer sensitivities extending to astrophysically signicant distances, gravitational wave detectors can now add value to even marginally signicant sources detected via electromagnetic or particle telescopes. The search success depends not only on the intrinsic GW luminosity of the source but also on whether something about the signal model is known. In this case we are able to reduce the amount of data to be searched taking advantage of the combination of sky location and timing information (triggers) from optical and neutrino observatories. Moreover, GW data associated with optical signals from a well defined population can be selected, setting a specialized limit that, with a sample of candidate events reach enough, can be signicantly better than the accuracy of an individual measurement [3].

The main example of this search approach is the analysis carried on in coincidence to the occurrence of Gamma Ray Bursts (GRB) [6], [4]. In particular, we refer to the interesting example of the GRB 070201 event, a short GRB whose error box intersects the spiral arms of the Andromeda galaxy (M31). Previous optical studies of short GRBs had suggested a neutron star merger model origin for many short GRBs. When the event occurred the LIGO detectors was in Science mode operation and, given the close proximity of M31, the GW analysis could rule out a binary merger origin at and beyond the distance to M31.

In addition, if the event location is M31, the event is more consistent with the known emission from magnetars. However, the search for completely unmodeled emission found no sign of gravitational waves, and the derived upper limit on the gravitational wave energy flux results to be orders of magnitude greater than the known optical flux.

Recently an extended LIGO-Virgo search for GW - GRB coincidences have been concluded: 212 GRB events occurred during the fifth scientific run of LIGO (S5) and the first Virgo science run (VSR1). 137 events were selected as double GW coincidence (any two of LIGO Hanford, LIGO Livingston, Virgo) but any of them can be flag as a GW detection. Thus, a lower limits on GRB event distance was established assuming an energy emission of $E_{GW} = 0.01$ $M_{\odot}c^2$.

This analysis can be regarded as a preliminary exercise. A coalescent binary GW signal at redshift z is indistinguishable from a local binary [7], and the future interferometers should detect several of them. From the binary events the luminosity distance D_L can be identified and by the γ ray counterpart the red shift z measured. Then, by the best fit of the $D_L(z)$ data it should be possible to infer the dark energy equation-of-state, dark matter and dark energy density parameters [8].

3 The search of continuous GW signals.

We know that potential sources of continuous waves (CW) exist. In particular we observed in the e.m. spectrum the signal emitted by $\sim 2,000$ pulsar. However, robust evolutionary models predict the existence of ~ 1 billion of neutron stars in the Galaxy and a fraction of these emit in the sensitivity band of the terrestrial interferometers. The braking index value of the pulsar $n = \nu\ddot{\nu}\dot{\nu}/\dot{\nu}^2$, which depends on the rotation frequency ν and its derivatives, constrains the gravitational wave emission.

The observed value $n = 2.5$ still is not well understood on theoretical grounds, but since quadrupolar radiation has $n = 5$ it implies that only a small fraction of the spin-down power is emitted in gravitational waves. C. Palomba estimates that the highest possible today h_0 is about 40 % of the spin-down limit [9].

For the CW signals the antenna network is less crucial for the detection. The Doppler effect due to the Earth motion permits the source localization and taking advantage of the signal persistence the observation can be extended over several independent time periods. On the other hand, because of the signal weakness, also in this case the coherent combination of data from different detectors improves the sensitivity and the statistical confidence. Moreover we can take advantage of the detector differences: Virgo and LIGO have different sky coverage and frequency dependence of their sensitivity curve. Thus, we can optimize the selection of the data produced by the network in function of the

targeted source.
During the current data taking (S6 and VSR2) a significant astrophysical constraint is expected to be obtained for the Vela pulsar, which should emit mainly at $\sim$ 22 Hz. In this case, the analysis, currently under way, will be based mainly on the Virgo data, which shows a higher sensitivity at that frequency. On the other hand , the GW signal from the Crab pulsar was searched in the LIGO S5 data.
Among the various local sources of continuous gravitational wave, the Crab pulsar (PSR B0531+21, PSR J0534+2200) has long been regarded as one of the most promising known local sources of gravitational wave emission. Its high spin-down rate, $\dot{\nu} = 3.7 \cdot 10^{-10} \quad Hz\ s^{-1}$, is due to a variety of mechanisms, including magnetic dipole radiation, particle acceleration in the magnetosphere, and gravitational radiation. If one assumes that all the energy is being radiated gravitationally, the gravitational wave tensor amplitude at Earth is $h_o = 1.4 \cdot 10^{-24}$ assuming a value for the momentum of inertia ($I = 10^{38} \quad kgm^2$) and a star distance of 2 kpc. This is the so called *spin down limit* for the gravitational wave emission.

Since 1978, Hirakawa and his group at the university of Tokyo carried on an experimental effort to detect the GW Crab signal [10]. This effort was pursuived by developing resonant antennas of different geometry and in May 1993 the Japanese group of KEK and Tokyo university collected data with a cryogenic torsional antenna setting a 1 σ upper limit of $h \leq 2 \cdot 10^{-22}$ at 60 Hz for an averaged time of 1900 hours [11].

The fifth LIGO science run started on 2005 November 4 and ended on 2007 October 1. The search of the Crab GW signal uses the known frequency and position of the Crab pulsar, as derived from the Jodrell Bank Crab Pulsar Monthly Ephemeris. Using this ephemeris and the assumption that the gravitational wave and electromagnetic phase track each other precisely, the signal phase evolution is predicted with negligible uncertainty. The analysis assumes that emission will be at or near twice the pulsars spin frequency, 59.56 Hz, which is the frequency of emission by a steadily rotating quadrupolar deformation, i.e., a triaxial star. In fact, the amplitude of the emitted signals depends on the star ellipticity ϵ, a parameter largely unknown. The 95 % upper limit on the gravitational wave amplitude, using uniform priors on all the parameters, is ${h_o}^{95\%} = 4.9 \cdot 10^{-25}$, a value well below the *spin down limit* of the pulsar. Assuming the previous cited values for the star distance and its momentum of inertia, this results corresponds to a neutron star eccentricity $\epsilon \leq 2 \cdot 10^{-4}$ [12].

4 The Stochastic GW background.

The detection of the stochastic GW background would have a profound impact on early Universe cosmology and on high-energy physics. Recently the Virgo and LIGO Scientific collaborations published limits on the amplitude of the stochastic gravitational-wave background [13]. The search based on the data from a two-year science run of LIGO was performed by cross-correlating interferometer strain from pairs of detectors. The result constrains the energy density of the stochastic gravitational wave background normalized by the critical energy density of the Universe, in the frequency band around 100 Hz, to be $\Omega_0^{gw}(\nu) < 6.9 \cdot 10^{-6}$ at 95 % confidence.

This limit surpassed already the so-called *standard big bang nucleosynthesis bound* setting new constraints on the primordial gravitational wave energy spectrum predicted by several models. The nucleosynthesis bound is derived from the fact that a large gravitational-wave energy density at the time of nucleosynthesis would alter the abundances of the light nuclei produced in the process. Hence, the model and observations constrain the total gravitational wave energy density at the time of nucleosynthesis.

The data rule out models of early Universe evolution with relatively large equation of state parameter, as well as cosmic string models with relatively small string tension that are favoured in some string theory models. This search for the stochastic background improves on the indirect limits from Big Bang nucleosynthesis and cosmic microwave background at 100 Hz.

This results is a significant example of the potential impact of the GW search on fundamental phyisics. All the informations on the Universe when an elementary particle was still in thermal equilibrium at the temperature T are obliterated by the successive interactions. The condition for thermal equilibrium is that the rate Ξ of the processes that maintain equilibrium be larger than the rate of expansion of the Universe H (the Hubble parameter). The weaker the interaction of a particle, the higher is the energy scale when they drop out of thermal equilibrium. Since for the graviton we have [14]

$$\frac{\Xi}{H} \sim \left(\frac{T}{M_{Planck}}\right)^3$$

where $M_{Planck} \sim 10^{19}$ Gev is the Planck mass, it follows that relic gravitational waves are a potential source of informations on the state of the very early Universe and therefore on physics at correspondingly high energies, which cannot be accessed experimentally in any other way. The relation between time and temperature during the radiation dominant phase then tells us how far back in time are we exploring the Universe at temperature T. Moreover, for causality reasons, we expect that the characteristic frequency of gravitons or any other particles produced at that time will be directly related to the horizon dimension of that epoch. Thus, assuming the standard Friedmann-Robertson-Walker cosmological model the temperature and time production can be related to the detection frequency of the GW signal (see fig. 1).

The satellite mission Planck [15] focused on the measurement of the cosmic

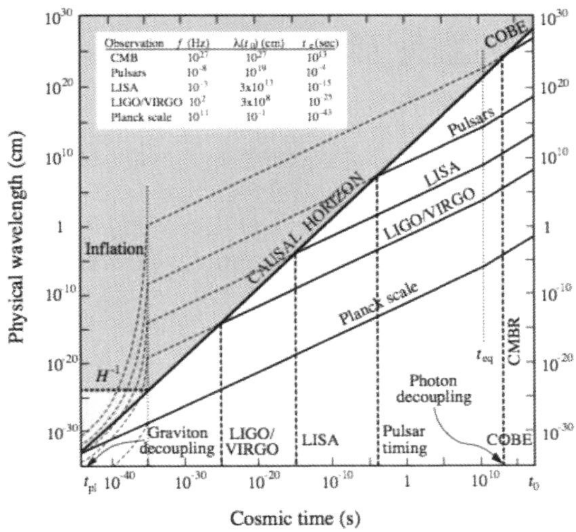

Figure 1: **Present GW radiation wavelength versus the cosmic emission time.**

microwave background (CMB), will either detect gravitational waves from inflation at very long wavelengths, or else rule out the simplest (and arguably the most compelling) inflationary models [16]. On shorter wavelengths pulsar timing techniques will reach far beyond the gravitational-wave sensitivity of individual pulsars [17], [18].

In the audio frequency range (0.1 - 1 kHz) of the laser interferometers on the Earth, the search is related to a production epoch ranging from 10^{-25} s to 10^{-27} s after the Big Bang, which corresponds to a high-energy scale that will never be reached with particle physics experiments. The future interferometric detectors on the Earth will be able to improve of orders of magnitude the limit already achieved, probing significantly the cosmological models near the Planck era.

5 Conclusion

Detecting gravitational waves is still one of the main challenges in physics for the years to come. At the present sensitivity level the GW network of detectors permist to contraint a variety of astrophysical and cosmological models. The analysis in coincidence with e.m. detectors for hunting GW signals associated to GRB, the pulsar observation via radio telescopes and GW detectors are examples of the astrophysical potentialities of the world wide network of GW detectors. Finally the new upper limit on the stochastic gravitational wave background below the barrier of nucleosynthesis bound, is the first example of a significant GW contribution in the fundamental physics domain.

Acknowledgments

The research was partially supported by the European Commission under the Framework Programme 7 (FP7, Grant Agreement 211743) and by the Italian minister of education under the financial program Prin 2007 (Grant prot. 2007NXMBHP 002).

References

[1] E. Amaldi et al., *Astron. and Astrophys.* 216, 325, (1989)
[2] P. Astone et al., *Phys. Rev.* **D** 68, 022001 (2003)
[3] L.S. Finn, S.D. Mohanty, and J.D. Romano, *Phys.Rev.* **D** 60, 121101 (1999)
[4] F. Acernese et al.,*Class. Quantum Grav.* 25, 225001 (2008)
[5] B. Abbott et al., *Astrophys. J.* 681, 1419 (2008)
[6] B. Abbott et al., preprint of astro-ph archive arXiv:0908.3824
[7] B. Schutz, *Nature* 323,310 (1986)
[8] B. Sathyaprakash, B. Schutz, C. van den Broeck, preprint *astro-ph* arXiv:0906.4151v1
[9] C. Palomba, *Astron. and Astrophys.* 354, 163 (2000)
[10] H. Hirakawa, K. Tsubono, and M. K. Fujimoto, *Phys.Rev.* **D** 17, 1919 (1978)
[11] T. Suzuki, in *"Gravitational Wave Experiments"* (Proceedings of the First Edoardo Amaldi Conference on Gravitational Wave Experiments, 14-17 June, 1994, Frascati (Rome), Italy), ed. by G. Pizzella, E. Coccia, F. Ronga World Scientific River Edge , 115, 1995.
[12] B. Abbott et al., *Astrophys. J. Lett.* 683, 45 (2008); Abbott et al. *Astrophys. J.* 706, L203 (2009) (erratum)
[13] B. P. Abbott et al., *Nature* 460, 990 (2009)
[14] M. Maggiore, *Phys. Reports* 331, 283 (2000)
[15] Planck Collaboration, http://www.rssd.esa.int/Planck/
[16] G. Efstathiouand, S. Gratton, *J. of Cosm. and Astropar. Phys.* 06, 011 (2009)
[17] M. Kramer, D. C. Backer, J. M. Cordes, T. J. W. Lazio, B. W. Stappers, and S. Johnston, *New Astron. Rev* 48, 993 (2004)
[18] F. A. Jenet et al., *Astrophys. J.* 653, 1571 (2006).

GRAVITATIONAL WAVE ASTRONOMY: AN EXPERIMENTAL OVERVIEW

Giles Hammond [a] for the LIGO Science Collaboration and Virgo Collaboration
Kelvin Building, Department of Physics and Astronomy, University of Glasgow, Glasgow, G12 8QQ, United Kingdom

Abstract. The current worldwide interferometer network comprises detectors in the US (LIGO) and Europe (GEO600 and VIRGO) as well as advanced facilities in Japan (TAMA, CLIO) and Australia (ACIGA). Detectors currently have the sensitivity to detect neutron star binary coalescences out to approximately 15Mpc (the Virgo supercluster). Furthermore the long baseline instruments (LIGO and VIRGO) will shortly be undergoing upgrades which will see their sensitivity increase by an order of magnitude by 2014. These detectors, together with an upgraded GEO detector, should make routine detections and open up the gravitational window on the universe. In the longer term, 3rd generation detectors operating post 2018 will further increase the sensitivity by an additional order of magnitude and will likely feature underground operation at cryogenic temperatures or operation in space.

1 Introduction

Gravitational waves are a prediction of general relativity and manifest themselves as a deformation of the metric of spacetime. Acceleration of mass creates ripples in the spacetime metric which carry information and energy from astrophysical sources at the speed of light. The effect of a gravitational wave at the Earth will be very weak. A neutron star binary coalescence in the Virgo supercluster (15Mpc) will result in a strain of approximately 10^{-21}. Over the baseline of one astronomical unit this equates to a displacement change of 0.1nm and thus incredibly sensitive detectors need to be utilised. Over the past couple of decades there has been a significant experimental and theoretical effort to make a positive detection of gravitational waves and open up a new, non-electromagnetic, window on the Universe. Although measurements of the orbital decay of the binary pulsar PSR1913+16 provide excellent indirect evidence for the existence of gravitational waves [1] the field still awaits the first positive detection. Ground-based detectors are currently operating in a worldwide network to achieve this goal. The LIGO [2] detectors in the US comprise a 2km/4km detector sharing an identical vacuum system in Hanford, WA and a 4km detector in Livingston, LA. In Europe, detectors include the joint German-British 600m GEO [3] detector in Ruthe, Germany and the 3km VIRGO [4] detector located just outside Pisa. Prototype detectors and advanced test facilities in Japan (TAMA and CLIO) [5] and Australia (ACIGA) [6] further complement the worldwide network. Long-baseline interferometers are the detector of choice as they combine good strain sensitivity over a wide bandwidth from approximately 10Hz-10kHz. All interferometers operate on a similar basis

[a] e-mail: g.hammond@physics.gla.ac.uk

and a typical detector topology is shown in figure 1. A stabilised laser (typically $\simeq$ 10W) passes through a mode cleaning cavity which is used to select the desired TEM mode. A 50:50 beamsplitter is then used to send the laser along orthogonal paths to end mirrors. In order to increase the interaction time of the light beam with the gravitational wave, Fabry-Perot cavities (or beam folding in GEO) are utilised in each arm. The finesse of the arm cavities can be used to increase the light storage time and LIGO uses a value of $\simeq$220 which results in an optimum sensitivity at approximately 150Hz. The beams are then recombined at the beamsplitter and a photodiode at the output port monitors the power exiting the interferometer. The detector is operated near a dark fringe in order to reduce the effect of laser fluctuations and this means that the majority of the light exits back towards the laser. A power recycling mirror is incorporated into the interferometer (between the mode cleaning cavity and beamsplitter) to reflect this light back into the detector and is equivalent to increasing the laser power.

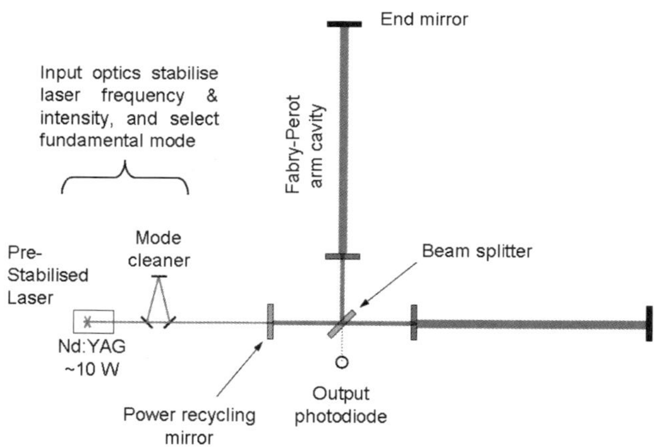

Figure 1: The topology of a large kilometre-scale interferometric gravitational wave detector.

Gravitational waves will give a totally new and complimentary view of the Universe[b]. Each detector has a broad angular sensitivity [2] and thus a network of detectors is required to be able to provide pointing information on the location of a source. This is accomplished through timing measurements in a similar way to long baseline interferometry in radio astronomy. An additional benefit of a network of detectors is also an improved veto to "noise events" through coincidence measurements. The LIGO, GEO and VIRGO detectors are operating

[b]an often cited example is that if an electromagnetic telescope is equivalent to "seeing" then a gravitational wave telescope is equivalent to "hearing"

at design sensitivity and have participated in five science runs to date (while a sixth run is currently underway). Figure 2 shows the sensitivity achieved by these detectors during the S5 science run. The curve comprises three main regions; (i) a low frequency region (<60Hz) which is dominated by Newtonian gravitational gradient noise (interaction of the interferometer optics with moving mass sources) and seismic noise. Seismic noise is reduced by suspending the interferometer mirrors from the ground on seismic isolation systems. These systems provide an isolation factor of $1/f^{2n}$ for an "n-stage" suspension. (ii) a mid-frequency region ($\simeq$100Hz) which is dominated by thermal noise. This can be in the form of Brownian thermal noise or thermoelastic noise (statistical temperature fluctuations which can couple into mirror displacement through the temperature dependence of the thermal expansion coefficient). The fluctuation dissipation theorem states that the thermal noise may be reduced by using materials with ultra-low dissipation or through the reduction of the system temperature. In the GEO detector, fused silica is used in the final stage suspension in order to fulfill the requirement of providing a suspension with low mechanical loss. (iii) a high frequency region (>100Hz) which results from shot noise in the interferometer. Reduction of the shot-noise-limited sensitivity requires an increase in the laser power or the use of advanced interferometric techniques such as squeezed light sources to go beyond the Standard Quantum Limit (SQL) [7, 8].

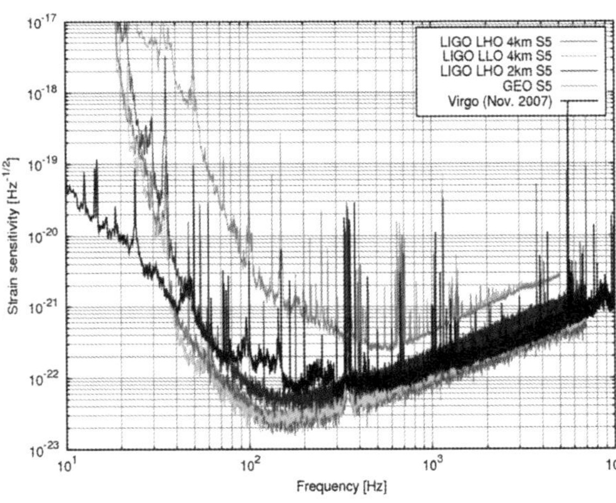

Figure 2: The sensitivity curves of LIGO, VIRGO and GEO600 during the S5 science run.

2 Second Generation Detectors

Current detectors are estimated to have detection rates of approximately 0.02 per year for neutron star binary coalescences [9, 10]. Thus at the current sensitivity there is no guarantee of a detection. As a result the next generation of interferometric gravitational wave detectors (2nd generation or "advanced detectors" [11]) will see an increase in strain sensitivity by an order of magnitude. This is equivalent to increasing the detected volume by a factor of one thousand and will allow neutron star binary coalescences to be detected out to several hundred Megaparsecs. Similarly event rates should be at the level of several events per year and will signal the beginning of Gravitational Wave astronomy. In order to achieve this improvement in strain sensitivity over all frequency bands requires research and development to be carried out on all aspects of the interferometer design. Over the next couple of years Enhanced LIGO, GEO-HF and VIRGO+ will increase sensitivity by approximately a factor of two through a combination of laser upgrades, a modified readout scheme for the output optics, the use of a squeezed light source and developments to the optic suspensions. This is seen as an intermediate step toward the advanced detectors (Advanced LIGO and Advanced VIRGO) which will follow in 2014 with a raft of further enhancements. The low frequency improvement will be achieved by sophisticated seismic isolation systems and suspensions which include a combination of passive (low frequency passive springs and pendulums) and active control (seismometers operated in closed loop). Figure 3 shows the Advanced LIGO mirror suspension which includes a seismic isolation system (SEI) comprising a two-stage active-passive platform. The SEI provides an isolation factor of approximately 1000 at 10Hz and also reduces the microseismic peak ($\simeq$ 160mHz) by an order of magnitude [12]. Attached to the underside of the SEI is a four stage quadruple pendulum [13] which incorporates the interferometer mirror as its lowest stage. The pendulum incorporates damping and control actuation (for beam steering) from a quiet reaction chain suspended behind the main pendulum. Coil-magnet actuators provide control forces to the upper portions of the suspension while an electrostatic drive provides actuation on the test mass itself. Three stages of vertical springs are incorporated into the pendulum to provide vertical isolation and the four stage pendulum provides a factor of 10^8 horizontal isolation at 10Hz. The test mass and penultimate mass are fabricated from fused silica and have a mass of 40kg. Fused silica attachment "ears" are silicate bonded onto the $\lambda/10$ polished sides of the test mass and allow for the attachment of four fused silica fibres. The fibres are pulled and welded to the silica ears with a CO_2 laser. The monolithic design of the lower stage is required to meet the thermal noise requirements for Advanced LIGO. In the mid frequency (40Hz-200Hz) the thermal noise of the mirror coatings is likely to be a limiting noise source. The coatings for the mirrors comprise alternating $\lambda/4$ stacks of high/low refractive index mate-

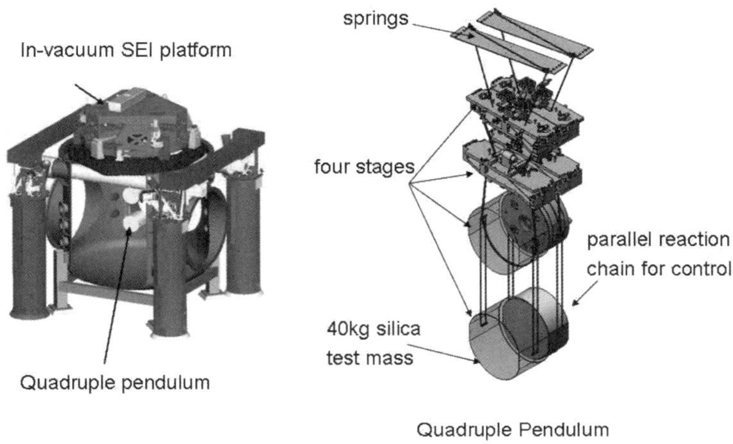

Figure 3: The seismic isolation system and quadruple pendulum suspension which will be used to suspend the mirrors and beamsplitter in Advanced LIGO.

rial (Ta_2O_5/SiO_2 respectively). Research into the dissipation of the coatings suggest that the Ta_2O_5 coatings are the dominant source of dissipation [14]. These measurements are typically performed by exciting a coated oscillator and measuring the ringdown time. It has further been shown that doping the coating material with TiO_2 can help to reduce the dissipation and thus reduce the coating thermal noise. Work is currently underway with coating vendors (CSIRO in Australia and LMA in France) to optimise the coatings for the advanced detectors and minimise the thermal noise contribution. In the high frequency regime shot noise will ultimately provide a limiting noise source. Statistical fluctuations in the number of photons in the laser beam result in a noise source which is essentially white with frequency. The shot-noise-limited sensitivity is typically reduced by increasing the laser power circulating in the detector. However, this also has the effect of increasing the radiation pressure noise which affects the low frequency performance of the interferometer. The combined effect of radiation pressure noise and shot noise is often termed quantum noise. In advanced detectors the circulating laser power in the cavities will be increased to $\simeq$800kW and the test masses increased from 10kg to 40kg in order to reduce the quantum noise in all frequency bands. Work is currently underway to test high power, low noise, lasers which have $>$ 95% of their power in the TEM_{00} mode. The lasers tyically comprise a low power 35W front end and a high power diode pumped output stage. The 35W first stage is currently being tested in Enhanced LIGO.

3 Third Generation Detectors

Third generation detectors will aim to provide a further order of magnitude increase in sensitivity over the advanced detectors in addition to increased bandwidth. Even though these detectors are scheduled for the post 2018 window there is already significant research and development underway. Low frequency operation (from 10^{-4}Hz-10^{-1}Hz) will be achieved by space-borne detectors such as LISA [15]. The mission features three drag-free spacecraft which orbit in a triangular configuration with armlengths of 5 million kilometres. Free-falling proof masses are shielded by the spacecraft and are used as the end mirrors of the interferometer. LISA will provide a complimentary view of the gravitational wave sky in the low frequency band. In order to prepare for this mission a technology demonstrator mission, LISA pathfinder [16], is currently being readied for launch in 2011. This mission will feature a scaled down version of a single spacecraft and will allow precision interferometry to be performed between two freely floating test masses. Research and development for a third generation ground-based detector is currently being directed toward the Einstein telescope [17]. This will most likely feature an underground detector with armlengths of approximately 10km. Underground operation is required in order to reduce the effect of gravitational gradient noise and will allow operation down to approximately 1Hz. Furthermore, the detector will be operated at low temperature in order to reduce the effect of thermal noise. A possible candidate material for the test masses and suspension elements may be silicon as this material has a null in its thermoelastic noise contribution at both 120K and 20K. High power lasers, in combination with test masses of several hundred kilogrammes, will also be required in order to push down the SQL in all frequencies of interest. Naturally this poses a number of technological challenges such as controlling cavities with high circulating power and developing effective strategies for inputting the laser power into the cavities without using transmissive optics [c].

4 Summary

We clearly stand at a very exciting time for Gravitational Wave astronomy. A worldwide network is currently operating at design sensitivity and producing important and interesting scientific results. Over the next 5 years the advanced detectors, with an increase in sensitivity by one order of magnitude, will come online (2014) and open up the gravitational window on the Universe. In the longer term (2018) third generation space-borne detectors (LISA) and ground-based detectors (Einstein Telescope) will yield high precision measurements of gravitational wave sources from 10^{-4}Hz to 10kHz.

[c] due to problems associated with thermal lensing

Acknowledgments

The author gratefully acknowledges the support of the United States National Science Foundation for the construction and operation of the LIGO Laboratory and the Science and Technology Facilities Council of the United Kingdom, the Max-Planck- Society, and the State of Niedersachsen/Germany for support of the construction and operation of the GEO600 detector. The author also gratefully acknowledges the support of the research by these agencies and by the Australian Research Council, the Council of Scientific and Industrial Research of India, the Istituto Nazionale di Fisica Nucleare of Italy, the Spanish Ministerio de Educación y Ciencia, the Conselleria d'Economia, Hisenda i Innovación of the Govern de les Illes Balears, the Royal Society, the Scottish Funding Council, the Scottish Universities Physics Al- liance, The National Aeronautics and Space Administration, the Carnegie Trust, the Leverhulme Trust, the David and Lucile Packard Foundation, the Research Corporation, and the Alfred P. Sloan Foundation. This article has document number LIGO P1000005-v1.

References

[1] J.H. Taylor and J.M. Weisberg, *Astrophysical Journal* 253, 908 (1982).
[2] B.P. Abbott *et al.*, *Rep. Prog. Phys.* 72, 076901 (2009).
[3] H. Grote, *Class. Quantum Grav.* 25, 114043 (2008).
[4] F. Acernese *et al.*, *Class. Quantum Grav.* 25, 114045 (2008).
[5] D. Tatsumi, *Class. Quantum Grav.* 24, S399S404 (2007).
[6] P. Barriga *et al.*,*Class. Quantum Grav.* 22, S199S208 (2005).
[7] V.B. Braginsky, M.L. Gorodetsky and F.Ya. Khalili, *Physics Letters A* 246, 458 (1998).
[8] R. Schnabel, J. Harms, K.A. Strain and K. Danzmann, *Class. Quantum Grav.* 21 S1045S1051 (2004).
[9] T. Regimbau, J. A. de Freitas Pacheco, A. Spallicci and S. Vincent, *Class. Quantum Grav.* 22, S935S941 (2005).
[10] I. Mandel *et al.*, *LIGO P0900125-v6* (2009).
[11] J.R. Smith, *Class. Quantum Grav.* 26, 114013 (2009).
[12] R. Abbott *et al.*, *Class. Quantum Grav.* 21, S915S921 (2004).
[13] N.A. Robertson *et al.*, *Class. Quantum Grav.* 19, 40434058 (2002).
[14] G.M. Harry *et al.*, *Class. Quantum Grav.* 24, 405415 (2007).
[15] O. Jennrich, *Class. Quantum Grav.* 26, 153001 (2009).
[16] M. Armano *et al.*, *Class. Quantum Grav.* 26, 094001 (2009).
[17] *http://www.et-gw.eu/*

FINE STRUCTURE IN THE COSMIC RAY ELECTRON SPECTRUM MEASURED BY ATIC

A. D. Panov [a], V. I. Zatsepin, N. V. Sokolskaya, G. L. Bashindzhagyan, E. N. Kouznetsov, M. I. Panasyuk, E. B. Postnikov
Skobeltsyn Institute of Nuclear Physics, Moscow State University, Moscow, Russia
J. H. Adams, Jr., M. Christl, J. Watts
Marshall Space Flight Center, Huntsville, AL, USA
H. S. Ahn, K. C. Kim, E. S. Seo, J. Wu
University of Maryland, Institute for Physical Science & Technology, College Park, MD, USA
J. Chang
Purple Mountain Observatory, Chinese Academy of Sciences, China
T. G. Guzik, J. Isbert, J. P. Wefel
Louisiana State University, Department of Physics and Astronomy, Baton Rouge, LA, USA

Abstract. New analysis to separate electrons from protons in the ATIC experiment has been performed. Five new discriminants were studied by different Monte Carlo programs. New electron spectrum, when compared with the published results [1], show good agreement in the most interesting region of energy (from 90 GeV to 600 GeV). It is argued that there is no disagreement between ATIC's results and Fermi-LAT ones. Finally, high-resolution electron spectrum is obtained and possible fine structure is found out in it.

The ATIC (Advanced Thin Ionization Calorimeter) balloon-borne spectrometer was designed to measure the energy spectra of elements from H to Fe with individual resolution of charges in primary cosmic rays for energy region from 50 GeV to 100 TeV. ATIC has had three successful flights around the South Pole in 2000–2001 (ATIC-1), 2002-2003 (ATIC-2) and 2007-2008 (ATIC-4). ATIC is comprised of a fully active bismuth germanate (BGO) calorimeter, a carbon target with embedded scintillator hodoscopes, and a silicon matrix that is used as the main charge detector. The calorimeter is comprised of 8 layers with 40 BGO crystalls in each for ATIC-1 and ATIC-2 and of 10 layers for ATIC-4. The details of the construction of the apparatus are described in the papers [2–4]. It was shown that it is possible also to measure the spectrum of cosmic ray electrons plus positrons [5] with ATIC (hereinafter we use 'electrons' for brevity). To separate electrons from the higher background of protons and other nuclei some differences in shower development for incident electrons and for nuclei are used. The spectrum of electrons measured with the ATIC spectrometer by this method was published in the paper [1]. The most notable detail of the electron spectrum reported was an 'excess' of electrons between energies of 300–800 GeV. The main purpose of this work is to investigate possible alternate techniques to separate electrons from hadrons and was curried

[a] e-mail: panov@dec1.sinp.msu.ru

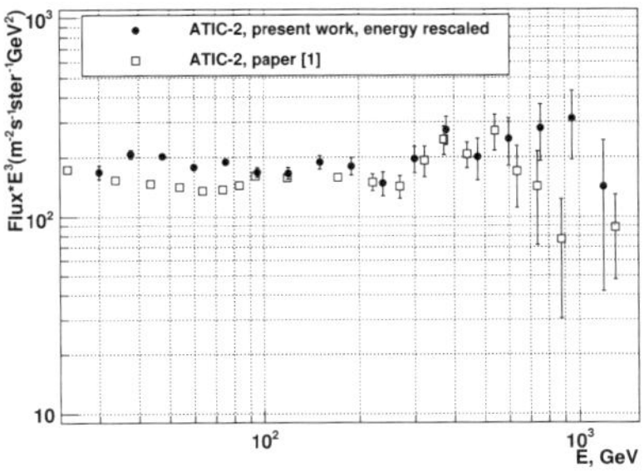

Figure 1: Comparison of the electron spectrum of ATIC-2 of present paper with the spectrum of the paper [1]. The proton background is not subtracted in both spectra.

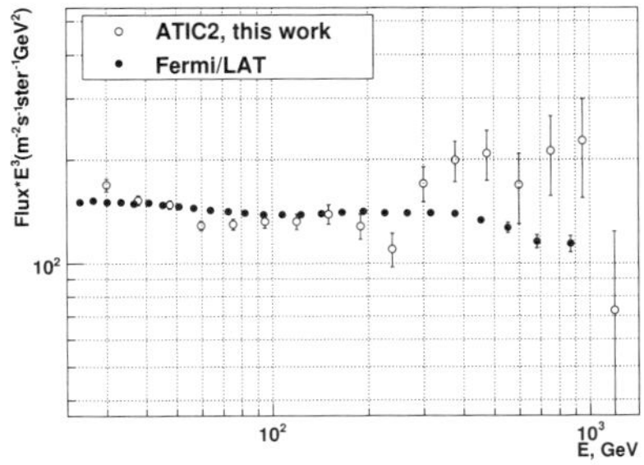

Figure 2: Comparison of the electron spectra of ATIC-2 (present paper) with the electron spectrum measured by Fermi-LAT [6].

out completely independent of the previous analysis with new discriminants to look at the resulting electron spectrum.

For better cross-checking of the results we constructed and worked with five new discriminants that permit to separate electron showers (by its shape) from proton showers. The first conclusion obtained during analysis was that the procedure of the proton background subtraction from the spectrum of the selected electron-like events may be a source of significant systematics. Therefore we did not rely on the proton background subtraction as was done in the paper [1], but, instead, we tried to reduce the proton contamination as much as possible. The things that may be studied are various sharp features in the behavior of the electron spectrum (like 'ATIC's excess').

We compare the obtained spectrum of electrons with the spectrum of [1] in Fig. 1. A problem was that the spectrum of [1] was corrected for the scattering of the electrons in the atmosphere by simple rescaling of the spectrum measured at the top of the apparatus, but we present the spectrum at the top of the apparatus as the main result. Therefore to compare the spectra we artificially normalize the energy of each event of our spectrum in Fig. 1 by factor 1.15 that corresponds to the mean loss of energy by the electrons in the atmosphere. The result of the comparison is shown in Fig. 1. It is seen that there exists very good agreement of both spectra in the region from 90 GeV to 600 GeV. We completely confirm 'ATIC's excess' measured in ATIC-2 in this region.

In Fig. 2 we compare our data for the ATIC-2 flight with recent results for the electron spectrum in the region from 20 GeV to 1 TeV measured by Fermi/LAT telescope [6]. The comparison of ATIC's results and the spectrum of Fermi/LAT shows no sign of disagreement between the experiments. Actually, the ATIC spectrum are shown without subtraction of proton back-

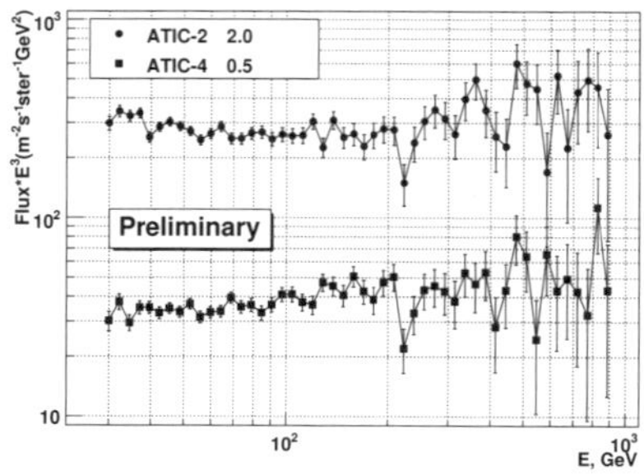

Figure 3: Fine structure in the spectra of electrons of ATIC-2 and ATIC-4.

ground. It is most probable that at the energies 30–100 GeV this background is not high, but above some hundreds of GeV the intensity of background is higher. Actually, the ATIC data agree with Fermi/LAT at the energies less than 200 GeV well and the deviation above 300 GeV may be related to unaccounted background. The only significant difference of the ATIC's spectrum from the Fermi/LAT's one is a sharp dip near 250 GeV but it may be explaind by lower energy resolution of Fermi/LAT relative to ATIC.

The energy resolution of the ATIC apparatus for electrons actually is very high and one could measure the electron spectrum with narrower energy bins than in the previous paper [1]. The electron spectra for ATIC-2 and ATIC-4 flight measured with bin width of 0.03 for decimal logarithm of energy are compared in Fig. 3. It is seen that between 200 GeV and 600 GeV some structure (three dips and three peaks) exists in ATIC-2 spectrum. One may think that it is only a play of statistics, but the preliminary spectrum measured by the ATIC-4 experiment in the same region completely confirms the structure. This is a sign that this fine structure may be real.

The work was supported in Russia by grant of RFBR 08-02-00238; in the US by NASA (NNG04WC12G at LSU), and in China by the Ministry of Science and Technology (N2002CB713905).

References

[1] J. Chang et al. *Nature*, 456, 362 (2008).
[2] T. G. Guzik et al. *Adv. Sp. Res.*, 33, 1763 (2004).
[3] V. I. Zatsepin et al. *Nucl. Instr. Meth. A*, 524, 195 (2004).
[4] A. D. Panov et al. *Instr. and Exp. Techn.*, 51, 665 (2008).
[5] J. Chang et al. *Adv. Sp. Res.*, 42, 431 (2008).
[6] A. A. Abdo et al. *Phys. Rev. Lett.*, 102, 181101 (2009).

THE COSMIC RAY SPECTRUM AT ULTRAHIGH ENERGIES

Mikhail Pravdin [a]
Yu.G. Shafer Institute of Cosmophysical Research and Aeronomy SB RAS, 677980 Yakutsk, Russia

Abstract. The energy spectrum obtained on the Yakutsk array is compared with the results obtained on the AGASA and HiRes arrays and the latest Pierre Auger Observatory data. The discrepancy in the intensity of the energy spectra obtained in different experiments can be explained by the presence of systematic errors in shower energy estimations.

Introdution

Study of the cosmic ray (CR) energy spectrum and establishment of its cutoff, predicted by Greisen [1] and Zatsepin and Kuzmin [2] (GZK cutoff), remain to be the most important problems for detecting CR sources at energies above 10^{19} eV. Four extensive air showers (EASs) with energies above the cutoff threshold have been detected on the Yakutsk EAS array [3]. The data of AGASA array (where 11 showers with energies above 10^{20} eV were registered) indicate the absence of cutoff in the spectrum . In the new Auger experiment, a record exposure has been reached and a spectrum cutoff was evidenced [5], which is in agreement with the HiRes results [6]. The spectra below 10^{20} eV obtained in all these experiments have a similar shape but significantly different intensities. This contradiction is most likely to be caused by the systematic difference in the energy estimates for individual showers in different experiments.

1 Determination of EAS energy

On EAS arrays, the primary-particle energy E_0 is estimated from the experimentally determined basic parameter. Such a parameter for the Yakutsk array is the density S_{600} at a distance of 600 m from the shower core. On the arrays similar to the Yakutsk array, the relation between the basic parameter and primary energy E_0 at the atmospheric depth X_0 , corresponding to vertical showers ($\theta = 0°$), is generally found. To estimate E_0 in events with zenith angles $\theta > 0°$, the found parameter value is recalculated to the vertical level from the zenith angle dependence.
Three main components are measured on the Yakutsk EAS array: charged-particle flux, Cherenkov light, and muon component. This circumstance allowed us to estimate energy by the calorimetric method and obtain a relation between S_{600} at $\theta = 0°$ and E_0 [7] from the experimental data with a minimum dependence on the development model. This method is based on experimental estimation of the energy dissipated by a shower above the observation level

[a] e-mail: m.i.pravdin@ikfia.ysn.ru

from the measured EAS Cherenkov light. This energy fraction exceeds 70% of the total energy.

For the events recorded during operation of Cherenkov light detectors, the parameter Q_{400} (Cherenkov flux density at a distance of 400 m from the shower core) was used to parameterize showers.

The error in determining the energy in individual events is affected by many factors. Some of them are random; however, there are systematic errors, which significantly influence determination of the energy spectrum intensity and shape. It was found in [8] that the relative systematic error in estimating energy on the Yakutsk array for vertical showers above 10^{19} eV is 25-26%. For a zenith angle of $\approx 60°$, it reaches 30%. The dominant contribution is from the uncertainty in absolute calibration of the Cherenkov light detectors and the error in determining the average transparency of atmosphere. However, this contribution is the same for all showers and does not affect the spectral shape but only leads to an energy shift of the entire spectrum.

A similar procedure for estimating the energy in showers is used on the AGASA array. The basic parameter is S_{600}. The relation of the parameter $S_{600}(0°)$ with the primary energy was determined from model simulation. It was found in [9] that the systematic error in determining energy may amount to 18%.

At the HiRes array, the fluorescence from ionization of air atoms by EAS particles is measured with telescopes. These data are used to reconstruct the cascade curve of electrons and find the number of particles in the shower maximum. The measured signal is proportional to the ionization loss of shower electrons. According to the HiRes collaboration estimates [6], the systematic error in determining energy is about 15%.

The Auger project uses both methods for recording showers. A ground-based array of thick water detectors occupies an area of $3000 km^2$. The distance between neighboring detectors is 1600 m and the area of each detector is $10 m^2$. The fluorescence produced by the showers falling in the ground-based array area is simultaneously measured with four telescopic systems. The energy is estimated from optical data in the same way as at the HiRes array. The response in units of an equivalent vertical muon at a distance of 1000 m from the shower core, $S(1000)$, is used as the basic parameter for the data from ground-based detectors [5]. The relationship between S_{1000} and the energy measured by fluorescence detectors is found from the common events. The systematic uncertainty of this energy scale is 22%.

2 Energy spectrum

Figure 1(a) shows the energy spectra registered on the Yakutsk [3], AGASA [4], Auger [5], and HiRes [6] arrays. The shapes of the energy spectra from different arrays are in good agreement; however their intensities significantly differ. The

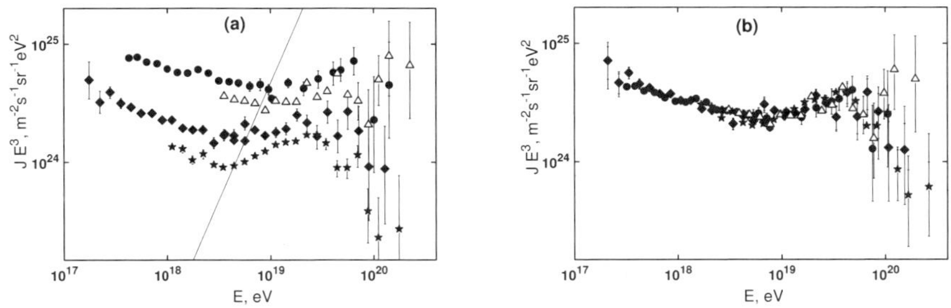

Figure 1: Differential energy spectrum according to the data of the (circles) Yakutsk [3], (triangles) AGASA [4], (asterisks) Auger [5], and (diamonds) HiRes [6] arrays. (a) Original data; (b) Energy spectrum after the energy correction: $E_C = K \times E_0; K = 0.75$ (Yakutsk), 0.87 (AGASA), 1.2 (HiRes), and 1.5 (Auger).

Auger and HiRes spectra are cut off for the most statistically provided data, a feature indicating the existence of the GZK cutoff. However, 11 events with energies above 10^{20} eV ($\theta < 45°$) have been detected on the AGASA array; i.e., the cutoff is absent. The Yakutsk array data are in better agreement with those of AGASA. Four events with energies above the cutoff threshold have been detected on the Yakutsk array.

It can be seen in Fig. 1(a) that the spectra of all arrays have characteristic features (dip and bump) and are in good agreement below 10^{20} eV. The oblique line shows the shift of a point on the plot in the coordinates used with a change in the estimated energy. It can be seen that dip position for all arrays only slightly deviates from this straight line. Figure 1(b) shows the spectra obtained by changing the energy estimates (multiplying the initial values by a coefficient K). The following values were used: K = 0.75 (Yakutsk), 0.87 (AGASA), 1.2 (HiRes), and 1.5 (Auger). The thus corrected results of the three experiments are in good agreement in the entire energy range, except for the AGASA data above 10^{20} eV. To put the intensities from all arrays into correspondence, a correction exceeding the specified systematic uncertainty by a factor of 2 was used for only the Auger data. The correction coefficients for other experiments were taken to be close to the systematic error in determining energy for each array.

According to the calculations of Berezinsky et al. [10], the observed features may arise in the energy spectrum if particles of extragalactic origin dominate in CRs above 10^{18} eV. Such a shape of the spectrum is explained by the energy loss on pion photoproduction and formation of electronpositron pairs on the relic radiation. In this case, protons should dominate in the spectrum. It was concluded in [10] that the most likely sources are active galactic nuclei.

3 Conclusions

The discrepancy in the intensities of the energy spectra obtained in different experiments can be explained by the presence of systematic errors in the estimated shower energies. The HiRes and Auger data indicate the relic cutoff of the CR spectrum. This fact and the presence of a dip in all experiments suggest that protons from extragalactic sources dominate in the spectrum even at energies above 10^{18} eV. However, this suggestion is inconsistent with the Yakutsk array data on the muon component behavior in the showers above 10^{19} eV [8]. The muon fraction in this region increases with an increase in energy. In the most inclined showers at $E_0 > 2 \cdot 10^{19}$ eV, the responses of the muon the same as ground-based detectors in a wide range of distances. These results may indicate the occurrence of new processes upon interactions of particles with such energies. If this is true, the estimated energy of the most intense showers may be incorrect for all arrays. The contradictictory data from different arrays on the existence of particles above the GZK cutoff can be related to unknown features of development of EASs produced by such particles.

Acknowledgments

This study was supported by the Russian Foundation for Basic Research, project No. 08-02-00348 and 09-02-12028.

References

[1] Greisen, K., Phys. Rev. Lett. , 1966, vol. 16, p. 748.
[2] Zatsepin, G.T. and Kuzmin, V.A., JETP Lett. 1966, vol. 4, p. 78.
[3] Egorova, V.P. et al., Nucl. Phys. B (Proc. Suppl.) , 2004, vol. 136, p. 3.
[4] Takeda, M., Hayashida, N., Honda, K., et al., Phys. Rev. Lett. , 1998, vol. 81, p. 1163; Sakaki, N. et al., Proc. 27th ICRC, Hamburg, 2001, vol. 1, p. 333.
[5] Abraham, J., Abreu, P., Aglietta, M., et al., Phys. Rev. Lett. , 2008, vol. 101, 061 101;
[6] Abbasi, R.U., Abu-Zayyad, T., Amman, J.F., et al., Preprint , 2005, astro-ph /0501317.
[7] Glushkov et al., Proc. 28th ICRC, Tsukuba, 2003, vol. 1, p. 393.; Ivanov, A.A., Knurenko, S.P., and Sleptsov, I.E. JETP 2007, vol. 131, p. 872.
[8] Pravdin, M.I. et al.Izv. Ross. Akad. Nauk, Ser. Fiz. , 2007, vol. 71, no. 4, p. 464.
[9] Takeda, M. et al. (AGASA Collaboration), Astropart. Phys., 2003, vol. 19, p. 447.
[10] Berezinsky, V., Gazizov, A., and Grigorieva, S., Phys.Rev. D, 2006, vol. 74, 043 005.

NEUTRINO ASYMMETRY AND THE GROWTH OF COSMOLOGICAL SEED HYPERMAGNETIC FIELDS

V.B. Semikoz[a]

Pushkov Institute of Terrestrial Magnetism,

Ionosphere and Radiowave Propagation of the Russian Academy of Sciences

IZMIRAN, Troitsk, Moscow region, 142190, Russia

Abstract. Primordial cosmological hypermagnetic fields polarize the early Universe plasma prior to the electroweak phase transition (EWPT). As a result of the long range parity violating gauge interaction present in the Standard Model their magnitude gets amplified, opening a new perturbative way of accounting for the observed intergalactic magnetic fields.

1 Introduction

The cosmological large scale magnetic fields can serve as the seed fields for observable galactic magnetic fields [1]. Primordial Maxwellian field appears in early universe after the electroweak phase transition (EWPT) as the combination $A_\mu = \sin\theta_W W^3_\mu + \cos\theta_W Y_\mu$. Below we consider hot plasma of early universe at high temperatures above EWPT, $T \gg T_{EW}$. The long-ranged non-Abelian magnetic fields (corresponding, for instance, to the color SU(3) or to the weak SU(2)) can not exist because at high temperatures the non-Abelian interactions induce a "magnetic" mass gap $\sim g^2T$ where g is the gauge coupling. The only long scale field that can exist for such conditions (restored $SU(2)\times U(1)_Y$ symmetry) is the hypercharge field Y_μ for which lattice studies confirmed non-perturbatively the absence of mass, $m_Y = 0$, as it is in SM for the massless electromagnetic field after EWPT obeying the boundary condition at T_{EW}, $\mathbf{A}_{em} = \cos\theta_W \mathbf{Y}$.

2 Generalized Maxwell equations for hypercharge fields

We start from the SM Lagrangian for hypercharge field Y_μ which includes manifestly parity violation terms for left (right) fermions and antifermions $f_{L,R} = [(1\mp\gamma_5)/2]f$:

$$L = -\frac{1}{4}Y_{\mu\nu}Y^{\mu\nu} + \sum_{l=e,\mu,\tau}\frac{g^{'}Y^\mu}{2}\left(-\bar{\nu}_{lL}\gamma_\mu\nu_{lL} - \bar{l}_L\gamma_\mu l_L - 2\bar{l}_R\gamma_\mu l_R\right) +$$

$$+\sum_i^N \frac{g^{'}Y^\mu}{2}\left[\frac{1}{3}\bar{U}_{iL}\gamma_\mu U_{iL} + \frac{1}{3}\bar{D}_{iL}\gamma_\mu D_{iL} + \frac{4}{3}\bar{U}_{iR}\gamma_\mu U_{iR} - -\frac{2}{3}\bar{D}_{iR}\gamma_\mu D_{iR}\right] +$$

$$+i\frac{g^{'}Y^\mu}{2}\left[\varphi^+ D_\mu\varphi - \left(D_\mu\varphi^+\right)\varphi\right], \quad (1)$$

[a]e-mail: semikoz@yandex.ru

where $l = e, \mu, \tau$, $U_i = u, c, t$, $D_i = d, s, b$ are leptons and quarks correspondingly, $\varphi = \left(\phi^+, \phi^{(0)}\right)^T$ - is the Higgs doublet.

From the Lagrangian (1) one can derive Maxwell equations for hypercharge fields $\mathbf{E}_Y = -\partial \mathbf{Y}/\partial t - \nabla Y_0$, $\mathbf{B}_Y = \nabla \times \mathbf{Y}$ which in hot *equilibrium* plasma at $T \gg T_{EW}$ take the form [2]:

$$\begin{aligned} &\nabla \cdot \mathbf{B}_Y = 0, \qquad \nabla \cdot \mathbf{E}_Y = 0, \\ &\frac{\partial \mathbf{B}_Y}{\partial t} = -\nabla \times \mathbf{E}_Y, \\ &-\frac{\partial \mathbf{E}_Y}{\partial t} + \nabla \times \mathbf{B}_Y = 4\pi \left[\mathbf{J}^Y(\mathbf{x}, t) + \frac{47}{378} \times \frac{g'^2 \mu_\nu}{\pi} \mathbf{B}_Y \right], \end{aligned} \tag{2}$$

where $\mu_\nu = \sum_l \mu_{\nu_L}^{(l)}$ is the net chemical potential for left neutrinos, $l = e, \mu, \tau$. The last pseudovector term ($\mathbf{J}_5^Y \sim \mu_\nu \mathbf{B}_Y$) originated by parity violation in SM is absent for Maxwellian fields in QED while namely it produces the amplification of a seed hypermagnetic field.

3 Faraday equation and α^2 -dynamo

In the rest frame $\mathbf{V} = 0$ of the isotropic early Universe plasma combining last Maxwell-like equations in Eq. (2) and the Ohm law $\mathbf{J}^Y = \sigma_{cond} \mathbf{E}_Y$ we can write the Faraday equation describing so-called α^2-dynamo [3] of hypermagnetic field:

$$\frac{\partial \mathbf{B}_Y}{\partial t} = \nabla \times \alpha \mathbf{B}_Y + \eta \nabla^2 \mathbf{B}_Y \ . \tag{3}$$

Here $\eta = (4\pi\sigma_{cond})^{-1}$ is the magnetic diffusion coefficent, the parameter α is the hypermagnetic helicity coefficent given as

$$\alpha = \frac{47 g'^2 \mu_\nu}{1512 \pi^2 \sigma_{cond}} \ . \tag{4}$$

It plays crucial role in the evolution of hypermagnetic field [2]. We can solve Eq. (3) through Foirier harmonics as $\mathbf{B}_Y(\mathbf{x}, t) = \int (d^3k/(2\pi)^3) \mathbf{B}_Y(\mathbf{k}, t) e^{i\mathbf{kx}}$ where $B_Y(k, t)$ is expressed as

$$B_Y(k, t) = B_0^Y \exp \left[\int_{t_0}^t [\alpha(t')k - \eta(t')k^2] dt' \right] \ . \tag{5}$$

For $0 < k < \alpha/\eta$, or correspondigly correlation length scales $\eta/\alpha < \Lambda < \infty$ such field gets exponentially amplified, but differently for different scales Λ. E.g. for the Fourier mode $k = \alpha/2\eta$ (or $\Lambda \simeq 2\eta/\alpha$) one gets the maximum amplification $\gamma = \alpha k - \eta k^2 = \alpha^2/4\eta$ [3,4]

$$B_Y(t) = B_0^Y \exp\left[\int_{t_0}^{t} \frac{\alpha^2(t')}{4\eta(t')} dt'\right] = B_0^Y \exp\left[32\int_{x}^{x_0} \frac{dx'}{x'^2}\left(\frac{\xi_\nu(x')}{0.001}\right)^2\right], \quad (6)$$

where we introduced the new variables $x = T/T_{EW}$, $\xi_\nu = \mu_\nu/T$ and B_0^Y is the assumed initial amplitude of the hypermagnetic field at $T \gg T_{Ew}$.

For larger scales, $\Lambda > 2\eta/\alpha$, the amplification factor in the exponent (6) becomes less than ~ 32, nevertheless, it is enough both for a strong enhancement of the initial hypermagnetic field and to survive against ohmic dissipation (magnetic field diffusion) if $\Lambda > l_{diff} = \sqrt{\eta l_H}$.

Let us comment on the physical interpretation of the new magnetic helicity term. The original seed field B_0^Y polarizes the fermions and antifermions (including neutrinos) propagating along the field in the main Landau level, $n = 0$. This polarization effect causes fermions and antifermions to move in opposite directions with a relative drift velocity proportional to the neutrino asymmetry $\mu_\nu = \sum_l \mu_{\nu_L}^{(l)}$. The existence of a basic parity violating hypercharge interaction in the SM induces a new term in the hypermagnetic field in Eq. (3) $\nabla \times \alpha \mathbf{B}_Y$ which winds around the rectilinear pseudovector hypercharge current $\mathbf{J}_5$ parallel to $\mathbf{B}_Y$. This term amplifies the seed hypermagnetic field B_0^Y according to Eq. (6).

The hypermagnetic helicity $H_Y = \int d^3x \mathbf{B}_Y \cdot \mathbf{Y}$ is another interesting object under study which is the important characteristic of the field structure as well as the magnetic helicity $H = \int d^3x \mathbf{B} \cdot \mathbf{A}$ of cosmological magnetic fields calculated in [5].

Acknowledgments

I acknowledge Dmitry Sokoloff and Jose Valle for fruitful discussions.

References

[1] For reviews see D. Grasso and H.R. Rubinstein, *Phys. Rept.* 348 (2001) ; M. Giovannini, *Int. J. Mod. Phys.* **D** 13, 391 (2004).
[2] V.B. Semikoz and J.W.F. Valle, *JHEP 03*, 067 (2008) .
[3] Ya.B. Zeldovich, A.A. Ruzmaikin, and D.D. Sokoloff, *Magnetic Fields in Astrophysics* (Gordon & Breach Publishers, New York), 1983.
[4] V.B. Semikoz and D.D. Sokoloff, *Phys. Rev. Lett.* 92, 131301 (2004).
[5] V.B. Semikoz, D.D. Sokoloff, *Astronomy and Astrophysics* 433, L53 (2005); V.B. Semikoz, D.D. Sokoloff, *Int. J. Mod. Phys* **D** 14, 1839 (2005).

REGULAR SOURCE OF THE KERR SPINNING PARTICLE

Alexander Burinskii [a]

Theor.Phys. Lab., NSI Russian Academy of Sciences, 115191 Moscow, Russia

*Abstract.*Superconducting source of the Kerr-Newman (KN) spinning particle is obtained as a consistent solution of the Einstein-Maxwell-Higgs system. The source forms an oblate rotating bubble filled by Higgs field which regularizes the KN electromagnetic (em) field, expelling it to the boundary of the bubble. External solution is KN one, while interior of the bubble is flat and extended analytically to flat 'negative' KN sheet, forming a holographic structure. Vortex of the em field forms a closed loop on the edge of bubble and quantizes spin of the solutions.

1.*Introduction.*Kerr-Newman (KN) solution for a charged and rotating Black-hole has $g = 2$ as that of the Dirac electron and paid attention as a classical model of electron coupled with gravity [1–5]. For parameters of electron (in the units $G = C = \hbar = 1$) $J = 1/2, m \sim 10^{-22}, e^2 \sim 137^{-1}$ the black-hole horizons disappear and there appears a naked source in the form of singular ring of the Compton radius $a = \hbar/2m \sim 10^{22}$. This ring is a branch line of the KN spacetime forming a gate to an "Alice" negative sheet of the Kerr geometry, which is the old mystery of the Kerr source. A consistent solution for the KN source was suggested recently [6] as a development of the bag-like model [5]. This solution is based on a supersymmetric version of the $U(1) \times \tilde{U}(1)$ field model which provides phase transition from external vacuum, with potential $V^{(ext)} = 0$, to an internal 'false' (superconducting) vacuum, $V^{(in)} = 0$. The Higgs mechanism of broken symmetry describes oscillating superconducting internal state and its interaction with the KN em field.

2. *Gravitational phase transition.* Following [5,6], for external region we use the exact KN solution in the Kerr-Schild (KS) form of metric

$$g_{\mu\nu} = \eta_{\mu\nu} + 2H k_\mu k_\nu \ , \tag{1}$$

where $\eta^{\mu\nu}$ is metric of the auxiliary Minkowski background in Cartesian coordinates $x^\mu = (t, x, y, z)$. Electromagnetic (em) KN field is given by vector potential

$$A^\mu_{KN} = Re \frac{e}{r + ia\cos\theta} k^\mu, \tag{2}$$

where $k^\mu(x^\mu)$ is the null vector field which is tangent to a vortex field of null geodesic lines, the Kerr principal null congruence (PNC). For the KN solution function H has the form $H = \frac{mr - e^2/2}{r^2 + a^2\cos^2\theta}$. The Kerr spheroidal oblate coordinates r, θ, ϕ_K, are related with the Cartesian coordinates as follows

$x + iy = (r + ia)e^{i\phi_K}\sin\theta, \ z = r\cos\theta$. Vector field k^μ has the form

$$k_\mu dx^\mu = dt - dr - a\sin^2\theta d\phi_K. \tag{3}$$

[a]e-mail: bur@ibrae.ac.ru

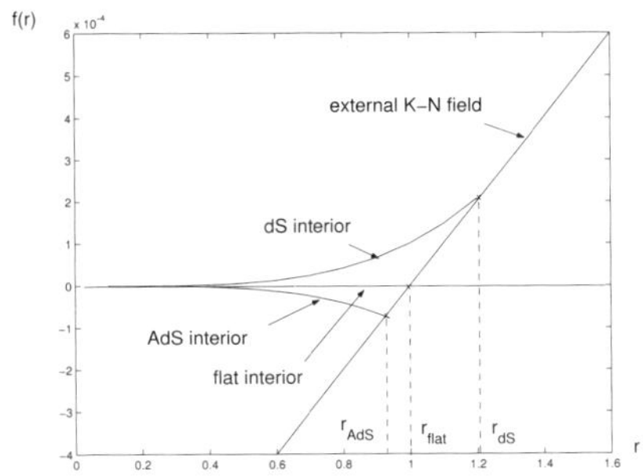

Figure 1: Matching of the metric for regular bubble interior with metric of external KN field.

For the metric *inside of the oblate bubble* we use the KS ansatz (1) in the form suggested by Gürses and Gürsey with $H = \frac{f(r)}{r^2+a^2\cos^2\theta}$. If we set for interior $f(r) = f_{int} = \alpha r^4$, the Kerr singularity is suppressed. For exterior, $r > r_0$, we use $f(r) = f_{KN} = mr - e^2/2$ corresponding to KN solution. Therefore, $f(r)$ describes a phase transition of the KS metric from 'true' to 'false' vacuum (see fig.1). We assume that the zone of phase transition, $r \approx r_0$, is very thin and metric is continuous (and maybe even smooth) there, so the point of intersection, $f_{int}(r_0) = f_{KN}(r_0)$, determines position of domain wall r_0 graphically and yields the 'balance matter equation' [5, 6], $m = m_{em}(r_0) + m_{mat}(r_0), \quad m_{em}(r_0) = \frac{e^2}{2r_0}$ which determines r_0 analytically. A small correction may be given by contribution to mass from intermediate zone. Interior has constant curvature, $\alpha = 8\pi\Lambda/6$. Following López, we fix the boundary of bubble at $r_0 = r_e = e^2/2m$, which yields flat interior, $\alpha = 0$. For parameters of electron the axis ratio of the ellipsoidal bubble is ~ 137, so it has the form of highly oblated disk.

3. *Chiral sector.* The corresponding phase transition is provided by Higgs model with two complex Higgs field Φ and Σ, two related gauge fields A^μ and B^μ, and one auxiliary real field Z. This is a given by Morris generalization of the $U(1)\times\tilde{U}(1)$ field model [5,6]. The potential $V(r) = \sum_i |\partial_i W|^2$, where $\partial_1 = \partial_\Phi$, $\partial_2 = \partial_Z$, $\partial_3 = \partial_\Sigma$, is determined by superpotential $W = \lambda Z(\Sigma\bar{\Sigma} - \eta^2) + (cZ + m)\Phi\bar{\Phi}$ and forms a domain wall interpolating between internal ('false') and external ('true') vacua. The vacuum states obey the conditions $\partial_i W = 0$ which yield $V = 0$ for 'false' vacuum $(r < r_0)$: $Z = -m/c; \Sigma = 0; |\Phi| = \eta\sqrt{\lambda/c}$, as well as for 'true' vacuum $(r > r_0) : Z = 0; \Phi = 0; \Sigma = \eta$.

4. *Higgs mechanism.* We set $B^\mu = 0$ and use only the gauge fields A^μ which interacts with the Higgs field $\Phi(x) = |\Phi(x)|e^{i\chi(x)}$ having a nonzero vev inside of the bubble, $|\Phi(x)|_{r<r_0} = \Phi_0$, and vanishing outside. The related part of the field model is a copy of the Nielsen-Olesen (NO) field model for a vortex string in superconducting media. However, there is a principal difference from

NO model, since we describe *superconducting interior* of the bubble, while the external em field has to be long-range. The Lagrangian $\mathcal{L}_I = -\frac{1}{4}F_{\mu\nu}F^{\mu\nu} + \frac{1}{2}(\mathcal{D}_\mu\Phi)(\mathcal{D}^\mu\Phi)^*$, where $\mathcal{D}_\mu = \nabla_\mu + ieA_\mu$, leads to the equations

$$\Box A_\mu = I_\mu = \frac{1}{2}e|\Phi|^2(\chi,_\mu + eA_\mu), \tag{4}$$

where we used flat d'Alemberian, so far as our metric is flat for $r < r_0$. In external region $|\Phi| = 0$ and the em field has to correspond to exact KN solution, $A^\mu \equiv A^\mu_{KN}$. and allows us to use flat We assume that current has to be expelled from interior of the bubble to its boundary, and we set in interior $I_\mu = 0$. The KN gauge field A_μ is given by (2) and (3). On the boundary and inside of the bubble the specific Kerr angular coordinate ϕ_K turns out to be inconsistent with Higgs field, and we transform the gauge field to the usual angular coordinate $\phi = -i\ln[(x+iy)/(x^2+y^2)^{1/2}]$. It reduces A_μ to the form

$$A_\mu dx^\mu = \frac{er}{r^2 + a^2\cos^2\theta}[dt - a\sin^2\theta d\phi] + d\ln(r^2 + a^2) \tag{5}$$

which shows that radial component is full differential. Near the boundary of bubble, $r = r_e = e^2/2m$, the value of potential is $A_\mu dx^\mu|_{r=r_e} = \frac{er_e}{r_e^2 + a^2\cos^2\theta}[dt - a\sin^2\theta d\phi] + d\ln(r^2 + a^2)$, and in equatorial plane $\cos\theta = 0$, we obtain

$$A^{(str)}_\mu dx^\mu = \frac{2m}{e}[dt - ad\phi] + d\ln(r_e^2 + a^2). \tag{6}$$

Tangent component of the gauge field at the border is $A^{(str)}_\phi = -2ma/e = -2J/e$ (since $J = ma$). Setting $J = \frac{n}{2}$ we find out that $A^{(str)}_\phi$ forms on the border of KN bubble a closed quantum loop $S = \oint eA^{(str)}_\phi d\phi = -n2\pi$, which has to be matched with angular periodicity of the Higgs field $\Phi \sim \exp in\phi$.

Inside the bubble, the condition $I_\mu = 0$ yields $\chi,_0 = -eA^{(in)}_0(r)$, which determines that $\chi,_0$ has to be a constant corresponding to the frequency of oscillation of the Higgs field, $\chi,_0 = \omega = -eA^{(str)}_0 = 2m$. Therefore, we have

$$A^{(in)}_0 = -\frac{2m}{e}; \; A^{(in)}_\phi = -\frac{2ma}{e}; \; A^{(in)}_r = \frac{2erdr}{(r^2 + a^2)}, \tag{7}$$

$$\Phi(x) = \Phi_0 \exp\{i\omega t - i\ln(r^2 + a^2) + 2i\phi\}. \tag{8}$$

For exclusion of the region of string-like loop at the equator, the time and ϕ components of the obtained vector field have a chock, crossing the boundary of the bubble, which generates circular currents. So, by $r \approx r_0$ we have to admit $I_\mu \neq 0$ and consider deviation of the em field from the obtained background solution (7), (8). Substituting $A^{(bound)}_\mu = A_\mu - A^{(in)}_\mu$ in (4) we obtain the equation

$$\Box A^{(M)}_\mu = I_\mu = e^2|\Phi|^2 A^{(M)}_\mu \tag{9}$$

which shows that a massive vector meson with mass $M = m_v = e|\Phi|$ may resides on the boundary of the bubble.

5. *Outlook.* For $a >> r_0$ the solutions of (9) form a massive rotating string at the edge of oblate bubble. It returns us to the old Kerr-Microgeon model [3], in which the Kerr ring was considered as a waveguide for circular motion of a photon. This string forms a similar waveguide for the em field. Being very close to the Kerr ring, it turns out to be relativistically rotating, $v_{tan} \approx c$, which should give a stringy contribution to mass. Besides, there have also to be the em excitations. It represents a strong analogy with the considered in [7] solvable model of soliton-antisoliton excitations moving on a rotating circle.

We should also note similarity of the KN source with the spinning Q-balls models of hadrons (oscillons).

Finally, the superconducting bubble source of the KN solution resolves the mysterious problem of the negative sheet of the KN solution which turns out to be related with a holographic structure [8,9] of the KN geometry adapted to its quantum treatment [10, 11]. The boundary of bubble forms a membrane separating the bulk regions of the in- and out-vacua and creates the Kerr/CFT (or the bulk/boundary) correspondence. The superconducting flat interior of the KN source may analytically be extended to 'negative' KN sheet, forming there a flat vacuum state. It allows one to identify the negative sheets of different KN sources and form a common vacuum state ensuring linear superposition of the vacuum fields from different sources.

References

[1] G.C. Debney, R.P. Kerr, A. Schild, *J. Math. Phys.* **10**, 1842 (1969).

[2] W. Israel, *Phys. Rev.* **D2**, 641 (1970).

[3] A.Ya. Burinskii, *Sov. Phys. JETP,* **39**, 193 (1974); *Russian Phys. J.* **17**, 1068 (1974), DOI 10.1007/BF00901591.

[4] C.A. López, *Phys. Rev.* **D30**, 313 (1984).

[5] A. Burinskii, *Grav. Cosmol.* **8**, 261 (2002), arXiv:hep-th/0110011; *J.Phys. A: Math.Gen.* **39**, 6209 (2006).

[6] A. Burinskii, in: Proc. of the XIII Workshop on High Energy Spin Physics DSPIN-09, ed. by A.Efremov, JINR 2009, arXiv:0910.5388[hep-th].

[7] R.F. Dashen, B.Hasslacher and A. Neveu, Phys. Rev. **D 11**, 3424 (1975).

[8] A. Burinskii, First Award of GRF 2009, *Gen. Rel. Grav.* **41** 2281, arXiv: 0903.3162[gr-qc].

[9] A. Burinskii, *Theor.Math.Phys.* 2010 (to appear); in: Proceedings of the 1th MCCQG, 2009 arXiv:1001.0332[gr-qc] .

[10] G. W. Gibbons, *Comm. Math. Phys.* **45** 191 (1975);

[11] C.R. Stephens , G. t' Hooft and B.F. Whiting, *Class. Quant. Grav.* **11** 621 (1994).

ATOMS AS RODS AND CLOCKS IN GRAVITATIONAL FIELD

Anatoly Nikishov [a]
I.E. Tamm Department of Physics, P.N.Lebedev Physical Institute, 117924, Leninsky Prospect 53, Moscow, Russia.

Abstract

We consider the implications of using an atom as a rod and a clock in a constant gravitational field.

1 Introduction and exposition of the approach

Almost 50 years ago the authors of [1] proposed to use, for example, the hydrogen atom as rod and clock in gravitational field and analyzed the consequences of this, see also [2]. The constant gravitational field modifies the Bohr radius r_b

$$r_b \to r_{b\phi} = \frac{r_b}{\sqrt{g_s}}, \quad r_b = \frac{\hbar^2}{me^2}. \tag{1}$$

[We use the notation

$$ds^2 = g_{00}c^2dt^2 + g_s(dx_1^2 + dx_2^2 + dx_3^2), \quad g_s \equiv g_{11} = g_{22} = g_{33}.] \tag{2}$$

Using the atom as a rod in gravitational field, we consider its size as the unchanged one. So the modified (or calculated [2]) radius $r_{b\phi}$ is multiplied by $\sqrt{g_s}$. In this way the space part of metric (2) appears. Similarly with g_{00} [1,2]. Other physical quantities such the velocity of light are also modified by gravitational field. But in the usual approach we consider them as unchanged and do not make any mistake [2]..This may be commented as follows: Modifications of physical quantities are determined by their dimensions [1].The terms of the Dirac equation (as any other equation) have the same dimension. So in constant gravitational field each term acquires the same factor and it can be dropped. In this way we get the observed (unchanged) quantities such as the atomic frequency, observed locally. An interesting exception is the Einstein gravitational constant

$$\kappa \to \kappa_\phi = \frac{\kappa}{\sqrt{|g_s g_{00}|}} \tag{3}$$

The modification is absent if $|g_s g_{00}| = 1$. An application of this approach is the deflection of light at finate distances from a gravitating body. Not far from the asymptotic behavior of the trajectory, where the deviation is small, it can be evaluated by purely mechanical means, see §4, Ch2 in Schwinger book [4]. This gives the angle of deviation [5]

$$\varphi = r_g\rho\int_x^\infty \frac{dx}{(x^2+\rho^2)^{3/2}} = \frac{r_g}{\rho}\left(1 - \frac{x\rho^{-1}}{\sqrt{1+(x\rho^{-1})^2}}\right). \tag{4}$$

[a] e-mail: nikishov@lpi.ru

Or in terms of

$$u = \frac{\rho}{r}, \quad \delta = \frac{r_g}{\rho}, \quad r_g = \frac{2GM}{c^2},$$

where ρ is impact parameter,

$$\varphi = \frac{1}{2}(u)^2\delta, . \tag{5}$$

The same result is obtained for the tangent of the trajectory in isotropic coordinate system. This is not the case for other coordinate systems [5].

2 Conclusions

It seems reasonable to expect that the considered approach should hold beyond the linear approximation as long as the tidal force inside the atom is negligible. It even seems reasonable to think that two functions g_{00} and g_s are functions of only Newtonian potential. This is the case for the isotropic coordinate system of the Schwarzschild field. It is inticing to think that the same form of metric should hold for any single body; only the Newtonian potentials are specific. But this agrees with general relativity only in linear approximation because only two functions g_{00} and g_s mast satisfy all Einstein equations. It would be also interesting to check experimentally that the relation $|g_s g_{00}| = 1$ holds only in linear approximation as predicted by general relativity.

Acknowledgments

The work was supported by Scientific Schools and Russian Fund for Fundamental Research (Grants 1615.2008.2 and 08-02-01118).

References

[1] H. Dehnen, H. Hönl, and K. Westpfahl, *Ann.der Phys.*, Bd 6, 7 Folge, Heft 7-8, 370 (1960).

[2] W.E. Thirring.,*Ann. Phys.* (N.Y.) **16**, 96 (1961).

[3] A.I. Nikishov, gr-qc/ 0710.4445,

[4] J.Schwinger, *Particles, Sources, and Fields.* V.1 Addison-Wesley (1970).

[5] A.I. Nikishov, *Physics of elementary particles and atomic nuclei*(Pis'ma) **6** 704 (2009).

NONHYDROGEN-LIKE GRAVIATOM RADIATION

Michael Fil'chenkov [a], Yuri Laptev
Institute of Gravitation and Cosmology, Peoples' Friendship University of Russia, 6 Miklukho-Maklay Street, 117198 Moscow, Russia

*Abstract.*Nonhydrogen-like graviatoms corrected for DeWitt's self-action, mini-hole rotation and particle spin have been considered.

1 Introduction

Graviatoms are bound quantum systems maintaining particles in orbit around mini-holes (primordial black holes) [1]. Hydrogen-like graviatoms satisfy Schrödinger's equation for the radial wave function as follows

$$\frac{1}{r^2}\frac{d}{dr}\left(r^2\frac{dR_{pl}}{dr}\right)-\frac{l(l+1)}{r^2}R_{pl}+\frac{2m}{\hbar^2}\left(E+\frac{mc^2r_g}{2r}\right)R_{pl}=0, \tag{1}$$

where m is the particle mass and $r_g=\frac{2GM}{c^2}$ is the mini-hole gravitational radius. The charged particles satisfying graviatom existence conditions are the electron, muon, taon, wino, pion and kaon. These conditions are geometrical, stability and indestructibility ones. Mini-hole and particle masses satisfy the relation $Mm=\alpha_g m_{pl}^2$ where $\alpha_g=0.5\div 0.6$ [2].

2 DeWitt's self-action corrections

The energy spectrum of a hydrogen-like atom has the form

$$E_n^{(0)}=-\frac{mc^2\alpha_g^2}{2n^2}, \tag{2}$$

where $\alpha_g=\frac{GMm}{\hbar c}$ is the fine-structure constant gravity equivalent. The perturbation being due to DeWitt's self-interaction [3]

$$V_q=\frac{mc^2r_gr_q}{4r^2}, \tag{3}$$

where $r_q=\frac{q^2}{mc^2}$ is the the classical radius if a particle withe the charge q. Corrections to the hydrogen-like spectrum

$$E_{nl}^{(1)}=\frac{mc^2\alpha_{eg}\alpha_g^2}{n^3\left(l+\frac{1}{2}\right)}, \tag{4}$$

where $\alpha_{eg}=\frac{e^2}{\hbar c}\alpha_g$.

[a]e-mail: fmichael@mail.ru

The intensity of the electric dipole radiation of a particle with mass and charge in the gravitational field of a mini-hole for the transition $2p \to 1s$ is

$$I_{10,21} = I^{(0)}_{10,21}\left(1 - \frac{46}{9}\right)^4 (1 - 6\alpha_{eg})^2, \tag{5}$$

where

$$I^{(0)}_{10,21} = \frac{2\hbar e^2[\omega^{(0)}_{12}]^3}{mc^3} f^{(0)}_{10,21} \tag{6}$$

is the intensity for a hydrogen-like graviatom, $\omega^{(0)}_{12} = \frac{E^{(0)}_1 - E^{(0)}_2}{\hbar}$, $f^{(0)}_{10,21} = \frac{2^{13}}{3^9}$ are the frequency and oscillator strength respectively. Thus, DeWitt's self-interaction diminishes both frequencies and in-tensities of the hydrogen-like graviatom.

3 Pauli's corrections

Taking account both DeWitt's self-action and Pauli's spin corrections [4], the graviatom energy spectrum takes the form

$$E = -\frac{mc^2\alpha_g^2}{2n^2} + \frac{mc^2\alpha_g^3}{2n^3}\left(\frac{2}{l + \frac{1}{2}}\frac{e^2}{\hbar c} - \frac{\alpha_g}{j + \frac{1}{2}}\right) + \frac{3mc^2\alpha_g^4}{8n^4}, \tag{7}$$

where $j = l + s$, $s = 0$ for mesons and $s = \pm\frac{1}{2}$ for leptons. The first nl_j levels: $1s_{1/2}, 2s_{1/2}, 2p_{1/2}, 2p_{3/2}$. The dipole transition $2p \to 1s$ splits into two transitions: $2p_{1/2} \to 1s_{1/2}$ with $\Delta j = 0$ and $2p_{3/2} \to 1s_{1/2}$ with $\Delta j = 1$. The intensity for the transition $2p_{3/2} \to 1s_{1/2}$ is

$$I_{10,21} = \frac{2}{3} I^{(0)}_{10,21}\left(1 + \frac{15}{48}\alpha_g^2\right)^4. \tag{8}$$

4 Graviatoms with slowly rotating miniholes

Lense-Thirring's metric for a slowly rotation black hole has the form [5]

$$ds^2 = \left(1 - \frac{r_g}{r}\right)c^2dt^2 - \frac{dr^2}{1 - \frac{r_g}{r}} - r^2(d\theta^2 + \sin^2\theta d\varphi^2) + \frac{2r_g a}{r}\sin^2\theta d\varphi dt, \tag{9}$$

where a the specific angular momentum. The perturbation being due to rotation

$$V_a = \frac{r_g a\hbar\sqrt{l(l+1)}}{r^3}\left(1 - \frac{r_g}{r}\right). \tag{10}$$

The intensity for the transition $2p \to 1s$

$$I_{21} = I^{(0)}_{21}\left(1 + 0.0349\frac{ma}{\hbar}\right)^4\left(1 - 1.1064\frac{ma}{\hbar}\right)^2. \tag{11}$$

The correction to the hydrogen-like spectrum of the graviatom taking account of DeWitt's self-force, particle spin and a slow rotation of the mini-hole reads

$$E_{nl}^{(1)} = -\frac{mc^2\alpha_g^2}{2n^2} + \frac{mc^2\alpha_g^3}{2n^3}\left(\frac{2}{l+\frac{1}{2}}\frac{e^2}{\hbar c} - \frac{\alpha_g}{j+\frac{1}{2}}\right) + \frac{3mc^2\alpha_g^4}{8n^4} + \frac{2\alpha_g^4 m^2 c^2 a}{\hbar n^3 (l+\frac{1}{2})^2}\left[1 - \frac{3n^2 - \left(l+\frac{1}{2}\right)^2}{n^2(l+\frac{1}{2})^2}\alpha_g^2\right]. \tag{12}$$

Providing $\frac{am}{\hbar}\alpha_g^2 \ll 1$, we obtain

$$\frac{E_{nl}^{(0)}}{E_n^{(0)}} = \frac{\alpha_n^2}{n}\left[\frac{1}{j+\frac{1}{2}} - \frac{4am}{\hbar\left(l+\frac{1}{2}\right)^2}\right]. \tag{13}$$

Hence $\frac{E_{nl}^{(0)}}{E_n^{(0)}} \ll 1$, if $\alpha_n^2 \ll n, \quad j+\frac{1}{2} \gg 1, \quad \frac{4am}{\hbar} \ll \left(l+\frac{1}{2}\right)^2$.

As a result, the perturbation theory is valid for a slow rotation of the minihole and higher levels of the particle being constituents of the graviatom.

5 Conclusion

The graviatoms can contain only leptons and mesons. The hydrogen-like graviatom is perturbed by DeWitt's self-action and minihole rotation diminishing, whereas particle spin enhancing the dipole radiation intensity.

Acknowledgments

We are grateful to the Organizing Committee and especially Prof. A.I. Studenikin for providing us the possibility of participating the 14th Lomonosov Conference.

References

[1] Yu.P. Laptev, M.L. Fil'chenkov, *Astronomical and Astrophysical Transactions* **25**, 33 (2006).

[2] M.L. Fil'chenkov, Yu.P. Laptev, S.V. Kopylov, in *"Particle Physics on the Eve of LHC"* (Proceedings of the 13th Lomonosov Conference on Elementary Particle Physics, August 23-29 2007, Moscow, Russia), ed. by A. Studenikin, World Scientific Singapore, 240, 2009.

[3] C.M. DeWitt, B.S. DeWitt, *Physics* **1**, 3 (1964).

[4] A.A. Sokolov, Yu.M. Loskutov, I.M. Ternov, *"Quantum Mechanics"*, (Prosveshchenie, Moscow) 353, 1965.

[5] J. Lense, H. Thirring, *Physikalische Zeitschrift* **19**, 156 (1918).

BRANE UNIVERSE: GLOBAL GEOMETRY

Alexey Smirnov [a]
Institute for Nuclear Research, Russian Academy of Sciences,60th October Anniversary Prospect, 7a, 117312, Moscow, Russia

Abstract
The global geometries of bulk vacuum space-times in the brane-universe models are investigated and classified in terms of geometrical invariants.

The aim of this talk is to unvestigate the possible global geometries of some "brane universe" scenarios in which "our Universe" is supposed to be embedded in a space-time of larger number of dimensions, "bulk".

In order to get the exact solutions we simplify the model as much as possible. Let us consider a $(N+1)$-dimensional space-time containing a single N-dimensional thin shell (brane) and supppose that this brane is time-like and has the cosmological symmetry, i.e., it is homogeneous and isotropic. Moreover, we demand the bulk to have the slicing with the same symmetry. Then the metric of the whole space-time can be written as

$$ds^2 = -dn^2 + B^2(n,t)dt^2 - A^2(n,t)\left[\frac{dr^2}{1-kr^2} + r^2 d\Omega^2_{N-2}\right], \tag{1}$$

Outside the shell the space-time is a vacuum with some cosmological constant Λ. Then the energy-momentum tensor in the right-hand side of the Einstein equations equals

$$T_{\mu\nu} = S_{\mu\nu}\delta(n) + \frac{1}{8\pi G}g_{\mu\nu}\Lambda, \quad S^n_n = S^n_i = 0, \quad S^2_2 = S^3_3 = S^N_N. \tag{2}$$

In what follows we will consider the so-called vacuum shells only, for which $S^0_0 = S^2_2 = const.$

Due to the proclaimed symmetry we are able to use yet another, more general, global approach, considering the metric in the form

$$ds^2 = \gamma_{AB}(x)dx^A dx^B - R^2(x)dl^2_{N-1} \quad A,B = 0,1\,, \tag{3}$$

which is essentially characterized by two invariant functions of two variables, $R(x)$ and $\Delta(x)$, where

$$\Delta = \gamma^{AB}\frac{\partial R}{\partial x^A} = -k + \frac{2}{N(N-1)}\Lambda R^2 \frac{\partial R}{\partial x^B} \tag{4}$$

is the square of the normal vector to the surfaces $R = const.$ Then

$$\begin{array}{lcl} \Delta < 0 & \Rightarrow & R_\pm\text{-regions} \\ \Delta > 0 & \Rightarrow & T_\pm\text{-regions} \\ \Delta = 0 & \Rightarrow & \text{Apparent horizons} \end{array}$$

[a]e-mail: smirnov@ms2.inr.ac.ru

It is the set of these regions and horizons together with the boundaries (infinities and singularities) that consists the global geometry.

Below we present Carter-Penrose conformal diagrams for all possible values (signs) of the cosmological constant Λ, spatial curvature $k = 0, \pm 1$ and the surface energy density S_0^0. In these figures the dashed curves represent surfaces of constant radii, and for the case $\Lambda < 0$, $k = -1$ we should distinguish between "heavy" shells with $|S_0^0| > \sqrt{\frac{(N-1)|\Lambda|}{2\pi G N}}$ and "light" shells with $|S_0^0| < \sqrt{\frac{(N-1)|\Lambda|}{2\pi G N}}$.

Note, that from the point of view of a possible genesis the positive values of the surface energy density $S_0^0 > 0$ is more plausible for positive cosmological constant $\Lambda > 0$, since it requires only one copy of the bulk space-time (and two copies if $S_0^0 < 0$). For $\Lambda < 0$, the negative $S_0^0 < 0$ seems more natural.

The local Z_2−symmetry can be preserved even in the case of two and more branes. This can be achieved when additional shells are placed on different sides of the Einstein-Rosen bridges in the space-times with horizons or beyond the infinities in the case of negative cosmological term because light rays in such a case reach he infinite value of radius in finite coordinate time interval.

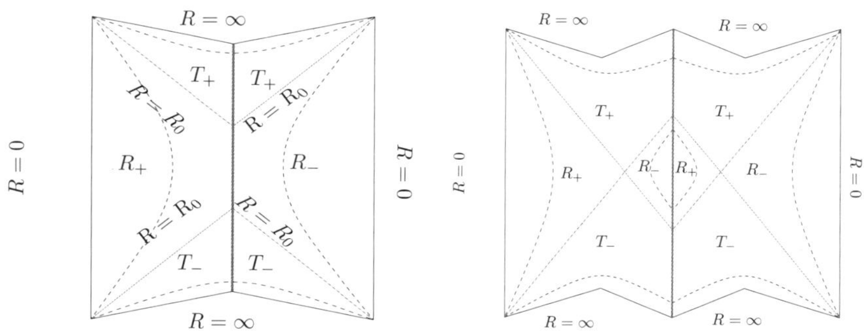

Figure 1: $\Lambda > 0$, $k = +1$. Left $S_0^0 > 0$. Right $S_0^0 < 0$

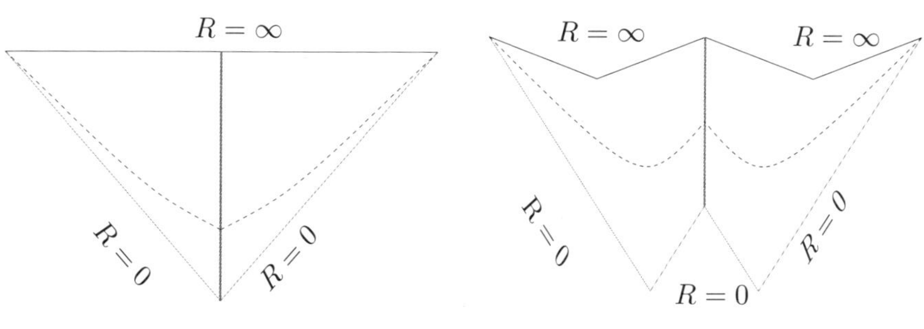

Figure 2: $\Lambda > 0$, $k = 0$. Left $S_0^0 > 0$. Right $S_0^0 < 0$

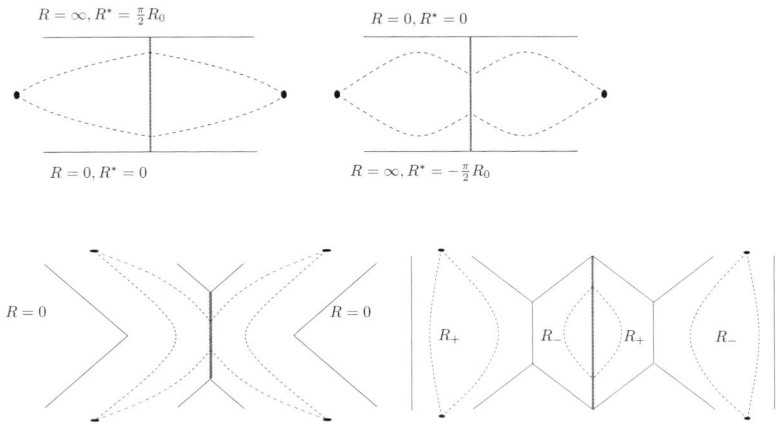

Figure 3: a) $\Lambda > 0$, $k = -1$, $S_0^0 > 0$ b) $\Lambda > 0$, $k = -1$, $S_0^0 < 0$ c) $\Lambda < 0$, $k = +1$, $S_0^0 > 0$ d) $\Lambda < 0$, $k = +1$, $S_0^0 < 0$

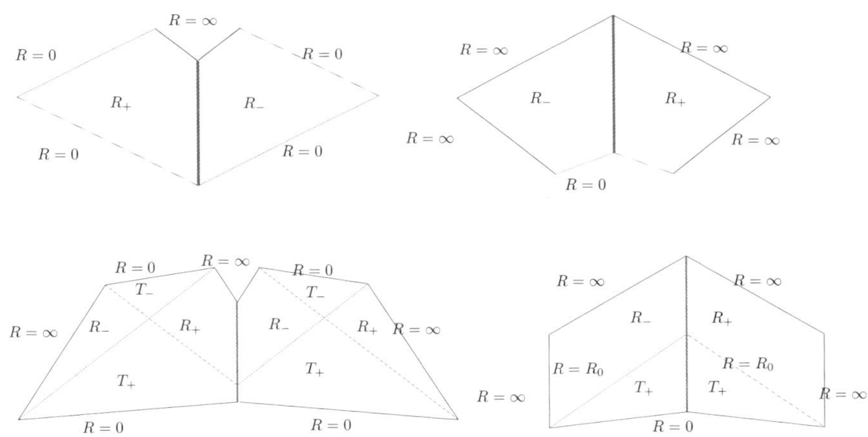

Figure 4: a) $\Lambda < 0$, $k = 0$, $S_0^0 > 0$ b) $\Lambda < 0$, $k = 0$, $S_0^0 < 0$ c) $\Lambda < 0$, $k = -1$, "heavy" shell, $S_0^0 > 0$ d) $\Lambda < 0$, $k = -1$, "heavy" shell, $S_0^0 < 0$

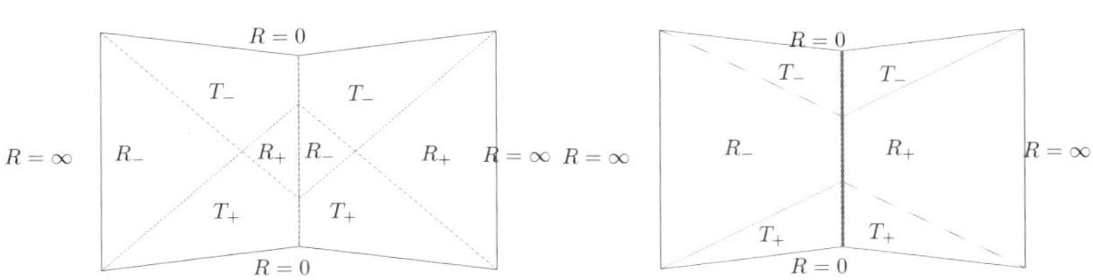

Figure 5: a)$\Lambda < 0$, $k = -1$, "light" shell, $S_0^0 > 0$ b) $\Lambda < 0$, $k = -1$, "light" shell, $S_0^0 < 0$

RADIATION EXPOSURE AND MISSION STRATEGIES FOR INTERPLANETARY MANNED MISSIONS AND INTERPLANETARY HABITATS

Piero Spillantini [a]

INFN and University, Firenze, Italy

Abstract. Active shieldings based on superconducting magnetic lenses were evaluated in the past in ESA supported studies. The present increasing interest on permanent space complexes, to be considered in the far future as 'Bases', rather than 'Stations', located in 'deep' space requires that this preliminary activity continues, envisaging the problem of the protection from Cosmic Ray (CR) action at a scale allowing long permanence in 'deep' space, besides to a relatively number of dedicated astronauts, also to citizens conducting there 'normal' activities. The realization of the magnetic protection of long permanence habitats by well-established nowadays materials and techniques is in principle possible, but not workable in practice for the huge required mass of the superconductor, the too low operating temperature and the corresponding required cooling power and thermal shielding. However the fast progress in the production of reliable High Temperature Superconducting (HTS) or MgB2 cables and of cryocoolers suitable for space operation opens the perspective of practicable solutions. Quantitative evaluations for the protection from Galactic CR of a large habitat of the 'Space Base' are presented. For possible solutions with outer diameter larger than that allowed by transportation to space vehicles it must in the meantime solved the problem of the assembling or deploying in space of the conductors for returning the electric current.

There are several ideas, initiatives and projects for defending astronauts from the most energetic solar cosmic rays in the most intense solar events, when they fly and operate far away from the protection of the terrestrial magnetic field. The projects of interplanetary spacecrafts ([1], [2], [3], [4]) foresee 'protected volumes' where astronauts can shelter in the unlikely but very dangerous case of an intense and energetic solar storm. These 'storm shelters' would be used in extreme cases, and only for a few hours, perhaps a few days, and are therefore small in volume ($\approx 1\text{m}^3$/astronaut) and Spartan: the protection is obtained by a water jacket and penalizes the mission with several tons of mass. A magnetic field sheath produced by superconducting coils could reduce the mass penalty for a factor 3 [5].

The criterion of 'shelter' cannot be used for protecting astronauts in long duration missions, either in spacecraft or in deep space permanent bases, where the protection is required to act during the whole duration of the mission against the action of the galactic cosmic rays (GCR). Astronauts must spend most of their time inside a volume much larger than a shelter, equipped for their life and activities, i.e. a large volume 'habitat' protected from GCR that are much more energetic that the solar ones emitted in the most energetic solar events. The dose absorbed by an astronaut in a few months overcomes the 'carrier limits' (normally expressed in dose absorbed in 10 years) recommended

[a]e-mail: spillantini@fi.infn.it

by the USA National Council on Radiation Protection for astronauts [6]. The volume to be protected should be 50-100m^3/astronaut, i.e. 300-600m^3 for a habitat for $\leq$6 men. Passive protection is not more possible, not only because of the required enormous mass, but also for the decreasing of its efficacy with the thickness due to the secondary production of particles by interaction of the cosmic ray on the material.

The problem must be set to this level of complexity because the plans of space agencies are evolving from sporadic manned missions of deep space exploration toward a strategy of exploitation of deep space, and not only by missions to celestial bodies of the solar system but also deep space bases for supporting them. This strategy is intended to attract private investments for digging out resources and implementing services from space, with the Space Agencies supplying the needed technical competences, guaranties and controls in conformity with the political indications of the respective governments. The growing of space tourism enterprises and of enterprises offering access to orbit (see for example the SpaceShipTwo successful spacecraft [7]), the studies for extracting useful materials from Moon [8] and asteroids, the growing awareness of the advantages of locating in Lagrange points permanent stations of transit and logistics and for transferring infrastructures ([9], [10], [11]) are signs of the growing of such a strategy.

In last decade several evaluations were made for protecting large volume habitats by magnetic fields produced by superconducting coil systems. In Europe ESA supported the work of a Topical Team study ([5], [12]) and an industrial tender [13] dedicated to the protection from ionizing radiation of astronauts in manned interplanetary missions. In both these studies the habitat was assumed to be protected by a sheath of magnetic field produced by superconducting coils in toroidal configuration (as schematically indicated in fig.1). The conclusions of both works were somewhat encouraging, supplying reasonable evaluations of the needed magnetic field intensities for mitigating by a factor 2-3 the dose of absorbed radiation inside a protected volume, assumed to be that typical of an ISS module (about 4m in diameter, see fig.2). In the conclusions of these studies the needed technologies to be developed were suggested to arrive to the design of a possible system.

In USA a design was presented [14], based, as the ESA evaluations, on toroidal coils and protecting a similar volume, but assuming a very high intensity magnetic field, what implied a huge mass of the system ($\approx$30t) for supporting the enormous ponderomotive forces.

The until now performed work on active shielding was reviewed at the last COSPAR in Montreal [15].

It can be regarded as 'pioneer' and must be updated for several reasons:

- (a) a diffuse wide experience was accumulated in realizing and operating huge volume and huge stored energy superconducting magnets for

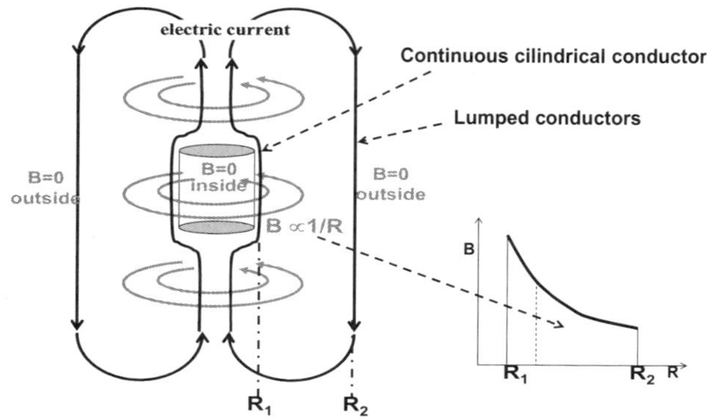

Figure 1: Schematics of a toroidal magnetic sheath for protecting a cylindrical volume inside

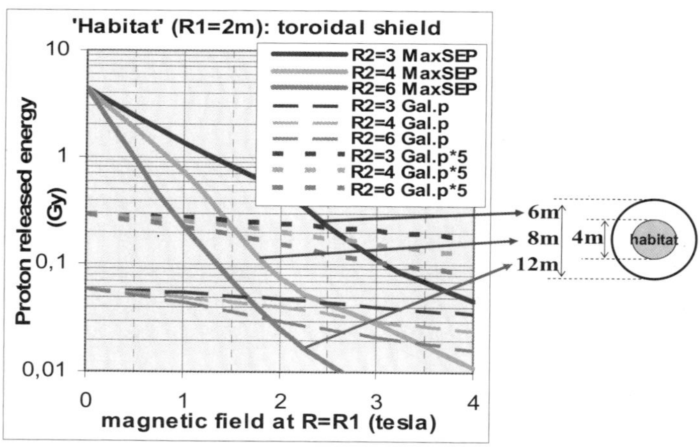

Figure 2: 'Dose' released in a ISS-like habitat evaluated for the three values 3, 4 and 6m of the outer radius R2 of the toroidal system as a function of the maximum field (at R=R1) of the system. For R2=6m the dose due to the maximum possible solar event (MaxSEP) is lower than that released by GCR in one year for a maximum magnetic field (at R=R1) of 0.8T. The GCR dose is reduced by a factor 2 (3) for maximum magnetic field of 2T(3T).

experiments at accelerators;

- (b) technical achievements on superconducting materials were obtained, with the production of High Temperature Superconductors (HTS) cables and of the medium temperature superconducting MgB2 cables ([16], [17]) and fast improvement of their parameters ([18], [19], [20], [21]), as well in more light and efficient cryocoolers [22];
- (c) the evolution toward the exploitation of the Solar System implies that in long duration missions to deep space will operate a not negligible number of citizen, besides astronauts, conducting there their activities (as for example personnel for running space hotels for tourists, or operating deep space stations, or assigned to transit and logistics);
- (d) plans are in preparation for establishing permanent bases on the surface of celestial bodies which also require to be protected from GCR, as well rovers for the mobility on their surface.

The protection from the ionizing radiation rises to a higher level than that until now afforded, either for the dimensions of the habitats to be protected or the possible long times of permanence or the possible new safety rules to comply with. The space bases must be equipped by one (or more) habitats offering a substantial reduction of GCR dose to the permanent staff during the whole permanence period and possibly acting as solar storm shelters easily and timely reachable by all the afferent personnel of the base.

The active system can be constituted by superconducting coils producing a high intensity magnetic field, protecting, as a sheath, the habitat. The most convenient configuration is the toroidal one (already shown in fig.1), also if other configurations can be considered. In order to allow in the habitat normal activities the residual magnetic field inside must be a few gauss at most, i.e. three or more order less than in the surrounding sheath. This is possible in principle if the inner current flows around the cylindrical external wall of the habitat uniformly in a continuous cylindrical surface; however the mechanical instability due to the magnetic buckling pressure must be taken into account in the safety systems. Preliminary tests and an accurate engineering are required to guarantee the cancellation of the magnetic field inside in any possible normal or emergency situation. A suitable distribution of high permeability materials (such as mu-metal sheets) can help to solve the problem.

Furthermore the design of the magnetic system must take in account the following points:

- \- it is advisable that the protection system be integrated with the habitat before its transportation to space;

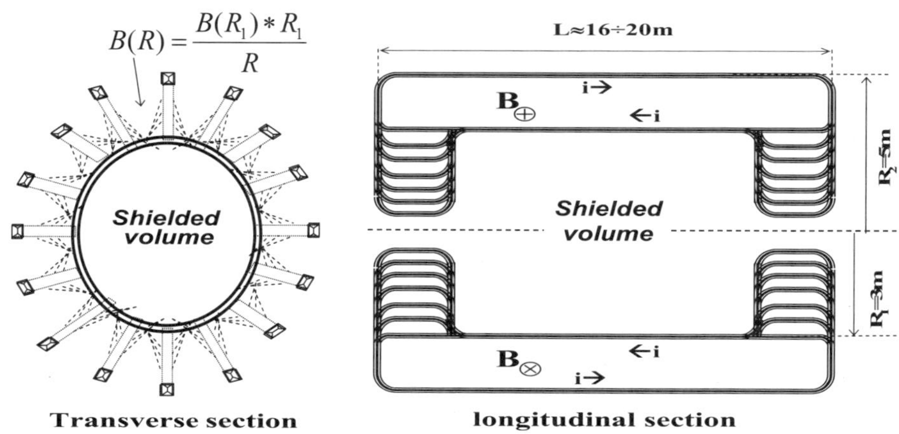

Figure 3: Configuration assumed to evaluate the protection of a 6m diameter cylindrical habitat

- - therefore its outer dimensions (diameter and length) and its mass must comply with the capability of the transport system;
- - the system, also if it should be transported already cooled, must be energized in its final location in space.

The sketch of the whole system shown in fig.3 is not a preliminary design but just a scheme for the subsequent evaluations, where it was assumed that:

- the total current is returned by a number (16 in the figure) of external conductors;
- the inner tube of current is supposed to be subdivided at the two ends in several tubes of smaller diameter, in order to protect down to small angles and meanwhile maintaining the magnetic field intensity not higher than that at R=R1 (this is important, because in a superconducting magnet the strictness of problems strongly depends from the maximum field reached in the system).

Taking in account the 10m maximum diameter until now foreseen for transport systems (that of the mammoth Ares V rocket for the next NASA mission to Moon) a habitat of 6m diameter can be protected by a 'sheath' of magnetic field not thicker that 2m. The dose in the habitat due to protons is reported in fig.4 as a function of the maximum magnetic field in the system for the two 'thicknesses' of 1m and 2m of the magnetic field sheath (dashed lines); in

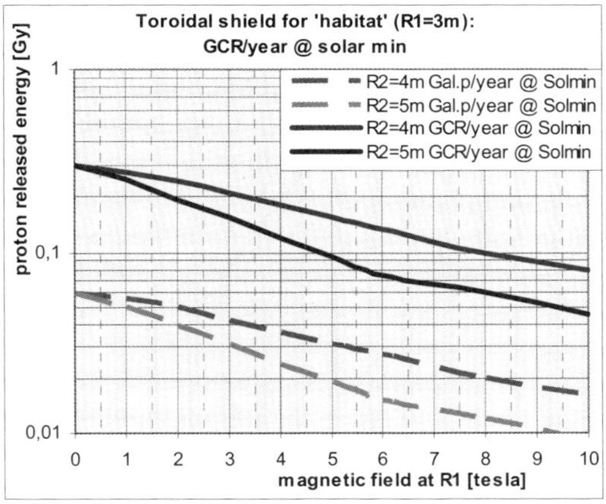

Figure 4: Reduction of the dose released by galactic protons as a function of the maximum magnetic field for two values of the outer diameter (8 and 10m) of the system (see fig.3). The reduction of the dose released by all the GCR is reported as indication, obtained by multiplying for 5 the dose released by the protons. For 10m outer diameter the reduction is about 70% (80%) for maximum field of 5 tesla (6 tesla). For the smaller 8m outer diameter the reduction of 70% is obtained with maximum field of 9 tesla.

the figure it is also indicated the dose due to all the GCR flux, obtained by multiplying by x5 the dose due to protons.

For the assumed geometry (diameter of the habitat 6m, external diameter of the whole system 10m, 2m thick magnetic sheath) the spectra of the GCR fluxes at solar minimum are reported in fig.5; at the bottom of the figure it is reported (in %) the reduction of the dose for different values of the maximum magnetic field on the superconductor. Reductions higher than 50% can be reached only for very high values of magnetic field at R1=3m: 6T (8T) for a reduction of 75% (80%).

It is difficult to conceive a system with such high fields without a very heavy and bulky mechanics. It must however observed that reductions about 3/4 of the GCR dose in periods of high solar activity (i.e. minimum GCR flux) can be regarded as acceptable by nowadays criteria for permanence in space of about one year. Because the GCR flux varies by a factor 2 according to the 11 year cycle of the solar activity, also this input must be taken into account in the planning of long duration permanence in deep space.

In order to evaluate the mass of the magnetic system the following criteria have been adopted:

- - the coils are maintained cool by cryocoolers at expenses of electrical power (Cryogen Free Superconducting Magnet concept [23]);

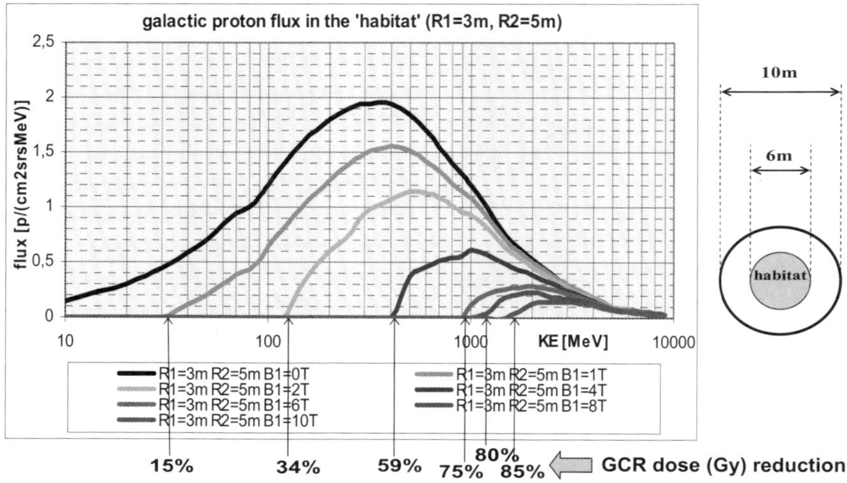

Figure 5: Reduction of the galactic proton flux inside the habitat. The corresponding reduction of the dose due to GCR flux is reported at the bottom of the figure for the values (1, 2, 4, 6, 8, 10T) of the maximum magnetic field of the system

- - the superconducting cable is constituted by MgB2 wires produced by the in-situ method in a titanium sheath ([18], [19]) stabilized outside in aluminium. This choice, proposed by Turin university and Alenia industry [21], seems at present the most promising: the cable can work at relatively high temperature (greater than 20K, a value that can be efficiently reached by cryocoolers), with a high current density (1kA/mm^2 at 2T, assumed diminishing with the inverse of the field for B(R1)$\geq$2T), its low density (3g/cm^3) guaranties an important mass saving, and it can be produced very thin [20], therefore less suffering current and temperature instability and better distributing current in the surrounding cables in case of bad functioning;

- - as observed commenting fig.5 it would be not expedient to approach the high values of the magnetic field needed for a substantial reduction of the GCR dose inside the habitat: the stored energy and the ponderomotive forces increase with the square of the field intensity, as well increase the possible instabilities of the superconducting regime of the current, critically depending from the point of higher field in the superconductor;

- - for this reason the mass of the magnetic system was evaluated also for outer diameters beyond the 10m supposed to be accepted by the transportation system, supposing that the conductors returning the electric

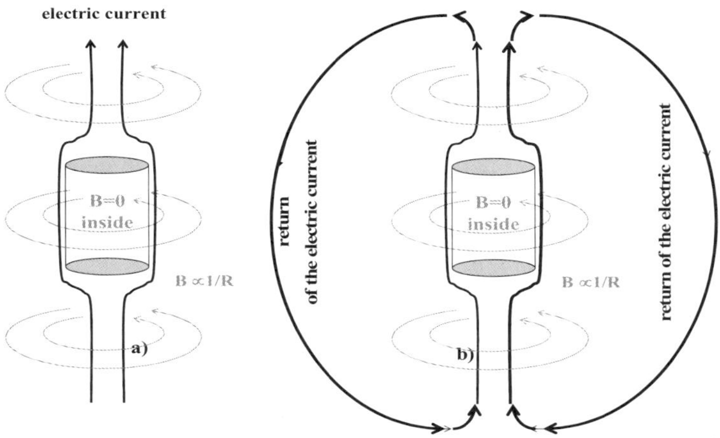

Figure 6: For large diameter magnetic systems the solenoidal configuration is not adequate, and must be adopted a toroidal configuration where the magnetic field diminishes at the increasing of the radius. The outer part of the system should be deployed or assembled in space

current at the outer radius could be either assembled or deployed in space in order to increase the 'thickness' of the magnetic sheath meanwhile allowing decrease its intensity fig.6.

For the 6m diameter habitat the needed cold mass of the system is represented in fig.7 as a function of the maximum magnetic field of the system for different values of the outer diameter (from 10m to 20m, corresponding to a magnetic field sheath thickness from 2m to 7m) and for 3 values of the GCR reduction. It must be underlined the strong dependence of the cold mass from the outer diameter: for a fixed GCR reduction the needed maximum field diminishes at the increasing the outer radius and the cold mass also decreases due to the increase of the allowed current density in the superconductor. The table in the figure is for a reduction of 82% of the GCR flux.

An assessment of the total mass is hardly possible without a detailed project, also considering that the mass of the supports strongly depends from the maximum field and the consequent ponderomotive forces. In the above mentioned ESA Topical Team study [5], taking in account that the system has a cylindrical symmetry, that in free space the vacuum tank (constituting most of the mass on the ground) is not needed and that is not necessary to contrast the gravitation field, it was assumed that the total mass could amount to 1.5 times the cold mass. Diminishing the maximum field in the superconductor also the ratio of the total mass to the cold mass could further diminish due to the decreasing of the ponderomotive forces.

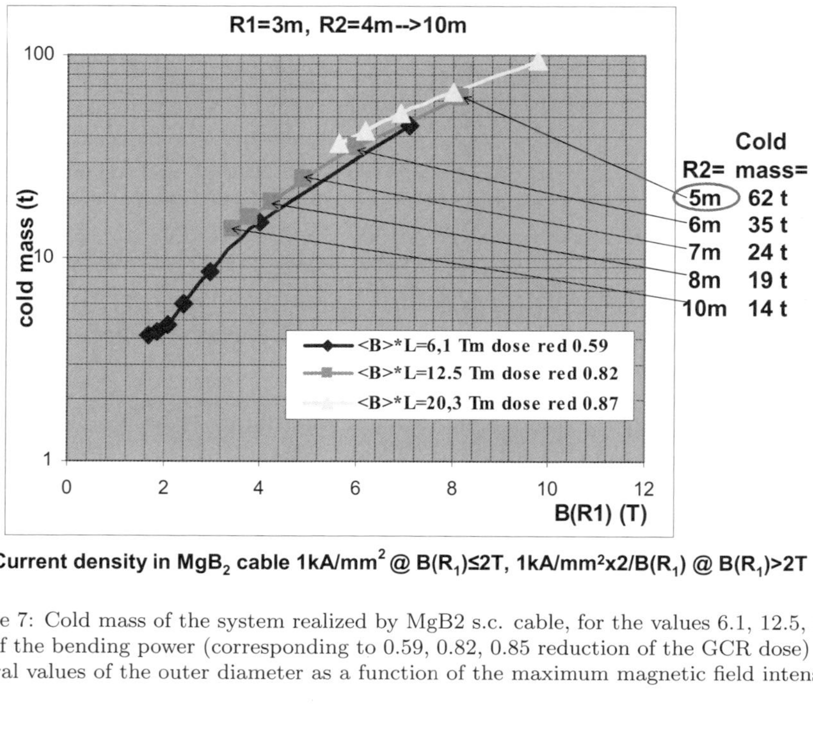

Current density in MgB_2 cable 1kA/mm^2 @ $B(R_1) \leq 2T$, 1kA/mm^{2}x2/$B(R_1)$ @ $B(R_1) > 2T$

Figure 7: Cold mass of the system realized by MgB2 s.c. cable, for the values 6.1, 12.5, 20.3 Tm of the bending power (corresponding to 0.59, 0.82, 0.85 reduction of the GCR dose) and several values of the outer diameter as a function of the maximum magnetic field intensity

These results underline the importance of developing systems for deploying or assembling in space the conductors returning the electric current running around the external cylindrical surface of the habitat, and the need of affording a specific development program, as recommended in the ESA studies.

An alternative solution to the limitedness of the maximum diameter of the transportation system could be the concept of a 'distributed habitat' in deep space: the protected volume could be subdivided in several protected sub-volumes (habitat modules), connected together for supplying to the personnel the total required living and working space (as it is now in the ISS). The 6m in diameter, 20m long habitat above considered could be composed by 8 habitat modules 3m in diameter, 10m long, each protected by its own toroidal field of 10m outer diameter. In this case a maximum magnetic field intensity less that 4T would be enough for their protection. Probably such architecture would be more expensive, but could be a useful approach for providing an artificial gravity gradient in the habitats (e.g. by putting them in rotation around a common axis).

References

[1] NASA Mars Exploration Study Team, Reference Mission version 3, Addendum to the Human Exploration of Mars: "The reference Mission of the NASA Mars Exploration Study Team", June 1998

[2] ISTC, "Preliminary project of the Manned Mars Expedition", project N 1172, 2000.

[3] Kennedy, K.J. Lessons from Transab: an Architect's Experience. NASA JSC report, AIAA 2002-6105, AIAA Space Architecture Symposium, Houston, Texas, 10-11 October 2002.

[4] HMM assessment study, "Human mission to Mars", CDF study report, CDF-20(A), March 2004.

[5] ESA Topical Team on "Shielding from the cosmic radiation for interplanetary missions: active and passive methods", 35th COSPAR Scientific Assembly, Paris, July 2004, and ESA sp-1281 report on: "Topical Teams in the Life and Physical Sciences: toward New Research Application in Space", pp.184-199, October 2005

[6] National Council on Radiation Protection and Measurement (NCRP), Recommendations on Dose Limits for Low Earth Orbit, NCRP report 132, edited by NCRP, Bethesda, MD., 2000

[7] Malik, T., Virgin Galactic unveils Suborbital Spaceliner Design', 23 January 2008, www.space.com/news/080123-virgingalactic-ss2-design.html.

[8] Moon Base - A challenge for humanity. Compagnone, F., Perozzi, E. (editors). Donzelli Editore, Roma 2007.

[9] Lo, M.W., Ross, S.D. The Lunar L1 gateway: portal to the stars and be-

yond, AIAA Space 2001 Conference, Albuquerque, New Mexico, August 2001.

[10] Casini, S. A quick look to a European strategy for solar system exploration (the wide vision). Moon Base: A Challenge for Humanity. 3rd Workshop, Moscow, November 2006.

[11] Thronson, H.A. Logistics in NASA Exploration Architecture and Major Science. Goals in Space. The Intern. Society of Logistics, 41, 4, 2007

[12] Spillantini, P., Casolino, M., Durante, M., Mueller-Mellin, R., Reitz, G., Rossi, L., Shurshakov, V., Sorbi, M. Shielding from the cosmic radiation for interplanetary missions: active and passive methods. Radiation Measurements 42, pp.14-23, 2007

[13] Cougnet, C., Crosby, N.B., Foullon, C., et al. Radiation Exposure and Mission Strategies for Interplanetary Manned Missions (REMSIM). Earth, Moon and Planets, 94, pp.279-285, 2005

[14] Chung, C.H., Fopp, S., Kim, Th., et al. AMS on ISS: Application of particle physics technology to manned interplanetary flight. AMS collaboration internal report, March 2005, to be submitted to NIM.

[15] Spillantini P. "Active shielding for long duration interplanetary manned missions". J.Adv. Space Res. (2010), doi:10.1016/j.asr.2010.01.025

[16] Nagamatsu, J., Nakagawa, N., Muranaka, T., Zenitani, Y., Akimitsu, J. Superconductivity at 39K in magnesium diboride. Nature 410, 2001.

[17] Dou, S.X., Collings, E.W., Shcherbakova, A. Aluminium stabilized superconductor. Superconductor Science and Technology 19, 333, 2006.

[18] Alessandrini, M., Fang, H., Hanna, M., Putman, P., Zhou, Y.X., Salama, K. High critical current of Ti-sheathed MgB2 wires for AC and weight-critical applications. Superc. Science and Technology 19, 129, 2006

[19] Kovac, P., Husek, I., Melisek, T., Holubek, T. Properties of stabilized MgB2 composite wire with Ti barrier. Superconductor Science and Technology 20, 771, 2007.

[20] Schlachter, S.I., Goldacker, W., Frank, A., Ringsdorf, B., Orshulko, H. Properties of mgb2 superconductors with regard to space applications. Cryogenics 46, 201, 2006

[21] Chevallard, J. Definizione di schermi magnetici per la radioprotezione degli astronauti. Turin University, degree thesis, September 2007

[22] Ross Jr, R.G., Boyle, R.F., Kittel, P. NASA Space cryocooler programs: a 2003 overview. TDW 2003: International Workshop on Thermal Detectors for Space Based Planetary, Solar and Earth Space Applications, 19 Jun 2003, Washington D.C., USA

[23] Rossi, L., CFSM (Cryogen Free Superconducting Magnet)'. Proposal to ASI for "Methodological Techniques in Space Systems and Subsystems", Alenia report SG-PP-Al-0864, ASI workshop ISS 2000, Naples.

CP Violation and Rare Decays

KLOE RECENT RESULTS: A REVIEW

Antonio De Santis [a] on behalf of the KLOE collaboration[b]
Sapienza Universitá di Roma & sez. INFN Roma

1 The KLOE experiment

The KLOE experiment [1] runs at the Frascati ϕ factory DAFNE, a high luminosity e^+e^- collider working at $\sqrt{s} \sim 1020$ MeV, corresponding to the ϕ meson mass.
The KLOE detector consists of a large cylindrical drift chamber [2], surrounded by a sampling lead-scintillating fiber electromagnetic calorimeter [3]. Both detectors operate inside a uniform magnetic field of ~ 0.5 T provided by a superconducting coil. In the whole data taking (2001 – 2006) KLOE has collected an integrated luminosity of 2.5 fb^{-1} corresponding to about 8 billions of ϕ produced. The KLOE trigger system [4] is highly efficient on most of the ϕ decay.

2 Kaon physics

CKM Unitarity and V_{us} [16] Precise measurements of semileptonic kaon decay rates provide the measurement of the V_{us} element of the CKM mixing matrix and information about lepton universality. Helicity-suppressed leptonic kaon decays provide an independent measurement of $|V_{us}|^2 / |V_{ud}|^2$, through the ratio $\Gamma(K \to \mu\nu)/\Gamma(\pi \to \mu\nu)$. These measurements, together with the result of $|V_{ud}|$ from nuclear β transitions, provide the most precise test of the unitarity of the CKM mixing matrix through the relation $|V_{ud}|^2 + |V_{us}|^2 + |V_{ub}|^2 = 1$. Unitarity can also be interpreted as a test of the universality of lepton and quark weak couplings, testing the relation $G_F^2 = G_{CKM}^2 = G_F^2(V_{ud}^2 + V_{us}^2 + V_{ub}^2)$, with G_F the Fermi coupling constant from the muon decay.
The kaon semileptonic decay rate is given by:

$$\Gamma(K_{l3}) = \frac{C_K^2 G_F^2 M_K^5}{192\pi^3} S_{EW} |V_{us}|^2 |f_+(0)|^2 I_{K,l}(\lambda)(1 + 2\Delta_K^{SU(2)} + 2\Delta_{K,l}^{EM}) \quad (1)$$

where $K = K^0, K^\pm$, $l = e, \mu$ and C_K is a Clebsch-Gordan coefficient, equal to 1/2 and 1 for $K^\pm$ and K^0, respectively. The decay width $\Gamma(K_{l3})$ is experi-

[a] e-mail: antonio.desantis@roma1.infn.it

[b] F. Ambrosino, A. Antonelli, M. Antonelli, F. Archilli, P. Beltrame, G. Bencivenni, C. Bini, C. Bloise, S. Bocchetta, F. Bossi, P. Branchini, G. Capon, T. Capussela, F. Ceradini, P. Ciambrone, E. De Lucia, A. De Santis, P. De Simone, G. De Zorzi, A. Denig, A. Di Domenico, C. Di Donato, B. Di Micco, M. Dreucci, G. Felici, S. Fiore, P. Franzini, C. Gatti, P. Gauzzi, S. Giovannella, E. Graziani, M. Jacewicz, W. Kluge, J. Lee-Franzini, M. Martini, P. Massarotti, S. Meola, S. Miscetti, M. Moulson, S. Müller, F. Murtas, M. Napolitano, F. Nguyen, M. Palutan, A. Passeri, V. Patera, P. Santangelo, B. Sciascia, T. Spadaro, L. Tortora, P. Valente, G. Venanzoni, R. Versaci, G. Xu

mentally determined by measuring the kaon lifetime and the semileptonic BRs totally inclusive of radiation. Several theoretical inputs are need: S_{EW}, $\Delta_K^{SU(2)}$, $\Delta_{K,l}^{EM}$ and $f_+(0) \equiv f_+^{K^0\pi^-}(0)$. The form factor dependence on the momentum transfer can be described by one or more slope parameters λ, measured from the decay spectra, and enters in the phase space integral $I_{K,l}(\lambda)$.

The KLOE experiment measure all the relevant inputs to extract V_{us} from K_{l3} decay rates of both charged and neutral kaons: BRs, lifetimes and form factors. The values of $|V_{us}|f_+(0)$ is determined from five decay modes: K_Le3, $K_L\mu3$, K_Se3, $K^{\pm}e3$, and $K^{\pm}\mu3$. The average is $|V_{us}|f_+(0) = 0.2157 \pm 0.0006$. Using the determination of $|V_{us}|f_+(0)$ from K_{l3} decays and the result $f_+(0) = 0.964(5)$ from [18], we get $|V_{us}| = 0.2237(13)$. With KLOE results it is also possible test the lepton universality by mean the ratio $r_{\mu e} = |f_+(0)\, V_{us}|^2_{\mu3}/|f_+(0)\, V_{us}|^2_{e3}$. Final result ($r_{\mu e} = 1.000(8)$) agrees with the SM expectation $r_{\mu e}^{SM} = 1$ with similar accuracy with respect to other precision measurements. To improve the V_{us} determination we extract the ratio V_{us}/V_{ud} by using $K^{\pm} \to \mu^{\pm}\nu(\gamma)$ [15] and $\pi^{\pm} \to \mu^{\pm}\nu(\gamma)$ and refitting together V_{us} and V_{ud}. The result of this fit is $V_{ud} = 0.97417(26)$ and $V_{us} = 0.2249(10)$, with $\chi^2/ndf = 2.34/1$ (Prob= 13%), from which we get $1 - V_{ud}^2 + V_{us}^2 + V_{ub}^2 = 4(7) \times 10^{-4}$ compatible with unitarity at 0.6σ level. Using these results, we then evaluate $G_{CKM} = G_F(V_{ud}^2 + V_{us}^2 + V_{ub}^2)^{1/2} = (1.16614 \pm 0.00040) \times 10^{-5}$ GeV^{-2} which is in perfect agreement with the measurement from the muon lifetime $G_F = (1.166371 \pm 0.000006) \times 10^{-5}$ GeV^{-2} and is competitive with the present accuracy of the measurements from tau-lepton decays and electroweak precision tests.

Furthermore we build the observable: $R_{\ell 23} = |V_{us}(K_{\ell 2})/V_{us}(K_{\ell 3}) \times V_{ud}(0^+ \to 0^+)/V_{ud}(\pi_{\mu 2})|$ as the ratio of the V_{us} values obtained from helicity suppressed and allowed decay modes, having SM expectation equal to 1. The presence of a scalar current due to a charged Higgs H^+ exchange is expected to lower the value of $R_{\ell 23}$ [19]. From our results and theoretical imputs we obtain $R_{\ell 23} = 1.008 \pm 0.008$ [16]. With this result we exclude region of parameters space coplementary to those excluded by the B results [20].

Kaon interferometry The decay of the ϕ produces a neutral kaon pair in a coherent quantum state with $J^{PC} = 1^{--}$:

$$|i\rangle = \frac{1}{\sqrt{2}}\{|K^0\rangle|\bar{K}^0\rangle - |\bar{K}^0\rangle|K^0\rangle\} = \frac{N}{\sqrt{2}}\{|K_S\rangle|K_L\rangle - |K_L\rangle|K_S\rangle\} \qquad (2)$$

The time evolution of the *entangled* system (2) has been observed for the first time by the KLOE [21] using the CP violating channel $K_SK_L \to \pi^+\pi^-\pi^+\pi^-$. To test the basic quantum mechanics (QM) principles that describe such evolution we heve introduced in the integrated decay amplitude an extra parameter

($\zeta_{0\bar{0}}$). It represents a loss of coherence in the entangled state and the QM prediction is zero. The analysis of a data sample corresponding to L$\sim$ 1.5fb^{-1} yields the following results [25]: $\zeta_{0\bar{0}} = (1.4 \pm 9.5_{\text{stat}} \pm 3.8_{\text{syst}}) \times 10^{-7}$ compatible with the prediction of QM. This result is well below the previous limit obtained with CPLEAR data [23] and better than the one obtained in the B meson system [22].

As discussed in [24], in a quantum gravity framework inducing decoherence, the CPT operator is *ill-defined*. This consideration might have intriguing consequences in correlated neutral kaon states, where the resulting loss of particle-antiparticle identity could induce a breakdown of the correlation in the initial state imposed by Bose statistics. As a result the odd initial state acquire a symmetric component: $|i\rangle = \frac{1}{\sqrt{2}}[|K^0\rangle|\bar{K}^0\rangle - |\bar{K}^0\rangle|K^0\rangle + \omega\left(|K^0\rangle|\bar{K}^0\rangle + |\bar{K}^0\rangle|K^0\rangle\right)]$, where ω is a complex parameter describing a completely novel CPT violation phenomenon, not included in previous analyses. Its order of magnitude could be at most $|\omega| \sim [(m_K^2/M_{\text{Planck}})/\Delta\Gamma]^{1/2} \sim 10^{-3}$. The analysis performed by KLOE on the integrated decay amplitude as a function of the decay time yields the measurement of the complex parameter ω [25]: $\Re(\omega) = \left(1.1^{+8.7}_{-5.3}{}_{\text{stat}} \pm 0.9_{\text{syst}}\right) \times 10^{-4}$, $\Im(\omega) = \left(3.4^{+4.8}_{-5.0}{}_{\text{stat}} \pm 0.6_{\text{syst}}\right) \times 10^{-4}$ with an accuracy that already reaches the interesting Planck scale region.

$\mathbf{K_{e2}/K_{\mu 2}}$ [31] The decay $K^\pm \to e^\pm\nu$ is strongly suppressed, $\mathcal{O}(\infty\prime^{-\nabla})$, because of conservation of angular momentum and the vector structure of the charged weak current. It therefore offers the possibility of detecting minute contributions from physics beyond the SM. This is enhanced in the ratio $R_K = \Gamma(K \to e\nu)/\Gamma(K \to \mu\nu)$ which, in the SM, is calculable without hadronic uncertainties. It has been shown in Ref. [30] that deviations of R_K of up to *a few percent* are possible in minimal supersymmetric extensions of the SM (MSSM) with non vanishing e-τ scalar lepton mixing. R_K is inclusive of IB, ignoring however DE contributions. To avoid high systematic error we define as signal events only those with energy of the radiated photon less than 10 MeV. Then we define the rate R_{10} as: $R_{10} = \Gamma(K \to e\nu(\gamma),\ E_\gamma < 10MeV)/\Gamma(K \to \mu\nu)$ from which we extract the R_K value. The measured kaon and decay particle momenta are used to compute the squared mass m_ℓ^2 of the lepton for the decay $K \to \ell\nu$ assuming zero missing mass. Since the decay particle impinge on the EMC we use all information about shower profile and total energy deposition by combining them with a neural network (NN) to improve the signal selection.

The number of $K \to e\nu(\gamma_{IB})$, is determined with a binned likelihood fit to the two-dimensional NN vs m_ℓ^2 distribution. Using the full KLOE dataset we obtain: $R_K = (2.493 \pm 0.025_{stat} \pm 0.019_{syst}) \times 10^{-5}$ in agreement with SM prediction. In the framework of MSSM with lepton-flavor violating (LFV) couplings, R_K can be used to set constraints in the space of relevant parameters.

3 Hadron physics

Pseudoscalars The ϕ meson decays about 1.3% of times into $\eta\gamma$ and then KLOE has collected one of the largest sample of η mesons in the world, about 10^8.

Among all decays we presented study of η decaying into four charged particles: $\pi^+\pi^-e^+e^-$ [12]. This decay allows us to probe the η internal structure by exploiting the conversion of the virtual photon into a leptons pair [13]. The invariant mass spectrum of the four particle is used to count signal events after background subtraction. Theese have been selected in the η mass region. the final result is: $BR(\eta \to \pi\pi ee) = (26.8 \pm 0.9_{Stat.} \pm 0.7_{Syst.}) \times 10^{-5}$. In [14] a non-CKM CP violating mechanism has been proposed. It should manifest as an angular asymmetry (A_ϕ), between the pion and the electron decay planes in the η rest frame. We measure: $A_\phi = (-0.6 \pm 2.5_{Stat.} \pm 1.8_{Syst.}) \times 10^{-2}$.

We have studied also the $\eta \to e^+e^-e^+e^-$ decay which is interesting, together with the $\eta \to \mu^+\mu^-e^+e^-$, for the η form factor since there are only leptons in the final state. The number of events is obtained fitting the data distribution of the 4 electron invariant mass, M_{eeee}, with signal and background shapes. From the fit we obtain 413 events.

Furthermore with the KLOE experiment it is possible to observe also the η' meson through the decay $\phi \to \eta'\gamma$ Using theese events we have measured the ration of the branching ratios $R_\phi = BR(\phi \to \eta'\gamma)/BR(\phi \to \eta\gamma)$. As described in [9] we have considered η and η' in the quark mixing base as ($|\eta'\rangle = X_{\eta'}|q\bar{q}\rangle + Y_{\eta'}|s\bar{s}\rangle + Z_G|G\rangle$). Recently we publish a new fit [8] to all data to constrain gluonium content in the η' meson. We use updated values for all imputs in the fit and the new KLOE results on the ω meson [11]. As a result we observe a 3σ evidence of the gluonium content in the η' meson: $|Z_G| = 0.12 \pm 0.04$.

Scalars The structure of the scalars below 1 GeV needs to be further clarified. Several models have been proposed to describe them (e.g. $q\bar{q}$, four quarks, $K\overline{K}$ molecules, etc.). The decay of the scalars into two pseudoscalars ($S \to PP'$) can be used to investigate their nature because the branching ratios and the invariant mass spectra are sensitive to the scalar structure.

At KLOE the scalar mesons are produced through $\phi \to S\gamma$ radiative decays. This allows to observe the f_0, σ and a_0 members of the light scalar multiplet.

The f_0 meson has been studied using charged and neutral pions pair. In both cases two different theoretical models have been used: "no-structure" (NS) [5] and the "kaon loop" (KL) [6].

The a_0 meson has been studied using the decay $\phi \to a_0\gamma \to \eta\pi^0\gamma$. The analysis has been performed for two different η final states, i.e. $\eta \to \gamma\gamma$ and $\eta \to \pi^+\pi^-\pi^0$. Also in this case the $\eta\pi^0(a_0)$ invariant mass distribution has been fitted with the NS and KL models after background subtraction. It is interesting to note that both models give a large coupling of the a_0 with the ϕ meson, indicating a sizable strange quark content in the a_0. The branching ratio obtained for the two different decay chains are in agreement: $BR(\phi \rightarrow \eta\pi^0\gamma) = (7.01 \pm 0.10_{stat.} \pm 0.20_{syst.}) \times 10^{-5}$ for the $\eta \rightarrow \gamma\gamma$ final state and $BR(\phi \rightarrow \eta\pi^0\gamma) = (7.12 \pm 0.13_{stat.} \pm 0.22_{syst.}) \times 10^{-5}$ for the $\eta \rightarrow \pi^+\pi^-\pi^0$ final state.

To complete the study of the scalar sector we observe the decay $\phi \rightarrow K^0\bar{K}^0\gamma$. This process should proceed through the intermediate f_0 and a_0 scalar mesons. The kaon pair is produced in a $J^{PC} = 0^{++}$ state, so the two kaons are both K_S or K_L. We limit our search on a final state with both K_S decaying to $\pi^+\pi^-$. The whole KLOE dataset, $\sim$ 2.2 fb^{-1}, has been used. We have observed 5 events in the data, with 3.2 ± 0.7 expected background events from MC. A 90% confidence level upper limit on the branching ratio has been obtained: $BR(\phi \rightarrow K^0\bar{K}^0\gamma) < 1.9 \times 10^{-8}$. This measurement excludes some of the theoretical predictions and is in agreement with expectations from other KLOE measurements.

Hadronic cross section [29] The anomalous magnetic moment of the muon has recently been measured to an accuracy of 0.54 ppm [26]. For the SM prediction the main source of uncertainty [27] is given by the hadronic contribution to the lowest order a_μ^{hlo} evaluated from the hadronic cross section. Among all contributors the dominant one is the $\pi^+\pi^-$ cross section.

At KLOE we extract the total cross section $\sigma_{\pi\pi} \equiv \sigma_{e^+e^- \rightarrow \pi^+\pi^-}$ from the differential spectrum of the $\pi^+\pi^-$ invariant mass, $M_{\pi\pi}$, from Initial State Radiation (ISR) events using [28]:

$$s \, \frac{d\sigma_{\pi\pi\gamma}}{dM^2_{\pi\pi}} = \sigma_{\pi\pi}(M^2_{\pi\pi}) \; H(M^2_{\pi\pi}) \; , \tag{3}$$

where H is the radiator function. This formula neglects Final State Radiation (FSR) terms (which are properly taken into account in the analysis). In the selected events the two charged pion tracks have $50^\circ < \theta_\pi < 130^\circ$. The ISR photon is not explicitly detected and its direction is reconstructed by closing the kinematics: $\vec{p}_\gamma \simeq \vec{p}_{miss} = -(\vec{p}_{\pi^+} + \vec{p}_{\pi^-})$.

We have measured the dipion contribution to the muon anomaly, $a_\mu^{\pi\pi}$, in the interval $0.592 < M_{\pi\pi} < 0.975$ GeV, with negligible statistical error and a full systematic uncertainty of 0.9%. Combining all errors KLOE gives: $a_\mu^{\pi\pi}(0.592 < M_{\pi\pi} < 0.975 \text{ GeV}) = (387.2 \pm 3.3) \times 10^{-10}$. This new result confirms the current disagreement between the standard model prediction for a_μ and the measured value.

References

[1] KLOE Coll., *LNF-92*/019 (IR) (1992) and *LNF-93*/002 (IR) (1993).
[2] KLOE Coll., M. Adinolfi *et al*, *Nucl.Ins.Met.* **A**488,51(2002).
[3] KLOE Coll., M. Adinolfi *et al*, *Nucl.Ins.Met.* **A**482,363(2002).
[4] KLOE Coll., M. Adinolfi *et al*, *Nucl.Ins.Met.* **A**492,134(2002).
[5] G. Isidori,*et al*,*JHEP* **05**,049(2006),hep-ph/0603241.
[6] N. N. Achasov, and A. V. Kiselev,*Phys.Rev.***D**68,014006(2003).
[7] KLOE Coll.,F. Ambrosino *et al*,*Phys.Lett.***B**679,10(2009).
[8] KLOE Coll., F. Ambrosino *et al*,*JHEP* 07,105(2009).
[9] J. L. Rosner,*Phys.Rev.***D**27,1101(1983).
[10] C. Amsler *et al*,*Phys.Lett.***B**667,1(2008).
[11] KLOE Coll.F. Ambrosino *et al*,*Phys.Lett.***B**669,223(2008),0807.4909.
[12] KLOE Coll.F. Ambrosino *et al*,*Phys.Lett.***B**675,283(2009), 0812.4830.
[13] L. G. Landsberg, *Phys. Rept.* **128**, 301–376 (1985).
[14] D.-N. Gao, *Mod. Phys. Lett.* **A17**, 1583–1588 (2002), hep-ph/0202002.
[15] KLOE Coll., F. Ambrosino *et al*,*Phys.Lett.***B**632,76(2006).
[16] KLOE Coll., F. Ambrosino *et al*,*JHEP* 04,059(2008).
[17] W.J. Marciano,*Phys.Rev.Lett.*93,231803(2004).
[18] HPQCD and UKQCD Coll., *Phys.Rev.Lett.*100,062002(2008).
[19] G. Isidori and P. Paradisi, *Phys. Lett.* B **639**, 499 (2006).
[20] Belle Coll., K. Ikado, *et al*,*Phys.Rev.Lett.*97,251802(2006);
BaBar Coll., B. Aubert *et al*0705.1820.
[21] KLOE Coll., F.Ambrosino *et al*,*Phys.Lett.***B**642,315(2006).
[22] Belle Coll., A. Go *et al*,*Phys.Rev.Lett.*,99,131802(2007).
[23] CPLEAR Coll., A. Apostolakis *et al*,*Phys.Lett.***B**422,339(1998).
[24] J. Bernabeu *et al*, *Phys.Rev.Lett.*92,131601(2004);
*Phys.Lett.***B**744,180(2006).
[25] A. Di Domenico *J.Phys.: Conf. Ser.***171**,012008,2009.
[26] G. W. Bennett *et al*,*Phys.Rev.***D**73,07200(2006).
[27] F. Jegerlehner,hep-ph/0703125
[28] S. Binner, *et al*,*Phys.Lett.***B**459,279(1999).
[29] KLOE Coll., F. Ambrosino *et al*,*Phys.Lett.***B**670,285(2009).
[30] A. Masiero, P. Paradisi and R. Petronzio, *JHEP* **0811** (2008) 042
[31] KLOE Coll., F. Ambrosino *et al*,*Eur.Phys.J.***C**64,627(2009).

ISR PHYSICS AT BABAR

Sergey Serednyakov [a]

Budker Institute of Nuclear Physics, Novosibirsk, Russia

Abstract. Experimental data from the PEP-II B-factory, obtained via initial-state radiation , are presented. Processes of e^+e^- annihilation into hadrons in the energy range up to 5 GeV are studied. From the measured cross sections the parameters of many resonances are improved, the baryons form factors are derived. New states, e.g; Y(4260) and Y(2175), for which the internal structure is not yet established, are observed.

In recent years many cross section measurements were performed with the Babar detector [1] using the initial-state radiation (ISR) approach [2]. ISR measurements of $e^+e^- \to hadrons$ cross sections are important because they are used in the calculation of the Standard model parameters such as the muon anomaly $a_\mu = \frac{g-2}{2}$ and the fine structure constant at Z-mass $\alpha_{em}(s = M_Z^2)$. At present, there is a considerable disagreement $\sim 3\sigma$ between the a_μ measurement and calculation; so new accurate cross section data are needed. In this review the recent Babar ISR results are presented.

Production of mesons. The channels $e^+e^- \to \pi^+\pi^-\pi^+\pi^-, \pi^+\pi^-\pi^0\pi^0$ give the largest contribution to the hadronic vacuum polarization term of a_μ above 1 GeV. Babar data give the considerable improvement of the accuracy above 1.4 GeV [3,4]. In these reactions the $\omega\pi^0$, πa_1, $\rho^+\rho^-$, $\rho^0 f^0$ intermediate states dominate. In the channel $e^+e^- \to \pi^+\pi^-\pi^0\pi^0$ a peak above 2 GeV (Fig.1) is seen, which can be a manifestation of the hypothetical $\rho(2150)$ or ρ''' state.

The cross sections of $e^+e^- \to 5\pi, 6\pi$ processes [5, 6] have been measured by Babar from the threshold up to 4.5 GeV in the final states $2(\pi^+\pi^-)\pi^0$, $2(\pi^+\pi^-)2\pi^0$, $3(\pi^+\pi^-)$. In the $e^+e^- \to 6\pi$ cross section some resonance structure below 2 GeV is seen (Fig.2).

In the $e^+e^- \to K^+K^-\pi^0$ and $e^+e^- \to K^\pm K_S\pi^\mp$ reactions [7] the $\phi(1650)$ isoscalar resonance dominates. The cross sections of the $\phi\eta$ and $\phi\pi^0$ final states have been measured for the first time. In the $\phi\eta$ final state the candidate for the Y(2175) or ϕ'' state is possibly seen (Fig.3). The processes $e^+e^- \to K^+K^-\pi^+\pi^-, K^+K^-\pi^0\pi^0$ [8] show domination of the $\phi(1650)$ in the total cross section and $K^\star(890)$, $K_1(1270)$, $\eta(1500)$ in intermediate states. The cross section of $e^+e^- \to K^+K^-\pi^+\pi^-\pi^0, K^+K^-\pi^+\pi^-\eta$ [5], $e^+e^- \to K^+K^-\pi^+\pi^-\pi^+\pi^-$ [6] and $e^+e^- \to K^+K^-K^+K^-$ [8] processes have been measured for the first time.

In the study of e^+e^- annihilation into D-mesons [9] the D-mesons ($D^\pm, D^0, \bar{D^0}$, $D^{*\pm}, D^{0*}, \bar{D^{0*}}$) are selected via modes $K\pi, K2\pi, K3\pi$. The main goal of this study is to see the Table charmonium states $\psi(4040), \psi(4160), \psi(4400)$ and to

[a] e-mail: seredn@inp.nsk.su

look for the recently observed $Y(4260)$ state in the $D\bar{D}$ mode. The measured cross sections are presented in Figs.4a,b,c for $D\bar{D}$, $D^*\bar{D}$, and $D^*\bar{D}^*$ modes respectively. While the Table charmonium states are seen, no sign of $Y(4260)$ is observed; so an upper limit on the $Y(4260)$ branching fraction into $D\bar{D}$ is set: $BF(Y(4260) \to D\bar{D}) < BF(Y(4260) \to J/\psi\pi^+\pi^-)$.

Production of baryons The proton electromagnetic form factor (FF) was measured by Babar in the reaction $e^+e^- \to p\bar{p}\gamma$ [10]. The Babar data (Fig.5) agree with the previous measurements, confirm the rise of the proton FF at the threshold and modestly satisfy the QCD fit formula [11] above 2.5 GeV. Structures at 2.15 and 2.9 GeV are possibly seen. Study of the angular distribution allows to extract the value of $|G_E/G_M|$ (the ratio of the electric and magnetic FFs) (Fig.6), which is found to be contradicting the only previous measurement at LEAR [12].

Later on the reactions $e^+e^- \to \Lambda\bar{\Lambda}\gamma$, $\Sigma\bar{\Sigma}\gamma$, $\Lambda\bar{\Sigma}\gamma(\Sigma\bar{\Lambda}\gamma)$ have been studied [13]. The most accurate FF measurement is done for Λ. Previously the Λ FF was measured by DM2 only [14] in a single energy point. The Babar data on the Λ and proton FF are shown in Fig.7. The asymptotic QCD prediction [15] for the ratio of the Λ and proton FF is 0.24. Figure 7 shows the strong disagreement between the data and prediction below 2.5 GeV. The disagreement at higher masses is less, but the measurements accuracy is not sufficient for the test at higher masses.

Resonance physics Using the ISR technique, parameters of vector states $\rho(1450)$, $\rho(1700)$, $\omega(1420)$, $\omega(1650)$ and $\phi(1650)$ have been improved (see Babar references cited above). Concerning the J/ψ and $\psi(2S)$ resonances, many decays have been observed for the first time ($J/\psi \to K^+K^-\pi^0\pi^0$, $K^+K^-\pi^+\pi^-\eta$, $K^{*0}\bar{K}^{*0}$, $\phi\pi^0$ ..., $\psi(2S) \to \pi^+\pi^-\pi^+\pi^-\eta$, $K^+K^-\pi^+\pi^-\pi^+\pi^-$ ...).

The most known new found state is the charmonium-like Y(4260) resonance [16], decaying into $J/\psi\pi^+\pi^-$ (Fig.8). It has low electron width $\Gamma_{ee} = 7.5 \pm 1.1\ eV$, which is much less than $\sim 500-800\ eV$ typical for the $\psi(4040)$, $\psi(4160)$,$\psi(4400)$ states. No decays of Y(4260) into $D\bar{D}$ is seen. For these reasons the Y(4260) is considered as a candidate for the exotic 4-quark $c\bar{c}n\bar{n}$ ($n = u, d$) or molecular state. One could mention, that the $\pi^+\pi^-$ mass spectrum in Y(4260) decay has a peak close to 1 GeV, that might be an indication on the contribution of $J/\psi f_0(980)$ intermediate state.

Another new state is the Y(4320) [17], found in $\psi(2S)\pi^+\pi^-$ channel. Judging by the presence of the charmonium in the final state and a small production cross section, this state might be of a similar nature as Y(4260).

One more exotic candidate is the Y(2175) [18], first seen in $K^+K^-f_0(980)$, $f_0 \to \pi^+\pi^-$, $\pi^0\pi^0$ final states (Fig.9) and then in the $\phi\eta$ (Fig.3) state. Similar to the Y(4260) it is relatively narrow, has a low electron width and is considered as a possible $s\bar{s}s\bar{s}$ exotic candidate or ϕ'' state.

Summarizing the isovector data one can suggest the existence of a new

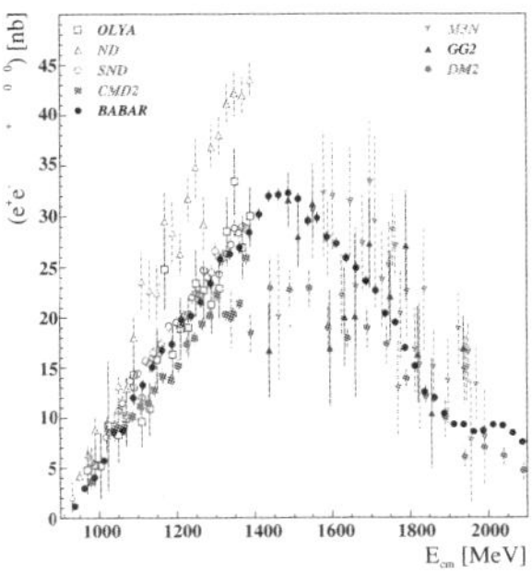

Figure 1: The $e^+e^- \to \pi^+\pi^-\pi^0\pi^0$ cross section

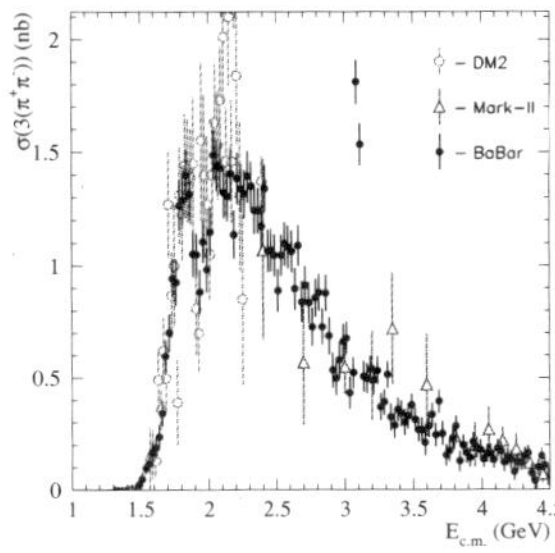

Figure 2: The $e^+e^- \to 3(\pi^+\pi^-)$ cross section.

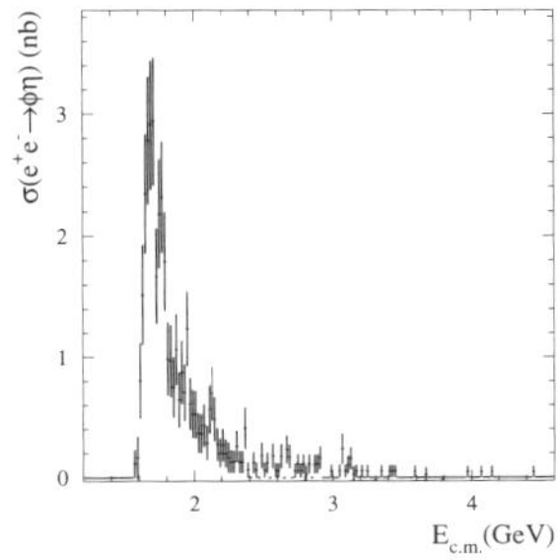

Figure 3: The $e^+e^- \to \phi\eta$ cross section.

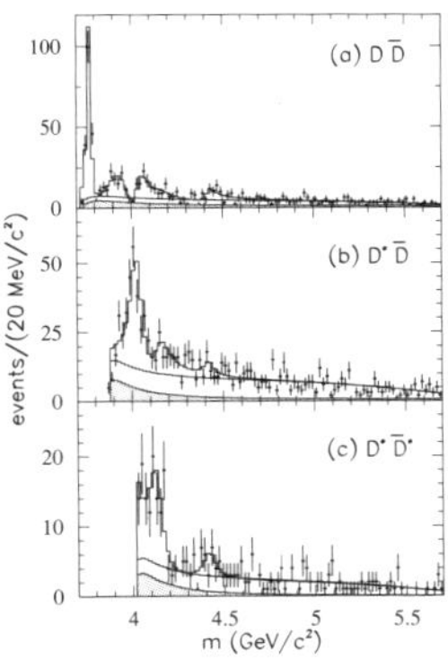

Figure 4: The $e^+e^- \to D\bar{D}$, $D^*\bar{D}$, and $D^*\bar{D}^*$ cross sections.

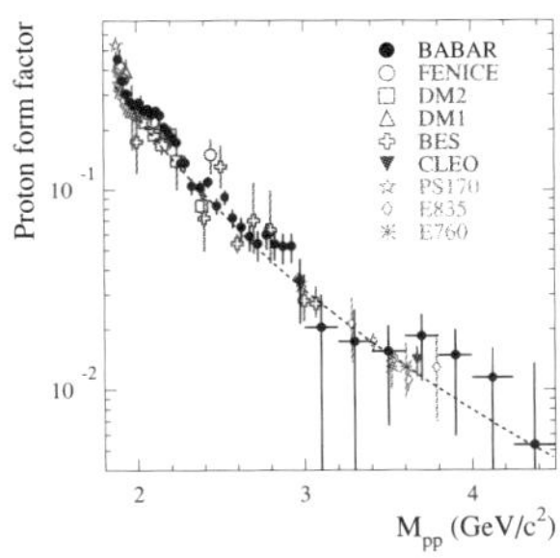

Figure 5: The proton form factor. The solid line is the QCD fit [11].

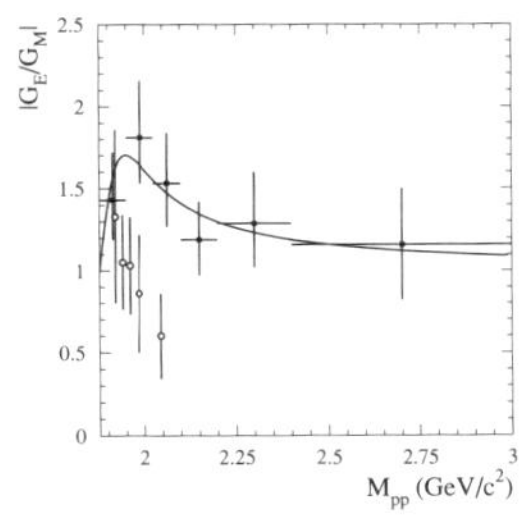

Figure 6: The Babar G_E/G_M ratio (black points), compared with LEAR (open circles).

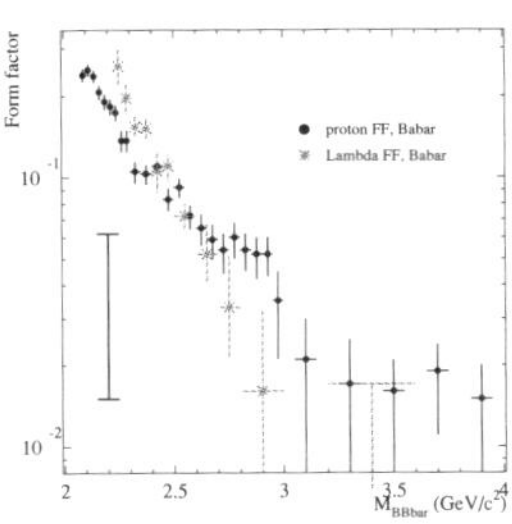

Figure 7: The comparison of proton and Λ form factors. The vertical line is the asymptotic ratio of proton and Λ form factors.

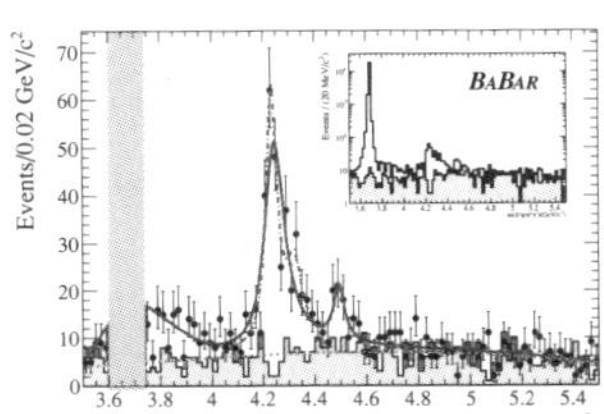

Figure 8: The $J/\psi\pi^+\pi^-$ mass spectrum in the $e^+e^- \to J/\psi\pi^+\pi^-\gamma$ reaction.

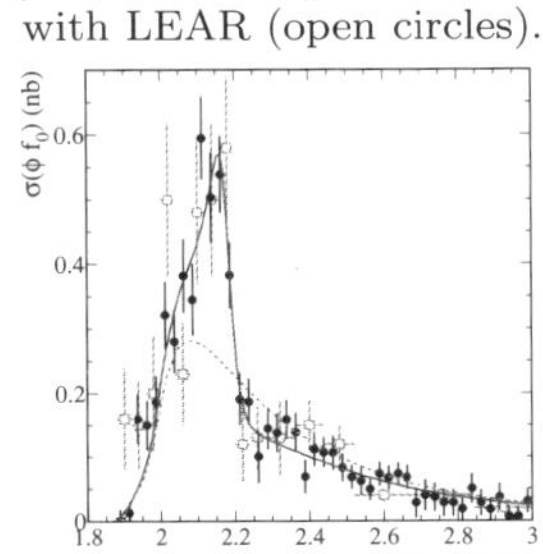

Figure 9: The $e^+e^- \to \phi f_0$ cross section, measured in $K^+K^-\pi^+\pi^-$ (filled circles) and $K^+K^-\pi^0\pi^0$ (squares) modes.

isovector state $\rho(2150)$ or ρ'''. This resonance is seen in $\pi^+\pi^-\pi^0\pi^0$ (Fig.1), $\eta'(958)\rho^0$ [5] and $p\bar{p}$ final states. In the latter case, we mean the step in the proton form factor near 2.15 GeV (Fig.5).

One could mention of an isovector candidate X(1880), seen by Babar as an interference pattern (Fig.2) in $e^+e^- \to 6\pi$ cross section [6] with the parameters M=1.88±0.03 GeV/c^2 and Γ=0.13±0.03 GeV/c^2. This state was seen earlier in DM2 [19] and FOCUS [20] experiments. The rise of the proton form factor at the threshold (Fig.5) can be due to this X(1880) state too.

Conclusions The $e^+e^- \to hadrons$ cross sections are measured at Babar in the energy range $1 \div 5$ GeV using the ISR method. Parameters of many vector mesons (ρ, ω, ϕ excitations and J/ψ and $\psi(2S)$ states) are improved. New resonances are observed via ISR: Y(4260), Y(4320), and Y(2175), which are candidates for exotic states. Also there are seen two new isovector candidates $\rho(2150)$ and X(1880).

[1] Babar Collaboration, B. Aubert *et al.*, Nucl. Instr. and Meth. A **479**, 1 (2002).
[2] M.Benayoun et al., Mod. Phys. Lett. **A14**, 2605 (1999).
[3] Babar Collaboration, B. Aubert *et al.*, Phys. Rev. D **71**, 052001 (2005).
[4] V. P. Druzhinin, arxiv: hep-ex/0710.3455 (2007).
[5] Babar Collaboration, B. Aubert *et al.*, Phys. Rev. D **76**, 092005 (2007); erratum-ibid, **77**, 119902 (2008).
[6] Babar Collaboration, B. Aubert *et al.*, Phys. Rev. D **73**, 052003 (2006).
[7] Babar Collaboration, B. Aubert *et al.*, Phys. Rev. D **77**, 092002 (2008).
[8] Babar Collaboration, B. Aubert *et al.*, Phys. Rev. D **74**, 091103 (2006).
[9] Babar Collaboration, B. Aubert *et al.*, Phys. Rev. D **79**, 092001 (2009).
[10] Babar Collaboration, B. Aubert *et al.*, Phys. Rev. D **73**, 012005 (2006); hep-ex/05112023.
[11] V. L. Chernyak, A. R. Zhitnitsky, JETP Lett. **25**, 510 (1977).
[12] PS170 Collaboration, G. Bardin *et al.*, Nucl. Phys. **B411**, 3 (1994).
[13] Babar Collaboration, B. Aubert *et al.*, Phys. Rev. D **76**, 092006 (2007); hep-ex/0709.1988.
[14] DM2 Collaboration, D. Bisello *et al.*, Z. Phys. C **48**, 23 (1990).
[15] V. L. Chernyak *et al.*, Z. Phys. C **42**, 569 (1989).
[16] Babar Collaboration, B. Aubert *et al.*, Phys. Rev. Lett. **95**, 142001 (2005).
[17] Babar Collaboration, B. Aubert *et al.*, Phys. Rev. Lett. **98**, 912001 (2007).
[18] Babar Collaboration, B. Aubert *et al.*, Phys. Rev. D **74**, 091103 (2006).
[19] A. B. Clegg, A. Donnachie, Z. Phys. C **45**, 677 (1990).
[20] FOCUS Collaboration, P. L. Frabetti *et al.*, Phys. Lett. **B514**, 240 (2000).

SEMILEPTONIC B-MESON DECAYS AT *BABAR*

Michael Sigamani [a]
Department of Physics, Queen Mary, University of London, Mile End Road, London, E1 4NS, UK

Abstract. Presented are selected results from semileptonic B-meson decays at *BABAR*. Two measurements of the Cabibbo-Kobayashi-Maskawa matrix element $|V_{cb}|$ are reported, using moments of the hadronic-mass spectrum in inclusive $\bar{B} \to X_c \ell^- \bar{\nu}$ decays, and also exclusive $\overline{B} \to D\ell^- \bar{\nu}_\ell$ decays. These results are based on data samples of 232 (inclusive $\bar{B} \to X_c \ell^- \bar{\nu}$) and 460 (exclusive $\overline{B} \to D\ell^- \bar{\nu}_\ell$) million $\Upsilon(4S) \to B\bar{B}$ decays recorded by the *BABAR* detector at the PEP-II e^+e^--storage rings. Semileptonic events are identified by requiring a lepton (e or μ) in events tagged by a full reconstruction of one of the B mesons in the $B\overline{B}$ pair.

1 Introduction

In the Standard Model (SM) of electroweak interactions, the coupling strength of the b to the c quark in the weak interaction is described by the Cabibbo-Kobayashi-Maskawa (CKM) [1] matrix element $|V_{cb}|$. A precise determination of $|V_{cb}|$ is therefore crucial for probing the CKM mechanism for quark mixing.

Experimentally, $|V_{cb}|$ is obtained using decays with two distinct approaches: An exclusive analysis, where the hadronic system X_c is reconstructed in a specific mode. Or, an inclusive analysis where X_c is not reconstructed, but summed over all possible hadronic final states. Theoretically, inclusive determinations rely on an Operator Product Expansion (OPE) in inverse powers of the b-quark mass m_b [2] which relates the total rate to $|V_{cb}|$. Whereas exclusive determinations use Form Factors (FF) to describe the hadronization process. Current measurements of $|V_{cb}|$ using inclusive and exclusive determinations generally differ by around two standard deviations, with the inclusive result being twice as precise as the exclusive [3].

2 Hadronic Reconstruction for the tagged B sample

The signal sample for both exclusive and inclusive measurements are selected using $B\overline{B}$ events whereby a full reconstruction for one of the B mesons is required (referred to as the B_{tag}). The B_{tag} [4] candidate is reconstructed using hadronic modes of the type $\overline{B} \to D^{(*)}Y$, where Y represents a collection of charmed hadrons. The remaining particles in the event are assumed to belong to the other B (referred to as the B_{recoil}), which leaves a very clean sample of semileptonic events.

[a] e-mail: sigamani@slac.stanford.edu

3 Exclusive determination of $|V_{cb}|$ using $\overline{B} \to D\ell^-\bar{\nu}_\ell$ decays

The $\overline{B} \to D\ell^-\bar{\nu}_\ell$ decay rate is proportional to the square of $|V_{cb}|$. In the limit of very small lepton masses we have the following relation

$$\frac{\mathrm{d}\Gamma(\overline{B} \to D\ell\nu)}{\mathrm{d}w} = \frac{G_F^2}{48\pi^3\hbar} M_D^3(M_B + M_D)^2(w^2 - 1)^{3/2} \mid V_{cb} \mid^2 \; \mathcal{G}^2(w), \qquad (1)$$

where G_F is the Fermi coupling constant, $\mathcal{G}(w)$ a FF which describes the effects of strong interactions in $\overline{B} \to D$ transitions, and M_B and M_D are the masses of the B and D mesons, respectively. The variable w denotes the product of the B and D meson four-velocities V_B and V_D, $w = V_B \cdot V_D = (M_B^2 + M_D^2 - q^2)/(2M_B M_D)$, where $q^2 \equiv (p_B - p_D)^2$, and p_B and p_D are the four-momenta of the B and D mesons. In the limit of infinite quark masses, $\mathcal{G}(w)$ coincides with the Isgur-Wise function [5]. This function is normalized to unity at zero recoil, where q^2 is maximum. Thus $|V_{cb}|$ can be extracted by extrapolating the differential decay rate to $w = 1$ using $\overline{B}^0 \to D^+\ell^-\bar{\nu}_\ell$ and $B^- \to D^0\ell^-\bar{\nu}_\ell$ decays [6]. Using the B_{tag} sample, the B_{recoil} is reconstructed by selecting a lepton with momentum in the CM frame p_ℓ^* larger than 0.6 GeV. Semileptonic B decays are identified by a missing-mass squared value,

$$m_{\mathrm{miss}}^2 = \left[p(\Upsilon(4S)) - p(B_{\mathrm{tag}}) - p(D) - p(\ell)\right]^2, \qquad (2)$$

defined in terms of the measured particle four-momenta. Signal events will peak at zero in the m_{miss}^2 distribution as there is only one associated missing particle. Other semileptonic B decays, such as $\overline{B} \to D^{(*,**)}\ell^-\bar{\nu}_\ell$ (feed-down) will yield larger values of m_{miss}^2 due to higher numbers of missing particles generated from secondary semi-leptonic decays of the $D^{(*,**)}$. A measurement of $|V_{cb}|$ is then made using a fit to the w distribution, where data and MC events are evaluated in ten equal-size bins in the interval $1 < w < 1.6$. Data samples are assumed to contain four different contributions: $\overline{B} \to D\ell^-\bar{\nu}_\ell$ signal events, feed-down from other semileptonic B decays, combinatorial $B\overline{B}$ and continuum background, and fake lepton events (predominantly from hadronic B decays with hadrons misidentified as leptons).

From the fit to the combined $\overline{B} \to D\ell^-\bar{\nu}_\ell$ sample, a measurement of $\mathcal{G}(1)|V_{cb}| = (43.0 \pm 1.9 \pm 1.4) \times 10^{-3}$ is made. Using an unquenched lattice calculation, corrected by a factor of 1.007 for QED effects, a value of $|V_{cb}| = (39.8 \pm 1.8 \pm 1.3 \pm 0.9_{FF}) \times 10^{-3}$ is obtained, where the third error is due to the theoretical uncertainty from $\mathcal{G}(1)$.

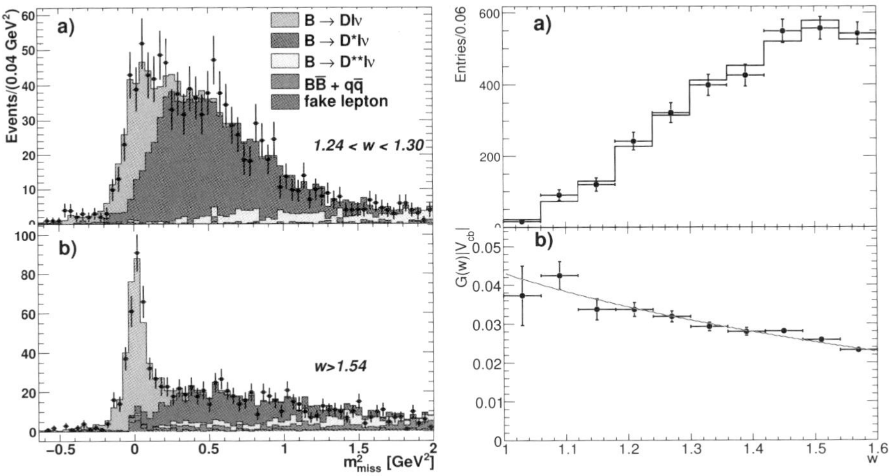

Figure 1: Left: Fit to the m^2_{miss} distribution, in two different w intervals, for $B^- \to D^0\ell^-\bar{\nu}_\ell$: the data (points with error bars) are compared to the results of the overall fit (sum of the solid histograms). The PDFs for the different fit components are stacked in the order shown in the legend. Right: (a) The w distribution obtained by summing the $B^- \to D^0\ell^-\bar{\nu}_\ell$ and $\overline{B}^0 \to D^+\ell^-\bar{\nu}_\ell$ yields. Data (points) are compared to the results of the overall fit (solid histogram). (b) The $\mathcal{G}(w)|V_{cb}|$ distribution corrected for the reconstruction efficiency, with the fit result superimposed.

4 Inclusive determination of $|V_{cb}|$ using hadronic-mass moments in $\bar{B} \to X_c\ell^-\bar{\nu}$ decays

We can also extract $|V_{cb}|$ using measurements of the hadronic mass moments $\langle m^k_X\rangle$ [7], with $k = 1, \ldots 6$ in semileptonic $\overline{B} \to X_c\ell^-\overline{\nu}$ decays. These moments are measured as functions of the lower limit on the lepton momentum, $p^*_{\ell,\text{min}}$, between 0.8 GeV/c and 1.9 GeV/c calculated in the rest frame of the B_{recoil}. The measured hadronic mass moments $\langle m^k_X\rangle$ are shown in Fig. 2 with $k = 1 \ldots 6$ as functions of the minimal lepton momentum $p^*_{\ell,\text{min}}$. The statistical uncertainty consists of contributions from the data statistics, and the statistics of the MC.

The fit method designed to extract the $|V_{cb}|$ from the moments measurements have been reported in [8] and is based on a χ^2 minimization technique. There are eight fit parameters namely: $|V_{cb}|$, the quark masses m_b and m_c, the total semileptonic branching fraction $\mathcal{B}(\overline{B} \to X_c\ell^-\overline{\nu})$, and the dominant non-perturbative HQE parameters μ^2_G and ρ^3_{LS}. This presentation only quotes results for $|V_{cb}|$. Combined fits to these moments and moments of the photon-energy spectrum in $B \to X_s\gamma$ decays [9] have resulted in: $|V_{cb}| = (42.0 \pm 0.2 \pm 0.7) \times 10^{-3}$.

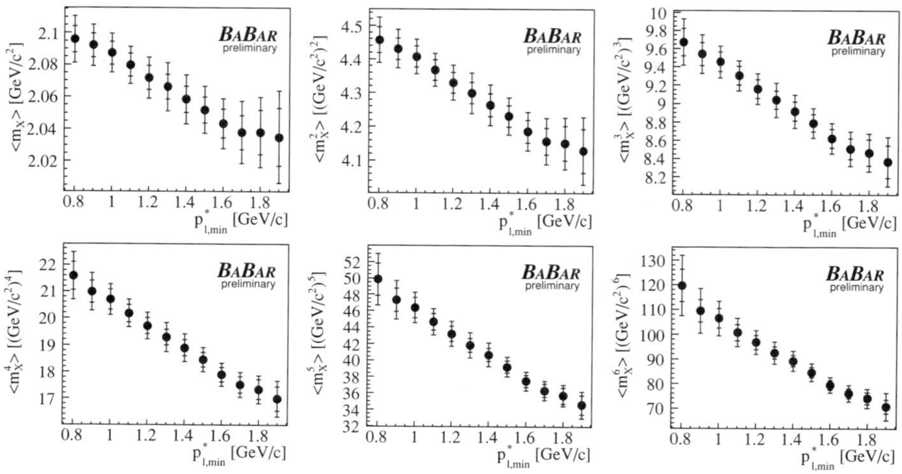

Figure 2: Measured hadronic mass moments $\langle m_X^k \rangle$ with $k = 1 \ldots 6$ for different selection criteria on the minimal lepton momentum $p^*_{\ell,\mathrm{min}}$. The inner error bars correspond to the statistical uncertainties while the full error bars correspond to the total uncertainties. All the moments are highly correlated since they share subsets of selected events.

5 Summary and Conclusion

Presented are measurements of $|V_{cb}|$ using exclusive $\overline{B} \to D\ell^-\bar{\nu}_\ell$ decays, and the moments of the hadronic mass distribution in inclusive $\overline{B} \to X_c\ell^-\bar{\nu}$ decays. Using the exclusive determination, the fit to the combined $\overline{B} \to D\ell^-\bar{\nu}_\ell$ sample, and an unquenched lattice calculation yields: $|V_{cb}| = (39.8 \pm 1.8 \pm 1.3 \pm 0.9_{FF}) \times 10^{-3}$. Using the hadronic mass moments $\langle m_X^k \rangle$ in the inclusive determination, and an extraction in the kinetic mass scheme yields: $|V_{cb}| = (42.0 \pm 0.2 \pm 0.7) \times 10^{-3}$.

[1] M. Kobayashi and T. Maskawa, Prog. Theor. Phys. **49** (1973) 652.

[2] I. I. Y. Bigi, M. A. Shifman, N. G. Uraltsev and A. I. Vainshtein, Phys. Rev. Lett. **71**, 496 (1993) [arXiv:hep-ph/9304225].

[3] C. Amsler *et al.*, Phys. Lett. B**667**, 1 (2008).

[4] B. Aubert *et al.* (*BABAR* Collaboration), Phys. Rev. Lett. **92**, 071802 (2004).

[5] N. Isgur and M. B. Wise, Phys. Lett. B**237**, 527 (1990).

[6] B. Aubert *et al.* [BABAR Collaboration], arXiv:0904.4063 [hep-ex].

[7] B. Aubert *et al.* [BABAR Collaboration], arXiv:0707.2670 [hep-ex].

[8] O. Buchmuller and H. Flacher, Phys. Rev. D **73**, 073008 (2006) [arXiv:hep-ph/0507253].

[9] B. Aubert *et al.* [BABAR Collaboration], Phys. Rev. Lett. **96** (2006) 221801 [arXiv:hep-ex/0601046].

BABAR : SEARCH FOR LEPTON FLAVOR VIOLATION IN TAU DECAY

Marcello A. Giorgi [a]
(on behalf of the BABAR Collaboration)
INFN sezione di Pisa and Dipartimento "Enrico Fermi" dell' Università di Pisa, 56127 Pisa, Italy

Abstract. The search for lepton Flavor Violation (LFV) in τ lepton decay in *BABAR* experiment at SLAC is presented. The data analyzed were collected at the e^+e^- asymmetric collider PEPII , operating at the center of mass energy corresponding to the mass of the Y_{4S} resonance. Although the machine and the *BABAR* detector were built for the discovery of the CP Violation in B meson decay and the measurements of angles of the Unitarity Triangle of the CKM matrix, a large sample of τ leptons was collected. The large statistics available has allowed to set the upper limits (ULs) to the Branching Ratios (BR) of decay channels showing LFV, that in the Standard Theory are either forbidden or permitted at an experimentally unreachable level.

1 Introduction

In this analysis we set upper limits for the branching fractions (BR) of LFV τ decays. Three categories of decays have been considered in this paper:
$\tau \to \ell\gamma$, $\tau \to \ell\ell\ell$ and $\tau \to \ell V^0$

where l could be either a μ or an electron.

LFV decays are extremely suppressed in the Standard Model (SM) even with neutrino mixing [1–3]

although not forbidden the expected BR (e.g. BR ($\tau \to \mu\gamma$) and BR($\tau \to \ell\ell\ell$) both estimated to be $< \mathcal{O}$ (10^{-54}) are well below the sensitivities of the present and even future experiments. Detection of these rare processes implies New Physics (NP) beyond SM. Many theoretical NP models predict higher (up to $\mathcal{O}$ (10^{-8})) τ LFV Branching Fraction, close to the limits obtainable in the experiments at the present e^+e^- factories or in the next generation machines with luminosity two order of magnitude higher, (see table 1).

So different models expect different processes leading to the LFV in charged lepton decays, therefore once the violation is observed the NP flavor structure may be investigated by looking at BR ratios of various LFV decay channels [11].

2 Measurement in *BABAR*

This paper contains the results of *BABAR* . The used data were collected at the asymmetric e^+e^- B-factory PEP-II operating at SLAC with the *BABAR* apparatus

[a]e-mail: marcello.giorgi@pi.infn.it

Table 1: Upper limits of τ violating decay Branching Fractions in different Theoretical scenarios.

Framework	Model	BR
SM	SM + right heavy Majorana ν [4]	$< 10^{-10}$
	SM + left-right neutral isosinglets [4]	$\sim 10^{-9}$
SUSY	MSSM+right heavy Majorana ν [5]	$\sim 10^{-9}$
	left-right SUSY [5]	$\sim 10^{-10}$
	SUSY+neutral Higgs [6]	$10^{-10} - 10^{-7}$
	SUSY+Higgs triplet [7]	$\sim 10^{-7}$
	MSSM+universal soft SUSY breaking [8]	$\sim 10^{-9}$
	MSSM+non-universal soft SUSY breaking [9]	$\sim 10^{-6}$
Other	Technicolor [10]	$\sim 10^{-8}$

. The apparatus is a general purpose detector, with a very precise vertex and tracking measurement, a very good particle identification for electrons, muons, pions and an electromagnetic calorimeter that in addition to the identification of the electrons, gives a very precise measurement of energy and direction of γ's. The machine has been run at different center of mass energies , mainly at the energy corresponding to the mass of $\Upsilon(4S)$, but with different settings corresponding to Y(3S) and Y(2S). In the search for $\tau \to \ell\gamma$ a statistics from an integrated luminosity of $516 fb^{-1}$ (of which $470 fb^{-1}$ at $\Upsilon(4S)$, $31 fb^{-1}$ at $\Upsilon(3S)$ and $15 fb^{-1}$ at $\Upsilon(2S)$Y, for $\tau \to \ell\ell\ell$ $470 fb^{-1}$ and for $\tau \to \ell V^0$ $451 fb^{-1}$, both at Y(4S) only.

3 Search For $\tau \to \ell\gamma$

In the search for $\tau \to \ell\gamma$ decays events were separated in five different samples, depending on the decay of the non signal τ lepton. The e-tag, μ-tag, and π-tag have only one charged track with no neutral energy deposits associated to the tag τ, the ρ-tag associates one charged track and neutral deposits consistent to be a π^0 to the τ, while the $3h$-tag presents three charged tracks. Further discrimination is obtained requesting charge conservation, reconstructed τ mass in the tag side to be compatible with the τ mass, and missing momentum and missing mass consistent to arise only from the non LFV tau decay. The main tool used for background reduction consist of a Neural Network (NN) selector, designed to reduce radiative QED backgrounds (e.g. $e^+e^- \to \mu^+\mu^-\gamma$) and radiative non-LFV τ decays (e.g. $\tau \to e\nu\bar{\nu}\gamma$). The NN receives six variables as input: the total tag side reconstructed momentum, the recoil angle of the tag side with respect to the reconstructed signal τ direction, the lepton-photon opening angle, the missing transverse momentum, and $\Delta E_\gamma = \frac{E_\gamma}{\sqrt{s}} \frac{\sin(\theta_1+\theta_2)}{\sin(\theta_1)+\sin(\theta_2)+|sin(\theta_1+\theta_2|}$, where θ_1 and θ_2 are the angles the photon momentum makes with the signal-track and the total observed tag-side momentum. The analysis is performed in a blind way, and the signal candidates are identified by two kinematic variables: $\Delta E = E_{\ell\gamma}^{CM} - sqrts/2$ and the beam constrained τ mass M_{ec}, the signal distributions peak at m_τ and small negative values for M_{ec} and ΔE respectively, the blind signal region is defined as the elliptical region within 3σ of the distribution center. The selection and NN is optimized to yeld the smallest UL. The background events are extrapolated from a fit to the MC and data distributions in the regions near the signal region, the fit to the background PDFs is validated looking at larger signal regions and both for each single tag and for the whole sample. The main backgrounds comes from non LFV τ decays for both the e channel $(70 \pm 15\%)$ and μ channel $(90 \pm 8\%)$. The main systematics affecting signal efficiency estimations arise from signal track M_{ec} and ΔE modeling, while the background estimation main source of uncertainties comes from the choice of PDFs used to model backgrounds. The ULs are obtained using Feldman and Cousin method [12] and are shown in Tab.2 [13], the results after the selection are shown in Fig.1.

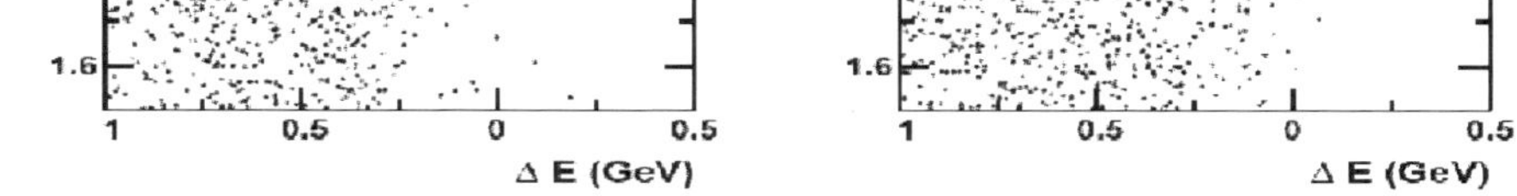

Figure 1: Selection for $\tau \rightarrow l\gamma$ decay channel

4 Search for $\tau \to \ell\ell\ell$

In $\tau \to \ell\ell\ell$ search all possible lepton combinations, consistent with charge conservation, are looked for, resulting in a total of six channels under study. After PID is applied, using a set of multivariate analysis selectors for muon and electron identification, two variables are defined: $\Delta E \equiv E^{\star}_{\rm rec} - E^{\star}_{\rm beam}$, where $E^{\star}_{\rm rec}$ is the total energy of the tracks observed in the 3-prong hemisphere and $E^{\star}_{\rm beam}$ is the beam energy, with both quantities measured in the c.m. frame, and $\Delta M_{ec} \equiv M_{\rm rec} - m_{\tau}$ where $M^2_{\rm rec} \equiv E^{\star 2}_{beam}/c^4 - |\vec{p}^{\,\star}_{3l}|^2/c^2$, where $|\vec{p}^{\,\star}_{3l}|^2$ is the squared momentum of the 3-prong tracks in the c.m. and $m_{\tau} = 1.77 GeV/c^2 7$ is the tau mass. Further background reduction is obtained using topological and kinematic variables (such as the angle between the 3-prong reconstructed momenta and the 1-prong momentum, the missing transverse momentum, and the minimum invariant mass of the two oppositely charged leptons in the 3 prong hemisphere). The selection procedure is blind: for each channel a different rectangular signal box is defined in the $(\Delta M_{ec}, \Delta E)$ plane, the dimensions of the signal box are chosen taking into account the smearing caused by experimental resolution and radiative effects. We also define a common Large Box region, identical for all channels, chosen so that all signal events lie in this region. The expected background events are estimated fitting the MC and data in the sideband region (defined as the region of the large box not including the signal box). Cross checks have been performed to validate the fit, by looking at expected and observed background events in the sideband regions adjacent to the signal box. The numbers of events observed and the background expectations in the signal box are shown in Tab. 2: no events were found in the signal box for all channels. ULs on the branching fractions are calculated using Cousin and Highland (CH) method with Barlow implementation. The main sources of systematic uncertainties are represented by the uncertainty on the number of τ-pairs produced, which is estimated to be 0.9%, uncertainties on the selection efficiency, dominated by the PID efficiency error in the 3-prong hemisphere, and uncertainties on the number of expected background events, which are dominated by the choice of the shapes used to model the MC distributions in the $(\Delta M_{ec}, \Delta E)$ plane. The selection on the $(\Delta M_{ec}, \Delta E)$ plane is shown in Fig. 2.

5 Search for $\tau \to \ell V^0$

For the eight $\tau \to \ell V^0$ ($\ell = \mu, e$ and $V^0 = \rho^0, K^{\star 0}, \bar{K^{\star 0}}, \Phi$) channels PID is not sufficient to completely remove backgrounds. The main sources of contamination come from low multiplicity $e^+e^- \to uds$ processes and an irreducible background constituted by $D \to \ell V^0 \nu$ processes. In order to reduce these backgrounds, a blind analysis procedure is performed using the same variables and technique used in $\tau \to \ell\ell\ell$ study. Further background reduction is ob-

Table 2: Efficiency, number of expected background events ($n_{\rm bgd}$), number of observed events ($N_{\rm obs}$), and observed UL at 90% CL for each decay mode in units of 10^{-8}.

Mode	$\epsilon(\%)$	$n_{\rm bkg}$	$N_{\rm obs}$	UL	Mode	$\epsilon(\%)$	$n_{\rm bkg}$	$N_{\rm obs}$	UL
$e\gamma$	3.91	1.6 ± 0.5	0	3.3	$\mu\gamma$	6.11	3.6 ± 0.6	2	4.4
$e^-e^+e^-$	8.6	0.12 ± 0.02	0	2.9	$\mu^-e^+e^-$	8.8	0.64 ± 0.19	0	2.2
$e^+\mu^-\mu^-$	10.2	0.03 ± 0.02	0	2.6	$\mu^+e^-e^-$	12.7	0.34 ± 0.12	0	1.8
$e^-\mu^+\mu^-$	6.4	0.54 ± 0.14	0	3.2	$\mu^-\mu^+\mu^-$	6.6	0.44 ± 0.17	0	3.3
$e\rho^0$	7.31	1.32 ± 0.19	1	4.3	$\mu\rho^0$	4.52	2.04 ± 0.21	0	0.8
eK^{*0}	8.00	1.64 ± 0.29	2	5.6	μK^{*0}	4.57	1.79 ± 0.25	4	16.7
$e\overline{K}^{*0}$	7.76	2.76 ± 0.30	2	4.0	$\mu\overline{K}^{*0}$	4.11	1.72 ± 0.18	1	6.4
$e\phi$	6.43	0.68 ± 0.14	0	3.1	$\mu\phi$	5.18	2.76 ± 0.21	6	18.2

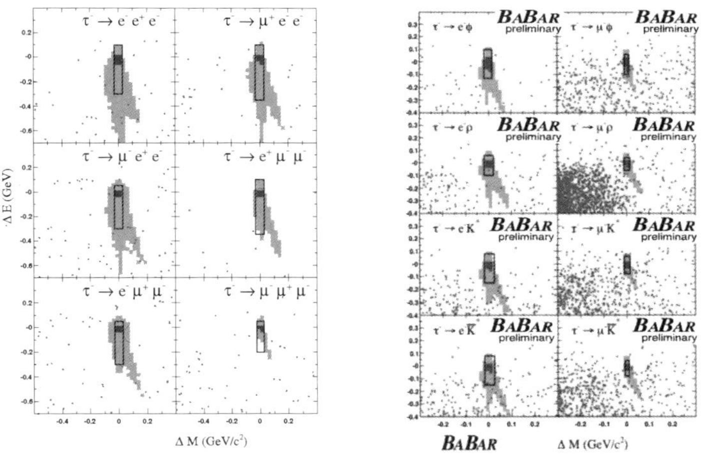

Figure 2: Selection for $\tau \to 3leptons$ decay channel on the left and selection for $\tau \to \ell V^0$ on the right

tained making tight requirement on the reconstructed mass of the resonances, moreover the whole decay tree is fitted requiring that, within reconstruction uncertainties, the hadrons from the meson decay form a vertex, and the lepton and the meson trajectory form a vertex close to the beam interaction region. As for $\tau \to \ell\ell\ell$ expected background is extrapolated from the sideband region of the $(\Delta M_{ec}, \Delta E)$ plane under exam. The main systematic uncertainties arise from PID in the signal side and the uncertainty on Φ branching fraction in $\tau \to \ell\Phi$ channel and range between 1.2% to 9.0%. Errors on background estimation are dominated by background modelization and finite data available in the GS region. The BR is estimated using CH method, and the measured

UL are reported in Tab. 2. Large background overfluctuations are observed in $\tau \to \mu K^{\star 0}$ and $\tau \to \mu\Phi$, resulting in poor ULs for the two channels [14].The selection on the $(\Delta M_{ec}, \Delta E)$ plane is shown in Fig. 2.

References

[1] W.J. Marciano, A.I. Sanda, *Phys. Lett.*, **B67**, 303, 1977.

[2] B.W. Lee, R.E. Shrock, *Phys. Rev.*, **D16**, 1444, 1977.

[3] T.P. Cheng, L.F. Li, *Phys. Rev.*, **D16**, 1425, 1977.

[4] G. Cvetic, C. Dib, C.S. Kim, J.D. Kim *Lepton flavor violation in tau decays*, PRD**66**, 034008, 2002.

[5] S.T. Petcov, S. Profumo, Y. Takanishi, C.E. Yaguna*Charged Lepton Flavor Violating Decays: Leading Logarithmic Approximation versus Full RG Results*, arXiv:hep-ph/0306195.

[6] A. Brignole, A. Rossi, *Lepton Flavor Violating Decays of Supersymmetric Higgs Boson*, arXiv:hep-ph/0304081.

[7] E. Ma, *Theoretical Expectations for Rare and Forbidden Tau Decays*, arXiv:hep-ph/0209170.

[8] J. Ellis, M.E. Gomez, G.K. Leontaris, S. Lola, D.V. Nanopoulos, *Charged-Lepton-Flavor Violation in the Light of the Super-Kamiokande Data*, arXiv:hep-ph/9911459.

[9] T.F. Feng, T. Huang, X.Q. Li, X.M. Zhang, S.M. Zhao, *Lepton Dipole Moments and Rare Decays in the CP-violating MSSM with Nonuniversal Soft-Supersymmetry Breking*, arXiv:hep-ph/0305290.

[10] C. Yue, Y. Zhang, Lanjun. Liu, *Non-universal gauge bosons Z' and lepton flavor-violation tau decays.*, arXiv:hep-ph/0209291.

[11] Andrea Brignole and Anna Rossi, *Anatomy and phenomenology of mu tau lepton flavour violation in the MSSM*, Nucl. Phys. **B701**:353, 2004, hep-ph/0404211.

[12] Robert D. Cousins and Virgil L. Highland, *Nucl. Instrum. Meth.*, **A320**, 331 (1992).

[13] *BABAR* Collaboration, B. Aubert *et al.*, Phys. Rev. Lett. **104**, 021802 (2010)

[14] *BABAR* Collaboration, B. Aubert *et al.*, arXiv:0904.0339v3.

PROSPECTS FOR RARE B-DECAYS AT THE LHCb

Nicola Serra [a]

Nationaal instituut voor subatomaire fysica (NIKHEF), Science Park 105 1098 XG Amsterdam , The Netherlands

*Abstract.*The search for physics beyond the Standard Model will be the primary task of the LHC experiments. At present, the most stringent constraints to new physics are given by Flavor Changing Neutral Current (FCNC) decays, such as processes that involve $b \to s\gamma$ transitions. The LHCb experiment has the possibility to consistently improve present results on rare FCNC decays of B-mesons. Here, the LHCb potential for the search of new physics in the rare decays $B_s \to \mu^+\mu^-$, $B_d \to K^*l^+l^-$, $B_{s,d} \to e^\pm\mu^\mp$ and $B_s \to \phi\gamma$ is discussed.

1 Introduction

Rare decays of B-mesons which proceed via Flavor Changing Neutral Currents (FCNC) are forbidden at the tree level and suppressed by the GIM mechanism in the Standard Model (SM). As a result, new physics (NP) contributions can enter at the same order as SM contributions. Moreover, since new particles contribute as virtual particles, the potential of discovering NP is not limited by the available energy, but only by our ability to make precise predictions and measurements. In particular FCNC involving $b \to d, s$ transitions are excellent indirect probes for NP, providing information about the masses and couplings of virtual NP particles entering into the loops. Here the decays $B_{s,d} \to \mu^+\mu^-$, $B_d \to K^*l^+l^-$, $B_s \to \phi\gamma$ and $B_{s,d} \to e^\pm\mu^\mp$ will be discussed. For all these decays the LHCb experiment [1] has the possibility to significantly improve the results of current experiments.

2 The rare decays: $B_s \to \mu^+\mu^-$ and $B_{s,d} \to e^\pm\mu^\mp$

The decay $B_s \to \mu^+\mu^-$ is a very rare decay. Its branching ratio is predicted to be $(3.35 \pm 0.32) \times 10^{-9}$ within the SM [2]. However, this observable is quite sensitive to NP involving new scalar/pseudo-scalar couplings. For instance models involving an extended Higgs sector, such as SUSY models, will change this branching ratio. In the Minimal Supersymmetric extension of the SM (MSSM) this branching ratio depends on the sixth power of the parameter $\tan\beta$. The decay $B_s \to \mu^+\mu^-$ has not yet been observed, and will be one of the key measurements for the LHCb experiment. Its most recent upper limit is 3.6×10^{-8}@$90\%CL$, set by the Tevatron experiments [3].
The LHCb event selection for this decay is based on a "soft" preselection to reject most of the background, followed by a multidimensional analysis based on three variables: the invariant mass of the di-muons, a geometrical likelihood and a particle identification likelihood. The geometrical likelihood combines

[a]e-mail: nicola.serra@cern.ch

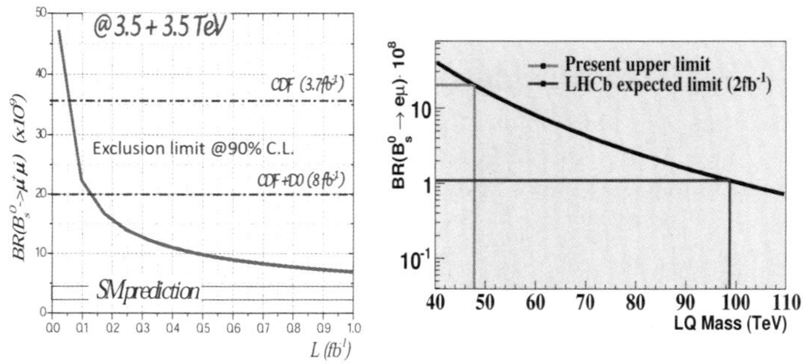

Figure 1: Exclusion limit at 90% CL for the measurement of $BR(B_s \to \mu^+\mu^-)$ at LHCb (left). Exclusion limit for the $BR(B_s \to e^\pm\mu^\mp)$ and corresponding limit for the Pati-Salam Leptoquark (right).

topological discriminating variables. For a given number of observed events the exclusion at $90\%CL$ or the 3σ signal evidence (5σ discovery) can be computed. This method is described in [4]. In Fig.1 the expectation for the measurement of the $Br(B_s \to \mu^+\mu^-)$ at LHCb, as a function of the integrated luminosity, is shown. The LHCb experiment will therefore be able of scanning the interesting region from the final Tevatron limit down to the SM expectation with the first two years of data taking in 2010 and 2011, corresponding to an integrated luminosity of 1fb^{-1}.
A similar analysis strategy is foreseen for the rare decays $B_{s,d} \to e^\pm\mu^\mp$. These decays are forbidden in the SM. The simplest model which allows these decays is the Pati-Salam model, based on the $SU(4)$ symmetry group [5]. In this model the lepton number is seen as the forth color of the theory. LHCb will further constrain the $BR(B \to e\mu)$ which in turn will increase the lower limit on the mass of the Pati-Salam leptoquark by a factor 2 to 100 TeV (Fig. 1) [6].

3 Angular observables in the $B_d \to K^*\mu\mu$ decay

The decay $B_d \to K^*\mu\mu$ is fully described by three angles (θ_l, θ_k, ϕ) [9] and the di-muon invariant mass q^2. The angular distribution of this decay gives access to a number of observables sensitive to NP [8]. The most promising of these observables for the search of NP is the Forward-Backward Asymmetry A_{Fb}. In particular the zero-crossing point of this asymmetry is well predicted in the SM and in a number of NP extensions of the SM. Recently, measurements of the A_{FB} were published by the BaBar, Belle and CDF experiments [7]. Although the statistics is too low for claiming for an evidence of NP, all the measurements show no evidence of a zero-crossing point, a distinctive sign of the SM. The expected sensitivity for LHCb to the A_{FB} for 2fb^{-1} of data taking, assuming the SM, is shown in Fig.2. The expected resolution in the zero-crossing point is expected to be $\sigma(s_0) = 0.5GeV^2$ [9].

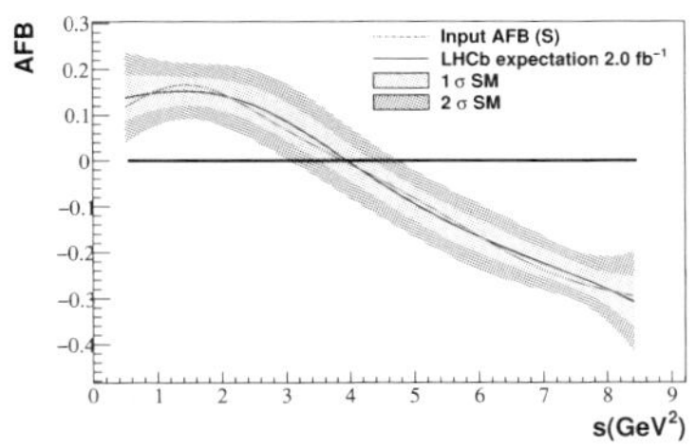

Figure 2: Expectation for the measurement of the A_{FB} at LHCb for 2fb^{-1} (right).

4 Measuring the polarization of the photon in $b \to s\gamma$ processes

One of the most stringent constraint on NP comes from measurements of $b \to s\gamma$ transitions [10]. In the SM the emitted photon is almost completely left-handed polarized due to the V-A structure of the SM. The right-handed part is of the order of $\frac{m_s}{m_b} \sim 10^{-2}$. However NP contributions (as shown in the right diagram of Fig. 3), could significantly enhance the right-handed contribution. At the LHCb, the helicity of the photon can be probed with the decays $B_s \to \phi\gamma$ and $B_d \to K^* e^+ e^-$. In the former decay the photon helicity can be extracted by performing a time dependent analysis. The decay rate can be written, for B_s and $\overline{B}_s$ respectively:

$$\begin{aligned} B(\overline{B})(t) &= B_0 e^{-\Gamma t} (\cosh(\frac{\Delta\Gamma t}{2}) - \\ &- H \sinh(\frac{\Delta\Gamma t}{2}) \pm C \cos(\Delta m_s t) \mp S \sin(\Delta m_s t)) \end{aligned} \tag{1}$$

where $H \sim (2\psi)$ and $tan\psi = \frac{A_R}{A_L}$. It is worth noticing that for measuring the parameters C and S the knowledge of the initial flavor of the B-meson is needed. To measure the parameter H, instead, no flavor tagging information is necessary, with a clear experimental advantage [11]. The LHCb sensitivity for this decay in a nominal year of data taking (2fb^{-1}), according to the MonteCarlo analysis, is $\sigma(\frac{A_R}{A_L}) \sim 0.1$ [9].
Another way for accessing the photon polarization in $b \to s\gamma$ transition is by measuring the virtual photon in the $B_d \to K^* e^+ e^-$ decay. The region of low di-lepton invariant mass, where the photon is quasi-real is dominated by the left diagram of Fig. 3 in the SM. The polarization of the photon is here accessible by performing an angular analysis.

According to MC simulation a sensitivity of $\frac{2A_R}{A_L} \sim 0.1$ is expected in a nominal year of data taking [12]. A similar sensitivity on the right-handed

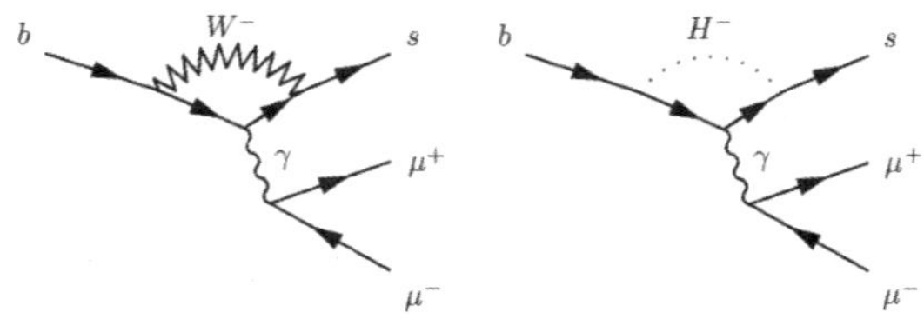

Figure 3: Possible diagrams for the $B_s \to K^* l^+ l^-$ decays, for the SM (left) and for a NP model (right).

component is expected for these two uncorrelated measurements, which allows for an independent cross-check.

References

[1] A. Augusto Alves Junior et al. [LHCb collaboration], *J.Instr.* **3** 4S08005 (2008).
[2] M. Blanke et al., *JHEP* 003, 0610 (2006).
[3] T. Aaltonen et al. [CDF collaboration], CDF note 9892 (2009). D collaboration, D note 5344-CONF (2007).
[4] D. Martnez, J.A. Hernando and F. Teubert, LHCb note 2007-033 (2007).
[5] J.Pati and A.Salam, *Phys. Rev.* **D** 10 275 (1974).
[6] N. Serra and W. Bonivento, LHCb note 2007-028 (2007).
[7] B. Aubert et al. [Babar collaboration], *Phys.Rev.* **D** 73, 092001 (2006). J.T. Wei et al. [Belle collaboration] *Phys.Rev.Lett.* **103** 171801 (2009). T. Aaltonen et al., *Phys.Rev.* **D** 79, 011104(R) (2009).
[8] W. Altmannshofer et al., *JHEP* 0901, 019 (2009). F. Kruger and J. Matias, *Phys. Rev.* **D**71, 094009 (2005). U. Egede et al, *JHEP* 032, 0811 (2008).
[9] B. Adeva et al, [LHCb collaboration], arXiv:0912.4179v2.
[10] K. Fujikawa and A. Yamada *Phys. Rev.* 49, 5890 (1994). J. Ellis et al., *JHEP* 063, 0605 (2006).
[11] F. Muheim, Y. Xie, R. Zwicky, *Phys.Lett.* **B** 664, 174-179 (2008)
[12] J. Lefranois, M.H. Schune, LHCb note 2009-008 (2009).

PROSPECTS OF CP VIOLATION MEASUREMENTS AT LHCb

Tristan du Pree [ab]

Nikhef, Postbus 41882, 1009 DB Amsterdam, The Netherlands

Abstract. In this contribution the prospects for CP violation measurements at the LHCb experiment are discussed. In particular, the expected improvement of the sensitivity to the CP-violating phases γ and β_s are given.

1 Introduction

The main subject of B-physics research is the study of flavor changing interactions of quarks $q_i \rightarrow Wq_j$. In the Standard Model (SM) these interactions are governed by the Cabibbo Kobayashi Maskawa (CKM) mechanism, which contains a (3×3) matrix with complex coupling constants V_{ij}, describing their relative strength and phase. These can be determined by measurements of the decay rates of B decays.

Only if a relative weak phase, i.e. a phase that flips sign under CP conjugation, is present in the different amplitudes contributing to the total decay amplitude, the decay width $\Gamma_{B\rightarrow f}$ and the rate of the CP-conjugated decay $\Gamma_{\bar{B}\rightarrow \bar{f}}$ can be unequal. Also a strong phase (that does not flip sign) is required. Then

$$A_{CP} = \frac{\Gamma_{B\rightarrow f} - \Gamma_{\bar{B}\rightarrow \bar{f}}}{\Gamma_{B\rightarrow f} + \Gamma_{\bar{B}\rightarrow \bar{f}}}, \tag{1}$$

the CP asymmetry, can be nonzero and CP-violating processes can occur.

Due to its unitarity, the CKM-matrix contains (only) four free parameters. Of these, only one is an irreducible (weak) phase, and all measurements of CP violation, albeit in kaon or B_d^0 or B_s^0 decays, constrain the same single irreducible phase that generates CP-violating processes. Inconsistencies between different probes of the same parameter can indicate contributions of new physics (NP) processes.

Phase-convention independent combinations of coupling constants in the form $V_{ij}V_{ik}^*/V_{lj}V_{lk}^*$ lead to different observable CP-violating phases in different B-decays, e.g. $\alpha, \beta, \gamma, \beta_s$. Their definitions and the current measurements that constrain these observables are shown in table 1. It follows that at the current level of precision, the direct and indirect constraints of the observable phases in B-decays are compatible, and the CKM mechanism provides the dominant source of CP violation in the B system.

The LHCb experiment aims to further constrain the two phases that are determined by direct measurements with the largest uncertainty, γ and β_s. It is situated at the LHC accelerator, which operates at a design center-of-mass energy of 14 TeV, at which the $b\bar{b}$ cross section is expected to be $\sim$ 500 μb.

[a] e-mail: tdupree@nikhef.nl

[b] On behalf of the LHCb collaboration.

CP angle	Indirect constraints (°)	Direct constraints (°)
$\alpha \equiv \arg[-V_{td}V_{tb}^*/V_{ud}V_{ub}^*]$	$95.6^{+3.3}_{-8.8}$	$89.0^{+4.4}_{-4.2}$
$\beta \equiv \arg[-V_{cd}V_{cb}^*/V_{td}V_{tb}^*]$	$27.4^{+1.3}_{-1.9}$	$21.07^{+0.90}_{-0.88}$
$\gamma \equiv \arg[-V_{ud}V_{ub}^*/V_{cd}V_{cb}^*]$	$67.8^{+4.2}_{-3.9}$	70^{+27}_{-30}
$\beta_s \equiv \arg[-V_{cs}^*V_{cb}/V_{ts}^*V_{tb}]$	$1.032^{+0.049}_{-0.046}$	22 ± 10 or 68 ± 10

Table 1: *The current constraints from indirect and direct measurements of α, β, and γ [1], and β_s [2].*

The design luminosity at LHCb is 2×10^{32} cm^{-2}s^{-1}, corresponding to an annual luminosity of $\mathcal{L} = 2.0$ fb^{-1}. Since $\sim$ 11% of the b quarks hadronize into B_s mesons, in a nominal year $\mathcal{O}(10^{11})$ B_s^0 mesons are expected to be produced at LHCb, of which $\sim$ 20% of the decays occur in the acceptance of the single-arm spectrometer [3].

2 Measurements of β_s

The decay leading to most sensitivity to β_s is the decay of a B_s^0 meson to a $c\bar{c}s\bar{s}$ final state. CP violation in this process originates from the interference of the mixing and the decay amplitude, leading to a time-dependent CP asymmetry[c]

$$A_{CP} \propto \sin\phi_s \sin\Delta m_s t. \tag{2}$$

In the SM $\phi_s = -2\beta_s \approx 2°$, but since the mixing originates from off-shell processes, this value can change due to contributions of NP processes. The LHCb proper time resolution of $\sim$ 40 fs meets the requirement to resolve the proper time dependent oscillations of the CP asymmetry, occuring with a frequency proportional to $\Delta m_s = 17.77$ ps^{-1}. Furthermore, to determine the CP asymmetry the tagging flavor (the flavor of the B_s^0 meson at production), needs to be estimated. The large power to do so originates from the particle identification capabilities of LHCb.

The $J/\psi\phi$ final state of the B_s^0 decay (a $c\bar{c}s\bar{s}$ state), however, is a superposition of states with odd and even orbital angular momentum. Hence, the final state is an admixture of states with positive and negative CP eigenvalue, which, at leading order, contribute with different signs to the CP asymmetry, diluting the observed CP asymmetry. In order to determine the sizes of the different polarizations, and hence the true CP violating phase, an angular analysis needs to be performed. The angular distributions of the different polarization amplitudes are given in [4].

For optimal sensitivity to ϕ_s, a multi-dimensional likelihood fit is performed, simultaneously fitting for the distributions of the proper time, three angles,

[c] Assuming the B_s^0 width difference $\Delta\Gamma = 0$. More general expressions can be found in [4].

tagging flavor, and invariant mass, taking into account inefficiencies, dilutions and backgrounds in all observables. With $\sim$ 100k $B_s^0 \to J/\psi(\mu^+\mu^-)\phi(K^+K^-)$ events expected to be selected, $\sigma(\phi_s) \approx 1.8°$ for 2.0 fb^{-1}. With $\mathcal{L} = 0.2$ fb^{-1} at 10 TeV, the expected first year of operation, the sensitivity of LHCb to ϕ_s is expected to be better than the sensitivity of the Tevatron experiments, extrapolating their current measurements [5] to a luminosity of 9 fb^{-1} each.

A similar measurement, for which a proper time dependent, flavor tagged, angular analysis needs to be performed, is the measurement of CP violation in the $B_s^0 \to \phi\phi$ channel. In this channel off-shell NP processes can contribute not only to the mixing amplitude, but also to loop-amplitudes in the decay. Both can alter the CP violating phase $\Phi_{\phi\phi} \approx 0$ in the SM. With 2.0 fb^{-1} a sensitivity to the CP-violating phase of $\sigma(\Phi_{\phi\phi}) \approx 6°$ is expected [6].

3 Measurements of γ

All analyses used to extract the CP-violating phase γ are performed in decays in which $b \to (u, c)$ quark transitions occur, followed by a decay into a fully hadronic final state. Since the final state of decays sensitive to γ consists of kaons and pions, both particle identification and mass resolution is crucial to distinguish signal and background.

One possibility to measure γ is from the analysis of $B_s \to D_s^{\mp}K^{\pm}$. As for the previous decays, this decay can occur both direct and via mixing. This leads to a proper time dependent CP asymmetry with an amplitude of $\sin(\Delta_s \pm (\gamma + \phi_s))$. Δ_s is a hadronic strong phase, whereas the weak phase ϕ_s originates from the mixing amplitude, and γ originates from the decay amplitudes. Performing a tagged proper time dependent analysis, the expected sensitivity for 2.0 fb^{-1} is $\sigma(\gamma + \phi_s) \approx 10°$ [7].

Secondly, sensitivity to γ (and ϕ_s) can originate from decays in which possible off-shell processes contribute not only to the mixing, but, as in $B_s^0 \to \phi\phi$, also to the decay amplitude. This occurs in the $B \to hh$ decays. The interference of different decay amplitudes leads to a sensitivity to γ, but since off-shell particles can also contribute via loops, there is sensitivity to possible NP in the decay amplitude. Furthermore, there is sensitivity to ϕ_s, from the mixing amplitude. Combining all measurements of $B \to hh$ channels, sensitivities of $\sigma(\gamma) \approx 7°$ and $\sigma(\phi_s) \approx 3°$ are expected for 2.0 fb^{-1}, with weak assumptions on U-spin symmetry [4].

Finally, a theoretically clean SM estimate of γ is possible from measuring the decay time independent decay rates of $B^- \to DK^-$. From the rates of the decay of the intermediate D into both a CP eigenstate and into D^0 and $\bar{D}^0$ specific states, the GLW method, the phase $\delta + \gamma$ can be determined from the sum of the three amplitudes. δ is the relative strong phase of the two amplitudes contributing to the CP-violating decay, and is beforehand unknown.

Measuring also the rates of all three CP conjugated $B^+ \to DK^+$ decay rates, $\delta - \gamma$ can be measured, which makes it possible to resolve the ambiguity and determine γ. The main amplitudes contributing to these decays originate from on-shell processes, whereas off-shell loop amplitudes are negligible. Therefore, this measurement serves as a control of SM processes. Combining the above method with all other methods to determine γ using on-shell processes only, a combined sensitivity of $\sigma(\gamma) \approx 5°$ is found [4].

4 Conclusions

The CKM model provides a successful description of CP violation in the SM, but the uncertainties of β_s and γ are still sizeable. LHCb will further constrain these phases in different ways, both with and without possible NP contributions in loop diagrams. Benefiting from the number of B_s^0 mesons, proper time

channel(s)	contribution(s)	sensitivity
$B_s^0 \to J/\psi\phi$	loops in mixing	$\sigma(\phi_s) \approx 1.8°$
$B_s^0 \to \phi\phi$	loops in mixing and decay	$\sigma(\Phi_{\phi\phi}) \approx 6°$
$B_s^0 \to D_s^{\pm}K^{\mp}$	loops in mixing	$\sigma(\gamma + \phi_s) \approx 8°$
$B \to hh$	loops in mixing and decay	$\sigma(\gamma) \approx 7°$,$\sigma(\phi_s) \approx 3°$
$B \to DK$	SM only	$\sigma(\gamma) \approx 5°$

Table 2: *Expected sensitivities of LHCb to CP violating phases from different direct measurements after 2.0 fb^{-1}. $\phi_s = -2\beta_s$ in the SM.*

and mass resolutions, and particle identification, LHCb will, as summarized in table 2, drastically improve the current constraints. Comparison of the indirect constraints and SM measurements with the direct measurements in which off-shell processes can contribute to loops in both mixing and decays, makes it possible to check the consistency of the CKM mechanism and helps to disentangle the type of possibly involved NP processes.

References

[1] CKMfitter group, *Eur.Phys.J.* **C** 41, 1-131 (2005).
[2] HFAG, http://www.slac.stanford.edu/xorg/hfag/osc/PDG_2009.
[3] The LHCb Collabration, *JINST* 3 S08005 (2008).
[4] The LHCb Collaboration, LHCb-PUB-2009-029 (2009).
[5] H. Fisk, this conference.
[6] Y. Xie et. al., LHCb-2007-047 (2007).
[7] S. Cohen et. al., LHCb-2007-041 (2007).

SEARCHES FOR NEW PHYSICS AT NA62

Vito Palladino [a] on Behalf of NA62 Coll.
Università e INFN, Dipartimento di Scienze Fisiche, Napoli, Italia

Abstract. We present the latest NA62 results in the search for physics beyond Standard Model (SM). NA62 aims to have indirect evidences of new physics, measuring rare K decays. NA62 phase I took place in 2007 when we collected data in order to measure the ratio $R_K = K_{e2}/K_{\mu 2}$ (were $K_l 2$ means $K \to l\nu_l$) at few per mill level. A brief experimental layout description will be followed by analysis strategy and preliminary results. The last part of present paper will be devoted to the description of NA62 phase II, which has the main goal of measuring the ultra-rare $K \to \pi^+ \nu\bar{\nu}$ decay Branching Ratio.

1 NA62 Phase I

1.1 R_K in Standard Model an beyond

The R_K ratio in SM framework is a very well determined quantity [1]:

$$R_K^{SM} = \frac{m_e^2}{m_\mu^2} \cdot \frac{m_k^2 - m_e^2}{m_k^2 - m_\mu^2} \cdot (1 + \delta R_K^{Rad.Corr.}) = (2.477 \pm 0.001) \times 10^{-5} \qquad (1)$$

Any significant deviation from this value could signal new physics. In minimal SUSY scenario R_K value is modified due to Lepton Flavor Violating (LFV) terms in charged Higgs exchange diagrams (Fig. 1.2). Using reasonable SUSY parameters values ($\Delta_{13} = 5 \times 10^{-4}$, $\tan(\beta)$ and $m_H = 500$ GeV) sizable deviations from SM value have been predicted [2]:

$$R_K^{LFV} = 2\frac{\Gamma_{SM}(K \to e\nu_e) + \Gamma_{LFV}(K \to e\nu_\tau)}{\Gamma(K \to \mu\nu_\mu)} = R_K^{SM}(1 + 0.013) \qquad (2)$$

1.2 Experimental Status

R_K PDG08 value is computed using three measurements dating back to the 70s:

$$R_K^{70s} = (2.45 \pm 0.11) \times 10^{-5} \qquad (3)$$

A recent new result from KLOE [3] experiment improved the measurement:

$$R_K^{KLOE} = (2.493 \pm 0.031) \times 10^{-5} \qquad .(4)$$

[a] e-mail: Vito.Palladino@cern.ch

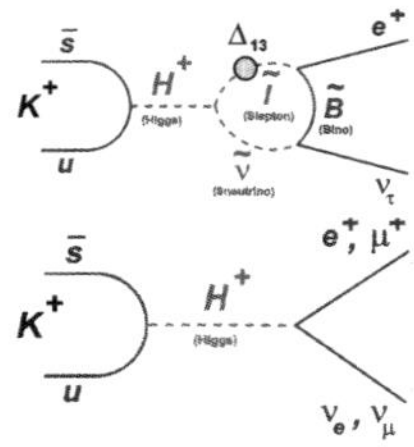

Figure 1: SUSY contributions to $BR(K \to e\nu_e)$.

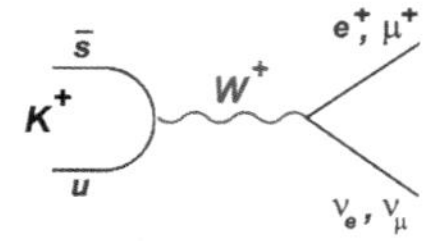

Figure 2: SM decay mode.

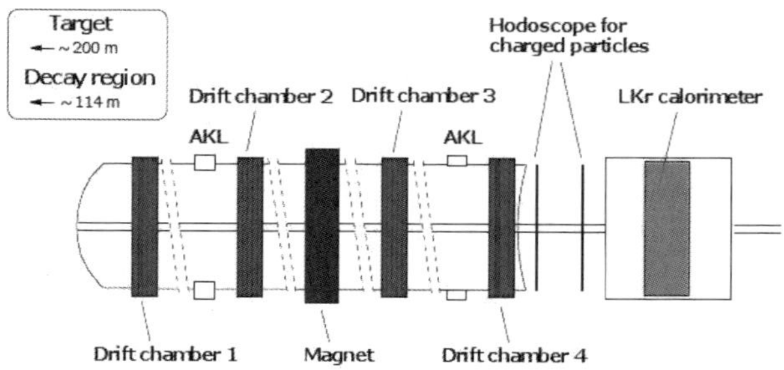

Figure 3: Scheme of apparatus for R_K measurement.

1.3 Apparatus, Trigger Logic and Measurement Strategy

Data have been taken in the June - October 2007 period. The apparatus used is reported in Fig. 1.2.

It is composed by a charged Hodoscope (called HODO) used as fast trigger, a Drift CHamber (DCH) spectrometer, a photon veto (called AKL) and the NA48 Liquid Kripton (LKr) calorimeter.

We used a minimum bias hardware trigger in order to select simultaneously K_{e2} and $K_{\mu2}$ events to minimize the systematics. The two samples only differ for energy release in LKr. Common logical conditions used are: activities in DCHs and energy release into both the hodoscope planes. K_{e2} events have to satisfy a further condition on energy released in LKr (higher than 10 GeV). Data taking goal was to collect about 150K events of K_{e2} in order to have an accuracy better than 0.5%.

The experimental quantity to be measured is:

$$R = \frac{N_{Ke2} - N_{Ke2}(BG)}{N_{K\mu2} - N_{K\mu2}(BG)} \cdot \frac{A_{K\mu2} \times \epsilon_{K\mu2} \times PID_{K\mu2}}{A_{Ke2} \times \epsilon_{Ke2} \times PID_{Ke2}} \tag{5}$$

where N_{Kl2} (l=e,μ) is the number of selected events, $N_{Kl2}(BG)$ is the number

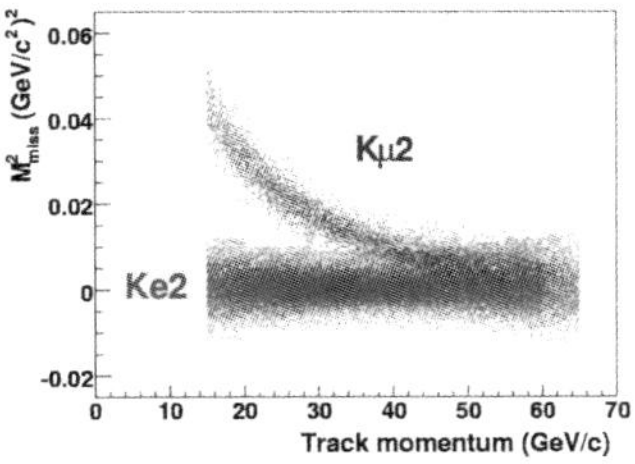

Figure 4: M^2_{miss} vs track momentum in electron mass hypothesis.

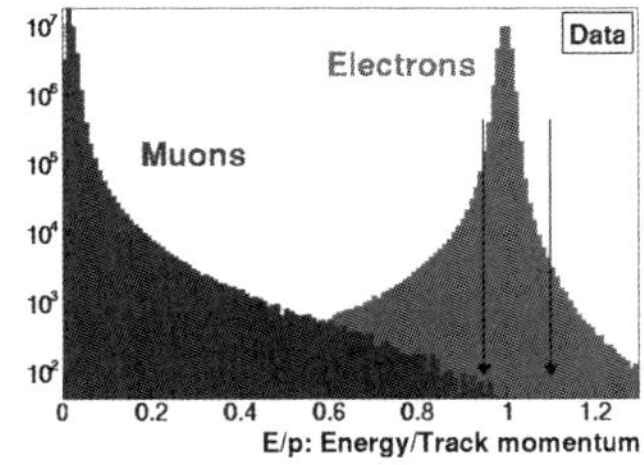

Figure 5: E/P distribution, the arrows define the signal cuts.

of background (BG) events, A_{Kl2} the geometrical acceptance, ϵ_{Kl2} and PID_{Kl2} the trigger and selection efficiencies respectively.

The ratio R has been evaluated in 10 momentum bins.

1.4 Signal Selection and Main Background

In order to separate the two samples we exploited the kinematic separation (using $M^2_{miss} = (p_K - p_l)^2$ see Fig. 1.3) which is optimal for tracks with energy up to 25 GeV and particle identification using E/p ratio (energy released in LKr/measured track momentum see Fig. 1.3).

The main background source for K_{e2} sample are $K_{\mu 2}$ events in which the muon loose all its energy into LKr (catastrophic bremsstrahlung) thus emulating an electron (therefore K_{e2} event). These events ere expected to contribute at O(10%) level to the final sample and we thus decided to directly measure their fraction in order to validate Monte Carlo estimates.

This measurement has been done using a pure (electron contamination was evaluated to be $\sim 10^{-7}$) muon sample obtained interposing a $10X_o$ deep lead wall between the two hodoscope planes. A MC simulation was made with and without the lead wall, the first was compared with data finding a very good agreement. The second was used to evaluate the real background contamination: $(6.28 \pm 0.17)\%$.

A preliminary result of R_K measurement computed on 40% of whole collected statistics is:

$$R_K = (2.005 \pm 0.012_{stat} \pm 0.011_{syst}) \times 10^{-5} = (2.500 \pm 0.016) \times 10^{-5} \quad (6)$$

the precision reached is 0.64%, with the whole sample we expect to reach $\sim 0.3\%$.

2 NA62 Phase II

2.1 Motivations and Experimental Status

During its phase II NA62 aims to measure the Branching Ratio (BR) of the ultra-rare Kaon decay $K^+ \rightarrow \pi^+\nu\bar{\nu}$ observing $\tilde{1}00$ signal events in about 2 years. This decay is a theoretically well described process in the SM and is very sensitive to SM extensions. The SM and minimal SUSY predictions and the current experimental status are:

$$BR(K \rightarrow \pi^+\nu\bar{\nu})_{SM} = (8.5 \pm 0.7) \times 10^{-11} \tag{7}$$

$$\frac{BR(K \rightarrow \pi^+\nu\bar{\nu})_{SM}}{BR(K \rightarrow \pi^+\nu\bar{\nu})_{MSUSY}} \in (0.65, 1.03) \tag{8}$$

$$BR(K \rightarrow \pi^+\nu\bar{\nu})_{exp} = (1.47^{+1.30}_{-1.05}) \times 10^{-10} \tag{9}$$

2.2 Apparatus Design

NA62 experiment [4] has been designed using the following guidelines: Kaons decay in flight in order to reduce the background and uncertainties introduced by stopping target, high beam momentum for improving background rejection, precise timing in order to associate the outgoing particle with the correct incoming Kaon, redundant $\pi - \mu$ particle ID, high hermeticity to photons to reduce π^0 induced background.

Almost all detectors have completed the *R&D* programs. Some of them are in the final construction phase.

3 Conclusions

NA62 phase I is stretching SM measuring R_K. At the end of this phase we expect to have a sample $\tilde{1}0$ times larger than the world integrated statistics collected up to now. The preliminary result does not show any discrepancy from SM prediction.

Concerning NA62 Phase II we expect to start our data taking in 2012.

References

[1] Cirigliano V. and Rossell I., *Phys. Lett.* 99, 231801 (2007).
[2] Masiero A., Paradisi P., Petronzio R. *Phys. Rev.* **D** 74, 01171 (2006).
[3] M. Antonelli et al., La Thuile '09.
[4] NA62 Technical design, http://na62.web.cern.ch.

QCD TESTS WITH KAON DECAYS

Cristina Biino [a]

Istituto Nazionale di Fisica Nucleare, via P. Giuria 1, 10125 Torino, Italy

Abstract. Kaon decays are an ideal laboratory to study strong interaction dynamics in the low energy regime. The NA48/2 fixed target experiment was dedicated to the study of CP violation (CPV) and rare decays in the kaon sector. In this paper the most recent results are reported. The analysis of more than one million $K^{\pm} \to \pi^{+}\pi^{-}e^{\pm}\nu$ (Ke4) rare decays allows a model independent approach to the study of low energy π-π scattering close to threshold providing an accurate test of Chiral Perturbation Theory (ChPT) predictions. This result is combined with the independent and complementary NA48/2 result obtained in the analysis of about 60 millions $K^{\pm} \to \pi^{0}\pi^{0}\pi^{\pm}$ (K3π), leading to an experimental measurement of a_0 and a_2, the isospin 0 and 2 s-wave π-π scattering lengths, of unprecedented precision. In the radiative decay $K^{\pm} \to \pi^{\pm}\pi^{0}\gamma$ about one million events were reconstructed leading to the first measurement of the interference between direct photon emission and inner bremsstrahlung amplitude. Finally, we report on precise measurements of the branching fractions and the form factors of the rare decays $K^{\pm} \to \pi^{\pm} lepton^{+} lepton^{-}$.

1 The NA48/2 experiment at CERN SPS

The main goal of NA48/2 experiment was to measure the CP violating asymmetry in the charged kaon decay in three pions. Thanks to the very high statistics accumulated many other interesting measurements have been performed with unprecedented precision. The NA48/2 experiment beamline [1] is designed to deliver simultaneously K^{+} and K^{-}. Charged particles with momentum 60 ± 3 GeV/c are selected by an achromatic system of dipole magnets, which splits the two beams in the vertical plane and then recombines them on a common axis. The ratio of K^{+} and K^{-} fluxes is about 1.8.

The NA48 detector [2] consists of a magnetic spectrometer made of four drift chambers and a dipole magnet in the middle. The momentum resolution is 1.4% for 20 GeV/c charged tracks. The spectrometer is followed by a plastic scintillator hodoscope used to produce fast trigger signals and precise time measurements for charged particles. Then there is a high resolution liquid krypton electromagnetic calorimeter used for photon detection and particle identification. It is an almost homogeneous ionization chamber, 27 X_0 deep, segmented transversally into 13,248 cells $2 \times 2cm^2$ each, with good time resolution and energy resolution of about 1% for 20 GeV electrons and photons. Further downstream there is a hadron calorimeter and a muon detector.

2 The $K^{\pm} \to \pi^{+}\pi^{-}e^{\pm}\nu$ (Ke4) Decay Analysis

The selection of the Ke4 semileptonic decays ($BR = 4.1 \times 10^{-5}$) requires three charged tracks forming a common vertex. Only one of them should

[a] e-mail: cristina.biino@to.infn.it

be consistent with the electron hypothesis, (associated energy deposit in the calorimeter consistent with the measured track momentum). The other two tracks should have opposite signs and should not give signal in the muon detector. The background, coming mainly from K3π decay with a pion misidentified as electron or with a pion decay $\pi \to e\nu$, is evaluated by studying the $K^{\pm} \to \pi^{+}\pi^{-}e^{\mp}\nu$ decays that are suppressed by the $\Delta S = \Delta Q$ rule. The total background level is at 0.5%. The 1.1 million event sample is distributed over a grid of $(10 \times 5 \times 5 \times 5 \times 12)$ equal population boxes in the five variables space that describes the decay kinematics. For each bin in $M_{\pi\pi}$, comparing data and MonteCarlo simulation, ten independent five parameters fits are performed. This allows to determine the axial and vectorial form factors in terms of partial s and p wave expansions which do not depend upon any particular model. The unprecedented precision of NA48/2 result triggered theoretical work to determine the effect of isospin symmetry breaking on phase shift. After subtracting isospin effects, numerical solution of Roy equations are used to extract scattering lengths from the phase measurements in a 2-parameter fit: $a_0 m_\pi = 0.2220 \pm 0.0128_{stat} \pm 0.0050_{syst} \pm 0.0037_{theor}$, $a_2 m_\pi = -0.0432 \pm 0.0086_{stat} \pm 0.0034_{syst} \pm 0.0028_{theor}$, with a correlation of 96.7%. Using the ChPT constraint we obtain: $a_0 m_\pi = 0.2206 \pm 0.0049_{stat} \pm 0.0018_{syst} \pm 0.0064_{theor}$, in excellent agreement with the ChPT prediction [3].

3 The $K^{\pm} \to \pi^0\pi^0\pi^{\pm}$ (K3π) Decays and the Cusp Analysis

The pion scattering lenghts can also be measured through the study of the *cusp* in the $\pi^0\pi^0$ invariant mass distribution (M^2_{00}) in K3π decay, see Fig.1 left-side. At $2m_\pi$ a singularity (cusp) appear. The existence of this threshold anomaly

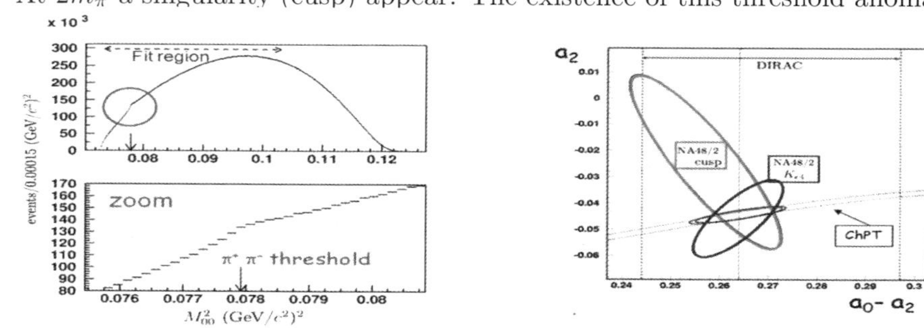

Figure 1: Left-side: Top: distribution of M^2_{00}. Bottom: enlargement of a narrow region centered at $M^2_{00} = (2m_\pi)^2$. Right-side: Ke4 (blue) and K3π (green) results as 2-par fit in the $(a_0 - a_2), a_2)$ plane. The lines correspond to DIRAC experiment result band.

had been first predicted in 1961 [4] as due to $\pi^+\pi^- \to \pi^0\pi^0$ strong rescattering, having different real and imaginary behaviour below and above the $2\pi^+$ production threshold. More recently different theoretical approaches using different formalisms and hypothesis have been developed in order to exploit the experimental precision obtained by NA48/2 thanks to the statistics collected (60.31×10^6 decays) and the excellent mass resolution M_{00}. The $K3\pi$ analysis is essentially based on the liquid krypton calorimeter to identify the γs from π^0 decays. We use in particular two models of the decay amplitude estimating one and two-loop rescattering effects [5] to fit our data. In both approaches, the M_{00}^2 distribution is fitted using the detector response matrix obtained from a large MonteCarlo simulation, to extract the scattering lengths a_0-a_2, a_2 and the Dalitz slopes. The results are in good agreement. The final results presented here are obtained fron the fits to the most complete description of rescattering effects (BFGKR [5]): $(a_0 - a_2)m_\pi = 0.2571 \pm 0.0048_{stat} \pm 0.0025_{syst} \pm 0.0014_{ext}$; $a_2 m_\pi = -0.024 \pm 0.013_{stat} \pm 0.009_{syst} \pm 0.002_{ext}$; and, using the ChPT constraint mentioned earlier, the 1-parameter fit gives: $(a_0 - a_2)m_\pi = 0.2633 \pm 0.0024_{stat} \pm 0.0014_{syst} \pm 0.0019_{ext}$.

4 Comparison between Ke4 and K3π Analysis

Two different approaches to measure the pion scattering lengths have been presented. From an experimental point of view the two processes are collected by different sub-detectors with different contributions to the systematic uncertainty. On the other hand also the theoretical inputs used to extract informations from the decay dynamics are totally different. In spite of these differences the results obtained are in good agreement, as shown in Fig.1 right-side. The combined results are: $(a_0 - a_2)m_\pi = 0.2639 \pm 0.0020_{stat} \pm 0.0004_{syst}$ and $a_2 m_\pi = -0.0429 \pm 0.0044_{stat} \pm 0.0016_{syst}$ and, assuming the ChPT constraint: $(a_0 - a_2)m_\pi = 0.264 \pm 0.0020_{stat} \pm 0.0017_{syst}$, in perfect agreement with the ChPT prediction [3]: $(a_0 - a_2)m_\pi = 0.265 \pm 0.004$.

5 The $K^\pm \to \pi^\pm \pi^0 \gamma$ Radiative Decay

Radiative kaon decays offer a unique opportunity to study ChPT in detail. The total amplitude of $K^\pm \to \pi^\pm \pi^0 \gamma$ decay is the sum of two terms: inner bremsstrahlung (IB), with the photon being emitted from the outgoing charged pion, and direct emission (DE), where the photon is emitted from the weak vertex. The IB component can be predicted from QED corrections to $K^\pm \to \pi^\pm \pi^0$ in a straightforward way. For the DE term, several studies within the framework of ChPT exist. At $O(p^4)$ChPT, DE can occur through both electric (XE, no definitive prediction exists) and magnetic (XM, calculable) and a direct amplitude (expected to be small). In addition to IB and DE contributions the total decay rate contains also the interference (INT) between IB and DE, which depend on XE and a possible CP violating phase. By measuring the INT term it is possible to disentangle the electric and magnetic amplitudes and to investigate possible CPV. For this measurement the full NA48/2 data set with about one million reconstructed decays was analyzed, which is a factor 30 larger than previous experiments. A strong suppression of background from $K^{\pm} \to \pi^0\pi^0\pi^{\pm}$ events is possible due to the excellent performance of the e.m. calorimeter. The extraction of the IB, DE and INT amplitudes was done with an extended maximum-likelihood fit of the MonteCarlo distributions of the single components to the data distribution. The high precision results are: $Frac(DE) = (3.32 \pm 0.15)\%$ and $Frac(INT) = (-2.35 \pm 0.35)\%$ being observed for the first time. The data distribution and DE result are shown in Fig.2 left-side. From this we also obtain $XE = (-24 \pm 4_{stat} \pm 4_{syst})Gev^{-4}$ and $XM = (254 \pm 11_{stat} \pm 11_{syst})$ very close to the WZW prediction of $271GeV^{-4}$. NA48/2 is the first experiment with both K^+ and K^- events and the CP violating asymmetry between K^+ and K^- has been obtained to be less than 10^{-3} in this channel.

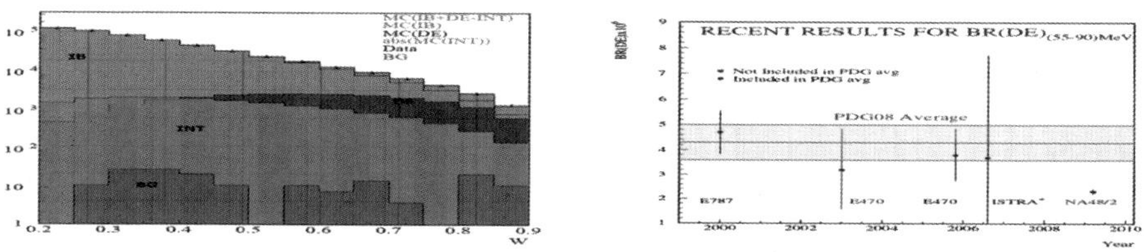

Figure 2: Left-side: Data distribution. Right-side: DE experimental results.

6 The $K^{\pm} \to \pi^{\pm} lepton^{+} lepton^{-}$ Rare Decays

The full NA48/2 data set has been analyzed, leading to more than 7,200 reconstructed events in the electronic and more than 3,000 events in the muonic channels, the latter exceeding the total existing statistics by more than a factor of five. For both channels the selected events are almost background free. From these events we have determined the branching fractions and form factors of the decays using different theoretical models. Our results of the two channels are in good agreement with each other and improve the existing world averages significantly. A paper is in preparation. We also measured the CP violating asymmetry between K^+ and K^- in this channels to be less than a few percent.

References

[1] J.R.Batley et al., *Phys.Lett.* **B** 638, 22 (2006).
[2] V.Fanti et al., *Nucl.Instr.Meth.* **A** 574, 433 (2007).
[3] G.Colangelo et al., *Nucl.Phys.* **B** 638, 187 (2006).
[4] P.Budini, L.Fonda, *Phys.Rev.Lett.* **6**, 419 (1961).
[5] N.Cabibbo, G.Isidori, *JHEP* 0503,021 (2005); M.Bissegger et al., *Phys.Lett.* **B** 638, 187 (2006); M.Bissegger et al. (BFGKR), *Nucl.Phys.* **B** 806, 178 (2009).

SCALED MOMENTUM SPECTRA IN DEEP INELASTIC SCATTERING AT HERA

Beata Brzozowska[a]
Faculty of Physics, University of Warsaw, 00681 Warsaw, Poland
(on behalf of the ZEUS collaboration)

Abstract. Particle production in *ep* deep inelastic scattering (DIS) has been studied in the ZEUS detector at HERA with an integrated luminosity of 0.44 fb^{-1}. The distributions of scaled particle momenta in the Breit frame are measured for charged particles in the current fragmentation region. The results are presented in a very wide kinematic range $10 < Q^2 < 41000$ GeV^2 and compared with theoretical predictions. The concept of quark-fragmentation universality is tested in comparison with results from e^+e^- annihilation.

1 Introduction

In this paper, a study of the fragmentation of struck quarks in neutral current NC *ep* DIS is presented. For the study, the inclusive spectrum of charge particle scaled momenta, $x_p = 2p^{Breit}/Q$, is used, where p^{Breit} denotes the momentum of a hadron measured in the Breit frame and Q is the virtuality of the exchanged boson. The Breit frame gives the best separation between the scattered quark (current region) and proton remnants. It is defined as the frame in which the four-vector of the exchanged photon becomes (0,0,0,-Q). The current region closely resembles a single hemisphere of an e^+e^- annihilation event, thus allowing a comparison to be made between results from *ep* and e^+e^- [1] experiments.

The data presented here were collected with the ZEUS detector at HERA between 1996-2007 and correspond to a luminosity of 0.44 fb^{-1} [2]. A detailed description of the ZEUS detector can be found elsewhere [4]. The reconstructed charged particle tracks used in the analysis had to be associated with the primary interaction vertex and were required to be in the region of high Central Tracking Detector (CTD) acceptance. The particles had to pass through at least three CTD superlayers and were required to have a transverse momentum, p_T^{track}, of more than 150 MeV.

2 Scaled momentum spectra

The scaled momentum distributions as functions of $1/\ln(x_p)$ were studied in (x, Q^2) bins for $0.002 < x < 0.75$ and $10 < Q^2 < 40960$ GeV^2. The distributions (see Fig. 1, left) are approximately Gaussian in shape with the mean charged multiplicity given by the integral of the distributions. As Q^2 increases, the multiplicity increases and, in addition, the peak of the distribution moves to larger values of $\ln(1/x_p)$.

[a]e-mail: beata.brzozowska@fuw.edu.pl

The data are compared with calculations based on the modified leading log approximation (MLLA) [5] which describes parton production in terms of the shower evolution. These calculations depend on two parameters only, the effective QCD scale, $\Lambda_{\rm eff}$, and the infrared cutoff scale, Q_0, at which the parton cascade is stopped. The calculations intrinsically include colour coherence and gluon interference effects. To connect predictions at the parton-level to experimental hadron-level data, the hypothesis of local parton-hadron duality (LPHD) [6] is assumed with the hadronisation constant $K_{\rm h}$.

The MLLA+LPHD predictions are calculated with parameter values obtained from a fit to LEP data including e^+e^- data with lower centre-of-mass energies: $Q_0 = \Lambda_{\rm eff} = 270 \pm 20$ MeV and $K_{\rm h} = 1.31 \pm 0.3$. The predictions give a reasonable description of the shape at $80 < Q^2 < 1280$ GeV^2, except for long tails at large values of $\ln(1/x_p)$ which are sensitive to mass corrections [7]. At low Q^2, the measured average multiplicity is lower than predicted by MLLA+LPHD. This can be explained by a significant migration of particles from the current to the target region of the Breit frame. As Q^2 increases, the peaks are shifted more than expected towards higher values of $\ln(1/x_p)$. Thus, the analytical MLLA+LPHD predictions with the LEP-tuned parameters cannot describe the data in the entire range of x_p and Q^2.

As the energy scale Q increases, α_s decreases and the phase space for soft-gluon radiation increases. This leads to a fast increase of the number of soft particles with small fractional momenta x_p. These scaling violations are seen when the data in bins of x_p are plotted as a function of the energy (see Fig. 1, right). The distributions rise with Q^2 at low x_p and fall off at high x_p and high Q^2 (the three lowest x_p bins are scaled by factors of 30, 5 and 2, respectively). Also shown is the comparison of recent ZEUS measurements with H1 [8] and e^+e^- data [1], a general trend of quark fragmentation features is observed.

3 Conclusions

Scaled momentum spectra in NC DIS have been measured for the current region in the Breit frame over a large range of Q^2 from 10 GeV^2 to 40960 GeV^2. Clear evidence for scaling violations is observed. Comparing the data to e^+e^- results from the LEP experiments supports the concept of quark-fragmentation universality. The MLLA+LPHD calculations fail to describe the data over the whole kinematic range.

Acknowledgments

I wish to thank the 14th Lomonosov Conference Organizers for their hospitality. I am grateful for discussions with W. Khoze, W. Ochs and R. Sassot. Special thanks go to S. Albino for providing the QCD calculations.

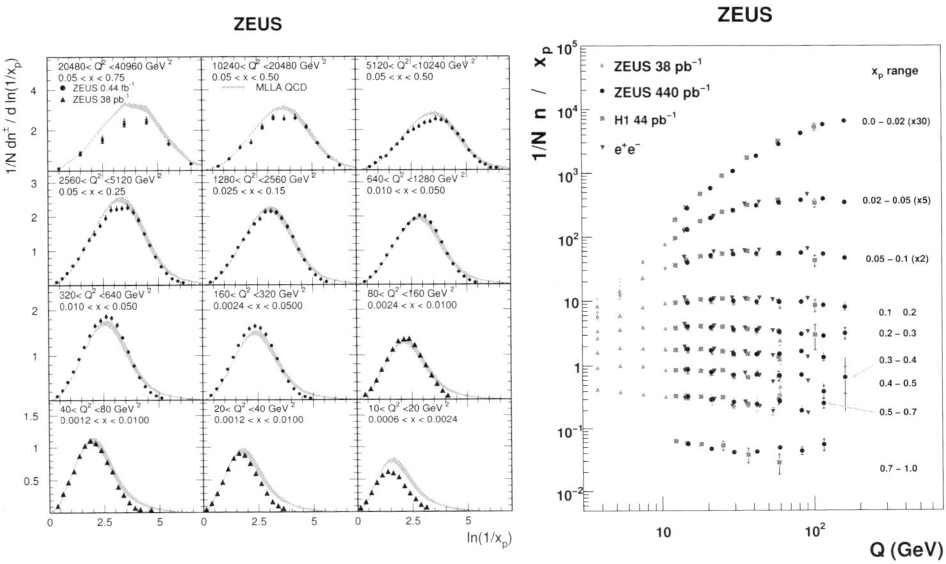

Figure 1: Left: scaled momentum spectra as functions of $\ln(1/x_p)$. The bands represent the range of the MLLA+LPHD predictions. Right: the normalized number of charged particles as a function of Q in x_p bins. The dots (triangles) represent the new (previous [3]) ZEUS measurement, the squares the H1 data [8] and the inverted triangles the e^+e^- data [1]. The inner error bars, where visible, indicate statistical uncertainties, the outer bars statistical and systematic uncertainties added in quadrature.

References

[1] TASSO Coll., W. Braunschweig et al., *Z. Phys.* **C** 47, 187 (1990)
MARK II Coll., A. Petersen et al., *Phys. Rev.* **D** 37, 1 (1988)
DELPHI Coll., P. Abreu et al., *Phys. Lett.* **B** 311, 408 (1993)
AMY Coll., Y.K. Li et al., *Phys. Rev.* **D** 41, 2675 (1990)

[2] ZEUS Coll., H. Abramowicz et al., DESY-09-229

[3] ZEUS Coll., J. Breitweg et al., *Eur. Phys. J.* **C** 11, 251 (1999).

[4] ZEUS Coll., U. Holm (ed.), *The ZEUS Detector.* Status Report (unpublished), DESY (1993, available on http://www-zeus.desy.de/bluebook/bluebook.html)

[5] V.A. Khoze and W. Ochs, *Int. J. Mod. Phys.* **A** 12, 2949 (1997)

[6] Y.I. Azimov et al., *Z. Phys.* **C** 27, 65 (1985)

[7] P. Dixon, D. Kant and G. Thompson, *Nucl. Part. Phys.* 25, 1453 (1999)

[8] H1 Coll., F.D. Aaron et al., *Phys. Lett.* **B** 654, 148 (2007)

HOW THE INTERACTION OF π MESONS IN THE FINAL STATE CHANGES A CORRELATION BETWEEN $K \to 2\pi$ AND $K \to 3\pi$ AMPLITUDES

Evgeny Shabalin [a]

Institute of Theoretical and Experimental Physics, 117218 Moscow, Russia

Abstract. The unitarization of the amplitudes of $K_1^0 \to \pi^+\pi^-$ and $K^+ \to \pi^+\pi^+\pi^-$ decays allows, using the parameters extracted from the data on $K \to 2\pi$ decays, to obtain a value of width of $K^+ \to \pi^+\pi^+\pi^-$ decay coinciding to a level of a few percent with its experimental magnitude.

The slope parameter g_{++-} calculated with taking into account the higher order terms in the momentum expansion of the $K^+ \to \pi^+\pi^+\pi^-$ amplitude turns out to be $g^{th}_{++-} = 02182$ in agreement with $g^{exp}_{++-} = 02154 \pm 0.0035$.

More than 40 years ago, the use of the algebra of currents allowed to find a correlation between the amplitudes of the $K \to 2\pi$ and $K \to 3\pi$ decays. As an example, let's consider the relation (see [1])

$$M(K^+ \to \pi^+(p_1)\pi^+(p_2)\pi^-(p_3)) = \frac{i}{3F_\pi} M(K_1^0 \to \pi^+\pi^-)[1 + y + 6\zeta], \quad (1)$$

$$y = 3E_3/m_K - 1, \qquad \zeta = -M(K^+ \to \pi^+\pi^0)/M(K_S \to \pi^+\pi^-). \quad (2)$$

This equatuion, beeing correct to leading order in momentum expansion of both amplitudes, however, gives a probability of $K^+ \to \pi^+\pi^+\pi^-$ decay by 1.575 times smaller than its experimental value, if the data on $K \to 2\pi$ decays are used. It means that the higher-order momentum corrections to these amplitudes are essential [2], [3] and, as it was found in [4], [5], the effect of final-state interaction (FSI) of pions, in different manner changing a value of these amplitudes, also has to be taken into account.

In [5], it was found that a repeated rescattering of final pions turns the arbitrary initial partial amplitude $a_l^{(I)}$, where I is the isotopic spin, into

$$\left(a_l^{(I)}\right)_{unitar} = \frac{a_l^{(I)}(s)}{1 - \Re\Pi(s)} \cos\delta_l^{(I)}(s) e^{i\delta_l^{(I)}(s)}, \quad (3)$$

where $\Re\Pi(s)$ is the real part of the loop integral over momenta of the intermediate particles. The reliable estimate of this part can be extracted from a fitting of $\delta_0^0(s)$ and $\delta_0^2(s)$ to data for these phase shifts [5], [6].

In terms of the Wilson coefficients c_i of the effective lagrangian of non-

[a] e-mail: shabalin@itep.ru

leptonic interactions with $\Delta S = 1$ [7] , to leading order of momentum expansion

$$\begin{aligned}
&M(K_1^0 \to \pi^+\pi^-) \\
&\quad = \tfrac{1}{\sqrt{2}} G_F F_\pi \sin\theta_C \cos\theta_C (m_{K^0}^2 - m_{\pi^+}^2)[(c_1 - c_2 - c_3 + \tfrac{32}{9}\beta\tilde{c}_5) - c_4], \\
&M(K_1^0 \to \pi^0\pi^0) \\
&\quad = \tfrac{1}{\sqrt{2}} G_F F_\pi \sin\theta_C \cos\theta_C (m_{K^0}^2 - m_{\pi^0}^2)[(c_1 - c_2 - c_3 + \tfrac{32}{9}\beta\tilde{c}_5) + 2c_4], \\
&M(K^+ \to \pi^+\pi^0) = \tfrac{1}{\sqrt{2}} G_F F_\pi \sin\theta_C \cos\theta_C (m_{K^+}^2 - \tilde{m}_\pi^2)\tfrac{3c_4}{2}, \\
&M(K^+(k) \to \pi^+(p_1)\pi^+(p_2)\pi^-(p_3)) = \tfrac{1}{2\sqrt{2}} G_F \sin\theta_C \cos\theta_C \cdot \\
&\left[(c_1 - c_2 - c_3 - c_4 + \tfrac{32}{9}\beta\tilde{c}_5)(s_1 + s_2 - 2\mu^2) + 9c_4(s_0 - s_3)\right], \\
&M(K^+(k) \to \pi^0(p_1)\pi^0(p_2)\pi^+(p_3)) = \tfrac{1}{2\sqrt{2}} G_F \sin\theta_C \cos\theta_C \cdot \\
&\left[(c_1 - c_2 - c_3 - c_4 + \tfrac{32}{9}\beta\tilde{c}_5)(s_3 - \mu^2) + \tfrac{9}{2}c_4(s_0 - s_3)\right].
\end{aligned} \tag{4}$$

In the above equations

$$\begin{aligned}
&\beta = 2\mu^4/[(m_d + m_u)^2\Lambda^2], \quad \Lambda^2 = m_{\sigma_\pi}^2 - \mu^2, \quad \tilde{c}_5 = c_5 + \tfrac{3}{16}c_6, \\
&\tilde{m}_\pi^2 = (m_{\pi^+}^2 + m_{\pi^0}^2)/2, \quad s_1 = (k - p_1)^2, \quad s_2 = (k - p_2)^2, \\
&s_3 = (k - p_3)^2, \quad s_0 = m_K^2/3 + \mu^2.
\end{aligned}$$

The higher-order momentum corrections can be found using the theory [6].As a result

$$\begin{aligned}
&M^{corr}(K_1^0 \to \pi^+\pi^-) = \tfrac{1}{\sqrt{2}} G_F \sin\theta_C \cos\theta_C (m_{K^0}^2 - m_{\pi^+}^2) \cdot [(c_1 - c_2 - c_3 + \\
&\tfrac{32}{9}\beta\tilde{c}_5 \cdot 1.3565) \cdot \cos\delta_0^0 \cdot \exp(i\delta_0^0)/(1 - \Re\Pi^0(m_K^2)) - c_4 \cos\delta_0^2 \cdot \exp(i\delta_0^2)]
\end{aligned} \tag{5}$$

The rest modes of $K \to 2\pi$ decays undergo the corresponding changes. Comparing the calculated widths of $K \to 2\pi$ decays with the experimental ones, we come to the result:

$$(c_1 - c_2 - c_3 + \frac{32}{9}\beta\tilde{c}_5) = -13.845; \quad c_4 = 0.3212 \quad \text{and} \quad \frac{32}{9}\beta\tilde{c}_5 = -8.0758, \tag{6}$$

if $(c_1 - c_2 - c_3) = -2.89$ (see [8]) is used.

The higher-order momentum corrections and rescattering ot final pions transform the initial $K^+ \to \pi^+\pi^+\pi^-$ amplitude into

$$\begin{aligned}
&M^{corr}(K^+ \to \pi^+\pi^+\pi^-) = \tfrac{1}{2\sqrt{2}} G_F \sin\theta_C \cos\theta_C \tfrac{2m_K^2}{3} 0.96[(c_1 - c_2 - c_3 - c_4) \\
&\cdot 1.67012(1 - 0.12694Y) + \tfrac{32}{9}\beta\tilde{c}_5 \cdot 1.001(1 - 0.12844Y) + 0.09150Y \\
&+9c_4 \tfrac{3\mu^2}{2m_K^2} Y] \exp(i8.53^\circ),
\end{aligned} \tag{7}$$

where $Y = (s_3 - s_0)/\mu^2$. At the end we obtain

$$\Gamma^{th}(K^+ \to \pi^+\pi^+\pi^-) = 2.8713 \cdot 10^{-9}\text{eV}. \tag{8}$$

Comparing this result with $\Gamma^{exp} = 2.9709 \cdot 10^{-9}$eV we come to conclusion that calculating the $K \to 3\pi$ amplitudes basing on data on $K \to 2\pi$ decays, it is necessary to take into account, besides the higher-order momentum terms in expansion of these amplitudes, the effect of FSI too. Our approach gave also the result g_{++-} =-0.2178 coinciding with the experimental one: $g^{exp}_{++-} = -(0.2154 \pm 0.0035)$.

[1] A.I. Vainshtein, V.I. Zakharov, *Usp. Fiz. Nauk* **100**, 225 (1970).
[2] E.P. Shabalin, *Yad. Fiz.* **48**, 271 (1988).
[3] J. Kambor, J. Missimer and D. Wyler, *Phys. Lett. B* **261**, 496 (1991); J. Bijnens, P. Dhonte and F. Persson, *Nucl. Phys. B* **648**, 317 (2003).
[4] T.N. Truong, *Phys. Lett. B* **207**, 495 (1988).
[5] E.P. Shabalin, *Phys. At. Nucl.* **71**, 1036 (2008).
[6] E.P. Shabalin, *Phys. At. Nucl.* **63**, 594 (2000).
[7] M.A. Shifman, A.I. Vainshtein and V.I. Zakharov, *Nucl.Phys. B* **210**, 316 (1977).
[8] L.B. Okun, *"Leptons and Quarks"*(North-Holland Publ.Co.) pp. 315, 323, 1982.

Hadron Physics

AXIAL ANOMALY AND STRANGE QUARKS

Oleg Teryaev [a]
Bogoliubov Laboratory of Theoretical Physics,
Joint Institute for Nuclear Research, 141980 Dubna, Russia

Abstract. The axial (related to axial anomaly) and vector currents of heavy quarks are considered. The special attention is payed to the strangeness polarization mediated by gluon anomaly and treatment of the strange quarks in a heavy ones in a multiscale nucleon. It is shown that the straightforward modification of Heisenberg-Euler effective lagrangian allows to calculate the vector current of strange quarks and describes an analog of Chiral Magnetic Effect for strange and heavy quarks.

1 Introduction

The spin structure of nucleon is a major problem since EMC Spin crisis (puzzle) emerged in 80's. The first observation [1] was related to the role of gluon anomaly which was interpreted as a (circular) gluon polarization. The extensive experimental investigations at HERMES, COMPASS and RHIC, however, did not find the significant polarization. Sometimes this is described as an "absence" of anomaly which is quite strange because of the fundamental character of this phenomenon. Because of this, I am going to discuss the manifestation of anomaly through strange quarks polarization mediated by very small polarizations of off-shell gluons. the similar consideration of vector currents leads to the description of charge separation of strangeness in heavy ions collisions.

2 Axial current of Strange quarks, gluonic anomaly and strangeness polarization

The divergence of the singlet axial current contains a normal and an anomalous piece,

$$\partial^{\mu} j_{5\mu}^{(0)} = 2i \sum_q m_q \bar{q} \gamma_5 q - \left(\frac{N_f \alpha_s}{4\pi} \right) G^a_{\mu\nu} \widetilde{G}^{\mu\nu,a}, \tag{1}$$

where N_f is the number of flavours. The two terms at the r.h.s. of the last equation are known to cancel in the limit of infinite quark mass. This is the so-called cancellation of physical and regulator fermions, related to the fact, that the anomaly may be regarded as a usual mass term in the infinite mass limit, up to a sign, resulting from the subtraction in the definition of the regularized operators. Consequently, one should expect, that the contribution of infinitely heavy quarks to the first moment of g_1 is zero. This is exactly what happens [2] in a perturbative calculation of the triangle anomaly graph. One may wonder, what is the size of this correction for large, but finite masses and how does

[a] e-mail: teryaev@theor.jinr.ru

it compare with the purely perturbative result. To answer this question, one should calculate the r.h.s. of (1) for heavy fermions. The leading coefficient is of the order m^{-2},

$$\partial^\mu j^c_{5\mu} = \frac{\alpha_s}{48\pi m_c^2}\partial^\mu R_\mu \tag{2}$$

where

$$R_\mu = \partial_\mu\left(G^a_{\rho\nu}\tilde{G}^{\rho\nu,a}\right) - 4\left(D_\alpha G^{\nu\alpha}\right)^a\tilde{G}^a_{\mu\nu}\,. \tag{3}$$

The contribution [3] of heavy (say, charm) quarks to the nucleon forward matrix element is

$$\langle N(p,\lambda)|j^{(c)}_{5\mu}(0)|N(p,\lambda)\rangle = \frac{\alpha_s}{48\pi m_c^2}\langle N(p,\lambda)|R_\mu(0)|N(p,\lambda)\rangle \tag{4}$$

Note that the first term in R_μ does not contribute to the forward matrix element because of its gradient form, while the contribution of the second one is rewritten, by making use of the equation of motion, as matrix element of the operator

$$\begin{aligned}\langle N(p,\lambda)|j^{(c)}_{5\mu}(0)|N(p,\lambda)\rangle &= \tfrac{\alpha_s}{12\pi m_c^2}\langle N(p,\lambda)|g\textstyle\sum_{\mathrm{f=u,d,s}}\bar{\psi}_f\gamma_\nu\tilde{G}_\mu{}^\nu\psi_f|N(p,\lambda)\rangle\\ &\equiv \tfrac{\alpha_s}{12\pi m_c^2}2m_N^3 s_\mu f_S^{(2)},\end{aligned} \tag{5}$$

The parameter $f_S^{(2)}$ appears in calculations of the power corrections to the first moment of the singlet part of g_1 part of which is given by exactly the quark-gluon-quark matrix element we got. The non-perturbative calculations are resulting in the estimate

$$\bar{G}^c_A(0) = -\frac{\alpha_s}{12\pi}f_S^{(2)}\left(\frac{m_N}{m_c}\right)^2 \approx -5\cdot 10^{-4}.$$

The seemingly naive application of this approach to the case of strange quarks was presented already ten years from now [3] giving for their contribution to the first moment of g_1 roughly $-5\cdot 10^{-2}$, which is compatible with the experimental data which is preserved also now despite the problem of matching DIS and SIDIS analyses.

At that time the reason for such a success which was mentioned in [3] was the possible applicability of a heavy quark expansions for strange quarks [4,5] in the case of the vacuum condensates of heavy quarks. That analysis was also related to the anomaly equation for heavy quarks, however, for the trace anomaly, rather than the axial one. The current lattice data provide somewhat smaller result of the same order. At the same time, the lattice simulations of scalar matrix elements [6] are in the case of strange quarks fact surprisingly close to what ones should get in the heavy quark limit. The current understanding may also include what I would call "multiscale" picture of nucleon with (squared) strange quark mass (and Λ) being much smaller than that of nucleon but much larger than (genuine)higher twist parameter. Whether the smallness of higher twists holds for consecutive terms in the series of higher twists may be checked by use of very accurate JLAB data for g_1. The simplest case of course is the non-singlet combination related to Bjorken Sum Rule. As higher twists are more pronounced at low Q^2, one should take care on the Landau singularities which may be achieved by use of Analytic Perturbation Theory (the main experts in which are here) or Simonov's soft freezing. The result [7] looks like a first terms of converging series of higher twists compatible with semiclassical picture.

It is instructive to compare the physical interpretation of gluonic anomaly for massless and massive quarks. While in the former, most popular, case it corresponds to the circular polarization of on-shell gluons (recall, that it is rather small, according to various experimental data), in the case of massive quarks one deals with very small (because of small higher twist strength) correlation of nucleon polarization and a sort of polarization of *off-shell* gluons. As soon as strange quark mass is not very large, it partially compensated the smallness of higher twist and this gluon polarization is transmitted to the non-negligible strange quark polarization. The role of gluonic anomaly is therefore to produce the "anomaly-mediated" strangeness polarization.

3 Vector current of Strange quarks and Chiral Magnetic Effect in Heavy Ions Collisions

The calculation of vector rather than axial current of heavy (and strange) quarks appears to be even easier. The answer is actually contained in the classical Heisenberg-Euler effective lagrangian for light-by light scattering. Calculating its variation with respect to the electromagnetic potential one immediately get the expression for the current. The transition from QED to QCD is performed by the substitution of three of quark-photon vertices by the quark-gluon ones. The C-parity ensures that the result contains only the symmetric $SU(3)$ structure d^{abc} and is proportional to Abelian one which was explored earlier [9]. The notion of this current allows to determine the strangeness contribution to the anomalous magnetic moment of the nucleon and to the mean square radius of the pion.

Another interesting manifestation of strange quark vector current emerges if one substitutes [8] the *two* rather than three quark-photon vertices by quark-gluon ones. The results describes the vector current induced by cooperative action of (two) gluonic and electromagnetic fields. Its physical realization corresponds, in particular, to heavy ions collisions where extremely strong magnetic fields are generated. The most interesting contributions comes from the term

$(F\tilde{F})^2$ in Heisenberg-Euler lagrangian, leading to the current

$$j^s_\mu = \frac{7\alpha\alpha_s}{45m_s^4}\tilde{F}_{\mu\nu}\partial^\nu(G\tilde{G}) \tag{6}$$

One may easily recognize here the analog of the famous Chiral Magnetic Effect (CME) (see [10] and Ref. therein). The correspondence is manifested by the substitution

$$\frac{1}{m_s^4}\partial^\nu(G\tilde{G}) \to \partial^\nu \int d^4z(G\tilde{G}) \to \partial^\nu\theta.$$

The latter requires the appearance of two scales, now in multiscale medium, (the ones corresponding to integration and taking the derivative) and leads to the derivative of topological field θ. The later property makes the interpretation of lattice simulations [11, 12] ambiguous. Note also that the calculated effect for heavy quarks is not directly related neither to topology nor to chirality. In particular, it is present also when all the three fields are electromagnetic ones, which may be of physical interest, as soon as the electromagnetic and chromodynamical fields are of the same order in heavy ions collisions.

Another possibility arises when coupling to electromagnetic field is substituted by the coupling to medium velocity $\vec{v}$, as suggested recently [13], which is accounted for by the substitution $e\vec{H} \to \mu rot\vec{v}$. This charge separation due to vorticity (appearing also for light quarks and described by the same substitution) may be called "Chiral vorticity effect" providing another source (vorticity of medium instead of magnetic field) of the same observable phenomenon.

4 Conclusions

The role of gluonic axial anomaly in nucleon spin structure may be quite different from what was expected before. Instead of inducing circular polarization of on-shell gluons it can mediate the nucleon polarization to strange quarks via the tiny high twist correlations of off-shell gluons with the nucleon spin. This implies the picture of multiscale nucleon, with strange quark mass (squared) being much smaller than that of nucleon and much larger than higher twist parameters.

In turn, the matrix elements of vector currents in external gluonic and magnetic (or vorticity) fields describe the analog of Chiral Magnetic (or Vorticity) Effect for strange quarks in heavy ions collisions.

Acknowledgments

I am indebted to G. Bali, P. Buividovich, D. Kharzeev, A. Moiseeva, M.Polyakov, M. Polikarpov, A. Schaefer and V.I. Zakharov for discussions. This work was supported in part by the Russian Foundation for Basic Research (grants No.

09-02-01149, 09-02-00732-), and the Russian Federation Ministry of Education and Science (grant No. MIREA 2.2.2.2.6546).

References

[1] A.V. Efremov and O.V. Teryaev, Report JINR-E2-88-287, Czech.Hadron Symp.1988, p.302.

[2] A.V. Efremov, J. Soffer and O.V. Teryaev, Nucl.Phys. B346 (1990) 97

[3] M. V. Polyakov, A. Schafer and O. V. Teryaev, Phys. Rev. D **60**, 051502 (1999) [arXiv:hep-ph/9812393].

[4] M.A. Shifman, A.I. Vainshtein and V.I. Zakharov, Nucl.Phys. B147 (1979) 385 (section 6.8);

[5] D.J. Broadhurst and S.C. Generalis, Phys. Lett. B139 (1984) 85.

[6] G. Bali, S. Collins and A. Schafer [QCDSF Collaboration], arXiv:0911.2407 [hep-lat].

[7] R. S. Pasechnik, D. V. Shirkov and O. V. Teryaev, Phys. Rev. D **78**, 071902 (2008) [arXiv:0808.0066 [hep-ph]]; R. S. Pasechnik et al., These Proceedings.

[8] O.V. Teryaev, Abstracts of the International Bogolyubov Conference (Moscow-Dubna, August 21-27 2009), p. 181.

[9] D. B. Kaplan and A. Manohar, Nucl. Phys. B **310**, 527 (1988).

[10] D. E. Kharzeev, arXiv:0911.3715 [hep-ph].

[11] P. V. Buividovich, E. V. Lushchevskaya, M. I. Polikarpov and M. N. Chernodub, JETP Lett. **90** (2009) 412 [Pisma Zh. Eksp. Teor. Fiz. **90** (2009) 456].

[12] P. V. Buividovich, M. N. Chernodub, E. V. Luschevskaya and M. I. Polikarpov, Nucl. Phys. B **826**, 313 (2010) [arXiv:0906.0488 [hep-lat]].

[13] D. T. Son and P. Surowka, Phys. Rev. Lett. **103**, 191601 (2009) [arXiv:0906.5044 [hep-th]].

STRONG MAGNETIC FIELDS IN LATTICE QCD

P. V. Buividovich [a], E. V. Luschevskaya [b], M. I. Polikarpov [c]
Institute for Theoretical and Experimental Physics, 117218 Russia, Moscow, B. Cheremushkinskaya str. 25
M. N. Chernodub [d]
CNRS, LMPT, Fédération Denis Poisson, Université de Tours, 37200 France
DMPA, University of Gent, Krijgslaan 281, S9, B-9000 Gent, Belgium
Institute for Theoretical and Experimental Physics, 117218 Russia, Moscow, B. Cheremushkinskaya str. 25

*Abstract.*Hadronic matter in very strong (hadron-scale) magnetic fields acquires many new interesting properties. Some of these properties can be studied with the help of lattice simulations in quenched lattice gauge theory. One of the interesting effects is the chiral magnetic effect, that is, the enhancement of the fluctuations of currents and charge densities by the magnetic field. By virtue of the Green-Kubo dispersion relations, this enhancement of fluctuations corresponds to the enhancement of the electric conductivity of hadronic matter. We review here the results of our recent studies of this phenomenon.

1 Introduction

Very strong magnetic fields of the hadronic scale can significantly modify the properties of strongly interacting matter: they change the order of the phase transition from the confinement phase to the quark-gluon plasma, shift the position of the transition line [1], etc. At the Relativistic Heavy Ion Collider (RHIC) in noncentral heavy-ion collisions the strong magnetic field arises due to the relative motion of the ions and products of the collision [2]. The induced magnetic field is perpendicular to the reaction plane. At first moments ($\tau \sim 1$ fm/c) of the collision the value of magnetic field at RHIC may reach the hadronic scale, $\sqrt{qB} \sim (10 - 100 \text{ MeV})$ [2,3]. Such strong magnetic fields can also be created in future at the ALICE experiment at LHC, at the Facility for Antiproton and Ion Research (FAIR) at GSI, and at the Nuclotron Ion Collider fAcility (NICA) in Dubna.

The so-called chiral magnetic effect (CME) is the generation of a local electric current in the direction of the external magnetic field in topologically nontrivial configurations of the gauge fields [2]. If we consider u and d-quarks as massless particles then the right-handed quarks should move in the direction of the magnetic field and the left-handed quarks should move in the opposite direction because in the external field magnetic moments of quarks are parallel to the direction of the field. Nonzero topological charge of gauge fields leads to a local imbalance between left-handed and right-handed quarks, which, in turn, gives

[a]e-mail: buividovich@itep.ru
[b]e-mail: luschevskaya@itep.ru
[c]e-mail: polykarp@itep.ru
[d]e-mail: maxim.chernodub@lmpt.univ-tours.fr

rise to a nonzero net electric current along the axis of the magnetic field. Since the topological charge of gauge fields fluctuates, with negative and positive values of the charge appearing with equal probabilities, the averaged net current is of course zero. However, such a local "$\mathcal{CP}$ - violation" can be observed as the enhancement of current fluctuations in the direction of the magnetic field. This enhancement can have several experimental consequences: one is the direct observation of the enhanced fluctuations in the numbers of positively and negatively charged particles in the direction perpendicular to the reaction plane [4], and the other is the enhanced production of soft photons and soft dilepton pairs [5].

A natural question to ask is whether this enhancement of electric current and charge fluctuations along the direction of magnetic field corresponds to a real flow of current, or is just caused by short-lived quantum fluctuations. This question can be answered by studying the current-current correlation functions. The currents which correspond to a real transport of charged particles should have long-range correlations in time, while quantum fluctuations are typically characterized by a finite correlation time [6]. Recalling Green-Kubo relations, one can see that this property is intimately related to the electric conductivity - namely, the real transport of charged particles can occur only in conducting media. In this paper we try to find out whether the currents induced due to the CME correspond to a real transport of charged particles by studying the tensor of electric conductivity of the vacuum. Here we consider the vacuum of quenched $SU(2)$ lattice gauge theory in external magnetic field. We find that the magnetic field induces nonzero electric conductivity along its direction, transforming the confining vacuum from an insulator into an anisotropic conductor.

2 Details of simulations

We generate statistically independent SU(2) gauge field configurations with the tadpole improved Symanzik action [7]. Using the chirally invariant Dirac operator [8], we solve the Dirac equation $D\psi_k = \lambda_k \psi_k$ numerically and determine the corresponding eigenfunctions ψ_k and eigenvalues λ_k. Here $D = \gamma^\mu(\partial_\mu - iA_\mu)$ is the massless Dirac operator in the gauge field A_μ. The uniform magnetic field is introduced into this operator as described in [9].

We performed the zero-temperature simulations on 14^4 lattice with the lattice spacing $a = 0.103$ fm and on 16^4 lattice with the lattice spacings and $a = 0.103$ fm and $a = 0.089$ fm. The values of the magnetic field are quantized due to the periodic boundary conditions imposed in finite lattice volume. In our simulations a minimal nonzero value of magnetic field is $qB_{min} = (348 \text{ MeV})^2$.

3 Local fluctuations of electromagnetic current

In order to study the simplest signatures of the chiral magnetic effect we first explore the *local* fluctuations of the electromagnetic current:

$$\langle j_\mu^2 \rangle_{IR}(B,T) = \frac{1}{V}\int_V d^4x \langle j_\mu^2(x)\rangle_{B,T} - \frac{1}{V}\int_V d^4x \langle j_\mu^2(x)\rangle_{B=0,T=0}. \qquad (1)$$

We estimate the local squares of the currents by applying the Wick theorem to fermion bilinears and by inverting the Dirac operator on the subspace spanned by some number $M \sim 20$ of its lowest eigenmodes. This method allows us to study the contributions of both connected quark diagrams and the "hairpin" diagrams. It should be stressed, however, that such method leads to uncontrollable systematic errors [10].

We checked on the lattice, that all components of the electric current (**??**) are on average equal to zero within statistical error bars. The component of the current along the direction of the magnetic field is called longitudinal. Transverse components are perpendicular to the direction of the field. At zero temperature only the magnetic field in the $\mu = 3$ direction breaks the rotational symmetry, so that we have: $\langle j_1^2 \rangle = \langle j_2^2 \rangle$ and $\langle j_0^2 \rangle = \langle j_3^2 \rangle$.

In Fig. 1 we show the average squares of all components of the electric current. All of them grow with the strength of the field. The longitudinal component of the electric current and the electric charge fluctuate stronger than the transverse ones. The fluctuations of the spatially transverse components grow with the field because the transverse momentum of a quark occupying the lowest Landau level increases with the strength of the magnetic field.

4 Current-current correlator and electric conductivity of the quenched $SU(2)$ lattice gauge theory

By virtue of the Green-Kubo relations, electric conductivity tensor $\sigma_{ij}(w)$ can be extracted from the low-frequency limit of the spectral function $\rho_{ij}(w)$ which corresponds the correlator of two vector currents $j_i(x) = \bar{q}(x)\gamma_i q(x)$:

$$G_{ij}(\tau) = \int d^3\vec{x} \langle j_i(\vec{0},0)\, j_j(\vec{x},\tau)\rangle, \qquad (2)$$

$$G_{ij}(\tau) = \int_0^{+\infty} \frac{dw}{2\pi} K(w,\tau)\,\rho_{ij}(w), \qquad (3)$$

$$K(w,\tau) = \frac{w}{2T}\frac{\cosh\left(w\left(\tau - \frac{1}{2T}\right)\right)}{\sinh\left(\frac{w}{2T}\right)}, \qquad (4)$$

$$\sigma_{ij} = \lim_{\omega\to 0}\frac{\rho_{ij}(\omega)}{4T}. \qquad (5)$$

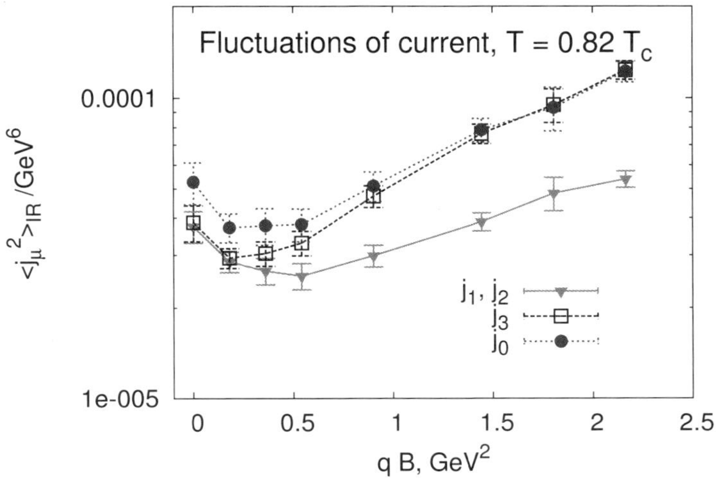

Figure 1: The squares of the longitudinal and transverse components of the electric currents vs the external magnetic field in the confinement phase.

The correlators $G_{ij}(\tau)$ in (2) are shown on Fig. 2 on the left and the corresponding spectral functions - on Fig. 2 on the right. Here we consider only the zero-temperature case, since above the deconfinement phase transition hadronic matter is a conductor even at zero magnetic field. In order to measure the correlators, we directly invert the Dirac operator by using the SHUMR algorithm proposed in [11].

It can be seen that the correlator of the longitudinal components of the current decays decays slower when the magnetic field is turned on. The corresponding spectral functions acquire nonzero value in the low-frequency limit, which, according to (5), corresponds to nonzero electric conductivity of the QCD vacuum.

5 Conclusion

We have studied the signatures of the chiral magnetic effect in lattice $SU(2)$ gauge theory with strong external magnetic field. The effect manifests itself both in the growth of the local fluctuations of the electromagnetic current, $\langle j_i(x) j_k(x) \rangle$, in the direction of the magnetic field, and in the enhancement of the electric conductivity in this direction.

Finally, let us comment on possible experimental consequences of the phenomenon described above. The expectation value $\langle j_k(x) j_l(y) \rangle$ is related to the polarization of soft photons and to the angular distribution of soft photons

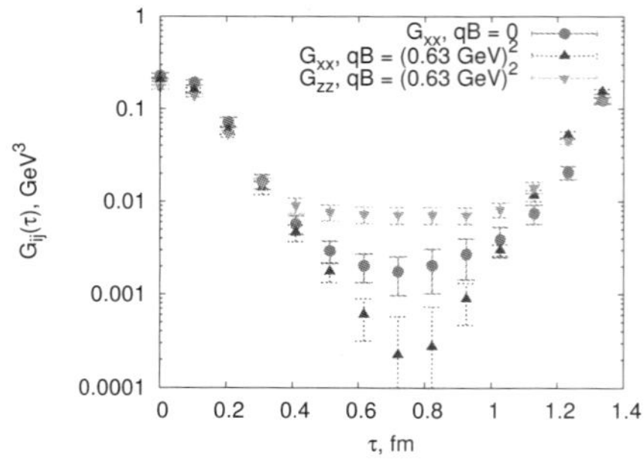

Figure 2:

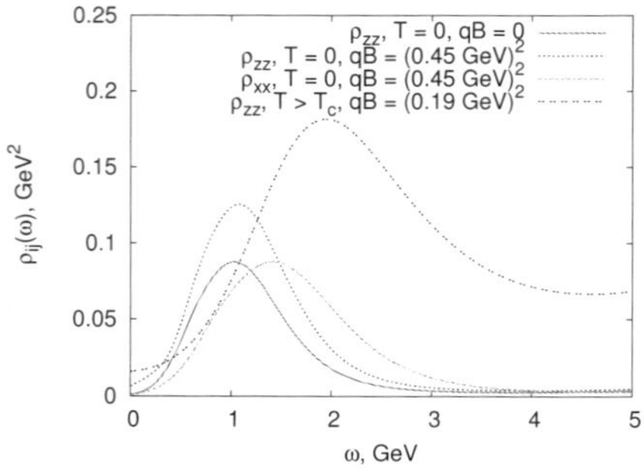

Figure 3:

and dilepton pairs emitted in the collision process [5,12,13]. One can therefore expect an enhancement of production rate in the direction perpendicular to the reaction plane, possibly resulting in a negative elliptic flow for soft photons and dilepton pairs.

Acknowledgments

The authors are grateful to Ph. Boucaud, V.G. Bornyakov, V.V. Braguta, A.S. Gorsky, B.L. Ioffe, B.O. Kerbikov, D. Kharzeev, A. Krikun, S.M. Morozov, V.A. Novikov, B. Pire, V.I. Shevchenko, M.I. Vysotsky, and V.I. Zakharov for interesting discussions and suggestions. This work was partly supported by Grants RFBR Nos. 06-02-04010-NNIO-a, 08-02-00661-a, 06-02-17012, 09-02-00338- and DFG-RFBR 436 RUS, a grant for scientific schools No. NSh-6260.2010.2, by the Russian Federal Agency for Nuclear Power, and by the STINT Institutional grant IG2004-2 025.s. P.B. was partially supported by a personal grant from the "Dynasty" foundation. The calculations were partially done on the MVS 50K at Moscow Joint Supercomputer Center.

References

[1] N. O. Agasian and S. M. Fedorov. Quark-hadron phase transition in a magnetic field. *Phys. Lett. B*, 663:445 – 449, 2008.

[2] D. E. Kharzeev, L. D. McLerran, and H. J. Warringa. The effects of topological charge change in heavy ion collisions: Event by event P and CP violation. *Nucl. Phys. A*, 803:227, 2008.

[3] V. Skokov, A. Illarionov, and V. Toneev. Estimate of the magnetic field strength in heavy-ion collisions. *Int. J. Mod. Phys. A*, 24:5925, 2009.

[4] S. A. Voloshin for the STAR Collaboration. Probe for the strong parity

violation effects at RHIC with three-particle correlations. Proceedings of Quark Matter 2008, 2008.

[5] E. L. Bratkovskaya, O. V. Teryaev, and V. D. Toneev. Anisotropy of dilepton emission from nuclear collisions. *Phys. Lett. B*, 348:283 – 289, 1995.

[6] L. P. Kadanoff and P. C. Martin. Hydrodynamic equations and correlation functions. *Ann. Phys.*, 24:19, 1963.

[7] V. G. Bornyakov, E. V. Luschevskaya, S. M. Morozov, M. I. Polikarpov, E.-M. Ilgenfritz, and M. Müller-Preussker. The topological structure of SU(2) gluodynamics at T>0 : an analysis using the Symanzik action and Neuberger overlap fermions. *Phys. Rev. D*, 79:054505, 2009.

[8] H. Neuberger. Exactly massless quarks on the lattice. *Phys. Lett. B*, 417:141, 1998.

[9] P. V. Buividovich, M. N. Chernodub, E. V. Luschevskaya, and M. I. Polikarpov. Numerical study of chiral symmetry breaking in non-abelian gauge theory with background magnetic field, 2008.

[10] P. Hasenfratz, S. Hauswirth, K. Holland, T. Jorg, and F. Niedermayer. Testing the fixed-point QCD action and the construction of chiral currents. *Nucl. Phys. B*, 643:280, 2002.

[11] A. Borici and A. Allkoci. Shifted unitary orthogonal methods for the overlap inversion. *PoS*, LAT2005:101, 2005.

[12] L. D. McLerran and T. Toimela. Photon and dilepton emission from the quark-gluon plasma: Some general considerations. *Phys. Rev. D*, 31:545 – 563, 1985.

[13] S. Gupta. The electrical conductivity and soft photon emissivity of the QCD plasma. *Phys. Lett. B*, 597:57, 2004.

HADRONIC EFFECTS IN LOW–ENERGY QCD: ADLER FUNCTION AND τ DECAY

A.V. Nesterenko

BLTPh, Joint Institute for Nuclear Research, Dubna, 141980, Russia

Abstract. It is shown that the nonperturbative effects due to hadronization play a crucial role in a number of strong interaction processes. In particular, these effects impose a stringent constraint on the infrared behavior of the Adler function and play an essential role in theoretical analysis of the inclusive τ lepton decay.

1 Introduction

The Adler function $D(Q^2)$ [1] plays a key role in elementary particle physics. Specifically, theoretical description of some strong interaction processes and hadronic contributions to some electroweak observables is inherently based on this function. Furthermore, theoretical analysis of timelike experimental data can be performed by making use of relevant dispersion relations only. In particular, dispersion relation for the Adler function [1]

$$D(Q^2) = Q^2 \int_{4m_\pi^2}^{\infty} \frac{R(s)}{(s+Q^2)^2}\, ds \tag{1}$$

represents the required link between experimentally measurable R–ratio of e^+e^-–annihilation into hadrons $R(s)$ and theoretically computable $D(Q^2)$.

The ultraviolet behavior of the Adler function can be approximated by the power series in the strong running coupling within perturbation theory [2]

$$D^{(\ell)}_{\text{pert}}(Q^2) = 1 + \sum_{j=1}^{\ell} d_j \left[\alpha_{\text{s}}^{(\ell)}(Q^2)\right]^j, \qquad Q^2 \to \infty \tag{2}$$

(the overall factor $3\sum_f Q_f^2$ is omitted throughout). Here $\alpha_{\text{s}}^{(\ell)}(Q^2)$ denotes the ℓ–loop perturbative QCD invariant charge, $\alpha_{\text{s}}^{(1)}(Q^2) = 4\pi/[\beta_0 \ln(Q^2/\Lambda^2)]$, $\beta_0 = 11 - 2n_{\text{f}}/3$, and $d_1 = 1/\pi$. However, expansion (2) is invalid at low energies and it is inconsistent with the dispersion relation (1) due to unphysical singularities of the strong running coupling $\alpha_{\text{s}}(Q^2)$ in the infrared domain.

2 Novel integral representation for the Adler function

Dispersion relation (1) imposes stringent physical nonperturbative constraints on the Adler function. Specifically, Eq. (1) implies that $D(Q^2)$ possesses the only cut $Q^2 \le -4m_\pi^2$ and that $D(Q^2)$ vanishes in the infrared limit $Q^2 \to 0$. These nonperturbative constraints have been merged with perturbative approximation (2) in Refs. [3, 4], that led to the following integral representations:

$$D(Q^2) = \frac{Q^2}{Q^2+4m_\pi^2}\left[1 + \int_{4m_\pi^2}^{\infty} \rho_{\text{D}}(\sigma)\frac{\sigma - 4m_\pi^2}{\sigma + Q^2}\frac{d\sigma}{\sigma}\right], \tag{3}$$

$$R(s) = \theta(s - 4m_\pi^2)\left[1 + \int_s^\infty \rho_{\rm D}(\sigma)\,\frac{d\sigma}{\sigma}\right]. \tag{4}$$

In Eqs. (3) and (4) $\rho_{\rm D}(\sigma)$ denotes the relevant spectral density and $\theta(x)$ is the unit step function. The developed approach [3,4] eliminates some intrinsic difficulties of perturbation theory and extends its range of applicability towards infrared domain. It is worth noting that in the massless limit ($m_\pi = 0$) expressions (3) and (4) become identical to those of the Analytic Perturbation Theory [5,6]. However, it is crucial to keep the pion mass nonvanishing, since it can be safely neglected only when one handles the strong interaction processes at high energies (see Refs. [3,4] for the details).

In Eqs. (3) and (4) the one–loop spectral function obtained in Ref. [7]

$$\rho^{(1)}(\sigma) = \frac{1 + \Lambda^2/\sigma}{\ln^2(\sigma/\Lambda^2) + \pi^2} \tag{5}$$

will be employed in what follows. The Adler function (3), corresponding to the spectral function (5), is presented in Fig. 1 by solid curve, whereas its one–loop perturbative approximation (2) is shown by the dot–dashed curve. As one may infer from Fig. 1, the obtained result for the Adler function is in a reasonable agreement with its experimental prediction (denoted by the shaded band) in the entire energy range $0 \le Q^2 < \infty$.

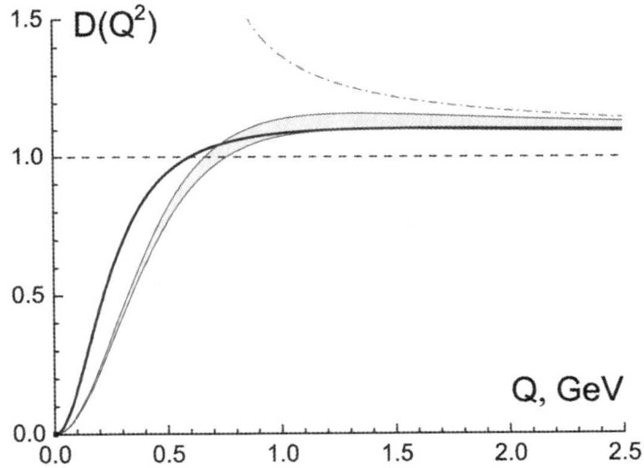

Figure 1: Adler function (3) corresponding to the spectral density (5) (solid curve) ($\Lambda = 441\,$MeV, $n_{\rm f} = 3$), its perturbative approximation (2) (dot–dashed curve), and its experimental prediction (shaded band).

3 Inclusive τ lepton decay

Let us proceed with the nonstrange part of the inclusive semileptonic branching ratio associated with the vector quark currents (see paper [8] for details)

$$R_{\tau,{\rm V}} = \frac{3}{2}\,|V_{\rm ud}|^2 S_{\rm EW}\,(\Delta_{\rm QCD} + \delta'_{\rm EW})\,. \tag{6}$$

Its experimental measurement [9] yields $R_{\tau,{\rm V}} = 1.764 \pm 0.016$. In Eq. (6) $|V_{\rm ud}| = 0.97418 \pm 0.00027$ denotes the CKM matrix element, $S_{\rm EW} = 1.0194 \pm 0.0050$ and $\delta'_{\rm EW} = 0.0010$ are the electroweak corrections [8,10], and

$$\Delta_{\rm QCD} = 2\int_0^{M_\tau^2}\left(1 - \frac{s}{M_\tau^2}\right)^2\left(1 + 2\,\frac{s}{M_\tau^2}\right)R(s)\,\frac{ds}{M_\tau^2}. \tag{7}$$

In the framework of the approach in hand at one–loop level Eq. (7) reads [4]

$$\Delta_{\rm QCD} = 1+d_1\alpha_{\rm TL}^{(1)}(M_\tau^2)-\delta_\Gamma-d_1\delta_\Gamma\alpha_{\rm TL}^{(1)}(m_\Gamma^2)+d_1\,\frac{4\pi}{\beta_0}\int_\chi^1 f(\xi)\,\rho^{(1)}(\xi M_\tau^2)\,d\xi, \quad (8)$$

where $f(\xi)=\xi^3-2\xi^2+2$, $\chi=m_\Gamma^2/M_\tau^2$, $\delta_\Gamma=\chi\,f(\chi)$, and

$$\alpha_{\rm TL}^{(1)}(s)=\frac{4\pi}{\beta_0}\,\theta(s-m_\Gamma^2)\int_s^\infty \rho^{(1)}(\sigma)\,\frac{d\sigma}{\sigma} \quad (9)$$

is the one–loop timelike effective coupling [3,4]. Here $m_\Gamma=m_{\pi^0}+m_{\pi^-}$ stands for the total mass of the lightest allowed hadronic decay mode of τ lepton. Eventually, Eq. (8) results in $\Lambda=(941\pm 86)\,$MeV for $n_{\rm f}=2$ active quarks, that is somewhat larger than the one–loop perturbative estimation $\Lambda=(678\pm 55)\,$MeV. It turns out that the effects due to the nonvanishing hadronic mass m_Γ play a substantial role herein. In particular, in the massless limit ($m_\Gamma=0$) Eq. (8) leads to $\Lambda=(493\pm 56)\,$MeV for $n_{\rm f}=2$ active quarks.

4 Conclusions

The infrared behavior of the Adler function is studied in the framework of Dispersive approach to QCD. It is shown that the effects due to the nonvanishing mass of the lightest hadron state play a substantial role in processing the experimental data on the inclusive τ lepton decay.

This work was partially supported by the grants RFBR-08-01-00686, BRFBR-JINR-F08D-001, and NS-1027.2008.2.

References

[1] S.L. Adler, *Phys. Rev.* **D** 10, 3714 (1974).
[2] P.A. Baikov, K.G. Chetyrkin, and J.H. Kuhn, *Phys. Rev. Lett.* 101, 012002 (2008).
[3] A.V. Nesterenko and J. Papavassiliou, *J. Phys.* **G** 32, 1025 (2006).
[4] A.V. Nesterenko, *Nucl. Phys. B (Proc. Suppl.)* 186, 207 (2009).
[5] D.V. Shirkov and I.L. Solovtsov, *Phys. Rev. Lett.* 79, 1209 (1997); *Theor. Math. Phys.* 120, 1220 (1999); 150, 132 (2007).
[6] K.A. Milton, I.L. Solovtsov, and O.P. Solovtsova, *Phys. Rev.* **D** 64, 016005 (2001); 65, 076009 (2002); *Mod. Phys. Lett.* **A** 21, 1355 (2006).
[7] A.V. Nesterenko, *Phys. Rev.* **D** 62, 094028 (2000); 64, 116009 (2001).
[8] E. Braaten, S. Narison, and A. Pich, *Nucl. Phys.* **B** 373, 581 (1992).
[9] K. Ackerstaff *et al.* (OPAL Collaboration), *Eur. Phys. J.* **C** 7, 571 (1999).
[10] W.J. Marciano and A. Sirlin, *Phys. Rev. Lett.* 61, 1815 (1988); E. Braaten and C.S. Li, *Phys. Rev.* **D** 42, 3888 (1990).

HELICITY QUARK DISTRIBUTIONS FROM DIS AND SIDIS MEASURED IN COMPASS

Dmitry Peshekhonov [a]
Laboratory of High Energy Physics, JINR, 141980 Dubna, Russia

Abstract

The COMPAS experiment at the CERN SPS has a broad physics program focused on the nucleon spin structure and hadron spectroscopy, using muon and hadron beams, respectively. In this report we present results from COMPASS on the inclusive longitudinal spin-dependent asymmetry A_1 at small and high Q^2, the spin-dependent structure function g1d and the first moment value, semi-inclusive asymmetries and distributions of polarized constituent and sea quarks obtained from 2002-2004 and 2006 data with a 160 GeV longitudinally polarised $^+$ beam and a ^{6}LiD target. Preliminary results on the analysis comprising 2007 data obtained with the same beam and the proton polarised target are discussed.

1 Introduction

The COMPASS experiment results on the spin-dependent asymmetry $A_{1,p,d}$ and structure functions $g_1^{p,d}$(x,Q^2) are presented. Data with deuteron target have been collected during 2002-2004 and 2006 years [1], proton target were used in 2007. COMPASS spectrometer, polarized target and muon beam are described in [2]. Inclusive and semi-inclusive DIS events were selected by cuts on the virtuality of the photon, $Q^2 > 1(\mathrm{GeV/c})^2$, and its fractional energy, $0.1 < \mathrm{y} < 0.9$. Proton and deuteron data have been combined in order to evaluate the non-singlet structure function g_1^{NS}(x,Q^2). The helicity quark distributions Δu, Δd, $\Delta \bar{u}$, $\Delta \bar{d}$ and $\Delta s(=\Delta \bar{s}$) have been obtained in combined analysis [3] of inclusive and semi-inclusive (with identified hadrons) asymmetries. Hadrons were identified as pions or kaons by RICH in the momenta region $10 < \mathrm{p} < 50$ GeV/c. To select the current fragmentation region the cut $\mathrm{z} > 0.2$ were used, and to separate scattered and secondary μ additional cut $\mathrm{z} < 0.85$ was applied. The total statistics of samples obtained with proton and deuteron targets for the inclusive, $\pi^+(\pi^-)$ and $\mathrm{K}^+(\mathrm{K}^-)$ events after all cuts are presented below in millions:
Proton target 92.5, 13.3, 11.8, 3.9, 2.6
Deuteron target 135.1, 22.8, 20.5, 4.8, 3.3

[a] e-mail: pdv@mail.cern.ch

2 Results on structure function $g_1^{d,p}$ and NLO QCD fit of g_1^{NS}

The longitudinal cross-section asymmetry can be decomposed into the virtual photon-deuteron(proton) asymmetry $A_1^{d,p}$ and $A_2^{d,p}$ as follows:

$$A_{LL} = D \cdot (A_1^{d(p)} + \eta \cdot A_2^{d(p)}) \approx D \cdot A_1^{d(p)} \tag{1}$$

where the photon depolarization factor D (as well asη) depends on the event kinematics. The term connected with A_2 can be neglected since its very small in the COMPASS kinematical range. The spin dependent structure function $g_1^{d(p)}$ is related to the asymmetry A_1 in the following way:

$$\frac{g_1^{d(p)} = F_2^{d(p)}}{2 \cdot x(1+R) \cdot A_1^{d(p)}} \tag{2}$$

where F_2 and R are spin-independent structure functions. $g_1^{d(p)}$ as a functions of x are shown in Fig.1 superposed to results of previous DIS experiments. The availability of g_1^d and g_1^p data with good and comparable precision at low x gives the opportunity to evaluate the non-singlet structure function g_1^{NS} defined as:

$$\frac{g_1^{NS}(x) = g_1^p(x) - g_1^n(x) = 2 \cdot [g_1^p(x) - g_1^d(x)/(1-3}{2w_D)]} \tag{3}$$

where w_D is the probability of the D state in the deuteron (w_D=0.050.01). An evaluation of the first moment of g_1^{NS}(x) provides a new test of the Bjorken sum rule which is considered as a fundamental result of QCD. The Q^2 dependence of g_1^{NS} is decoupled from the QCD evolution of $\Delta\Sigma$ and ΔG and the fit to the Q^2 evolution requires only small number of parameters to describe the shape of g_1^{NS}(x) at some reference Q^2 The values of $x \cdot g_1^{NS}$(x) at Q^2=3(GeV/c)^2 together with NLO QCD fit is shown in Fig.2. The first moment value:

$$g_1^{NS}(Q^2) = \frac{1}{2}\frac{g_A}{g_V} \cdot C^{NS}(Q^2) \tag{4}$$

leads the value of g_A/g_V=1.30 ± 0.07(stat.) ± 0.10(syst.) at Q^2=3 (GeV/c)^2 it is in perfect agreement with the one derived from neuteron β decay (1.2690.003 [4]). The dominant systematic error due to the uncertainty of the beam polarization (5%).

3 LO QCD analysis for polarized quark densities

The combined analysis of inclusive and of semi-inclusive asymmetries for identified hadrons (pions and kaons) measured on proton and deuteron targets allows to evaluate the helicity distributions for different quark flavours: Δu, Δd, $\Delta\bar{u}$, $\Delta\bar{d}$ and Δs assumed to be equal to $\Delta\bar{s}$. As in the previous COMPASS

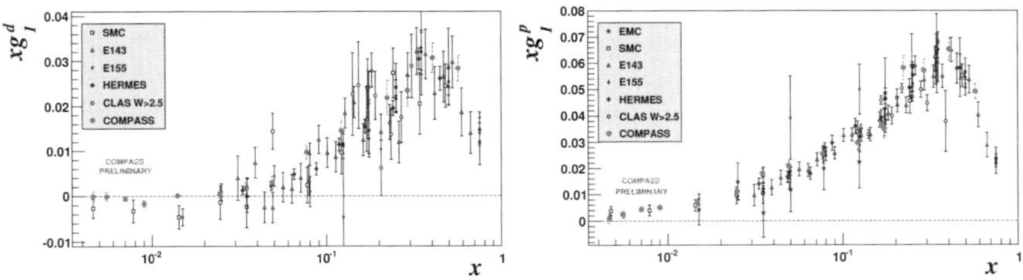

Figure 1: Figure 1: COMPASS result on $xg_1^{d(p)}$ as a function of x Bjorken. A good agreement with previous measurements are seen. Experimental points are shown in measured Q^2.

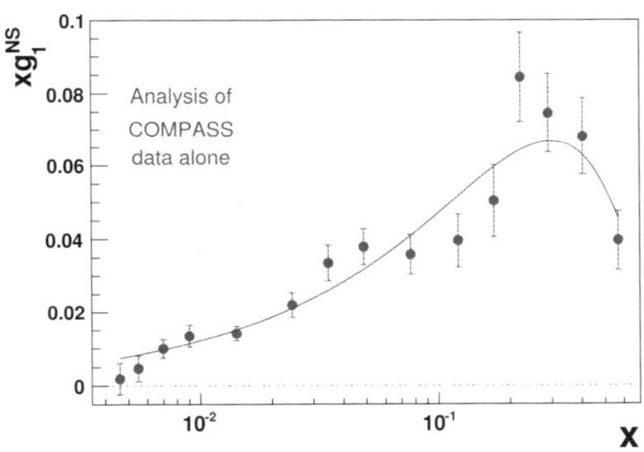

Figure 2: Figure 2: Preliminary values of $g_1^{NS}(x)$ at $Q^2 = 3(GeV/c)^2$. The curve represents the NLO QCD fit to the COMPASS data.

LO analysis [3] assumed that hadrons in the current fragmentation region are produced independent quark fragmentation and their spin asymmetries can be written in terms of PDFs q(x,Q^2) and Δq(x,Q^2) and fragmentation functions D_q^h(z,Q^2):

$$\frac{A_h(x,Q^2) = \Sigma_q e_q^2 \Delta q(x,Q^2) \int D_q^h(z,Q^2)dz}{\Sigma_q e_q^2 q(x,Q^2) \int D_q^h(z,Q^2)dz} \tag{5}$$

Unpolarized PDFs were taken from MRST [5] and the DSS parametrization of the fragmentation functions at LO DCQ was used [6]. The quark helicity distributions are shown in Fig.3. The curves obtained with LO DNS [7] parametrization of polarized PDFs are also shown. The Δu, Δd, $\Delta\bar{u}$, $\Delta\bar{d}$ curves from DNS parametrization fit well our points. However one can see discrepancy in the Δs graph. SIDIS measurement do not support fit behaviour, at least with DSS FFs. More detailes on the distribution Δs(x) are given in the Ref.[3]. The present analysis including 2007 proton data confirms the result obtained with only deuteron data and reduces significantly the statistical errors.

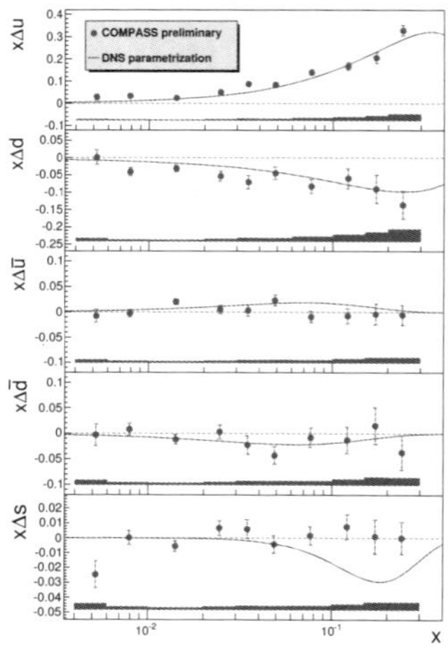

Figure 3: Figure 3: The quark helicity distributions evaluated at common value $Q^2 = 3(GeV/c)^2$ as a function of x Bjorken. Bands at bottom of each graph present systematic uncertainties. Solid markers and bands correspond to COMPASS data. For comparison the NLO DNS parametrization [7] is shown.

4 Conclusion

COMPASS data on the measured inclusive and identified hadron spin asymmetries on the proton and deuteron targets allow to evaluate the non-singlet structure function g_1^{NS}(x,Q^2). The first moment of this function confirms the validity of the Bkorken sum rule with better than one σ precision. The LO QCD evaluation of polarized quark densities was also presented. The full set of polarized PDFs Δu, Δd, $\Delta \bar{u}$, $\Delta \bar{d}$ and Δs were extracted. All sea quark densities are found to be compatible with zero in the full range of measurements. The discrepancy in a shape of Δs(x) with the one obtained in a typical qlobal QCD fit was found.

References

[1] V.Yu.Alexakhin et al., *Phys.Lett.* **B** 647, 330 (2007).
[2] P.Abbon et al., *Nucl.Instr.Meth.* **A** 577, 455 (2007).
[3] M.Alexeev et al., *hep-e[:0905.2828.*
[4] W.M.Yao et al., *Journal of Physics* **G** 33, 1 (2006).
[5] A.Martin, W.J.Stirling and R.S.Thorne, *Phys.Lett.* **B** 636, 259 (2006).
[6] D.de Florian, R.Sassot, M.Stratmann, *Phys.Rev.* **D** 75, 114010 (2007);**D** 76, 074033(2007).
[7] D.de Florian, R.Sassot, M.Stratmann, *Phys.Rev.* **D** 75, 094018 (2005).

SEMI-INCLUSIVE PION ELECTROPRODUCTION WITH CLAS

Mikhail Osipenko [a]
INFN, sez. di Genova, 16146 Genova, Italy,
SINP, Moscow State University, 119991 Moscow, Russia

Abstract. Measurement of the semi-inclusive π^+ electroproduction off the proton, performed with CLAS detector at Jefferson Lab, has been presented. The obtained fully-differential cross sections, including the azimuthal angle between hadronic and leptonic planes, ϕ, allowed us to separate the ϕ-dependent terms. While, the ϕ-independent part of the cross section was found to be in good agreement with current fragmentation pQCD calculations.

The semi-inclusive electroproduction of hadrons is an important tool for studying the nucleon structure in the perturbative Quantum Chromodynamics (pQCD) framework at medium energies. Indeed, the detection of a hadron produced by the struck quark or by nucleon spectator fragments provides an information about the orbital momentum of the quark in the initial state. Meanwhile, the undetected hadronic system allows to apply the optical theorem, reducing the energy necessary for the convergence towards basic pQCD processes.

CLAS has measured the semi-inclusive electroproduction of π^+ on the proton at the beam energy of 6 GeV. The measurement span over a wide, continuous 5-dimensional domain, which allows for a detailed study of ϕ and p_T behaviors. In this article we will focus on one particular aspect of the obtained results. This aspect deals with the comparison of the data, integrated in ϕ and transverse momentum p_T, expressed in terms of the structure function H_2:

$$\frac{d^3\sigma}{dxdQ^2dz} = \frac{4\pi\alpha^2}{xQ^4}\left[xy^2H_1(x,z,Q^2) + \left(1 - y - \frac{M^2x^2y^2}{Q^2}\right)H_2(x,z,Q^2)\right] \quad , \quad (1)$$

with pQCD calculations [4]. These calculations describe the structure function H_2 as the convolution of the parton density function $f(x,Q^2)$ obtained in inclusive processes and the parton fragmentation function $D^h(z,Q^2)$ measured in e^+e^- collisions:

$$H_2(x,z,Q^2) = \sum_i e_i^2 x f_i(x,Q^2) \otimes D_i^h(z,Q^2) \; , \quad (2)$$

where the sum runs over quark flavors i and e_i is the charge of ith flavor quark (we neglect the gluon contribution here).

In particular, we are interested in the difference between data-theory comparisons made using the structure function H_2 and structure function ratio H_2/F_2. The latter ratio, where the F_2 is the inclusive structure function, represents in

---

[a] e-mail: osipenko@ge.infn.it

DIS limit the widely used multiplicity observable, and can be calculated in pQCD as following:

$$\frac{H_2(x,z,Q^2)}{F_2(x,Q^2)} = \frac{\sum_i e_i^2 x f_i(x,Q^2) \otimes D_i^h(z,Q^2)}{\sum_i e_i^2 x f_i(x,Q^2)} . \tag{3}$$

Given the relatively low beam energy of Jefferson Lab one may expect a manifestation of visible deviations from pQCD calculations at the low-Q^2 end of the covered interval. Such deviations should indicate the contribution of higher twists in the semi-inclusive electroproduction. The detailed calculations of these higher twists are not available due to their complexity. However, one may phenomenologically divide them in two types: Initial State Interactions (ISI) of the current quark and Final State Interactions (FSI) of struck quark or produced hadron. If the dominant contribution to the total higher twist term would be due to ISI, one could expect a partial cancellation of them in H_2/F_2 ratio. Hence, it is possible that H_2/F_2 ratio agrees with pQCD calculations better that the H_2 structure function alone.

In Fig. 1 comparisons of the measured structure function H_2 and H_2/F_2 ratio to LO and NLO pQCD calculations are shown for two values of z. As one can see, the NLO calculations describe very well the data at $z = 0.45$ for both observables, while at $z = 0.11$ some deviation in the Q^2-slope is evident. This deviation can be due to the higher twist contribution.

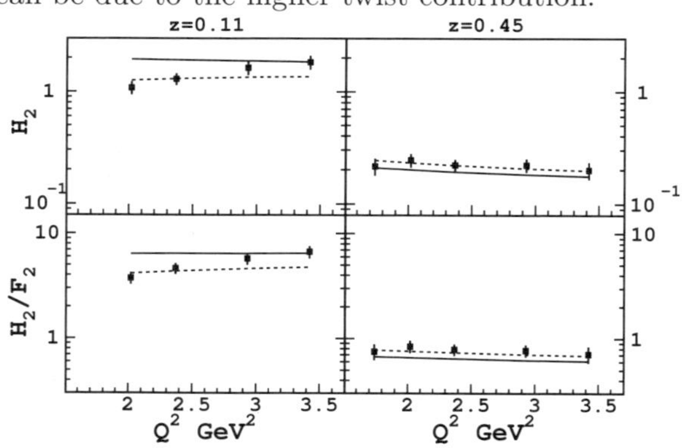

Figure 1: The Q^2-evolution of the data on structure function H_2 and ratio H_2/F_2 at $x = 0.34$ and two values of z, in comparison to pQCD calculations: LO - solid line, NLO - dashed line. The calculations use CTEQ 5 parton distributions [2], and Kretzer fragmentation functions [3]. The error bars give statistical and systematic uncertainties combined in quadrature.

In Fig. 2 the same comparison is shown for the ratio of the data over NLO calculations. In the amplified scale the deviation of the low-z data from the expected NLO evolution can be quantified. Both H_2 and H_2/F_2 comparisons show the deviation rising with Q^2 from -10% up to 40%. The difference between H_2 and H_2/F_2 ratios results in a few percent overall shift, well below the systematic uncertainties of the measurement (about 15% in average) and the theory.

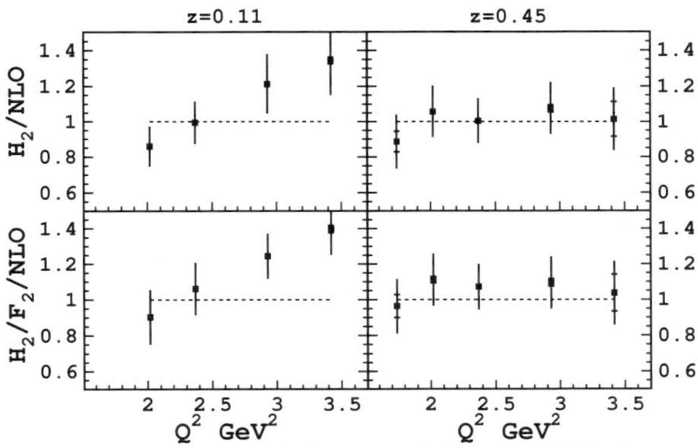

Figure 2: Same as Fig. 1 except with the ratio of H_2 and H_2/F_2 data to the NLO calculations. The dashed line indicates the unity. The inner error bars (mostly smaller than the symbol size) give statistical uncertainties.

Summarizing, the use of H_2 structure function or H_2/F_2 ratio in the comparison of the experimental data to pQCD calculations is equivalent in JLab energy domain. If the deviations of the data from NLO pQCD calculations, observed at low-z, are due to the higher twist contribution, they are probably related to the FSI mechanism. However, given the low-z values at which the difference is observed, an alternative explanation due to the mixing between current and target fragmentation evolutions is more likely. The pQCD calculations described above are due to the current fragmentation only. The target fragmentation mechanism [5], expected to play role in the low-z domain, is still poorly established. In particular, the corresponding partonic functions, fracture functions for the pions are completely unknown. This encourages further pQCD studies of semi-inclusive reactions at Jefferson Lab.

Acknowledgments

Author would like to express his gratitude to Prof. A.Kataev for kind invitation and Prof.B.S.Ishkhanov for the help with the present proceeding.

References

[1] M.Osipenko *et al.*, *Phys.Rev.* **D** 80, 032004 (2009).
[2] H.L.Lai *et al.*, *Eur.Phys.J.* **C** 12, 375 (2000).
[3] S.Kretzer, *et al.*, *Phys.Rev.* **D** 62, 054001 (2000).
[4] W.Furmanski and R.Petronzio, *Z.Phys.* **C** 11, 293 (1982).
[5] L.Trentadue and G.Veneziano, *Phys.Lett.* **B** 323, 201 (1994).

SELF-SIMILARITY OF JET PRODUCTION AND QCD

M. Tokarev [a] and T. Dedovich
Joint Institute for Nuclear Research,
141980 Dubna, Moscow region, Russia

Abstract. Results of the next-to-leading order QCD calculations of inclusive cross sections of jet production in $\bar{p}p$ and pp collisions at energies $\sqrt{s} = 63 - 14000$ GeV are used to test the z-scaling and compare with available experimental data. The dependence of the spectra in p_T- and z-presentations on the collision energy $\sqrt{s}$ and parton distributions is studied. It is shown that self-similar features of jet cross sections dictated by the z-scaling give strong restriction on the asymptotic behavior of $\psi(z)$ predicted by perturbative QCD. The new constraint on the gluon distribution function based on the properties of z-scaling for jet production is suggested.

1 Introduction

Search for new scaling regularities in high energy particle collisions is always to be a subject of intense investigations. Such regularities could give experimental indications on new physics phenomena and new insight in theory. Jets are traditionally considered as a best probe of constituent interactions and most suitable objects for precise test of the perturbative QCD at high energies. New data on inclusive cross sections of jet production in $\bar{p}p$ collisions at Tevatron [1,2] are analysed to verify properties of z-scaling [3]. Transverse jet spectra in the next-to-next leading order QCD in the z-presentation are calculated and compared with results of analysis of experimental data.

The method of phenomenological description of high-p_T particle production cross sections in inclusive reactions ($z-scaling$) was developed in [4]. It is based on the principles of locality, self-similarity and fractality reflecting properties of particle structure, their constituent interactions and particle formation. For the first time it has been applied in [5] for analysis of the jet production. A more sophisticated method of analysis of inclusive cross sections of jet production in z-presentation was proposed in [6].

2 z-Scaling

The collision of extended objects like hadrons and nuclei at sufficiently high energies is assumed to be an ensemble of individual self-similar interactions of their constituents. The structures of the colliding objects (M_1) and (M_2) are characterized by the parameters δ_1 and δ_2. The interacting constituents carry the fractions x_1, x_2 of the incoming momenta P_1, P_2. The inclusive particle (m_1) carries the momentum p.

[a] e-mail:tokarev@sunhe.jinr.ru

The elementary sub-process is considered as a binary collision of the constituents (x_1M_1) and (x_2M_2) resulting in the scattered (m_1) and recoil (M_X) objects in the final state. The momentum conservation law of the sub-process is written as follows

$$(x_1P_1 + x_2P_2 - p)^2 = M_X^2, \tag{1}$$

where $M_X = x_1M_1 + x_2M_2 + m_2$. This equation is expression of the locality of the hadron interaction at a constituent level. It represents a constraint on the fractions x_1, and x_2. The structural parameters δ_1, δ_2 are connected with the corresponding momentum fractions by the function

$$\Omega(x_1, x_2) = (1 - x_1)^{\delta_1}(1 - x_2)^{\delta_2}. \tag{2}$$

The quantity Ω is proportional to relative number of all such constituent configurations which contain the state defined by the fractions x_1 and x_2. We use $\delta_1 = \delta_2 \equiv \delta$, $M_1 = M_2$, and $m_2 = m_1 \equiv 0$ in the analysis of jet production in proton-(anti)proton interactions. The parameters δ_1 and δ_2 were found to be constant values [5]. They are interpreted as the fractal dimensions in the space of the momentum fractions $\{x_1, x_2\}$.

The maximal value of Ω is used in the definition of the scaling variable z which has the form

$$z = z_0\Omega^{-1}. \tag{3}$$

The quantity $z_0 = s_\perp^{1/2}/(dN/d\eta|_0)$ is proportional to the transverse kinetic energy $s_\perp^{1/2}$ of the sub-process consumed on the production of (m_1) and (m_2). The multiplicity density $dN/d\eta|_0$ is taken at $\eta = 0$. The scaling function

$$\psi(z) = -\frac{\pi s}{(dN/d\eta)\sigma_{in}} J^{-1} E\frac{d^3\sigma}{dp^3}. \tag{4}$$

is expressed in terms of the inclusive cross section, multiplicity density $dN/d\eta$, and total inelastic cross section σ_{in}. Here s is the square of the center-of-mass energy, and J is the corresponding Jacobian. The function $\psi(z)$ is normalized to unity.

3 Jet cross section data in z-presentation

The inclusive jet cross section measures the probability of observing a hadronic jet produced in a hadron-hadron collisions. A jet is a direct evidence of hard interaction of quarks and gluons. Experimental data on inclusive cross sections of jet production at different collision energies obtained by CDF [1] and D0 [2] Collaborations at Tevatron allow us to study the energy dependence of the function $\psi(z)$.

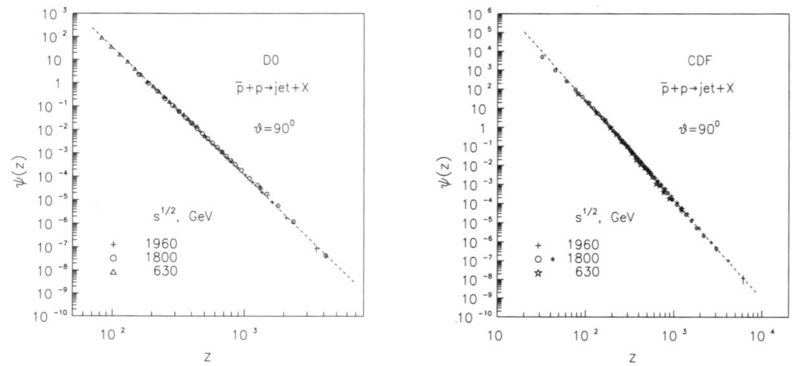

a) b)

Figure 1. Inclusive spectra of jet production in $\bar{p}p$ collisions at $\sqrt{s} = 630, 1800$ and 1960 GeV and $\theta \simeq 90^0$ measured by the D0 [7,8] (a) and CDF [9,10] (b) Collaborations in z-presentation.

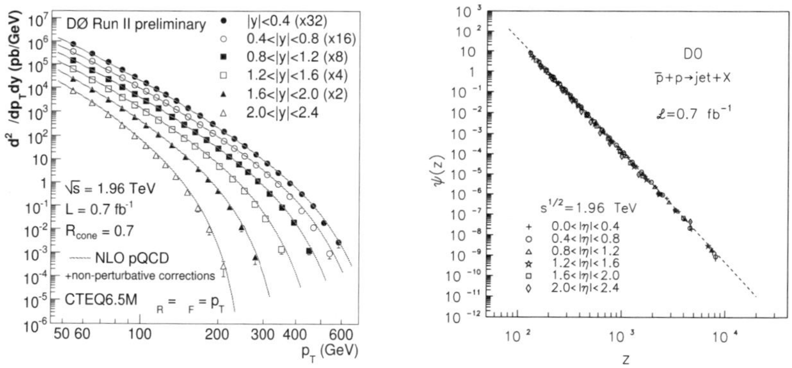

a) b)

Figure 2. Inclusive spectra of jet production in $\bar{p}p$ collisions at $\sqrt{s} = 1960$ GeV and different pseudorapidity intervals in p_T- (a) and z- (b) presentations. Experimental data obtained by the D0 Collaboration are taken from [2].

Figure 1 shows[b] the z-presentation of jet spectra. All data demonstrate the energy independence of the scaling function. As seen from Fig. 1(a) the power behavior of $\psi(z)$ is valid over a wide range of z. The function changes more than on twelve orders. The obtained results mean that mechanism of jet production manifests a property of self-similarity. Note, that this feature is most distinctly visible in z-representation in contrast to p_T-presentation.

Figure 2 demonstrates inclusive spectra [2] of jet production in $\bar{p}p$ collisions at a center-of-mass energy $\sqrt{s} = 1960$ GeV, over the momentum $p_T = 50 - 600$ GeV/c and pseudorapodity $|\eta| < 2.4$ ranges in p_T- (a) and z- (b)

[b]The version of z-scaling considered in [5] was used for the present data analysis.

presentations. The collected by the D0 Collaboration the data correspond to the integrated luminosity of 0.7 fb^{-1}. As seen from Fig. 2(a) the spectra measured for different pseudorapidity intervals demonstrate a strong angular dependence. It is enhanced with increasing of the transverse momentum. Jet yields decrease in the measured momentum range with p_T by more than ten orders. The theoretical calculations of the jet spectra in the NLO QCD (see [2] and references therein) are shown by the solid lines. Figure 2(b) demonstrates the angular independence of $\psi(z)$ over a wide kinematical range. The function is described by the power law, $\psi(z) \sim z^{-\beta}$, with a constant value of the slope parameter $\beta = 5.48 \pm 0.02$.

4 QCD test of z-scaling

We use z-scaling for analysis of inclusive cross sections of jet production in $\bar{p}p$ collisions calculated in the NLO QCD and comparison of the corresponding scaling function with that obtained from analysis of experimental data in the framework of the z-presentation. The code from Ellis-Kunszt-Soper [11] for calculations of jet transverse momentum spectra in the next-to-leading order QCD and the parton distribution functions the CTEQ6M [14] and MRST01 [15] are used in the present analysis.

The inclusive differential cross section for hadron production with the transverse momentum p_T and the pseudorapidity η is written in the form

$$E_C \frac{d^3\sigma}{dp_C^3} = \sum_{abc} \int dx_a dx_b \frac{dz}{z} f_{a/A}(x_a, \mu_F) f_{b/B}(x_b, \mu_F) \epsilon_c \times \\ \times \frac{d^3\sigma}{dk_c^3}(p_c/z\sqrt{s}, \mu_R) D_{C/c}(z, \mu_H) \quad (5)$$

The sum is taken over the various flavors of partons that can participate in the hard scattering process. The functions $f_{a/A}(x_a, \mu_F)$ and $f_{b/B}(x_b, \mu_F)$ describe the distributions of a parton a and b in the hadron A and B on the momentum fraction x_a and x_b at the factorization scale μ_F, respectively. The fragmentation of a parton c into the hadron C is described by the function $D_c^C(z, \mu_H)$. The parton momentum fraction carried by the hadron C at the fragmentation scale μ_H is equal to z. Inclusive cross section $\epsilon_c\ d^3\sigma/dk_c^3$ of the parton subprocess $a + b \to c + X$ is calculated perturbatively in the NLO QCD. In our case we have $c \equiv jet$. It is assumed that fragmentation does not destroy the transverse distribution of jet and $z = 1$. The $\overline{\text{MS}}$ scheme is used to subtract final state collinear singularities. The strong coupling $\alpha_S(\mu_R)$ is defined in the $\overline{\text{MS}}$ renormalization scheme at the scale μ_R.

The jet spectra as a function of the collision energy $\sqrt{s}$ and transverse momentum p_T calculated in the NLO QCD are shown in Fig. 3. The results calculationed with the parton distribution functions MRST01 [15] are depicted by the dashed lines. The renormalization, factorization and fragmentation

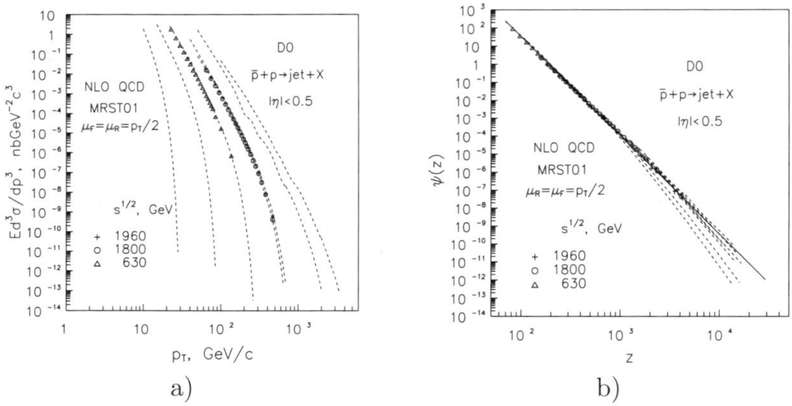

Figure 3. Inclusive spectra of jet production in $\bar{p}p$ collisions as a function of the transverse momentum p_T. Dashed lines are calculated results in NLO QCD with (a) MRST01 [15] PDFs at $\sqrt{s} = 63, 200, 630, 1800, 1960, 7000$ and 14000 GeV. Symbols are the D0 data taken from [7,8]. (b) The same data in z-presentation.

scales were set to be equal each other $\mu_R = \mu_F = \zeta\ p_T$, where ζ is the scale factor usualy varied in the range $\zeta = 0.5, 2$.

The spectra were calculated over a wide range of the transverse momentum $p_T = 10 - 1000$ GeV/c and the energy $\sqrt{s} = 63 - 14000$ GeV at $\theta \simeq 90^0$. As seen from Fig. 3(a) the strong dependence of the spectra on the collision energy increases with p_T. The D0 data [7,8] on cross sections at $\sqrt{s} = 630, 1800$ and 1960 GeV are shown in Fig. 3 by the symbols. The same data are plotted in the dependence on the variable z are shown in Fig. 3(b) as well. The solid line represents the asymptotic behavior of $\psi(z)$ obtained from analysis of experimental data. The z-presentation of the NLO QCD calculation results of jet spectra is shown by the dashed lines. As seen the NLO QCD predictions demonstrate dramatic deviation from the asymptotic behavior of $\psi(z)$ predicted by the z-scaling as the collision energy and transverse momentum increase. The quark and gluon distribution functions and fragmentation function could modify the asymptotic behavior of ψ. We assume that large uncertainties of the gluon distribution function could be essentially restricted by the power behavior of $\psi(z)$ established over a wide range of $\sqrt{s}$ and high-p_T. It could give us the new additional constraint on PDFs and FFs which are phenomenological quantities and weakly controlled by the perturbative QCD.

5 Conclusions

New data on jet spectra at Tevatron up to highest jet transverse momentum $p_T \simeq 600$ GeV/c confirm the energy and angular independence of the z-scaling. The power behavior of the scaling function, $\psi(z) \sim z^{-\beta}$, is observed. It means that properties of mechanism of jet production reflect the self-similarity, locality and fractality of the hadron interactions at a constituent level. We conclude that self-similar features of jet production dictated by the z-scaling give strong restriction on the asymptotic behavior of $\psi(z)$. The behavior is not reproduced by the NLO QCD evolution of cross sections with the phenomenological parton distribution functions used in the present analysis. The asymptotic behavior of $\psi(z)$ could be used as the additional constraint on a gluon distribution function in the global QCD analysis of experimental data.

Acknowledgments

The investigations have been partially supported by the special program of the Ministry of Science and Education of the Russian Federation, grant RNP.2.1.1.2512.

References

[1] A. Abulencia et al., *Phys.Rev.* **D** 75, 092006 (2007).
[2] V.M. Abazov et al., *Phys.Rev.Lett.* 101, 062001 (2008).
[3] I. Zborovský, M.V. Tokarev, *Int.J.Mod.Phys.* **A** 24, 1417 (2009).
[4] I. Zborovský, M.V. Tokarev, *Phys.Rev.* **D** 75, 094008 (2007).
[5] M.V. Tokarev, T.G. Dedovich, *Int.J.Mod.Phys.* **A** 15, 3495 (2000).
[6] I. Zborovský, M.V. Tokarev, *Part.Nucl.Lett.* 3, 312 (2006).
[7] B. Abbott et al., *Phys.Rev.Lett.* 82, 2451 (1999).
B. Abbott et al., *Phys.Rev.* **D** 64, 032003 (2001).
D. Elvira, Ph.D Thesis Universodad de Buenos Aires, Argentina (1995).
V.M. Abazov et al., *Phys.Lett.* **B** 525, 211 (2002).
[8] M. Begel et al. (D0 Collaboration), hep-ex/0305072.
M. Voutilainen (for D0 Collaboration) XIV Interbational Workshops on Deep-Inelastic Scattering and Related Subjects (DIS2006), April 20-24, 2006, Tsukuba, Japan; http://www-conf.kek.jp/dis06/
J. Commin (for D0 Collaboration) XV Interbational Workshops on Deep-Inelastic Scattering and Related Subjects (DIS2007), April 16-20, 2007, Munich, Germany; http://www.mppmu.mpg.de/dis2007/
[9] F. Abe et al., *Phys.Rev.Lett.* 77, 438 (1996).
T. Affolder et al., *Phys.Rev.* **D** 64, 032001 (2001).
[10] A. Abulencia et al., *Phys.Rev.Lett.* 96, 122001 (2006).
A. Abulencia et al., *Phys.Rev.* **D** 74, 071103 (2006).
[11] S.D. Ellis, Z. Kunszt, and D. Soper, *Phys.Rev.Lett.* 69, 1496 (1992); http://zebu.uoregon.edu/~soper/soper.html
[12] S.D. Ellis, Z. Kunszt, and D. Soper, *Phys.Rev.* **D** 40, 2188 (1989).
[13] Z. Kunszt, D. Soper, *Phys.Rev.* **D** 46, 192 (1992).
[14] J. Pumplin et al., *JHEP* **0207**, 012 (2002); hep-ph/0201195; http://www.phys.psu.edu/~cteq
[15] A.D. Martin et al., *Eur.Phys.J.* **C** 23, 73 (2002); hep-ph/0110215; http://www.durpdg.dur.ac.uk/hepdata/mrs

ELECTROMAGNETIC STRUCTURE FUNCTIONS OF NUCLEONS IN THE REGION OF VERY SMALL X Y

E.V. Bugaev
Institute for Nuclear Research of the Russian Academy of Sciences, 7a, 60th October Anniversary prospect, Moscow 117312, Russia
B.V. Mangazeev[a]
Irkutsk State University, 1, Karl Marx Street, Irkutsk 664003, Russia

Abstract. A two component model describing the electromagnetic nucleon structure functions in the low-x region, based on generalized vector dominance and color dipole approaches is briefly described.

1 Introduction

A starting point of the present model of deep inelastic lepton-nucleon scattering is generalized vector dominance (GVD) idea [1], the main ingredient of which is a double dispersion relation for the total cross section for the process $\gamma^* p \to X$,

$$\sigma = \sum_q \int \frac{dM^2}{M^2+Q^2} \int \frac{dM'^2}{M'^2+Q^2} \rho(s, M^2, M'^2) \frac{1}{s} A_{q\bar{q}\to p}(s, M^2, M'^2). \quad (1)$$

Here, M and M' are invariant masses of the incoming and outgoing $q\bar{q}$-pair in $\gamma^* p$ elastic amplitude A.

It is assumed here that at high energies the virtual photon γ^* fluctuates in a hadronic system ($q\bar{q}$-pair, to the leading order) well before the interaction with the target nucleon and, at the second step, $q\bar{q}$-pair interacts with the target. Later, in nineties [2-3], the relation (1) was used, in diagonal approximation, i.e., with $\rho \sim \delta\left(M^2 - M'^2\right)$, for a separation of the "soft" and "hard" interactions, and the soft part was described by the Vector Dominance Model (VDM).

The present two-component model of deep inelastic scattering differs (in a description of the soft part) from those developed in [2-3] in two respects: we do not use diagonal approximation and, most important, we do not limit ourselves by accounting of low mass vector meson (ρ) and, instead, take into account ρ-family (containing, except of ρ, the radial excitations ρ', ρ",...). The use of the real mass spectrum of vector mesons is more natural than the introduction, in [3], the additive quark model component and, besides, allows to introduce the non-diagonal transitions (and check the importance of them).

---

[a]Work partially supported by the program "Development of Scientific Potential in Higher Schools of Russia" (project 2.2.1.1/1483, 2.1.1/1539)".

2 A two component model

In our previous papers [4], [5] we formulated the two-component model of electromagnetic structure functions of the nucleon. The nonperturbative (soft) component of the structure functions is described by the off-diagonal GVD with vector mesons having properties of usual hadrons. It was shown in [4] that the approach of the off-diagonal GVD alone cannot describe the experimental data if the destructive interference effects and corresponding cancellations of VN-V'N amplitudes inside of GVD sums are small (and they are really small if the vector mesons in the tower have the properties of usual hadrons). It was shown, as a result, that two modifications of the off-diagonal GVD scheme are needed: 1) cut-off factors reducing the probability of initial γ V transitions must be introduced and 2) a "hard component" must be added to describe the perturbative QCD part of the total process of the virtual photon-nucleon interaction. For a calculation of the cut-off factor we use the following basic assumption: an interaction of the initial qq-pair with the nucleon is meson dominated if (and only if) the transverse momenta of pair's quarks are not too high; only in this case are confinement forces effective. It can be easily shown that the relative part of pair's phase volume for pairs with quarks having transverse momenta, p^T, in the limits $(m_q \div \mathrm{p}^T_{\max})$ is given, approximately, by $\eta \approx 3\left(p^T_{\max}/M_{q\bar{q}}\right)^2$, if $M^2_{q\bar{q}} >> \left(p^T_{\max}\right)^2$. By definition, this value is just the required cut-off factor. Here, the value $p^T_{\max}$ is the model parameter. The average transverse size of the qq-pair is, at not very large Q^2, inversely proportional to p^T. From comparison with experimental data on F_2 we obtained for this parameter the value 0.385 GeV.

The simplest model of VN-scattering had been used [4], based on two-gluon exchange approximation and relativistic constituent quark model. Wave functions of the vector mesons were obtained from a solution of the Bethe-Salpeter (BS) equation using quasipotential formalism in the light-front form. The kernel of the BS-equation has the confining term of the harmonic oscillator type. The vector meson mass spectrum is of the form $m^2_n = m^2_\rho(1+2n)$, for the ρ-meson family (only this family has been taken into account). For the hard component, we used the color dipole model and the parameterizations of the dipole cross section $\sigma\left(r_\perp, s\right)$ (perturbative QCD part) from the work by J. Forshaw et al [6].

As one can see from fig.1, there is rather good agreement of the model predictions with the available data on the electromagnetic structure function F_2 in the region $x < 0.05$ and $Q^2 < 10^2 GeV^2$. The relative contribution in F_2 of the soft (GVD) component strongly depends on the values of the kinematic variables x, Q^2. At not very small x the contribution of the soft component is dominant. With a decrease of the x-value the interval of Q^2 in which soft component is dominant is reduced. For example, at $x = 10^{-4}$ the soft component

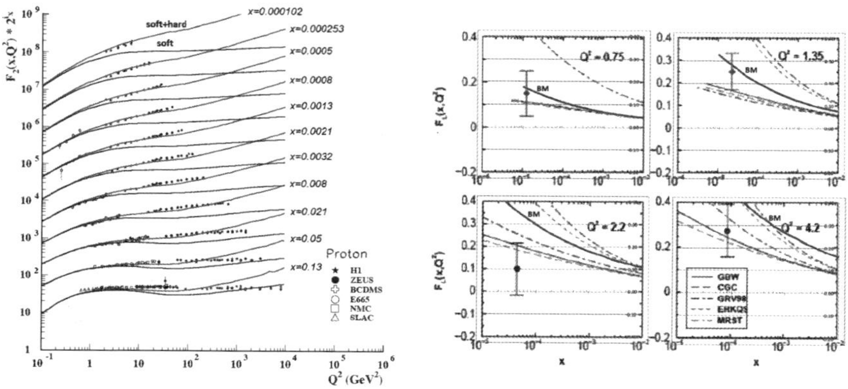

a b

Figure 1: a) The proton electromagnetic structure function F_2 measured in experiments H1, ZEUS, BCDMS, E665, NMC, SLAC [7]. Data of each bin of fixed x has been multiplied by 2^i, where i is the odd number of the bin, ranging from i=7 (x=0.13) to i=27 (x=0.000102). The red lines are predictions of the present model (soft+hard), blue lines are the contributions of the soft (GVD). b) The results for F_L as a function of x at fixed Q^2. The data are taken from H1 Collaboration. Solid brown curves (BM) are results of the present model. Other curves are predictions of different theoretical models (see [8]).

is relatively large only in the region of very small $Q^2, 0 < Q^2 < 1GeV^2$.

References

[1] V.N. Gribov, *Sov. Phys. JETP.* **30**, 709 (1970).
[2] E. Gotsman et al, *Eur. Phys. J.* **C5**, 303 (1998).
[3] A.D. Martin et al, *Eur. Phys. J.* **C7**, 643 (1999).
[4] E. Bugaev, B. Mangazeev and Y. Shlepin, *hep-ph/9912384* (1999).
[5] E. Bugaev and Y. Shlepin, *Phys. Rev.* **D67**, 034027 (2003).
[6] J. Forshaw, G. Kerley and G. Shaw, *Phys. Rev.* **D60,** 074012 (1999).
[7] *http://pdg.lbl.gov/2007/reviews/strucfunfigrpp.pdf* (2007).
[8] V.Gonsales, M.Machado, *hep-ph/0406230* (2004).

RESUMMATION IN QCD FRACTIONAL ANALYTIC PERTURBATION THEORY

Alexander P. Bakulev[a]

Bogoliubov Laboratory of Theoretical Physics, JINR, 141980 Dubna, Russia

Abstract. In my report I describe the generalization of Analytic Perturbation Theory (APT) for QCD observables, initiated by Radyushkin, Krasnikov, Pivovarov, Shirkov and Solovtsov [1–3], to fractional powers of coupling — the so called Fractional APT (FAPT) [4, 5]. The basic aspects of FAPT are shortly summarized. After that I discuss how to treat heavy-quark thresholds in FAPT [6] and then show how to resum perturbative series in both the one-loop APT and FAPT, provided that the generation function $P(t)$ of perturbative coefficients d_n is known [6–8]. As an application I consider FAPT description of the Higgs boson decay $H^0 \to b\bar{b}$ [4,6,9]. My main conclusion is: To achieve an accuracy of the order of 1% it is enough to take into account up to the third correction—in complete agreement with Kataev–Kim analysis in [10]. The d_4 coefficient value calculated in [9] is needed only to estimate the corresponding generating function $P(t)$.

The full version of this report has been published in [11]. Here in Fig. 1 we show only the main result for the width $\Gamma_{H\to b\bar{b}}$: The width of the shaded strip takes into account the overall uncertainties due to the resummation procedure and the renormgroup-invariant b-quark mass, $\hat{m}_b$.

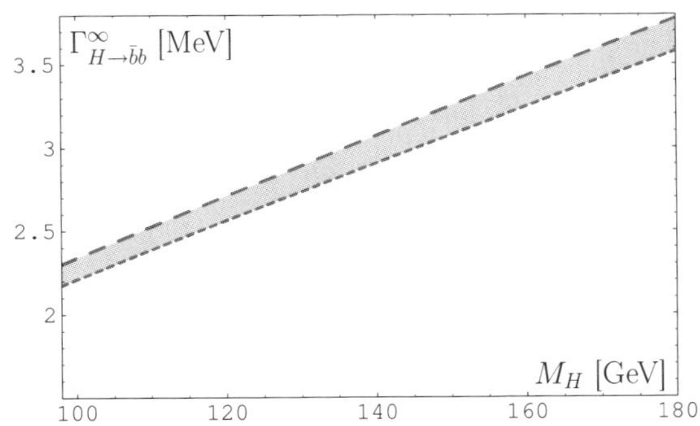

Figure 1: The width $\Gamma_{H\to b\bar{b}}$ as a function of the Higgs boson mass M_H in the resummed FAPT.

Acknowledgments

This work was supported in part by the Russian Foundation for Fundamental Research (grant No. 08-01-00686) and the Heisenberg–Landau Programme (2009–2010).

[a]e-mail: bakulev@theor.jinr.ru

References

[1] A. V. Radyushkin, *JINR Rapid Commun.* **78**, 96 (1996); [*JINR Preprint*, E2-82-159, 26 Febr. 1982; arXiv: hep-ph/9907228].

[2] N. V. Krasnikov and A. A. Pivovarov, *Phys. Lett.* **B116**, 168 (1982).

[3] D. V. Shirkov and I. L. Solovtsov, *JINR Rapid Commun.* **2** [76], 5 (1996); *Phys. Rev. Lett.* **79**, 1209 (1997); *Theor. Math. Phys.* **150**, 132 (2007).

[4] A. P. Bakulev, S. V. Mikhailov, and N. G. Stefanis, *Phys. Rev. D* **72**, 074014, 119908(E) (2005); *ibid.* **75**, 056005 (2007); **77**, 079901(E) (2008).

[5] A. P. Bakulev, A. I. Karanikas, and N. G. Stefanis, *Phys. Rev. D* **72**, 074015 (2005).

[6] A. P. Bakulev *Phys. Elem. Part. Nucl.* **40**, 715 (2009).

[7] S. V. Mikhailov, *JHEP* **06**, 009 (2007).

[8] A. P. Bakulev and S. V. Mikhailov, in *Proc. Int. Seminar on Contemp. Probl. of Part. Phys., dedicated to the memory of I. L. Solovtsov, Dubna, Jan. 17–18, 2008.*, Eds. A. P. Bakulev *et al.* (JINR, Dubna, 2008), pp. 119–133 (arXiv:0803.3013).

[9] P. A. Baikov, K. G. Chetyrkin, and J. H. Kühn, *Phys. Rev. Lett.* **96**, 012003 (2006).

[10] A. L. Kataev and V. T. Kim, *PoS* **ACAT08**, 004 (2009).

[11] Alexander P. Bakulev, *Ukr. J. Phys.* **50**, 15 (2010).

SCALING BEHAVIOR OF FACTORIAL MOMENTS OF THE MULTIPLICITY DIFFERENCE AND PHASE TRANSITION QGP TO HADRONS

Babichev L.F. [a], Khmialeuski A.N. [b]

Joint Institute for Power and Nuclear Research – Sosny, 220109 Minsk, Belarus

*Abstract.*The multiplicity difference correlators between two well-separated bins in high energy heavy-ion collisions are studied using Monte-Carlo simulations. Numerical value of the universal scaling exponents γ is calculated and compared whit results obtained within the framework of Ginzburg-Landau theory. Simulations of phase transitions using Monte Carlo generators are discussed.

1 Introductions

Experiments of elementary particles or nuclei collision at high energy show various collective phenomena, e.g. multiplicity fluctuations of secondary particles. At high temperatures or energy densities it should be a new superdense state of matter — quark-gluon plasma (QGP). The heavy ion collisions are most promising for formation QGP. But detection of QGP is a more difficult. Firstly, there is a large background due to the strong interaction between nucleons. Secondly, the duration of the evolution quark-gluon phase of the nuclear system is a small fraction of the total time of evolution. If the QGP was formed in a collision, then there should be observed a phase transition (PT) QGP→hadrons. Such PT can be defined by the study of the particles multiplicities fluctuations. The availability of sufficiently large fluctuations can be a indication of PT and correspondingly QGP formation. It was proposed to study such fluctuations using the factorial moments [1].

2 Scaling behavior of multiplicity difference correlators

The factorial moments are good tool for the study of the multiplicities fluctuations. But they are not effective for the study of the heavy ion collisions. Factorial moments of the multiplicity difference was introduced to overcome the problem of hight multiplicity per bin in heavy ion collisions [3]. They are multiplicity difference correlators (MDC) and a form of hybrid of factorial correlators and wavelets. MDC are

$$\mathcal{F}_q(\delta,\Delta) = \frac{\sum_{m=q}^{\infty} m(m-1)...(m-q+1)Q_m(\delta,\Delta)}{(\sum_{m=q}^{\infty} mQ_m(\delta,\Delta))^q} \tag{1}$$

where $m = m(\delta,\Delta) = |n_1 - n_2|$ is a multiplicity difference in two bins separated Δ, $Q_m(\delta,\Delta)$ is a multiplicity difference distribution m, q is order of MDC. The numerical value was calculated by averaging of MDC over several bins as in [4].

[a]e-mail: babichev@sosny.bas-net.by

[b]e-mail: akhmelevskiy@sosny.bas-net.by

If there are the nonstatistical fluctuations of multiplicity of secondary particles in the events, then there should be observed power dependence of the type $\mathcal{F}_q(\delta) \propto \delta^{-\alpha_q}$. If the system is characterized by the phenomenon intermittency, then perhaps scaling behavior of MDC $\mathcal{F}_q(\delta) \propto \mathcal{F}_2^{-\beta_q}$. Slope coefficients β_q can be obtained using the following expression: $\beta_q = (q-1)^\gamma$. γ is universal scaling exponent for MDC. The numerical value of the scaling exponent γ was obtained in the framework of the Ginzburg-Landau general theory of the phase transitions. The value γ for the second order PT is 1.09±0.02 [3]. The value of γ in the absence of PT is 1.33±0.02 [3].

For the study of the multiplicity fluctuations we have simulated 5% most of the central events of the Pb ions collisions for energies from 500 to 1400 GeV. For calculations we have selected 10^5 events of the secondary charge particles. The Monte Carlo generator HIJING (Heavy Ion Jet INteraction Generator) [2] was used for simulation. The multiplicity distribution for all the phase variables (pseudorapidity η, azimuthal angle ϕ and transverse momentum $p_\perp$) was calculated. The value of scaling exponents γ for different energies and the phase variables is calculated and shown in table 1.

Table 1: Value of scaling exponents γ for different energies and the phase variables

$\sqrt{s}$, GeV	γ_η, $[0;5,0]$, $\Delta=0,42$	γ_ϕ, $[0;3,0]$, $\Delta=0,19$	$\gamma_{p_\perp}$, $[0;2,0]$, $\Delta=0,21$	$\sqrt{s}$, GeV	γ_η, $[0;5,0]$, $\Delta=0,42$	γ_ϕ, $[0;3,0]$, $\Delta=0,19$	$\gamma_{p_\perp}$, $[0;2,0]$, $\Delta=0,21$
500	1,21 ± 0,03	0,86 ± 0,03	1,22 ± 0,16	600	1,22 ± 0,03	0,86 ± 0,06	1,22 ± 0,16
700	1,23 ± 0,03	0,86 ± 0,07	1,22 ± 0,17	800	1,24 ± 0,03	0,86 ± 0,08	1,21 ± 0,20
900	1,25 ± 0,04	0,85 ± 0,08	1,21 ± 0,20	1000	1,28 ± 0,05	0,84 ± 0,11	1,20 ± 0,32
1100	1,26 ± 0,04	0,85 ± 0,10	1,21 ± 0,25	1200	1,27 ± 0,04	0,84 ± 0.10	1.24 ± 0.29
1300	1,28 ± 0,05	0.85 ± 0.11	1.20 ± 0.33	1400	1,29 ± 0,05	0.85 ± 0.10	1.20 ± 0.32

3 Conclusion

Comparing the results for γ with values calculated in the framework of the Ginzburg-Landau general theory, can say that this method can be used for determining of the PT. Based on data from the table 1, can make the assumption that in the heavy ion collisions may be observed first order phase transition.

References

[1] Bialas A., Peschanski P. *Nucl.Phys.* **B** 273 (1986) 703, 308 (1988) 857.

[2] Wang X.-N., Gyulassy M. *Phys.Rev.* **D**. 44 (1991) 3501.

[3] R.C. Hwa. *Phys.Rev.* **D** 57 (1998) 1831-1837.

[4] De Wolf E.A., Dremin I. M., Kittel W. *Phys. Rept.* 270 (1996) 1-141.

HEAVY HADRONS IN QUARK-GLUON PLASMA

I.V. Narodetskii[a], Yu.A.Simonov[b], A.I.Veselov[c]
ITEP, 117218 Moscow, Russia

Abstract. We use the screened Coulomb potential with r-dependent coupling constant and the non–perturbative quark–antiquark potential derived within Field Correlator Method (FCM) to calculate J/ψ, Υ, Ω_c and Ω_b binding energies and melting temperatures in the deconfined phase of quark-gluon plasma.

1 Introduction

Since 1986, the gold-plated signature of deconfinement was thought to be J/ψ suppression. If Debye screening of the Coulomb potential above T_c is strong enough then J/ψ production in A+A collisions will be suppressed. But this is not the full story.

There is a significant change of views on physical properties and underlying dynamics of quark–gluon plasma (QGP), produced at RHIC, see *e.g.* [1] and references there in. Instead of behaving like a gas of free quasiparticles – quarks and gluons, the matter created in RHIC interacts much more strongly than originally expected. It is more appropriate to describe the non-perturbative (NP) properties of the QCD phase close to T_c in terms of the NP part of the QCD force rather than a strongly coupled Coulomb force. In the QCD vacuum, the NP quark-antiquark potential is $V = \sigma r$. At $T \geq T_c$, $\sigma = 0$, but that does not mean that the NP potential disappears.

In a recent paper [2] we calculated binding energies for the lowest $Q\overline{Q}$ and QQQ eigenstates ($Q = c, b$) above T_c using the NP $Q\overline{Q}$ potential derived in the Field Correlator Method (FCM) [3] and the screened Coulomb potential with the strong coupling constant $\alpha_s = 0.35$. In this contribution we extend our analysis to the case of the running $\alpha_s(r)$ [4]. Technical details and the full list of references can be found in [3], [2].

2 Results

We employ the current masses $m_c = 1.4$ GeV and $m_b = 4.8$ GeV. The melting temperatures for heavy meson and baryon ground states are shown in Table 1.

At $T = T_c$ we obtain the weakly bound $c\bar{c}$ state that disappears at $T \sim 1.3\,T_c$. Note that immediately above T_c the mass of the $c\bar{c}$ state is about 0.2 GeV higher than that of J/ψ. As expected, the 1S bottomoniium $T \sim 2Tc$ the Υ state remains intact up to the larger temperatures, $T \sim 2.3\,T_c$. The masses of the Υ lie in the interval 9.7–9.8 GeV, about 0.2–0.3 GeV higher than 9.460 GeV,

[a]e-mail: naro@itep.ru
[b]e-mail: simonov@itep.ru
[c]e-mail: veselov@itep.ru

Table 1: The melting points (in units of T_c) for J/ψ, Υ, Ω_c, and Ω_b

	J/ψ	Υ	Ω_c	Ω_b
This work	1.2	2.3		
Ref. [2]	1.3	2.3	1.3	2.3
Ref. [5]	1.2	2.3		
Ref. [6]	2	$(4-6)$		

the mass of $\Upsilon(1S)$ at $T = 0$. At $T = T_c$ we predict the $b\bar{b}$ separation $r_0 = 0.25$ fm that compatible with $r_0 = 0.28$ fm at $T = 0$. The results agree with those found previously for a constant $\alpha_s = 0.35$ [2]. We also mention that the melting temperatures for Ω_c and Ω_b calculated in [2] practically coincide with those for J/ψ and Υ.

Our results for $1S(J/\psi)$ are qualitatively agree with those of Refs. [5], [6] based on phenomenological $Q\overline{Q}$ potentials identified with the free quark-antiquark energy measured on the lattice while our melting temperature for $1S(\Upsilon)$ is much smaller than $T \sim (4-6)\, T_c$ found in Ref. [6].

Acknowledgements

This work was supported in part by RFBR Grants # 08-02-00657, # 08-02-00677, # 09-02-00629 and by the grant for scientific schools # NSh.4961.2008.2.

References

[1] M. J. Tannenbaum, Rep. Prog. Phys. **69**, 2005 (2006)

[2] I. M. Narodetskiy, Yu. A. Simonov, A. I. Veselov, JETP Lett. **90**, 232 (2009) [Pisma Zh. Eksp. Teor. Fiz. **90**, 254 (2009)]; arXiv: hep-ph/0907.5176

[3] A. V. Nefediev *et al.*, Int. J. Mod. Phys. **E18**, 549 (2009)

[4] I. M. Narodetskiy, Yu. A. Simonov, A. I. Veselov, arXiv: hep-ph/1001355

[5] D. Blaschke, O. Kaczmarek, E. Laermann, V. Yudichev, Eur. Phys. J. C**43**, 81 (2005)

[6] W. M. Alberico, A. Beraudo, A. De Pace, and A. Molinari, arXiv: hep-ph/0507084

New Developments in Quantum Field Theory

NEW BOUND STATES OF HEAVY QUARKS AT LHC AND TEVATRON

C.R. Das
Centre for Theoretical Particle Physics, IST, Lisbon, Portugal
C.D. Froggatt
Department of Physics and Astronomy, Glasgow University, Glasgow, Scotland
L.V. Laperashvili
Institute of Theoretical and Experimental Physics, Moscow, Russia
H.B. Nielsen
The Niels Bohr Institute, Copenhagen, Denmark

1 Introduction

The Higgs boson couples more strongly to the heavy top quarks than to the light ones. As a result, the Higgs exchanges between top quarks produce new type of bound states [1]. The present talk is devoted to the properties of the new bound states (NBS) [2]: 1) There exists a scalar $1S$–bound state of $6t+6\bar{t}$. The forces which bind these top-quarks are so strong that almost completely compensate the mass of the 12 top-quarks forming this bound state. 2) There exists a new bound state $6t+5\bar{t}$, which is a fermion similar to the quark of the fourth generation having quantum numbers of top quark. 3) Theory also predicts the existence of new bound states with b-quark replaced the t-quark [2]: for example, NBS $n_b b + (6t + 6\bar{t} - n_b t)$, etc., where $n_b = 1, ...6$. A new (earlier unknown) bound state $6t+6\bar{t}$, which is a color singlet (that is, 'white' state), was first suggested by Froggatt and Nielsen in Ref. [3]. Now all these NBS are named T-balls, or T-fireballs.

2 Higgs and gluon interactions of quarks

If the Higgs particle exists, then between quarks qq, quarks and anti-quarks $q\bar{q}$, and also between anti-quarks $\bar{q}\bar{q}$ there exist virtual exchanges by Higgs bosons, leading only to the attractive forces. It is well-known that the bound state $t\bar{t}$ – so called toponium – is obliged to the gluon virtual exchanges. In the case of the toponium the contributions of the Higgs scalar particles are essential, but less than gluon interactions. However, putting more and more top and anti-top quarks together in the lowest energy bound states, we notice that the attractive Higgs forces continue to increase. Simultaneously gluon (attractive and repulsive) forces first begin to compensate themselves, but then begin to decrease relatively to the Higgs effect with growth of the number of top-anti-top constituents in the NBS. The maximum of the binding energy value corresponds to the $1S$-wave state of the NBS $6t+6\bar{t}$ [1].

3 T-ball mass estimate

The VEV of the Higgs field in the EW-vacuum is $v = \langle|\Phi_H|\rangle = 246$ GeV. However, if the NBS is a bubble in the EW-vacuum and contains $N_{\rm const.}$ top-like constituents, then the Higgs field can modify its VEV insight the bubble (bag) [1], and we have: $v_0 = \langle|\Phi_h|\rangle$, where $\xi = v_0/v < 1$. Then the effective Higgs mass m_h insight the T-ball is smaller than the corresponding experimental mass M_H: $m_h = \xi M_H$ (and $m_t = \xi M_t$). In this case the attraction between the two top (or anti-top) quarks is presented by the Yukawa type of potential $V(r) = -(g_t^2/2)/(4\pi r)(\exp(-m_h r))$. Assuming that the radius R_0 of the bubble is small ($m_h R_0 \ll 1$), we obtain the Coulomb-like potential $V(r) \simeq -(g_t^2/2)/(4\pi r)$, describing the attraction between any pairs tt, $t\bar{t}$, $\bar{t}\bar{t}$.

By analogy with Bohr Hydrogen-atom-like model, the binding energy of a single top-quark relatively to the nucleus, containing $Z = N_{\rm const.}-1$ top-quarks, have been estimated in Refs. [1–3]. The total potential energy for the NBS with $Z = 11$ is: $V_{tot}(r) = -11(g_t^2/2)/(4\pi r)$. Here we would like to comment that the value of the mass m_h, which belongs to the Higgs field insight the NBS $6t + 6\bar{t}$, can just coincide with estimates given by Refs. [4].

Considering a set of Feynman diagrams (the Bethe-Salpeter equation) and including the contributions of all (s-,t- and u-) channels for the Higgs and gluon exchange forces, we obtain the following Taylor expansion:

$$M_T^2 = (N_{\rm const.}M_t)^2 \times \left\{1 - 2(N_{\rm const.} - 1)\left(\frac{N_{\rm const.}}{12}\right)^2 \left(\frac{g_t^2 + \frac{1}{6}g_s^2}{\pi}\right)^2 + \ldots.\right\}. \tag{1}$$

Here the QCD coupling constant g_s is given by its fine structure constant value at the EW-scale [5]: $\alpha_s(M_Z) = g_s^2(M_Z)/4\pi \approx 0.118$. Now the value of the total binding energy for arbitrary $N_{\rm const.}$ is equal to:

$$E_T = N_{\rm const.}(N_{\rm const.} - 1)\left(\frac{N_{\rm const.}}{12}\right)^2 \left(\frac{g_t^2 + \frac{1}{6}g_s^2}{\pi}\right)^2 m_t. \tag{2}$$

The mass of T-ball is: $M_T = N_{\rm const.}m_t - E_T$, and approximately this dependence is described by the following expression:

$$M_T = N_{\rm const.}m_t \left\{1 - (N_{\rm const.} - 1)\left(\frac{N_{\rm const.}}{12}\right)^2 \left(\frac{g_t^2 + \frac{1}{6}g_s^2}{\pi}\right)^2\right\}. \tag{3}$$

Below we shall use the following notations: T_s-ball is a scalar NBS $6t + 6\bar{t}$, having the spin $S = 0$, and T_f-ball presents the NBS $6t+5\bar{t}$, which is a fermion: $\overline{T_f} = 5t+6\bar{t}$. Let us consider now the condition: $(11/\pi^2)(g_t^2+(1/6)g_s^2)^2 = 1$. In this case the binding energy E_T compensates the NBS mass $12m_t$ so strongly that the mass of the scalar T_s-ball becomes zero:

$$M_{T_s} = 11m_t \left\{ 1 - \frac{11}{\pi^2} \cdot (g_t^2 + \frac{1}{6} g_s^2)^2 \right\} = 0. \tag{4}$$

Fig. 1 shows the dependence of T-ball masses on the number of NBS constituents $N_{\text{const.}}$. In the case when $M_{T_s} = 0$, we have:

$$M_T = N_{\text{const.}} m_t \left\{ 1 - \frac{(N_{\text{const.}} - 1)}{11} \frac{N_{\text{const.}}^2}{12^2} \right\}. \tag{5}$$

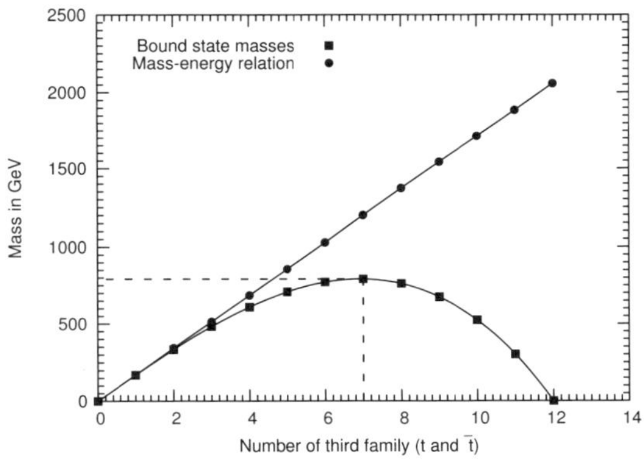

We easily see that the light scalar Higgs bosons with mass $m_h < M_H$ can bind the 12 top-like quarks so strongly that the mass M_{T_s} becomes almost zero, and even tachyonic: $M_{T_s}^2 < 0$. In the last case we obtain the Bose-Einstein condensate of T-balls – a new vacuum at the EW-scale.

References

[1] C.D. Froggatt, H.B. Nielsen and L.V. Laperashvili, Int.J.Mod.Phys. **A** 20, 1268 (2005) [ArXiv: hep-ph/0406110]; in: Proceedings of 13th International Seminar on High-Energy Physics: *QUARKS-2004*, Pushkinskie Gory, Russia, 24-30 May 2004 [ArXiv: hep-ph/0410243].

[2] C.D. Froggatt, L.V. Laperashvili, R.B. Nevzorov, H.B. Nielsen and C.R. Das, ArXiv: 0804.4506 [hep-ph].

[3] C.D. Froggatt and H.B. Nielsen, Surveys High Energy Phys **18**, 55 (2003) [ArXiv: hep-ph/0308144]; ArXiv: 0811.2089 [hep-ph].

[4] M.Yu. Kuchiev, V.V. Flambaum and E. Shuryak, Phys.Rev. **D** 78, 077502 (2008) [ArXiv: 0808.3632 [hep-ph]], ArXiv: 0811.1387.
Jean-Marc Richard, Few Body Syst. **45**, 65 (2009) [ArXiv: 0811.2711 [hep-ph]].

[5] Particle Data Group, C.Amster *et al.*, Phys.Lett. **B** 667, (2008).

BOUNDS ON NEW LIGHT PARTICLES FROM HIGH ENERGY AND VERY SMALL MOMENTUM TRANSFER np ELASTIC SCATTERING DATA

Mikhail Vysotsky [a]

ITEP, 117218 Moscow, Russia

*Abstract.*We found that vector and axial vector new light particle exchanges are strongly bounded by high energy data; the analogous bound for a scalar particle is considerably weaker, while for pseudo-scalar exchange particles there are no bounds at all.

1 Introduction

The talk is based on paper [1].

The Standard Model of three fundamental forces describes interactions of elementary particle very well. While the electromagnetic force has a long interaction range, the short radius of the "weak force" $\sim 1/1000$ fermi is determined by the heavy masses of mediating W and Z bosons ($\sim$ 100 GeV). The QCD forces are contained typically within a confinement radius of $\sim$ 1 fermi. The effects of the long-range gravity force can usually be neglected in elementary particle scattering experiments. The search for new forces of nature is a major goal of experiments at high-energy colliders. Rare transitions and decays of fundamental particles can also shed light on new interactions. These experiments are probing new forces at distances shorter than 1/1000 fermi. However, the existence of new forces at distances larger than the confinement radius of a nucleon $\sim$ 1 fermi can also be probed and constrained by sensitive experiments.

In the present paper we reanalyze the experimental small-angle np-elastic scattering data at high energy in terms of bounds on the existence of new forces expressed as S, P, V, or A covariant interactions. The data were obtained in the NA-6 experiment [2] performed at CERN SPS a quarter century ago. Incident neutron energy in the experiment was 100–400 GeV, while the square of 4-momentum transfer $|t|$ was varied in the range $6 \cdot 10^{-6}$ to $5 \cdot 10^{-1}$ GeV^2. The data of this experiment are consistent with extrapolation of the hadronic amplitude from higher $|t|$ values, while at $|t| < 10^{-4}$ GeV^2 the differential cross section rises due to Schwinger scattering, which is the interaction of the neutron's magnetic moment with the Coulomb field of the proton and electron. The purpose of NA-6 [2] was to measure hadronic interactions at high s in the region of momentum transfer $\sim |t| < 10^{-2}$ GeV^2 that was usually unaccessible in the scattering of charged hadrons due to Coulomb interactions. This is the region where the effect of a new force mediated by a light particle can be present.

[a]e-mail: vysotsky@itep.ru

np elastic scattering data in this experiment are well described by the following formula:

$$\frac{d\sigma}{dt} = A\exp[bt] - 2\left(\frac{\alpha k_n}{m_n}\right)^2 \frac{\pi}{t} \ , \tag{1}$$

where $A = (79.78 \pm 0.26) mb/\mathrm{GeV}^2$, $b = (11.63 \pm 0.08)$ GeV^{-2} were determined from the fit to the data, m_n is the neutron mass, and $k_n = -1.91$ originates from the neutron's magnetic moment. This description works rather satisfactorily with χ^2=41.5 for 31 degrees of freedom. We will call it the "zero model" since no new force contributions are included here.

2 Bounds from high energy np scattering

Let us suppose that a new light particle with mass μ exists which interacts with the neutron and proton with couplings g_n and g_p correspondingly. Assuming scalar, pseudoscalar, vector, and axial vector couplings of this particle with nucleons, we obtain the following addition to expression (1):

$$\frac{d\sigma_i}{dt}(g,\mu)|_{\mathrm{new}} = \frac{|A_i|^2}{16\pi s(s-4m^2)} \ , \tag{2}$$

where $s = (p_n + p_p)^2$ is the invariant energy square and m is the nucleon mass.

We use the following amplitude squares for different couplings:

$$|A_S|^2 = \frac{g_S^4}{(t-\mu^2)^2}(4m^2 - t)^2 \ , \tag{3}$$

$$|A_P|^2 = \frac{g_P^4 t^2}{(t-\mu^2)^2} \ , \tag{4}$$

$$|A_V|^2 = \frac{4g_V^4}{(t-\mu^2)^2}[s^2 - 4m^2 s + 4m^4 + st + \frac{1}{2}t^2] \ , \tag{5}$$

$$|A_A|^2 = \frac{4g_A^4}{(t-\mu^2)^2}[s^2 + 4m^2 s + 4m^4 + st + \frac{1}{2}t^2 + \frac{4m^4t^2}{\mu^4} + \frac{8m^4 t}{\mu^2}] \ , \tag{6}$$

where coupling constants $g_i^2 \equiv g_p^i g_n^i$.

Though the strong interaction np scattering amplitude can not be determined theoretically from the first principles, our confidence that $1/t$ dependence is absent in strong interactions for $|t| < m_\pi^2$ opens the road to bounding the light particle exchange if its mass is smaller than that of the π-meson.

Our bounds on the parameters g_V^2 and g_A^2 are rather strong; say for $\mu = 10$, MeV $g_{V,A} < 5 \cdot 10^{-3}$ at 90% C.L., which corresponds to

$$g_N^{V,A} < 0.071 \ , \tag{7}$$

four times smaller than the QED coupling constant $\sqrt{4\pi\alpha} \simeq 0.3$. For scalar exchange, taking $\mu = 10$ MeV, we get a much weaker bound, $g_S^2 < 1.4$.

More stringent constraints on axial couplings were obtained in [3] from the data on $K^+ \to \pi^+ +$ invisible decays.

3 Conclusions

Summarizing, we see that both low energy $n -^{208}$ Pb scattering data and high energy np scattering data lead to similar upper bounds on the coupling constants for ≈ 10 MeV scalar and vector bosons, though upper bounds from $n -$ Pb scattering on the coupling constants $g_N^{V,S}$ are 30 times lower and close to the bounds from $\pi^0 \to$ invisible and $\pi^0 \to \gamma +$ invisible decays on the vector coupling constants with quarks.

Our bounds provide exclusion of new forces at distances from 5 to 200 fermi, which corresponds to exchanged particle masses of 1 MeV to 40 MeV. These bounds are extracted in a covariant approach as an alternative to the bounds on couplings at larger distances extracted as a deviation from classical Newtonian gravitational law.

I am grateful to the organizers of the stimulating Lomonosov Conference and in particular to Alexander Studenikin. My work was partially supported by RFBR grant 08-02-00494 and State contract No. 02.740.11.5158.

[1] Yu. Kamyshkov, J. Tithof, M. Vysotsky, *Phys. Rev. D* **78**, 114029 (2008).
[2] A. Arefiev et al., *Nucl. Phys.* **B232**, 365 (1984).
[3] P. Fayet, *Phys. Rev. D* **74**, 054034 (2006); *Phys. Rev. D* **75**, 115017 (2007).

IS THE FIFTH FAMILY PREDICTED BY THE "APPROACH UNIFYING SPIN AND CHARGES" WHAT FORMS THE DARK MATTER?

Norma Susana Mankoč Borštnik [a]

Faculty of Mathematics and Physics, University of Ljubljana, 1000 Ljubljana, Slovenia

Abstract. The *approach unifying spin and charges* [1–4], which predicts that all the internal degrees of freedom—the spin, all the charges and the families—originate in only two kinds of spins in $d > (1+3)$, offers a new way to understanding the appearance of the charges and the families. A simple starting Lagrange density for gauge fields and for spinors in $d > (1+3)$, which carry nothing but two kinds of spins—the Dirac one and the additional one anticommuting with the Dirac one—and interact with only the gravitational field through the vielbeins and the two kinds of the spin connection fields (the gauge fields of the two kinds of the Clifford algebra objects) manifests (after particular breaks of the starting symmetry) in $d = (1+3)$ the properties of fermions and bosons as postulated by the *standard model of the electroweak and colour interactions*, with the Yukawa couplings included. The *approach* predicts the fourth family with the masses to be possibly seen at the LHC or at somewhat higher energies and the fifth family, which decouples in the Yukawa couplings from the lower four families. The properties of this fifth family members through the evolution of the universe up to today are estimated, predicting that they are what the dark matter is made out of.

1 Introduction

The *standard model of the electroweak and colour interactions* (extended by the right handed neutrinos), fitting with around 25 assumptions and parameters all the existing experimental data, leaves unanswered many open questions, among which are also the questions about the origin of the charges $(U(1), SU(2), SU(3))$, of the families, and correspondingly of the Yukawa couplings of quarks and leptons and of the Higgs mechanism. Answering the question about the origin of families and their masses is the most promising way leading beyond the today's knowledge about the elementary fermionic and bosonic fields.

The simple action in $d = (1+13)$-dimensional space of the *approach unifying spins and charges* [1–4]

$$S = \int d^d x\, E\, \mathcal{L}_f + \int d^d x\, E\, \mathcal{L}_g \tag{1}$$

with the Lagrange density for the two kinds of gauge fields linear in the curvature

$$\begin{aligned}\mathcal{L}_g &= E\,(\alpha\, R + \tilde{\alpha}\tilde{R}),\\ R &= f^{\alpha[a}f^{\beta b]}\,(\omega_{ab\alpha,\beta} - \omega_{ca\alpha}\omega^{c}{}_{b\beta}),\, \tilde{R} = f^{\alpha[a}f^{\beta b]}\,(\tilde{\omega}_{ab\alpha,\beta} - \tilde{\omega}_{ca\alpha}\tilde{\omega}^{c}{}_{b\beta}),\end{aligned} \tag{2}$$

[a] e-mail: norma.mankoc@fmf.uni-lj.si

and for a spinor, which carries in $d = (1+13)$ two kinds of the spin—the Dirac one and the additional one—represented by the two kinds of the Clifford algebra objects [1–4]

$$\begin{aligned} S^{ab} &= \frac{i}{4}(\gamma^a\gamma^b - \gamma^b\gamma^a), \quad \tilde{S}^{ab} = \frac{i}{4}(\tilde{\gamma}^a\tilde{\gamma}^b - \tilde{\gamma}^b\tilde{\gamma}^a), \\ \{\gamma^a,\gamma^b\}_+ &= 2\eta^{ab} = \{\tilde{\gamma}^a,\tilde{\gamma}^b\}_+, \quad \{\gamma^a,\tilde{\gamma}^b\}_+ = 0, \quad \{S^{ab},\tilde{S}^{cd}\}_- = 0, \end{aligned} \tag{3}$$

(there are no other kinds of the Clifford algebra objects) and no charges, and interacts correspondingly only with the vielbeins and the two kinds of spin connection fields

$$\begin{aligned} \mathcal{L}_f &= \frac{1}{2}(E\bar{\psi}\,\gamma^a p_{0a}\psi) + h.c. \\ p_{0a} &= f^{\alpha}{}_a p_{0\alpha}, \quad p_{0\alpha} = p_\alpha - \frac{1}{2}S^{ab}\omega_{ab\alpha} - \frac{1}{2}\tilde{S}^{ab}\tilde{\omega}_{ab\alpha} \end{aligned} \tag{4}$$

offers a real possibility to explain the assumptions of the *standard model*[b].

The *approach* predicts an even number of families, among which is the fourth family, which might be seen at the LHC [2,4,6] or at somewhat higher energies and the fifth family neutrinos and the fifth family baryons with masses several hundred TeV/c^2 forming the dark matter [5].

The *approach* confronted and still confronts several problems (among them are the problems common to all the Kaluza-Klein-like theories), which we[c] are studying step by step when searching for possible ways of spontaneous breaking of the starting symmetries, which lead to the observed properties of families of fermions and of gauge and scalar fields, and looking for predictions the *approach* is making [3,5,6].

I kindly ask the reader to look for more detailed presentation of the *approach* in the refs. [2,4,6].

The action of Eq. (1) starts with the massless spinor which through two kinds of spins interacts with the two kinds of the spin connection fields. The Dirac kind of the Clifford algebra objects (γ^a) determines, when the group $SO(1,13)$ is analysed with respect to the *standard model* groups in $d = (1+3)$ the spin and all the charges, manifesting the left handed quarks and leptons carrying the weak charge and the right handed weak chargeless quarks and leptons [2,4,6]. Accordingly the Lagrange density $\mathcal{L}_f$ manifests after the appropriate breaks of the symmetries all the properties of one family of fermions as assumed by

[b]This is the only theory in the literature to my knowledge, which does not explain the appearance of families by just postulating their numbers on one or another way, through the choice of a group, for example, but by offering the mechanism for generating families.

[c]I started the project named the *approach unifying spins and charges* fifteen years ago, proving alone or together with collaborators step by step, that such a theory has the chance to answer the open questions of the Standard model. The names of the collaborators and students can be found in the cited papers.

the *standard model*, with the three kinds of charges coupling fermions to the corresponding gauge fields, as presented bellow in the first term on the right hand side of Eq.(5).

The second kind ($\tilde{\gamma}^a$) of the Clifford algebra objects (defining the equivalent representations with respect to the Dirac one) determines families. Accordingly manifests the spinor Lagrange density, after the spontaneous breaks of the starting symmetry ($SO(1,13)$ into $SO(1,7) \times U(1) \times SU(3)$ and further into $SO(1,3) \times SU(2) \times SU(2) \times U(1) \times SU(3)$) the *standard model-like* Lagrange density for massless spinors of (four + four) families (defined by $2^{8/2-1} = 8$ spinor states for each member of one family). After two successive breaks first the upper four families (in the Yukawa couplings decoupled from the lower four families) and in the final break (leading to $SO(1,3) \times U(1) \times SU(3)$) the last four families obtain masses through the Yukawa couplings, presented in the second term on the right hand side of Eq.(5) bellow. The third term ("the rest") is unobservable at low energies

$$\mathcal{L}_f = \bar{\psi}\gamma^m (p_m - \sum_{A,i} g^A \tau^{Ai} A^{Ai}_m)\psi + \sum_{s=7,8} \bar{\psi}\gamma^s p_{0s}\, \psi + \text{the rest}. \qquad (5)$$

Here τ^{Ai} $(= \sum_{a,b} c^{Ai}{}_{ab}\, S^{ab})$ determine the hyper ($A = 1$), the weak ($A = 2$) and the colour ($A = 3$) charge: $\{\tau^{Ai}, \tau^{Bj}\}_- = i\delta^{AB} f^{Aijk}\tau^{Ak}$, $f^{1ijk} = 0$, $f^{2ijk} = \varepsilon^{ijk}$, f^{3ijk} is the $SU(3)$ structure tensor.

Correspondingly manifests the $\mathcal{L}_g$ at observable energies all the three known gauge fields. The scalar fields and the gauge fields manifest (after the breaks of symmetries, for which are *responsible* vielbeins and the two kinds of the spin connection fields with the indices $\alpha \neq 0,1,2,3$) through the vielbeins

$$e^a{}_\alpha = \begin{pmatrix} \delta^m{}_\mu & e^m{}_\sigma = 0 \\ e^s{}_\mu = e^s{}_\sigma E^\sigma{}_{Ai} A^{Ai}_\mu & e^s{}_\sigma \end{pmatrix}, \qquad (6)$$

with $E^\sigma{}_{Ai} = \tau^{Ai} x^\sigma$. ($a$ and α define the index in the tangent space and the Einstein index, respectively, m and μ determine the "observed" $(1+3)$-dimensional space. The break from $SO(1,3) \times SU(2) \times SU(2) \times U(1) \times SU(3)$ to $SO(1,3) \times SU(2) \times U(1) \times SU(3)$, appearing at around 10^{13} GeV, leaves the four lower families massless (and mass protected), while the upper four families, with no Yukawa couplings to the lower four families, obtain the masses. The quarks with the lowest mass among the upper four families, clustered into fifth family baryons, are, together with the fifth family neutrinos, the candidates to form the dark matter.

At the last break the action manifests the massive quarks and leptons of the lower four families and the scalar field, which corresponds to the Higgs field postulated by the *standard model*. Making estimations on the tree level we were not able to tell the masses of the fourth family [d]. We are now studying the properties of the Yukawa couplings for the upper four and the lower four families bellow the tree level.

2 The fifth family as the candidate for forming the dark matter clusters

In what follows I briefly present the estimations of the main properties of the fifth family members under the assumption that the fifth family neutron is the lightest fifth family baryon and that the neutrino is the lightest fifth family lepton. Other possibilities are under consideration. For known masses of quarks and leptons and for known breaks of symmetries the behaviour of the fifth family members in the evolution of the universe as well as when the clusters of galaxies are formed and when they interact with the ordinary matter follow (although the calculations are extremely demanded). We have not yet studied the masses and the mixing matrices for the upper four families. From what it is presented it follows that the masses of the fifth family members are above 1 TeV/c^2.

We assume no baryon/antibaryon asymmetry, which anyhow does not influence much the results for large enough masses (above several tens TeV/c^2) of the fifth family quarks. For high enough masses the one gluon exchange determines the properties of quarks in baryons as well as the quarks' interaction with the cosmic plasma when the temperature is above 1 TeV/k_b and correspondingly their freezing out of the plasma and their forming the fifth family neutrons during the expansion. The fifth family baryons and antibaryons decoupling from the plasma before (mostly) or during the phase transition interact among themselves through the "fifth family nuclear force", which is for the masses of the fifth family quarks of the order of several hundred TeV/c^2 for a factor 10^{-10} smaller than the ordinary (first family) nuclear force [5]. Correspondingly the fifth family baryons and antibaryons interact in the dark matter clouds in the galaxies and among the galaxies, when they scatter, dominantly with the weak force. With the ordinary matter they interact through the "fifth family nuclear force" and are obviously not WIMPS (weakly interacting massive particles, if they are meant to interact only with the weak force).

We estimated that the fifth family neutrinos with masses above TeV/c^2 and bellow 200 TeV/c^2 contribute to the dark matter and to the direct measurements less than the fifth family neutrons.

---

[d] For a particular choice of the additionally assumed symmetry on the Yukawa couplings we predicted the masses of the fourth family members as follows: $m_{u_i}/\mathrm{GeV} = (0.0034, 1.15, 176.5, 285.2)$, $m_{d_i}/\mathrm{GeV} = (0.0046, 0.11, 4.4, 224.0)$ and $m_{\nu_i}/GeV = (1\ 10^{-12}, 1\ 10^{-11}, 5\ 10^{-11}, 84.0)$, $m_{e_i}/GeV = (0.0005, 0.106, 1.8, 169.2)$ and the corresponding mixing matrices [4]. Without the additionally imposed symmetry we were able on the tree level only to evaluate the mixing matrices for the assumed fourth family masses.

The fifth family quarks in the expanding universe: To solve the coupled Boltzmann equations for the number density of the fifth family quarks and the colourless clusters of the quarks in the plasma of all the other fermions (quarks, leptons) and bosons (gauge fields) in the thermal equilibrium in the expanding universe we estimate [5] the cross sections for the annihilation of quarks with antiquarks and for forming clusters. We do this within some uncertainty intervals, which take into account the roughness of our estimations. Knowing only the interval for possible values of masses of the fifth family members we solve the Boltzmann equations for several values of quark masses, following the decoupling of the fifth family quarks and the fifth family neutrons out of the plasma down to the temperature 1 GeV/k_b when the colour phase transition starts.

The fifth family neutrons and antineutrons, packed into very tiny clusters so that they are totally decoupled from the plasma, do not feel the colour phase transition, while the fifth family quarks and coloured clusters (and antiquarks) do. Their scattering cross sections grow due to the nonperturbative behaviour of gluons (as do the scattering cross sections of all the other quarks and antiquarks). While the three of the lowest four families decay into the first family quarks, due to the corresponding Yukawa couplings, the fifth family quarks can not. Having the binding energy a few orders of magnitude larger than 1 GeV and moving in the rest of plasma of the first family quarks and antiquarks and of the other three families and gluons as very heavy objects with large scattering cross section, the fifth family coloured objects annihilate with their partners or form the colourless clusters (which result in the decoupling from the plasma) long before the temperature falls bellow a few MeV/k_b when the first family quarks start to form the bound states.

Following further the fifth family quarks in the expanding universe up to today and equating the today's dark matter density with the calculated one, we estimated the mass interval of the fifth family quarks to be 10TeV $< m_{q_5}\, c^2 <$ a few hundreds TeV The detailed calculations with all the needed explanations can be found in ref. [5].

Dynamics of a heavy family baryons in our galaxy and the direct measurements: Although the average properties of the dark matter in the Milky way are pretty well known (the average dark matter density at the position of our Sun is expected to be $\rho_0 \approx 0.3\,\mathrm{GeV}/(c^2\,\mathrm{cm}^3)$, and the average velocity of the dark matter constituents around the centre of our galaxy is expected to be approximately velocity of our Sun), their real local properties are known much less accurate, within the factor of 10 [e]. When evaluating [5] the number of

[e] In a simple model that all the clusters at any radius r from the centre of our galaxy travel in all possible circles around the centre so that the paths are spherically symmetrically distributed, the velocity of a cluster at the position of the Earth is equal to v_S, the velocity of our Sun in the absolute value, but has all possible orientations perpendicular to the radius

events which our fifth family members (or any stable heavy family baryons or neutrinos) trigger in the direct measurements of DAMA [8] and CDMS [9] experiments, we take into account all these uncertainties as well as the uncertainties in the theoretical estimation and the experimental treatments. Let the dark matter member hits the Earth with the velocity $\vec{v}_{dm\,i}$. The velocity of the Earth around the centre of the galaxy is equal to: $\vec{v}_E = \vec{v}_S + \vec{v}_{ES}$, with $v_{ES} = 30$ km/s and $\frac{\vec{v}_S \cdot \vec{v}_{ES}}{v_S v_{ES}} \approx \cos\theta, \theta = 60^0$, $v_S = (100 - 270)$ km/s. The dark matter cluster of the i- th velocity class hits the Earth with the velocity: $\vec{v}_{dmE\,i} = \vec{v}_{dm\,i} - \vec{v}_E$. Then the flux of our dark matter clusters hitting the Earth is: $\Phi_{dm} = \sum_i \frac{\rho_{dmi}}{m_{c_5}} |\vec{v}_{dm\,i} - \vec{v}_E|$, which can be approximated by $\Phi_{dm} = \frac{\rho_0\, \varepsilon_\rho}{m_{c_5}} \{\varepsilon_{v_{dmS}}\, v_S +, \; \varepsilon_{v_{dmES}}\, v_{ES} \cos\theta \, \sin\omega t\}$. The last term determines the annual modulations observed by DAMA [8]. We estimate (due to experimental data and our theoretical evaluations) that $\frac{1}{3} < \varepsilon_{v_{dmS}} < 3$ and $\frac{1}{3} < \frac{\varepsilon_{v_{dmES}}}{\varepsilon_{v_{dmS}}} < 3$.

The cross section for our fifth family baryon to elastically scatter on an ordinary nucleus with A nucleons is $\sigma_A \approx \frac{1}{\pi \hbar^2} < |M_{c_5 A}| >^2 \; m_A^2$, where m_A is the mass of the ordinary nucleus. Since scattering is expected to be coherent the cross section is almost independent of the recoil velocity of the nucleus. Accordingly is the cross section $\sigma(A) \approx \sigma_0\, A^4\, \varepsilon_\sigma$, with $\sigma_0\, \varepsilon_\sigma$, which is $9\,\pi r_{c_5}^2\, \varepsilon_{\sigma_{nucl}}$, with $\frac{1}{30} < \varepsilon_{\sigma_{nucl}} < 30$ (taking into account the roughness with which we treat our heavy baryon's properties and the scattering procedure) when the "nuclear force" dominates. In all the expressions the index c_5 denotes the fifth family cluster, while the index *nucl* denotes the ordinary nucleus.

Then the number of events per second (R_A) taking place in N_A nuclei of some experiment is due to the flux Φ_{dm} equal to $R_A\, \varepsilon_{cut} = \varepsilon_{cut}\, N_A \, \frac{\rho_0}{m_{c_5}}\, \sigma_0\, A^4\, v_S\, \varepsilon\, (1 + \frac{\varepsilon_{v_{dmES}}}{\varepsilon_{v_{dmS}}}\, \frac{v_{ES}}{v_S} \cos\theta\, \sin\omega t)$, where we estimate that $\frac{1}{300} < \varepsilon < 300$ demonstrates the uncertainties in the knowledge about the dark matter dynamics in our galaxy and our approximate treating of the dark matter properties, while ε_{cut} determines the uncertainties in the detections.

Taking these evaluations into account we predict that if DAMA [8] is measuring our fifth family (any heavy stable) baryons then CDMS [9] (or some other experiment) will measure in a few years these events as well [5] provided that our fifth family quarks masses are higher than $m_{q_5} \geq 200 \frac{\mathrm{TeV}}{c^2}$.

3 Concluding remarks

I demonstrated in my talk that the *approach unifying spin and charges*, which assumes in $d = (1+13)-$dimensional space a simple action for a gravitational

---

r with equal probability. In the model that the clusters only oscillate through the centre of the galaxy, the velocities of the dark matter clusters at the Earth position have values from zero to the escape velocity, each one weighted so that all the contributions give ρ_{dm}.

field and massless fermions carrying only two kinds of spin, no charges, and interacting with a gravitational field through vielbeins and the two kinds of spin connection fields, shows the new way beyond the standard model. The action namely manifests at low (observable) energies after particular breaks of symmetries eight families of left handed weak charged quarks and leptons and right handed weak chargeless quarks and leptons, as well as after additional breaks of symmetries, the Yukawa couplings among the lower four families and among the upper four families, while the Yukawa couplings among the lower and the upper four families are zero (in comparison with the age of the universe).

The *approach* predicts the fourth family, to be possibly observed at the LHC or at somewhat higher energies and the stable fifth family, whose baryons and neutrinos might form the dark matter. I presented the properties of the lower four families, estimated at the tree level. Following the history of the fifth family members in the expanding universe up to today and estimating also the scattering properties of this fifth family on the ordinary matter, the evaluated masses of the fifth family quarks, under the assumption that the lowest mass fifth family baryon is the fifth family neutron, are in the interval

$$200\,\mathrm{TeV} < m_{q_5}\, c^2 < 10^5\,\mathrm{TeV}. \tag{7}$$

The fifth family neutrino mass is estimated to be in the interval: a few TeV $< m_{\nu_5}\, c^2 <$ a few hundreds TeV.

References

[1] N.S. Mankoč Borštnik,*Phys. Lett.* **B** 292, 25 (1992), *J. Math. Phys.* 34, 3731 (1993), *Modern Phys. Lett.* **A** 10, 587 (1995), *Int. J. Theor. Phys.* 40, 315 (2001), hep-ph/0711.4681, p.94-113, 53-113, hep-ph/0401043, hep-ph/0401055, hep-ph/0301029.

[2] A. Borštnik Bračič, N.S. Mankoč Borštnik, *Phys Rev.* **D** 74,073013 (2006).

[3] N. S. Mankoč Borštnik, H. B. Nielsen, *Phys. Rev.* **D** 62, 04010 (2000), *J. of Math. Phys.* 43, 5782 (2002), *J. of Math. Phys.* 44, 4817 (2003), *Phys. Lett.* **B** 633, 771 (2006), *Phys. Lett.* **B** 644, 198 (2007), *Phys. Lett.* **B** 110, 1016 (2008).

[4] G. Bregar, M. Breskvar, D. Lukman, N.S. Mankoč Borštnik, *New J. of Phys.* 10, 093002 (2008).

[5] G. Bregar, N.S. Mankoč Borštnik, *Phys. Rev.* **D** 80, 083534 (2009).

[6] N.S. Mankoč Borštnik, arXiv:0912.4532, p.119-135.

[7] D. Lukman, N.S. Mankoč Borštnik, H.B. Nielsen, arXiv:1001.4679.

[8] R. Bernabei at al,*Int. J. Mod. Phys.* **D** 13, 2127 (2004).

[9] Z. Ahmed et al., astro-ph/0802.3530. hep-ph/0711.4681,p.195-201.

60 YEARS OF BROKEN SYMMETRIES IN QUANTUM PHYSICS

FROM THE BOGOLIUBOV THEORY OF SUPERFLUIDITY TO THE STANDARD MODEL

Dmitry Shirkov [a]
Joint Institute for Nuclear Research,
141980 Dubna, Moscow region, Russia

"Phase transition in quantum system, as a rule,
is accompanied by Spontaneous Symmetry Breaking"
Folklore of the middle of the XX century

A retrospective historical overview of the phenomenon of spontaneous symmetry breaking (SSB) in quantum theory, the issue that has been implemented in particle physics in the form of the Higgs mechanism. The main items are:
– The Bogoliubov's microscopical theory of superfluidity (1946);
– The BCS-Bogoliubov theory of superconductivity (1957);
– Superconductivity as a superfluidity of Cooper pairs (Bogoliubov - 1958);
– Transfer of the SSB into the QFT models (early 60s);
– The Higgs model triumph in the electro-weak theory (early 80s);
The role of the Higgs mechanism and its status in the current Standard Model is also touched upon.
This talk follows the line of an overview published in

[1] D. Shirkov, *Phys.Usp.* 52, 549-557 (2009); e-Print: arXiv:0903.3194 [physics.hist-ph]

[a] shirkovd@theor.jinr.ru

UNSTABLE-PARTICLES PAIR PRODUCTION IN MPT APPROACH IN NNLO

M.L.Nekrasov

Institute for High Energy Physics, 142284 Protvino, Russia

*Abstract.*We consider pair production and decay of fundamental unstable particles in the framework of a modified perturbation theory treating resonant contributions in the sense of distributions. Outcomes of a numerical simulation up to the NNLO with taking into account universal massless-particles contributions are presented.

In realistic descriptions of the processes of production and decays of fundamental unstable particles the gauge cancellations and high accuracy of the calculation of resonant contributions must be provided. A modified perturbation theory (MPT) [1, 2] is one of the approaches intended for the solution of these problems. Its main feature is the direct expansion of the probability instead of the amplitude in powers in the coupling constant, on the basis of distribution-theory methods. As the object to be expanded (the cross-section) is gauge invariant, the gauge cancellations in the MPT must be automatically maintained. However the accuracy of the calculation of resonant contributions remains vague. For clearing up this question numerical simulations are required.

In this report I present results of calculations in the MPT up to the NNLO in a model defined as the improved Born approximation for the process $e^+e^- \to \gamma, Z \to t\bar{t} \to W^+b\,W^-\bar{b}$. The total cross-section of the process is

$$\sigma(s) = \int\limits_{s_{\min}}^{s} \frac{\mathrm{d}s'}{s}\, \phi\Big(\frac{s'}{s};s\Big) \iint\limits_{\substack{s_{1\min}\\ s_{2\min}}}^{\infty\ \infty} \mathrm{d}s_1\, \mathrm{d}s_2\;\; \hat{\sigma}(s'\,;s_1,s_2)\,(1+\delta_c)\,, \tag{1}$$

where $\phi(s'/s;s)$ is the flux function, $\hat{\sigma}(s\,;s_1,s_2)$ is the exclusive cross-section, δ_c stands for soft massless-particles contributions, s_1 and s_2 are virtualities of the top quarks. With massless b quarks $s_{1\,\min} = s_{2\,\min} = M_W^2$, $s_{\min} = 4M_W^2$. The ϕ we consider in the leading log approximation. By δ_c the Coulomb factor is meant, taken into consideration in the one-gluon approximation with conventional resummation, not affecting the Breit-Wigner (BW) factors (see details in [2]).

In $\hat{\sigma}(s\,;s_1,s_2)$ the kinematic and BW factors can be extracted,

$$\hat{\sigma}(s\,;s_1,s_2) = \frac{1}{s^2}\,\theta(\sqrt{s}-\sqrt{s_1}-\sqrt{s_2}\,)\sqrt{\lambda(s,s_1,s_2)}\;\Phi(s;s_1,s_2)\,\rho(s_1)\,\rho(s_2)\,. \tag{2}$$

Further, $\Phi(s;s_1,s_2)$ may be determined in the conventional perturbation theory, but the BW factors $\rho(s_1)\rho(s_2)$ must be determined as asymptotic series in the distributions sense in powers in the coupling constant α. For isolated $\rho(s)$ in the case of a smooth weight, the series up to the NNLO looks as follows [1]:

$$\rho(s) \;=\; \delta(s-M^2) \,+\, PV\,\mathrm{Taylor}\big\{\rho(s)\big\} + \sum_{n=0}^{2} c_n\, \delta^{(n)}(s-M^2) + O(\alpha^3)\,. \tag{3}$$

Here M is the top quark mass, $\delta(\cdots)$ is the δ-function, $\delta^{(n)}$ is its n-th derivative. The 'PV Taylor' means the naive Taylor expansion with principal-value presc-

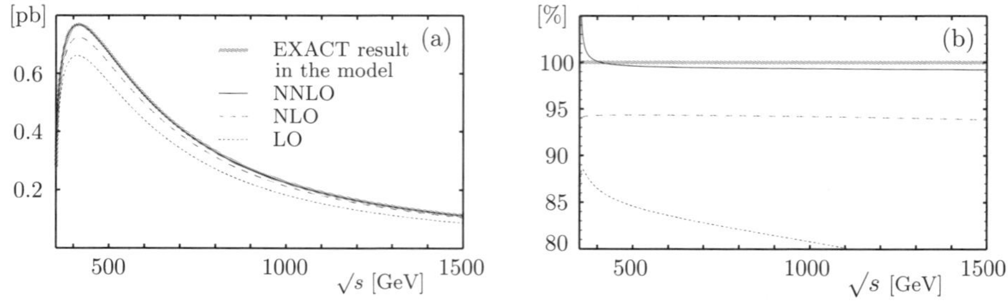

Table 1: The total cross-section in pb and in % with respect to the EXACT result.

$\sqrt{s}$ (TeV)	σ	σ_{LO}	σ_{NLO}	σ_{NNLO}
0.5	0.6724	0.5687	0.6344	0.6698(7)
	100%	84.6%	94.3%	99.6(1)%
1	0.2255	0.1821	0.2124	0.2240(2)
	100%	80.8%	94.2%	99.3(1)%

ription for the poles. Coefficients c_n within the NNLO are determined by three-loop self-energy contributions (for explicit expressions see [2]).

Unfortunately, the weight in (2) is not smooth because of the θ-function and the square root in the kinematic factor. However it can be analytically regularized by substitution $[\lambda(s,s_1,s_2)]^{1/2} \to [\lambda(s,s_1,s_2)]^{\nu}$. The singular integrals then can be analytically calculated and after removing the regularization the results become finite. Moreover, the expansion remains asymptotic [2].

The further numerical calculations, I carry out by means of rather general FORTRAN code with double precision. Parameters of the model I determine as follows: $M = 175$ GeV, $M_W = 80.4$ GeV. The results for the total cross-section above the threshold are shown in the figure (notation is obvious). The results in numerical form at typical ILC energies are presented in the table.

It should be emphasized a very stable behavior of the NLO and NNLO approximations at the energies above and near the maximum of the cross-section (this result to a large extent is model-independent [3]). Near the maximum the discrepancy of the NNLO is within $\pm 0.5\%$, which seemingly is sufficient for the ILC in the case of the top quarks. The further increase of the precision is possible at the proceeding to the NNNLO [2]. Another way for boosting the precision may be based on the combined use of the MPT, when the loop corrections are handled in the MPT framework but Born contribution in the conventional approach with the Dyson-resummation in the unstable-particles propagators (on the analogy of the actual practice of application of DPA).

[1] F.V. Tkachov, in Proc. of the 32nd PNPI Winter School, St.Petersburg, ed. by Ya.I.Azimov et al., PNPI, 166, 1999 [hep-ph/9802307].
[2] M.L.Nekrasov, *Int.Mod.Phys.* **A**24, 6071 (2009) [arXiv:0709.3046].
[3] M.L.Nekrasov, arXiv:0912.1025.

REVEALING STRUCTURE OF QUANTUM CORRECTIONS IN $N = 1$ SUPERSYMMETRIC THEORIES USING THE SCHWINGER–DYSON EQUATIONS

Konstantin Stepanyantz[a]

Faculty of Physics, Moscow State University, 119991 Moscow, Russia

*Abstract.*Some calculations show that all integrals, defining the β-function in supersymmetric theories, are integrals of total derivatives. Here we try to explain this substituting solutions of Slavnov–Taylor identities into Schwinger–Dyson equations.

Most calculations of quantum correction in supersymmetric theories [1] were made with the dimensional reduction [2]. However, it is well known that the dimensional reduction is not self-consistent. A self-consistent regularization, which does not break the supersymmetry, is the higher covariant derivative regularization [3], which was generalized to the supersymmetric case in Ref. [4]. Application of the higher covariant derivative regularization to calculation of quantum corrections in the N=1 supersymmetric electrodynamics in two and three loops [5, 6] reveals an interesting feature of quantum corrections: all integrals, defining the β-function appear to be integrals of total derivatives and can be easily calculated. For example, the two-loop β-function for the general renormalizable N=1 supersymmetric Yang–Mills theory was calculated in Ref. [6]. In the massless case this theory is described by the action

$$S = \frac{1}{2e^2}\mathrm{Re}\,\mathrm{tr}\int d^4x\, d^2\theta\, W_a C^{ab} W_b + \frac{1}{4}\int d^4x\, d^4\theta\, (\phi^*)^i (e^{2V})_i{}^j \phi_j + \\ +\Big(\frac{1}{6}\int d^4x\, d^2\theta\, \lambda^{ijk}\phi_i\phi_j\phi_k + \mathrm{h.c.}\Big). \qquad (1)$$

In order to find the β-function we consider

$$\frac{d}{d\ln\Lambda}\Big(d^{-1}(\alpha_0, \lambda_0, \Lambda/p) - \alpha_0^{-1}\Big)\Big|_{p=0} = -\frac{d\alpha_0^{-1}}{d\ln\Lambda} = \frac{\beta(\alpha_0)}{\alpha_0^2}. \qquad (2)$$

Then the calculations, made with the higher derivative regularization show that the integrals, defining the β-function, have the following structure [6]:

$$I = \int \frac{d^4q}{(2\pi)^4}\Big(\frac{1}{q^2}\frac{d}{dq^2}A(q) + \frac{\partial}{\partial q^\mu}A_\mu(q)\Big), \qquad (3)$$

where $A(q)$ and $A_\mu(q)$ are regular in the limit $q \to 0$. So, in the four dimensional spherical coordinates all these integrals are integrals of total derivatives.

[a]e-mail: stepan@phys.msu.ru

Therefore, they can be easily calculated analytically. The result in the considered approximation agrees with the exact NSVZ β-function [7] and with the results of calculations made with the dimensional reduction.

One can try to explain the factorization of integrands into total derivatives substituting solutions of the Slavnov–Taylor identities into the Schwinger–Dyson equations [8]. Here we consider only a contribution of the matter superfields. The Schwinger–Dyson equations for the considered theory can be written in the following graphical form:

$$\frac{\delta^2\Gamma}{\delta\mathbf{V}_x\delta\mathbf{V}_y} =$$ 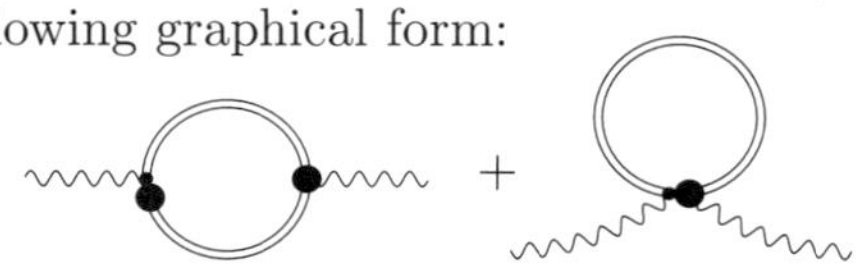 +gauge contribution

Vertexes here contain derivatives w.r.t. auxiliary sources ϕ_0 and φ, which are introduced as follows:

$$S_{\text{Source}} = \int d^4x\, d^2\theta\, (\varphi)_j(\mathbf{U}_1)^j + \frac{1}{4}\int d^8x\, (\phi_0^*)^i(e^{2V'}\phi + \mathbf{U}_0^*)_i + \text{h.c.}, \quad \text{where}$$

$$(\mathbf{U}_0)^i \equiv \frac{1}{6}\lambda^{ijk}\Big[\frac{D^2}{2\partial^2}\Big(\phi_j\phi_k\Big) + 2\Big(\frac{D^2}{2\partial^2}\phi_j\Big)\phi_k\Big]; \qquad (\mathbf{U}_1)^i \equiv \frac{1}{6}\lambda^{ijk}\phi_j\phi_k. \tag{4}$$

These vertexes are restricted by the Slavnov–Taylor identities. Substituting solution of these identities we obtain

$$\frac{d}{d\ln\Lambda}\Big(d^{-1}(\alpha_0,\lambda_0,\Lambda/p) - \alpha_0^{-1}\Big)\Big|_{p=0} = 4\pi T(R)\frac{d}{d\ln\Lambda}\int\frac{d^4q}{(2\pi)^4}\times$$

$$\times\frac{1}{q^2}\frac{d}{dq^2}\Big(\ln(q^2G^2) + 2K_\varphi\Big) +$$ 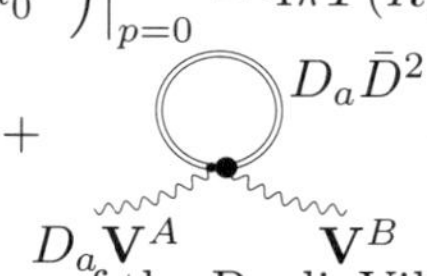

$$-(\text{PV},\text{reg})$$

where (PV,reg) denotes a contribution of the Pauli–Villars fields and a contribution coming from a regularizing term (for simplicity we do not write them explicitly), and the function K_φ is defined by

$$\frac{\delta^2\Gamma}{\delta(\mathbf{j}_y)^j\delta(\varphi_w)_i} \equiv -\frac{1}{2}(K_\varphi)_j{}^i\bar{D}_y^2\delta^8_{yw}. \tag{5}$$

For a theory with cubic superpotential this function is nontrivial starting from the two-loop approximation.

A term with the logarithm is an integral of a total derivative and exactly produces the contribution to the NSVZ β-function, containing the anomalous dimension of the matter superfield. However, it is necessary to explain why the additional contribution, containing the function K_φ and the remaining effective diagram, is an integral of a total derivative and vanishes, at least in the lowest loops. For this purpose we rewrite the last diagram as a sum of two-loop effective diagrams. In the next figure we present only a part of this sum, which corresponds to the triple matter vertex.

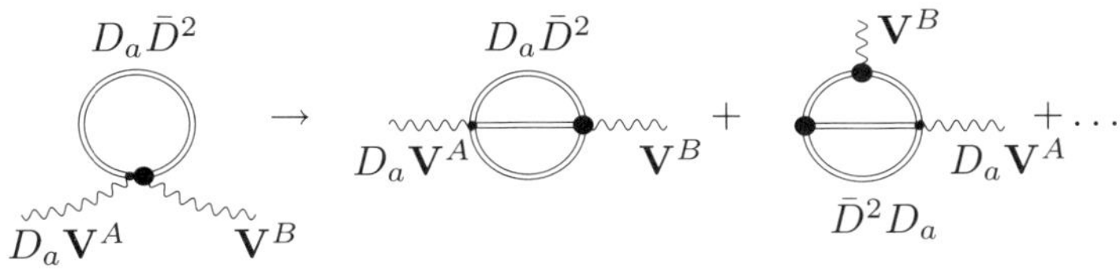

Substituting solutions of the Slavnov–Taylor identities into these diagrams we obtain that some contributions are really factorized into integrals of total derivatives. In particular, this assures vanishing of the additional contribution in the two-loop approximation. However, there is a contribution, containing a function, which can not be found from the Slavnov–Taylor identities. Therefore, the problem is shifted to higher loops. However, we obtained a method, by which one could possibly explain factorization of integrands into total derivatives. To conclude, now it is clear that higher loops in the Schwinger–Dyson equations are essential for explanation of this factorization.

Acknowledgments

This work was partially supported by RFBR grant No 08-01-00281a.

References

[1] L.V.Avdeev, O.V.Tarasov, *Phys.Lett.* **112 B**, 356 (1982); L.F.Abbott, M.T.Grisary, D.Zanon, *Nucl.Phys.* **B244**, 454 (1984); A.Parkes, P.West, *Phys.Lett.* **138B**, 99 (1983); I.Jack, D.R.T.Jones, C.G.North, *Phys.Lett* **B386**, 138 (1996); Nucl.Phys. **B473**, 308 (1996); *Nucl.Phys.* **B486**, 479 (1997); I.Jack, D.R.T.Jones, A.Pickering, *Phys.Lett.* **B435**, 61 (1998).

[2] W.Siegel, *Phys.Lett.* **84 B**, 193 (1979); **94B**, 37 (1980).

[3] A.A.Slavnov, *Nucl.Phys.*, **B31**, 301 (1971); *Theor.Math.Phys.* **13**, 1064 (1972).

[4] V.K.Krivoshchekov, *Theor.Math.Phys.* **36**, 745 (1978); P.West, *Nucl.Phys.* **B 268**, 113 (1986).

[5] A.A.Soloshenko, K.V.Stepanyantz, hep-th/0304083; A.B.Pimenov, E.S.Shevtsova, A.A.Soloshenko, K.V.Stepanyantz, *Russ.Phys.J.*, **51**, 444 (2008).

[6] A.B.Pimenov, E.S.Shevtsova, K.V.Stepanyantz, ArXiv:0912.5191 (hep-th).

[7] V.Novikov, M.Shifman, A.Vainstein, V.Zakharov, *Nucl.Phys.* **B 229**, 381 (1983); *Phys.Lett.* **166B**, 329 (1985); Shifman M.A., Vainshtein A.I., *Nucl.Phys.* **B 277**, 456 (1986).

[8] K.V.Stepanyantz, *Theor.Math.Phys.* **142**, 29 (2005); **150**, 377 (2007).

NEAR-THRESHOLD BOSON-PAIR PRODUCTION IN THE MODEL OF UNSTABLE PARTICLES

Vladimir Kuksa [a], Roman Pasechnik [b]
Institute of Physics, Southern Federal University, 119991 Rostov-on-Don, Russia
Bogoliubov Laboratory of Theoretical Physics, JINR, Dubna 141980, Russia
Abstract. Near-threshold production of boson pairs is considered within the framework of the model of unstable particles with smeared mass. The results of calculations are in good agreement with LEP II data and Monte-Carlo simulations.

1 Boson-pair Production in the Model of Unstable Particles with a Smeared Mass

Near-threshold production of the unstable particles (UP's) is the most suitable process to observe the finite-width effects (FWE). The measurements of boson-pair production at threshold energy (LEP II experiments) have provided us with an important information about the masses of bosons and non-abelian triple gauge-boson couplings. To extract the exact information from boson-pair production we have to calculate the radiative corrections (RC's), which give a noticeable contribution to the cross-section. However, in practice the problem of calculation of full RCs to the processes $e^+e^- \to ZZ$; ZH; $W^+W^- \to \sum_f 4f$ is very complicated [1] and can not be considered analytically. Therefore, the Monte-Carlo simulation, effective approaches and some phenomenological methods are used. In this paper, we describe FWE's in the near-threshold boson-pairs production within the framework of the model of UP's with a smeared masses [2]. The model provides simple and transparent analytical expressions for cross-section of the boson-pair production.

In the case of the boson-pair production $e^+e^- \to ZZ, W^+W^-, ZH$ model cross-section has double-convolution form:

$$\sigma^{tr}(s) = \int\int \sigma^{tr}(s, m_1, m_2)\rho(m_1)\rho(m_2)\, dm_1^2\, dm_2^2, \tag{1}$$

where m_a are the variable masses of bosons and probability density is:

$$\rho(m) = \frac{1}{\pi}\frac{m\Gamma(m)}{[m^2 - M^2(m)]^2 + [m\Gamma(m)]^2}\,. \tag{2}$$

In the case of one UP in the final state, for instance in the process $e^+e^- \to \gamma Z$, we have one integration over variable m_Z. In the calculations we take into account ISR and the main part of radiative corrections with help of the universal expression:

$$\sigma(s, m_1, m_2) = k_{QCD}\int_0^{k_{max}} \rho_\gamma(k)\sigma^{tr}(s(1-k), m_1, m_2)\, dk\,, \tag{3}$$

[a] e-mail: kuksa@list.ru
[b] e-mail:rpasech@theor.jinr.ru

where $\rho_\gamma(k)$ is the photon spectrum which describes soft, hard and virtual radiation. Some vertex corrections are accounted with the help of effective coupling $\alpha(M_W)$ and factor k_{QCD} (for more details, see Refs. [3]- [5]).

Results and Conclusion

The model cross-sections $\sigma(e^+e^- \to ZZ)$ and $\sigma(e^+e^- \to ZZ)$ including above mentioned corrections is represented in Fig. 1 (the solid lines) together with the result of Monte-Carlo simulation (the dashed lines) and LEP data points. Both results are consistent with the data within the error bars. We note that results of Monte-Carlo simulation and model calculations coincide one with another with very high precision. The formalism have been applied, also, to

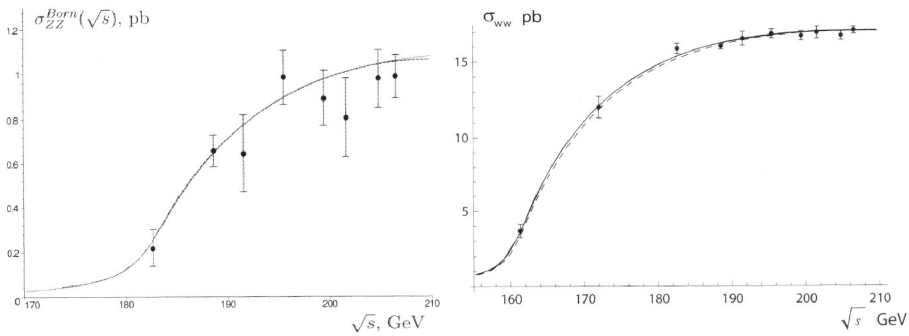

Figure 1: *Total cross-sections of the production ZZ (left panel) and WW (right panel).*

the processes $e^+e^- \to \gamma Z; HZ$ with reasonable results (see Ref. [5]). From our analysis it follows, that the approach under consideration is convenient, simple and transparent framework for the description of boson-pair production which significantly simplifies calculations with respect to standard perturbative one.

References

[1] A. Denner *et al.*, *Nucl. Phys.* **B** 724, 247 (2005).
[2] V. I. Kuksa, *Int. J. Mod. Phys.* **A** 24, 1185 (2009).
[3] V. I. Kuksa, R. S. Pasechnik, *Int. J. Mod. Phys.* **A** 23, 4125 (2008).
[4] V. I. Kuksa, R. S. Pasechnik, *Int. J. Mod. Phys.* **A** (in print) (2009);
[5] V. I. Kuksa, R. S. Pasechnik, *Phys. Atom. Nucl. (in print)*; arXiv:0910.4644.

RADIATIVE AND HADRONIC DECAYS OF VECTOR MESONS IN THE GAUGE MODEL OF QUARK-MESON INTERACTIONS

Vitaly Beylin [a], Vladimir Kuksa [b], Gregory Vereshkov [c]
Southern Federal University, 344090 Rostov-on-Don, Russia

Abstract. Vector meson radiative and hadronic decays are considered within the framework of $U_0(1) \times U(1) \times SU(2)$ gauge symmetry approach. The model parameters are fixed to describe an electromagnetic and strong meson decays in a good agreement with the data.

Failing an exact nonperturbative theory, hadron interactions at low energies are considered in terms of effective Lagrangians, following both from fundamental theory (QCD) and dynamical symmetry concepts. The last approach allows to account for meson-meson and (constituent) quark-meson interactions combined with the VMD ideas in the generalized linear sigma-model (LσM) [1]. To formulate the gauge description of low-energy hadron phenomenology, we used the LσM inspired scheme with $U_0(1) \times U(1) \times SU(2)$ gauge symmetry. It is the simplest group to analyze (non-chiral) electromagnetic and strong meson-meson and quark-meson interactions. Light vector mesons ρ, ω and photon arise in the model as gauge fields realizing tree level VMD and electromagnetic interaction due to extra $U(1)$ groups. To consider the strong effects, $SU(2)$ group, i.e. diagonal sum of $SU_{L,R}(2)$ subgroups of the global chiral group, is localized. After the spontaneous breaking of the $U_0(1) \times U(1) \times SU(2)$ local symmetry, residual scalar degrees of freedom (Higgs fields) can be associated with the scalar isotriplet $a_0(980)$ and singlet $f_0(980)$, and σ- meson is interpreted as $f_0(600)$ state (for details see [2]). Their mass spectrum and decay properties are reasonably described due to free parameters in the scalar sector.

In the model, vector fields couplings universality bounds in the gauge sector the number of free parameters (couplings, mixing angles), which are fixed by the data on two-particle decays ($\Gamma(\rho^+ \to \pi^+\pi^0)$, $\Gamma(\omega \to \pi^+\pi^-)$) and masses of vector mesons [3]. Namely, we get $g_0^2/4\pi = 7.32 \cdot 10^{-3}$, $g_1^2/4\pi = 2.86$, $g_2^2/4\pi = 2.81$, $\sin\phi = 0.031$, $\sin\theta = 0.051$. These values were used for the calculation of the vector meson decay widths to verify the gauge VMD approach in radiative decays $\rho^0 \to \pi^+\pi^-\gamma$ and $\omega \to \pi^+\pi^-\gamma$ at the tree level of meson-meson sector. The model quark-meson sector is tested in processes with quark loops: $\rho \to \pi^0\gamma$, $\omega \to \pi^0\gamma$ and $\omega(\rho) \to \pi^0\pi^+\pi^-$. Note, in calculations we use strong couplings as effective final values, without any loop corrections. But electromagnetic vertices are renormalized by the strong interactions.

The differential width of $\rho^0 \to \pi^+\pi^-\gamma$ can be presented in the form:

$$d\Gamma(E_\gamma)/dE_\gamma = \frac{\alpha_{em} \cdot g_2^2 \cos^2\theta \cdot \cos^2\phi/24\pi^2}{\kappa}\left(F_1(\kappa,\mu) + F_2(\kappa,\mu)\ln F_3(\kappa,\mu)\right),$$

[a] e-mail: vbey@rambler.ru
[b] e-mail: kuksa@list.ru
[c] e-mail: gveresh@gmail.com

where $\kappa = E_\gamma/m_\rho$ $\mu = m_\pi^2/m_\rho^2$ and $F_{1,2,3}(\kappa,\mu)$ have been calculated in [2].

This theoretical spectrum of photons (normalized by the total width) is in a good agreement with the experimental curve [4]. Integrating the spectrum from $E_\gamma^{min} = 50\,\text{MeV}$ up to $E_\gamma^{max} = m_\rho(1-4\mu)/2$, we get the branching $B^{theor}(\rho^0 \to \pi^+\pi^-\gamma) = 1.17 \cdot 10^{-2}$ which is very close to the experimental value $B^{exp} = (0.99 \pm 0.16) \cdot 10^{-2}$ [3]. A phenomenological account of the loop contributions leads to $B^{phen} = (1.22 \pm 0.02) \cdot 10^{-2}$ [5]. An excess of B^{theor} over B^{exp} can be caused by uncertainties in the experimental spectrum at $E_\gamma \leq 100\,\text{MeV}$. Analogously, with the replacement $\cos\phi \to \sin\phi \approx 0.034$; $m_\rho \to m_\omega$ for the decay $\omega \to \pi^+\pi^-\gamma$ we get: $B(\omega \to \pi^+\pi^-\gamma) = 4.0 \cdot 10^{-4}$ and $B(\omega \to \pi^+\pi^-\gamma) = 2.6 \cdot 10^{-4}$ for $E_\gamma^{min} = 30\,\text{MeV}$ and $50\,\text{MeV}$, respectively. Loop corrections can increase $B(\omega \to \pi^+\pi^-\gamma)$ up to $(2-3) \cdot 10^{-3}$, which does not contradict to the experimental limit, $B^{exp}(\omega \to \pi^+\pi^-\gamma) \leq 3.6 \cdot 10^{-3}$ [3].

In the model, radiative decays ρ^0, $\omega \to \pi^0\gamma$ and hadronic decays $\omega, \rho \to 3\pi$ occur via quark loops only with the gauge vertices. Supposing $m_u \approx m_d = m_q$ and using for the πqq coupling the Goldberger-Treiman relation, $g_{\pi qq} \approx m_q/f_\pi$, we have: $\Gamma^{theor}(\omega \to \pi^0\gamma) = 0.74 \pm 0.02\,\text{MeV}$, $(\Gamma^{exp}(\omega \to \pi^0\gamma) = 0.76 \pm 0.02\,\text{MeV})$, $\Gamma^{theor}(\rho^0 \to \pi^0\gamma) = 0.081 \pm 0.003\,\text{MeV}$, $(\Gamma^{exp}(\rho^0 \to \pi^0\gamma) = 0.090 \pm 0.012\,\text{MeV})$. Thus, both the widths agree with the data [3], and the only free parameter - effective quark mass - has the value $m_q = 175 \pm 5\,\text{MeV}$.

Analysis of $\omega(\rho) \to 3\pi$ decays is more cumbersome because of interfering box and $\rho-$ mediated quark loop diagrams. Due to non-negligible parameter m_π^2/m_q^2, the phase space integration in the case must be exact (cf. with [6]). Then $\Gamma^{theor}(\omega \to 3\pi) = 7.8 \pm 0.4(7.6 \pm 0.3)\,\text{MeV}$ for $m_q = 175(200)\,\text{MeV}$. Replacing parameters (mixing angles and masses), we have $\Gamma^{theor}(\rho \to 3\pi) = (0.66 \pm 0.06) \cdot 10^{-3}\,\text{MeV}$. These results are in agreement with the modern data: $\Gamma^{exp}(\omega \to 3\pi) = 7.5 \pm 0.1\,\text{MeV}$ and $\Gamma^{exp}(\rho \to 3\pi) = (0.5 - 2.9) \cdot 10^{-3}\,\text{MeV}$ [3].

So, the gauge quantum field approach can be effectively used to describe the radiative and hadronic decay modes of light vector mesons in the scheme of meson-meson and quark-meson interactions. Remained Higgs degrees of freedom can be associated with known scalar states $a_0(980)$ and $f_0(980)$. Relatively light effective quark can be interpreted as a nucleon component separated from gluon component, which is simulated by σ- meson ($f_0(600)$).

References

[1] M.D.Scadron et al., *Nucl. Phys.* **A** 724, 391 (2003).

[2] V.Beylin, V.Kuksa, G.Vereshkov, ArXiv: hep-ph/0803.0061 (2008).

[3] C.Amsler et al. (Particle Data Group), *Phys. Lett.* **B** 667, 1 (2008).

[4] S.I.Dolinsky et al., *Phys. Rept.* 202, 99 (1991).

[5] A.Gokalp, S.Solmaz, O. Yilmaz, *Phys. Rev.* **D** 67, 073007 (2003).

[6] J.L.Lucio et al., *Phys. Rev.* **D** 61, 034013 (2000).

ENERGY OF UNSTABLE STATES AT LONG TIMES [a]

K. Urbanowski [b], J. Piskorski [c]
University of Zielona Góra, Institute of Physics, ul. Prof. Z. Szafrana 4a, 65–516 Zielona Góra, Poland

Abstract. An effect generated by the nonexponential behavior of the survival amplitude of an unstable state in the long time region is considered. We find that the instantaneous energy of the unstable state for a large class of models of unstable states tends to the minimal energy of the system $\mathcal{E}_{min}$ as $t \to \infty$ which is much smaller than the energy of this state for t of the order of the lifetime of the considered state. Analyzing the transition time region between exponential and non-exponential form of the survival amplitude we find that the instantaneous energy of the considered unstable state can take large values, much larger than the energy of this state for t from the exponential time region. Taking into account results obtained for a model considered, it is hypothesized that this purely quantum mechanical effect may be responsible for the properties of broad resonances such as σ meson as well as having astrophysical and cosmological consequences.

1 Introduction

Searching for the properties of unstable states $|\phi\rangle \in \mathcal{H}$ (where $\mathcal{H}$ is the Hilbert space of states of the considered system) one analyzes their decay law. The decay law, $\mathcal{P}_{\phi}(t)$ of an unstable state $|\phi\rangle$ decaying in vacuum is defined as follows

$$\mathcal{P}_{\phi}(t) = |a(t)|^2, \tag{1}$$

where $a(t)$ is the probability amplitude of finding the system at the time t in the initial state $|\phi\rangle$ prepared at time $t_0 = 0$, $a(t) = \langle\phi|\phi(t)\rangle$ and $|\phi(t)\rangle$ is the solution of the Schrödinger equation for the initial condition $|\phi(0)\rangle = |\phi\rangle$. From basic principles of quantum theory it is known that the amplitude $a(t)$, and thus the decay law $\mathcal{P}_{\phi}(t)$ of the unstable state $|\phi\rangle$, are completely determined by the density of the energy distribution $\omega(\mathcal{E})$ for the system in this state [1–3]

$$a(t) = \int_{Spec.(H)} \omega(\mathcal{E})\, e^{-\frac{i}{\hbar}\,\mathcal{E}\,t}\, d\mathcal{E}. \tag{2}$$

where $\omega(\mathcal{E}) \geq 0$ and $a(0) = 1$.

In [4] assuming that the spectrum of H must be bounded from below, ($Spec.(H) > -\infty$) it was proved that in the case of unstable states there must be $|a(t)| \geq A \exp\left[-b\,t^q\right]$ for $|t| \to \infty$ (where $A > 0$, $b > 0$ and $0 < q < 1$).

Recently the experimental evidence of deviations from the exponential decay law at long times was reported in [5]. This result gives rise to another problem which seems to be important: If (and how) deviations from the exponential

---

[a] This paper is a shortened version of [1].

[b] e-mail: K.Urbanowski@proton.if.uz.zgora.pl; K.Urbanowski@if.uz.zgora.pl

[c] e-mail: J.Piskorski@proton.if.uz.zgora.pl

decay law at long times affect the energy of the unstable state and its decay rate at this time region.

Note that in fact the amplitude $a(t)$ contains information about the decay law $\mathcal{P}_\phi(t)$ of the state $|\phi\rangle$, that is about the decay rate γ_ϕ^0 of this state, as well as the energy $\mathcal{E}_\phi^0$ of the system in this state. This information can be extracted from $a(t)$. Indeed, if $|\phi\rangle$ is an unstable (a quasi–stationary) state then, there is $\mathcal{E}_\phi^0 - \frac{i}{2}\gamma_\phi^0 \equiv i\hbar \frac{\partial a_0(t)}{\partial t}\,\frac{1}{a_0(t)}$, for $t \sim \tau_\phi$, (where $a_0(t) = \exp\,[-\frac{i}{\hbar}(\mathcal{E}_\phi^0 - \frac{i}{2}\gamma_\phi^0)t] \simeq a(t)$ for $t \sim \tau_\phi$, $\tau_\phi = \frac{\hbar}{\gamma_\phi^0}$ and γ_ϕ^0 is the decay rate of $|\phi\rangle$). Taking the above into account one can define the "effective Hamiltonian", h_ϕ, for the one–dimensional subspace of states $\mathcal{H}_{||}$ spanned by the normalized vector $|\phi\rangle$ as follows (see, eg. [6]),

$$h_\phi \stackrel{\rm def}{=} i\hbar \frac{\partial a(t)}{\partial t}\,\frac{1}{a(t)}. \tag{3}$$

So starting with $a(t)$ and using the expression (3) one can calculate the effective Hamiltonian $h_\phi(t)$ in a general case for every t. Thus, one finds the following expressions for the energy and the decay rate of the system in the state $|\phi\rangle$ under considerations, to be more precise for the instantaneous energy and the instantaneous decay rate, (for details see: [7,8]),

$$\mathcal{E}_\phi \equiv \mathcal{E}_\phi(t) = \Re\,(h_\phi(t)), \qquad \gamma_\phi \equiv \gamma_\phi(t) = -\,2\,\Im\,(h_\phi(t)), \tag{4}$$

where $\Re\,(z)$ and $\Im\,(z)$ denote the real and imaginary parts of z, respectively.

The aim of this paper is to discuss the long time behaviour of $\mathcal{E}_\phi(t)$ using $a(t)$ calculated for the given density $\omega(\mathcal{E})$.

2 The model

Let us assume that $Spec.(H) = [\mathcal{E}_{min}, \infty)$, (where, $\mathcal{E}_{min} > -\infty$), and let us choose $\omega(\mathcal{E})$ as follows

$$\omega(\mathcal{E}) \equiv \omega_{BW}(\mathcal{E}, \mathcal{E}_{min}) = \frac{N}{2\pi}\,\Theta(\mathcal{E} - \mathcal{E}_{min})\,\frac{\gamma_\phi^0}{(\mathcal{E} - \mathcal{E}_\phi^0)^2 + (\frac{\gamma_\phi^0}{2})^2}, \tag{5}$$

where N is a normalization constant and $\Theta(\mathcal{E}) = \{1$ for $\mathcal{E} \geq 0$, and 0 for $\mathcal{E} < 0\}$. An substitution of this $\omega_{BW}(\mathcal{E}, \mathcal{E}_{min})$ into (2) leads to the result (see also [9])

$$\begin{aligned} a(t) \;\; &= \;\; N\,e^{-\frac{i}{\hbar}(\mathcal{E}_\phi^0 - i\frac{\gamma_\phi^0}{2})t}\,\Big\{1 - \frac{i}{2\pi}\Big[e^{\frac{\gamma_\phi^0 t}{\hbar}}\,E_1\Big(-\frac{i}{\hbar}(\mathcal{E}_\phi^0 - \mathcal{E}_{min} + \frac{i}{2}\gamma_\phi^0)t\Big) \\ &\quad - E_1\Big(-\frac{i}{\hbar}(\mathcal{E}_\phi^0 - \mathcal{E}_{min} - \frac{i}{2}\gamma_\phi^0)t\Big)\Big]\Big\}, \end{aligned} \tag{6}$$

where $E_1(x)$ denotes the integral–exponential function [9, 10].

In general one has

$$a(t) \equiv a_{exp}(t) + a_{non}(t), \tag{7}$$

where $a_{exp}(t) = N \exp\left[-\frac{i}{\hbar}(\mathcal{E}_\phi^0 - i\frac{\gamma_\phi^0}{2})t\right]$ and $a_{non}(t) = a(t) - a_{exp}(t)$.

Making use of the asymptotic expansion of $E_1(x)$ [10] and (3) one finds

$$h_\phi(t)|_{t\to\infty} \;\simeq\; \mathcal{E}_{min} - i\,\frac{\hbar}{t} - 2\,\frac{\mathcal{E}_\phi^0 - \mathcal{E}_{min}}{|\,h_\phi^0 - \mathcal{E}_{min}\,|^2}\left(\frac{\hbar}{t}\right)^2 + \ldots \tag{8}$$

for the considered case (5) of $\omega_{BW}(\mathcal{E}, \mathcal{E}_{min})$, (for details see [1, 7]). Here $h_\phi^0 = \mathcal{E}_\phi^0 - \frac{i}{2}\gamma_\phi^0$. ¿From (8) it follows that

$$\Re\left(h_\phi(t)|_{t\to\infty}\right) \stackrel{\text{def}}{=} \mathcal{E}_\phi^\infty(t) \simeq \mathcal{E}_{min} - 2\,\frac{\mathcal{E}_\phi^0 - \mathcal{E}_{min}}{|\,h_\phi^0 - \mathcal{E}_{min}\,|^2}\left(\frac{\hbar}{t}\right)^2 \underset{t\to\infty}{\longrightarrow} \mathcal{E}_{min}, \tag{9}$$

where $\mathcal{E}_\phi^\infty(t) = \mathcal{E}_\phi(t)|_{\,t\to\infty}$, and $\Im\left(h_\phi(t)|_{t\to\infty}\right) \simeq -\frac{\hbar}{t}$ for $t \to \infty$.

Relations (8), (9) become important for times $t > t_{as}$, where t_{as} denotes the time t at which contributions to $\left|a(t)|_{t\to\infty}\right|^2$ from the exponential component and from the nonexponential component proportional to $\frac{1}{t^2}$ are comparable, that is (see (7)),

$$|a_{exp}(t)|^2 \simeq |a_{non}(t)|^2 \tag{10}$$

for $t \to \infty$.

3 Numerical calculations

The model considered in Sec. 2 and defined by the density $\omega_{BW}(\mathcal{E}, \mathcal{E}_{min})$, (5), allows one to find numerically the decay curves and the instantaneous energy $\varepsilon_\phi(t)$ as a function of time t. The results presented in this Section have been obtained assuming for simplicity that the minimal energy $\mathcal{E}_{min}$ appearing in the formula (5) is equal to zero, $\mathcal{E}_{min} = 0$. So, all numerical calculations were performed for the density $\tilde{\omega}_{BW}(\mathcal{E}) \equiv \omega_{BW}(\mathcal{E}, \mathcal{E}_{min} = 0)$ for some chosen $\frac{\mathcal{E}_\phi^0}{\gamma_\phi^0}$. While performing calculations particular attention was paid to the form of the probability $|a(t)|^2$, i. e. of the decay curve, and of the instantaneous energy $\varepsilon_\phi(t)$ for times t belonging to the most interesting transition time-region between exponential and nonexponential parts of $|a(t)|^2$, where the property (10) is true and t is of order t_{as}. Some of the result discussed in [1] are presented graphically below:

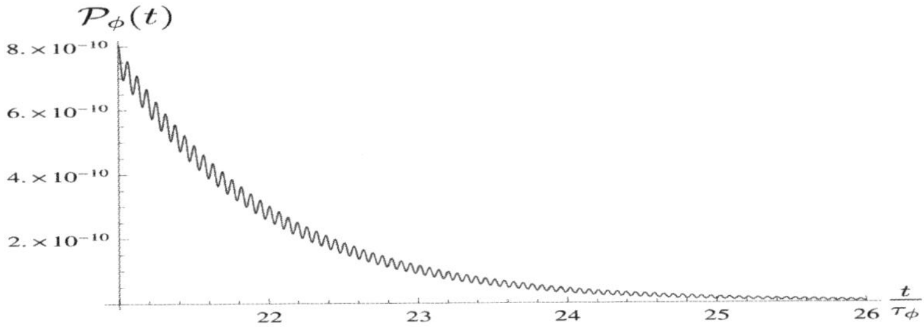

Figure 1: Survival probability $\mathcal{P}_\phi(t)$ in the transition time region. The case $\frac{\mathcal{E}_\phi^0}{\gamma_\phi^0} = 100$.

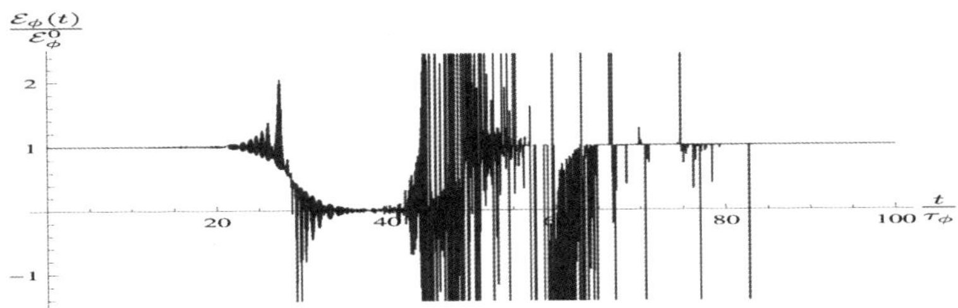

Figure 2: Instantaneous energy $\mathcal{E}_\phi(t)$ in the transition time region. The case $\frac{\mathcal{E}_\phi^0}{\gamma_\phi^0} = 100$.

4 Final remarks.

Decay curves of shapes similar to that in Fig. (1) are found in a large class of models (see [3, 11]). The characteristic feature of all these decay curves is the presence of sharp and frequent oscillations at the transition times region (see Fig. (1)). This means that derivatives of the amplitude $a(t)$ may reach extremely large values for some times from the transition time region and the modulus of these derivatives is much larger than the modulus of $a(t)$, which is very small for these times. This explains why in this time region the real and imaginary parts of $h_\phi(t) \equiv \mathcal{E}_\phi(t) - \frac{i}{2}\gamma_\phi(t)$ reach values much larger than the energy $\mathcal{E}_\phi^0$ of the the unstable state measured at times for which the decay curve has the exponential form.

The question is whether and where this effect can manifest itself. There are two possibilities to observe the above long time properties of unstable states: The first one is that one should analyze the properties of unstable states characterized by t_{as} which are not too long. The second one is finding a possibility to observe a suitably large number of events, i.e. unstable particles, created by the same source.

The problem with understanding the properties of broad resonances in the scalar sector (σ meson problem [12]) discussed in [13, 14], where the hypothesis was formulated that this problem could be connected with properties of the decay amplitude in the transition time region, seems to be possible manifestations of this effect and this problem refers to the first possibility mentioned above (see [1]).

Astrophysical and cosmological processes in which extremely huge numbers of unstable particles are created seem to be another possibility for the above discussed effect to become manifest. The probability $\mathcal{P}_{\phi}(t) = |a(t)|^2$ that an unstable particle, say ϕ, survives up to time $t \sim t_{as}$ is extremely small. Let $\mathcal{P}_{\phi}(t)$ be $\mathcal{P}_{\phi}(t)\big|_{t \sim t_{as}} \sim 10^{-k}$, where $k \gg 1$, then there is a chance to observe some of particles ϕ survived at $t \sim t_{as}$ only if there is a source creating these particles in $\mathcal{N}_{\phi}$ number such that $\mathcal{P}_{\phi}(t)\big|_{t \sim t_{as}} \mathcal{N}_{\phi} \gg 1$. So if a source exists that creates a flux containing $\mathcal{N}_{\phi} \sim 10^{\,l}$, unstable particles and $l \gg k$ then the probability theory states that the number N_{surv} unstable particles $N_{surv} = \mathcal{P}_{\phi}(t)\big|_{t \sim t_{as}} \mathcal{N}_{\phi} \sim 10^{l-k} \gg 0$, has to survive up to time $t \sim t_{as}$. Sources creating such numbers of unstable particles are known from cosmology and astrophysics.

So let us assume that we have an astrophysical source creating a sufficiently large number of unstable particles in a unit of time and emitting a flux of these particles and that this flux is constant or slowly varying in time. Consider as an example a flux of neutrons. There is $t_{as}^{n} \sim (250\tau_n - 300\tau_n)$ for the neutron (here $\tau_n \simeq 886$ [s]). If the energies of these neutrons are of order 30×10^{17} [eV] then during time $t \sim t_{as}^{n}$ they can reach a distance $d_n \sim 25000$ [ly], that is the distance of about a half of the Milky Way radius. Now if in a unit of time a suitably large number of neutrons $\mathcal{N}_n$ of the energies mentioned is created by this source then in the distance d_n from the source a number of spherically symmetric space areas (halos) surrounding the source, where neutron instantaneous energies $\mathcal{E}_n(t)$ are much larger than their energy $\mathcal{E}_n^0 = \frac{m_n^0\, c^2}{\sqrt{1-(\frac{v_n}{c})^2}}$, ($m_n^0$ is the neutron rest mass and v_n denotes its velocity) have to appear. Of course this conclusion holds also for other unstable particles ϕ_α produced by this source. Unstable particles ϕ_α forming such halos and having instantaneous energies $\mathcal{E}_{\phi_\alpha}(t) \gg \mathcal{E}_{\phi_\alpha}^0 = m_{\phi_\alpha}^0\, c^2$ have to interact gravitationally with objects outside of these halos as particles of masses $m_{\phi_\alpha}(t) = \frac{1}{c^2}\, \mathcal{E}_{\phi_\alpha}(t) \gg m_{\phi_\alpha}^0$. The possible observable effects depend on the astrophysical source of these particles considered.

If the halos are formed by unstable particles emitted as a result of internal star processes then in the case of very young stars cosmic dust and gases should be attracted by these halos as a result of a gravitational attraction. So, the halos should be places where the dust and gases condensate. Thus in this case one may consider the halos as the places where planets are born. On the other hand, in the case of much older stars the presence of halos should manifest itself in tiny changes of velocities and accelerations of objects moving in the considered planetary star system relative to those calculated without taking into account the presence of the halos.

If the halos are formed by unstable particles emitted by a galaxy core and these particles are such that the ratio $\frac{\mathcal{E}_\phi(t)}{\mathcal{E}_\phi^0}$ is suitably large inside the halos, then rotational velocities of stars circling the galaxy center outside the halos should differ from those calculated without taking into account the halos. Thus the halos may affect the form of rotation curves of galaxies.

All the above mentioned effects are the simple consequence of the fact that the instantaneous energy $\mathcal{E}_{\phi_\alpha}(t)$ of unstable particles becomes large compared with $\mathcal{E}_{\phi_\alpha}^0$ and for some times even extremely large. On the other hand, this property of $\mathcal{E}_{\phi_\alpha}(t)$ results from the rigorous analysis of properties of the quantum mechanical survival probability $a(t)$ and from the assumption that the energy spectrum is bounded from below.

References

[1] K. Urbanowski and J. Piskorski, arXiv: 0908.2219.

[2] S. Krylov, V. A. Fock, *Zh. Eksp. Teor. Fiz.* **17**, 93 (1947).

[3] L. Fonda, G. C. Ghirardii and A. Rimini, *Rep. on Prog. in Phys.* **41**, 587 (1978).

[4] L. A. Khalfin, *Zh. Eksp. Teor. Fiz.* **33**, 1371 (1957) [*Sov. Phys. — JETP* **6**, 1053 (1958)].

[5] C. Rothe, S. I. Hintschich and A. P. Monkman, *Phys. Rev. Lett.* **96**, 163601 (2006).

[6] K. Urbanowski, *Phys. Rev.* **A 50**, 2847 (1994).

[7] K. Urbanowski, *Cent. Eur. J. Phys.* **7**, 696 (2009).

[8] K. Urbanowski, *Eur. Phys. J.* **D 54**, 25 (2009).

[9] K. M. Sluis, E. A. Gislason, *Phys. Rev.* **A 43**, 4581 (1991).

[10] *Handbook of Mathematical Functions*, Natl. Bur. Stand. Appl. Math. Ser. No 55, eds. M. Abramowitz nad I. A. Stegun (U.S. GPO, Washington, D.C., 1964).

[11] N. G. Kelkar, M. Nowakowski, K. P. Khemchadani, *Phys. Rev.* **C 70**, 024601 (2004).

[12] C. Amsler *at al*, *Phys. Lett.* **B 667**, 1 (2008).

[13] M. Nowakowski, N. G. Kelkar, *Nishiharima 2004, Penataquark — Proceedings of International Workshop on PENATAQUARK 04, Spring – 8, Hyogo, Japan, 23 –24 July 2004*, pp. 182 – 189; arXiv: hep–ph/0411317.

[14] M. Nowakowski, N. G. Kelkar, *AIP Conf. Proc.* **1030**, 250 – 255 (2008); ArXiv:0807.5103.

THE QUANTUM DEFORMATION OF THE WEAKLY EXCITED ELECTRONS SYNCHROTRON RADIATION ANGULAR DISTRIBUTIONS

V. G. Bagrov [a], A. N. Bourimova [b]

Department of Physics, Tomsk State University, Lenin Avenue, Tomsk, Russia; Institute of High Current Electronics of the Siberian Branch of the Russian Academy of Sciences, Tomsk, Russia

Abstract. We give the precise analysis of first excited state electron synchrotron radiation (SR) power using quantum theory methods.

1 Introduction

The precise analytical expressions for the basic synchrotron radiation (SR) characteristics such as spectral-angular distribution, spectral and angular distribution, total output, radiation polarization are entirely formulated in terms of classical theory. These expressions could be found in monographic and educational literature [1–5].

As for the quantum theory the only expressions which could be represented analytically at any quantum factor value are the expressions for the spectral-angular and spectral distributions and for the total SR polarization components output in ultrarelativistic limit [1–5] when we can successfully use the quasi-classical approximation method modification. However even in ultrarelativistic limit the precise analytical expression for the angular distribution at the arbitrary quantum factor value is unknown. Still we can obtain the precise analytical expressions for SR characteristics in terms of quantum theory e. g. in case of initial state charge being in low energy level. The results of the precise analytical and numerical analysis of the first excited state electron SR characteristics are represented below.

2 Basic theoretical expressions

We study the synchrotron radiation (SR) angular distribution structure of an electron in the first excited state (charge $-e, \quad e>0$).

In this case the only possible quantum transition is the transition from the first excited state to the ground state. So, the electron radiates one harmonic with the frequency

$$\omega = \frac{mc^2}{\hbar}\frac{\gamma^2-1}{\gamma+\sqrt{\gamma^2\cos^2\theta+\sin^2\theta}} = \frac{E}{\hbar}\frac{\beta^2}{1+\sqrt{1-\beta^2\sin^2\theta}}, \quad 0\le\theta\le\pi. \quad (1)$$

[a] e-mail: bagrov@phys.tsu.ru

[b] e-mail: llefrith@yandex.ru

Here m - is the rest mass of an electron, c is the velocity of light, $\hbar$ is the Planck constant, $\gamma = E/mc^2 = (1-\beta^2)^{-1/2}$ is the relativistic factor, E is the electron energy, $\beta = v/c$, v is the particle velocity, θ is the angle between the photon's impulse direction and the magnetic field. For the first excited state we have the relation

$$\gamma = \sqrt{1 + 2\frac{H}{H_0}}, \quad H_0 = \frac{m^2c^3}{e\hbar} \text{ – the Schwinger field,} \tag{2}$$

where $H > 0$ - is the magnetic field strength.

Let the quantum number $\zeta = \pm 1$ specify the initial state electron spin orientation with respect to the magnetic field direction, ($\zeta = 1$ means that electron spin and the magnetic field do have the same direction, $\zeta = -1$ signifies that they have opposite directions). The final (ground) state of the electron is distinguished by that the electron being in it has the spin opposite to the field. So, the jump from the initial state $\zeta = 1$ to the ground state occurs with a spin reorientation (spin - flip) only , the transition from the initial state $\zeta = -1$ to the ground state happens without the spin-flip.

In the case under consideration the SR power W and its polarization components angular distribution can be represented as follows

$$\begin{gathered} \frac{dW_i}{d\Omega} = QMG_i, \quad Q = \frac{e^2m^2c^2}{\hbar^2}, \quad d\Omega = \sin\theta d\theta, \\ M = \frac{\beta^6(1+\sqrt{1-\beta^2})}{64(1-\beta^2)} = \frac{2}{(1-x_0^2)^2}\left(\frac{x_0}{1+x_0}\right)^3 = \frac{(\gamma+1)(\gamma^2-1)^3}{64\gamma^5}, \\ x_0 = \frac{1-\sqrt{1-\beta^2}}{1+\sqrt{1-\beta^2}} = \frac{\gamma-1}{\gamma+1}, \quad 0 \le x_0 < 1. \end{gathered} \tag{3}$$

where the index i denotes the radiation polarization: $i = 2$ corresponds to the radiation power of the "σ" - linear polarization component, $i = 3$ is for the "π" - linear polarization component, $i = 0$ is for the total (summed over polarizations) output, $i = l$ ($l = \pm 1$) is for the right ($l = 1$) and left ($l = -1$) circular polarizations. The angular distribution structure is completely defined by the functions $G_i = G_i(\zeta, \beta, \theta)$, given below

$$\begin{gathered} G_2(\zeta, \beta, \theta) = (1-\zeta)R_1(\beta, \theta) + (1+\zeta)x_0R_2(\beta, \theta), \\ G_3(\zeta, \beta, \theta) = (\gamma-\zeta)(\gamma+\zeta)^{-1}\, G_2(-\zeta, \beta, \theta), \\ G_0(\zeta, \beta, \theta) = 2[1 + x_0 - \zeta(1-x_0)]R_0(\beta, \theta), \\ 2R_0(\beta, \theta) = R_1(\beta, \theta) + R_2(\beta, \theta), \\ G_l(\zeta, \beta, \theta) = [1 + x_0 - \zeta(1-x_0)]\, S_l(\beta, \theta), \\ S_l(\beta, \theta) = R_0(\beta, \theta) + l\, S(\beta, \theta)\cos\theta. \end{gathered} \tag{4}$$

Here we introduce the notations

$$R_1(\beta,\,\theta)=\frac{(1-x_0x)(1+x)^3}{1-x}\,e^{-x},\quad R_2(\beta,\,\theta)=\frac{(x_0-x)(1+x)^3}{x_0(1-x)}\,e^{-x},$$

$$R_0(\beta,\,\theta)=\frac{[2\,x_0-(1+x_0^2)x](1+x)^3}{2\,x_0(1-x)}\,e^{-x},\quad S(\beta,\,\theta)=\frac{(1+x)^4}{1-x}\,e^{-x},\qquad(5)$$

$$x=\frac{1-\sqrt{1-\beta^2\sin^2\theta}}{1+\sqrt{1-\beta^2\sin^2\theta}}=\frac{\gamma-\sqrt{\gamma^2\cos^2\theta+\sin^2\theta}}{\gamma+\sqrt{\gamma^2\cos^2\theta+\sin^2\theta}},\quad 0\le x\le x_0<1\,.$$

Thus we can see that in our case the synchrotron radiation angular distribution structure is completely defined by the functions $R_s(\beta,\,\theta),\;\; s=0,\,1,\,2;\;\; S_l(\beta,\,\theta)$.

3 SR polarization components angular distribution structure of the first excited state electron

At first we mention that the functions $R_s(\beta,\,\theta),\quad S_l(\beta,\,\theta)$ are finite at any argument value $0\le\beta\le1$ (including the point $\beta=1$). In particular, at the boundary points of the segment $0\le\beta\le1$ segment we have

$$R_1(0,\,\theta)=1,\; R_2(0,\,\theta)=\cos^2\theta,\; R_0(0,\,\theta)=\frac{1+\cos^2\theta}{2},S_l(0,\,\theta)=\frac{(1+l\cos\theta)^2}{2}.$$
$$R_s(1,\,\theta)=(1+p)^3e^{-p},\;\; S_l(1,\,\theta)=\left(1+l\frac{\cos\theta}{|\cos\theta|}\right)(1+p)^3e^{-p};\;\; p=\frac{1-|\cos\theta|}{1+|\cos\theta|}.\qquad(6)$$

It is not difficult to obtain the relations

$$\begin{aligned}&R_s(\beta,\,0;\,\pi)=1,\quad S_l(\beta,\,0;)=1+l,\quad S_l(\beta,\,\pi)=1-l;\quad R_1(\beta,\,\pi/2)=\\&=2R_0(\beta,\,\pi/2)=2S_l(\beta,\,\pi/2)=(1+x_0)^4e^{-x_0},\quad R_2(\beta,\,\pi/2)=0.\end{aligned}\qquad(7)$$

It follows from (4) and (6) that in the ultrarelativistic limit ($\gamma\to\infty$)the functions $G_i=G_i(\zeta,\,\beta,\,\theta)$ do not longer depend on the spin orientation ζ. It means that in ultrarelativistic limit the spin-flip quantum jumps and the jumps without spin reorientation become equiprobable.

The second expression in (4) implies that the angular distribution structure of the "π" -linear polarization component for an electron with spin ζ and the "σ"-linear polarization component for an electron with spin $-\zeta$ are the same (the corresponding expressions differs only by the factors which do not depend on θ).

The "π" -linear polarization component does not vanish at $\theta=\pi/2$ for an electron with spin $\zeta=1$ and this is an essentially quantum effect.

As it ensues from (6) in the ultrarelativistic case ($\gamma\to\infty$)the only right circular polarization is radiated in the upper half plane ($\theta<\pi/2$),while the lower half plane($\theta > \pi/2$) contains a pure left circular polarization. It is a characteristic feature of the transitions to the ground state.

The analysis of the functions $R_s(\beta,\,\theta)$, $S_1(\beta,\,\theta)$ behavior shows up that the function $R_1(\beta,\,\theta)$ is a monotone increasing function of θ on $0 < \theta < \pi/2$ at any β.

The function $R_2(\beta,\,\theta)$ is monotone decreasing function of θ on $0 < \theta < \pi/2$ and while $\beta^2 \leq 3/4$ it has its maximum at some inner point $\theta = \theta_2^{(max)}(\beta)$ of this interval.

The functions $R_0(\beta,\,\theta)$ on $0 < \theta < \pi/2$ and $S_1(\beta,\,\theta)$ on $0 < \theta < \pi$ at $\beta^2 \leq 1/2$ are monotone decreasing functions of θ. At $\beta^2 > 1/2$ they reach maxima on $0 < \theta < \pi/2$ at the points $\theta = \theta_0^{(max)}(\beta)$ and $\theta = \theta_1^{(max)}(\beta)$, respectively.

The figures illustrate the graphs of the functions $R_s(\beta,\,\theta)$, $S_1(\beta,\,\theta)$ for different β and the graphs of $\theta_n^{(max)}(\beta), (n = 0,\,2)$

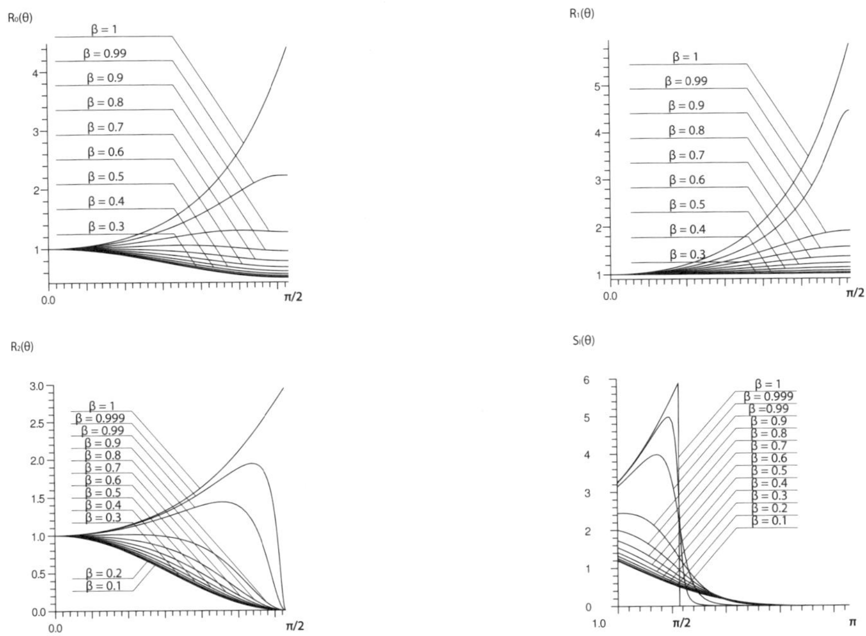

c) d)

Figure 1: This functions define the character of the polarization components angular distribution $R_0(\theta)$ (a), $R_1(\theta)$ (b), $R_2(\theta)$ (c) and $S_l(\theta)$ $(l = 1)$ (d).

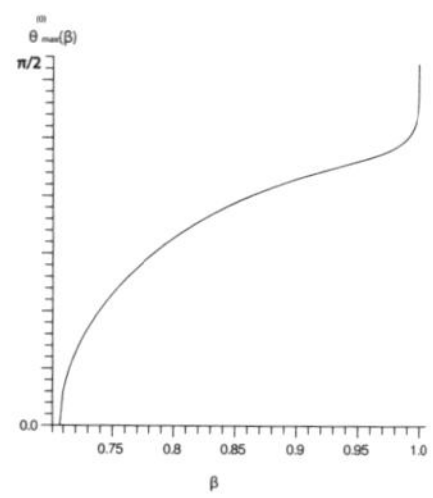

a)

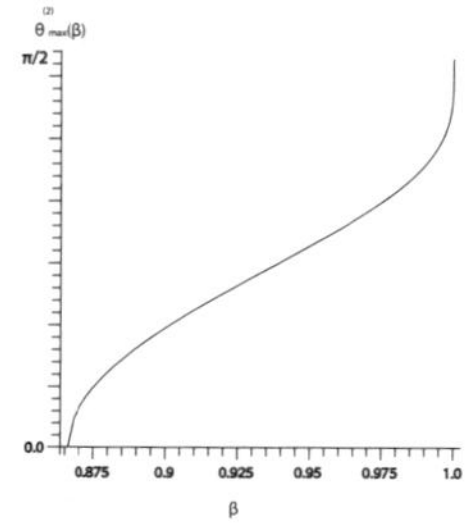

b)

Figure 2: Here the functions $\theta^{(0)}_{max}(\beta)$ (a) and $\theta^{(2)}_{max}(\beta)$ (b) are represented.

4 The total output

For the total output radiated in the upper half plane (integrated over θ from $0 \le \theta \le \pi/2$) we obtain

$$\begin{gathered} W_i = Q\overline{M}P_i(\zeta,\,\beta), \\ \overline{M} = \frac{\beta^6}{32(1-\beta^2)(1+\sqrt{1-\beta^2})} = \frac{1}{(1-x_0)^2}\left(\frac{x_0}{1+x_0}\right)^3 = \frac{(\gamma^2-1)^3}{32\gamma^3(\gamma+1)}, \\ P_2(\zeta,\,\beta) = (1-\zeta)f_1(\beta) + (1+\zeta)x_0 f_2(\beta), \\ P_3(\zeta,\,\beta) = (\gamma-\zeta)(\gamma+\zeta)^{-1}\,P_2(-\zeta,\,\beta), \\ P_0(\zeta,\,\beta) = [1+x_0-\zeta(1-x_0)]f_0(\beta), \quad f_0(\beta) = f_1(\beta)+f_2(\beta), \\ 2P_l(\zeta,\,\beta) = [1+x_0-\zeta(1-x_0)][f_0(\beta)+2l\phi(\beta)], \end{gathered} \tag{8}$$

In (8) the following notations are used

$$\begin{gathered} f_1(\beta) = \int_0^1 (1+x_0 y)\exp(-x_0 y)\sqrt{\frac{1-x_0^2 y}{1-y}}\,dy, \\ f_2(\beta) = \int_0^1 (1+x_0 y)\exp(-x_0 y)\sqrt{\frac{1-y}{1-x_0^2 y}}\,dy, \\ \phi(\beta) = \frac{2-(2+x_0)\exp(-x_0)}{x_0}. \end{gathered} \tag{9}$$

Substituting $l \longrightarrow -l$ in (8) we get the total output radiated in the lower half plane (integrated over θ from $\pi/2 \le \theta \le \pi$).

The functions defined by (9) are positive and bounded, the functions $f_0(\beta)$,

$f_1(\beta)$ and $\phi(\beta)$ are monotone decreasing and $f_2(\beta)$ is monotone increasing on the segment $0 \le \beta \le 1$. It is easy to calculate the values of the function (9) at the segment boundary points

$$f_1(0) = 3f_2(0) = 2\phi(0) = 2, \quad f_1(1) = f_2(1) = \phi(1) = 2 - \frac{3}{e}. \tag{10}$$

At $\beta \ll 1$, $(x_0 \ll 1)$ we find for the functions $f_n(\beta)$

$$f_1(\beta) \approx 2\left(1 - \frac{13}{15}x_0^2\right), \quad f_2(\beta) \approx \frac{2}{3}\left(1 + \frac{11}{35}x_0^2\right), \quad f_0(\beta) \approx \frac{8}{3}\left(1 - \frac{4}{7}x_0^2\right), \tag{11}$$

and for $\phi(\beta)$ we have the convergent at any x_0 series expansion

$$\phi(\beta) = \sum_{n=0}^{\infty} \frac{(-1)^n(1-n)}{(n+1)!}x_0^n \approx 1 - \frac{x_0^2}{6}. \tag{12}$$

For the functions $f_s(\beta)$ $(s = 0,\, 1,\, 2)$ and $\phi(\beta)$ at $\beta \approx 1\, (\gamma \gg 1)$ we find

$$\begin{aligned} f_0(\beta) &\approx 2\left(2 - \tfrac{3}{e} + \tfrac{2}{e}\tfrac{\ln\gamma}{\gamma}\right) + O\left(\tfrac{1}{\gamma}\right), \quad f_1(\beta) \approx 2 - \tfrac{3}{e} + \tfrac{8}{e}\tfrac{\ln\gamma}{\gamma} + O\left(\tfrac{1}{\gamma}\right), \\ f_2(\beta) &\approx 2 - \tfrac{3}{e} - \tfrac{4}{e}\tfrac{\ln\gamma}{\gamma} + O\left(\tfrac{1}{\gamma}\right), \quad \phi(\beta) \approx 2 - \tfrac{3}{e} + 2\left(2 - \tfrac{5}{e}\right)\tfrac{1}{\gamma} + O\left(\tfrac{1}{\gamma^2}\right). \end{aligned} \tag{13}$$

Acknowledgments

The work is partially supported by Russian Science and Innovations Federal Agency under contract No 02.740.11.0238 and Russia President grant SS-871.2008.2.

References

[1] A. A. Sokolov and I. M. Ternov. Synchrotron Radiation. Akademie – Verlag, Berlin; Pergamon Press, New York, 1968.

[2] A. A. Sokolov, I. M. Ternov. Radiation from relativistic electrons. American Institute of Physics, New York, 1986.

[3] I. M. Ternov, V. V. Mikhailin. Synchrotron Radiation. Theory and Experiment. Energoatomizdat, Moskow, 1986.

[4] V. G. Bagrov, G. S. Bisnovatyi – Kogan, V. A. Bordovitsyn et all. Synhrotron Radiation Theory and Its Development. Editor V. A. Bordovitsyn. World Scientific. Singapore. 1999.

[5] V. G. Bagrov, G. S. Bisnovaty – Kogan, V. A. Bordovitsyn et all. Radiation Theory of Relativistic Particles. Editor: V. A. Bordovitsyn. Moscow, Fizmatlit. 2002.

CASIMIR PRESSURE REGULARIZATION AND RENORMALIZATION IN TWO-DIMENSIONAL SCALAR FIELD MODEL

Yu.S. Voronina [a], P.K. Silaev [b]

Faculty of Physics, Moscow State University, 119991 Moscow, Russia

Abstract. A method of calculation Casimir pressure using Green surface function is considered in two-dimensional case. A method of approximate calculation of the regular part of the Green surface function with the help of a Born-type series is suggested.

1 Introduction

In this work we consider the massive scalar field defined on an area with zero boundary conditions. It is worth stressing that there are many various methods of regularization. One of the most effective one is regularization with the use of the Green surface function. The advantage of the method in question is in the fact that the Green function, through which we define the pressure, can be found via several different methods. In particular, the Green function can be found with the help of approximate methods. We consider the pressure rather than the energy as under renormalization of the pressure the choice of the normalization point is simplified, because the pressure, unlike the energy, which is defined up to a constant, has a straightforward physical meaning [2].

The pressure at the area boundary C at the point $\vec{\xi}$ can be found through spatial components of the energy-momentum tensor $p(\vec{\xi}) = \langle 0|n^i n^j T_{ij}(\vec{\xi})|0\rangle$, $i, j = 1, 2, 3$. We consider area D having axial symmetry. The pressure at the intersection point of symmetry axis and boundary is defined by expression [1,3]

$$p(\vec{\xi}) = -\frac{1}{2\pi}\int\limits_{m}^{\infty} \frac{\kappa}{\sqrt{\kappa^2 - m^2}} \left(\vec{n}, \nabla_{\vec{x}} S(\vec{x}, \vec{y})\right)\Big|_{\vec{x}=\vec{y}=\vec{\xi}} d\kappa, \qquad (1)$$

As one can see the pressure is defined through Green surface function $S(\vec{x}, \vec{y})$ of inner Dirichlet problem in area D for the Helmholtz equation. Thus, the pressure regularization and renormalization problem is reduced not only to the definition of the divergent part in the integral with respect to κ in (1) but also to the definition of the regular part in the corresponding Green surface function. In the general case the calculation of Green surface function is rather difficult, due to the nontriviality of the considered area. However, the regular part of this function can be calculated approximately.

[a] e-mail: voronina-yulya@yandex.ru
[b] e-mail: silaev@bog.msu.ru

2 The construction of a Born-type series

We divide the Green surface function into regular and singular parts

$$S(\vec{x},\vec{y}) = S^{(r)}(\vec{x},\vec{y}) + S^{(s)}(\vec{x},\vec{y}) \tag{2}$$

in such a way that the singular part would be defined only by the point $\vec{y} = \vec{\xi}$ and wouldn't depend on other points of the area, and with the regular part defined by the rest of the area. Also we can consider the case, when singular part is defined by the boundary curvature at the considered point. The division into regular and singular parts is ambiguous, and this ambiguity is eliminated after the selection of a normalization point for pressure.

One can show regular part of the Green surface function can be presented as

$$S^{(r)}(\vec{x},\vec{y}) = \sum_{m=1}^{\infty} \underbrace{\oint_C \cdots \oint_C}_{m} S^{(s)}(\vec{x},\vec{z}_1) S^{(r)}(\vec{z}_1,\vec{z}_2) \ldots$$

$$\ldots S^{(r)}(\vec{z}_{m-1},\vec{z}_m) S^{(r)}(\vec{z}_m,\vec{y}) dl_{z_1} \ldots dl_{z_m} \tag{3}$$

By calculating few ferst terms of this series we can obtain regular part of the Green surface function in resoective order of accuracy. The rate of Born series convergence is defined by the division (2). We examine the correctness and computational efficiency of the formula (3) using the example of the boundary problem for the circle, for which the precise analytical solution of the Green surface functions regular part is known [3].

3 Conclusion

So we suggested an algorithm for the approximate calculation of the Green surface function with the help of a Born-type series that is applicable for problems in wich the precise definition of the Green surface function is difficult, but singular part can be chosen in a certain arbitrary way. In the current work, no definite type of the subtracted Green surface function's singular part was determined. It should be noted that this function should be defined based on physical considerations. Thus, after the selection of the normalization point, the problem of the renormalized pressure calculation considered here will ultimately be solved.

References

[1] M. Bordag, U. Mohideen, V.M. Mostepanenko, *Phys.Rep.* **353**, 1 (2001).

[2] Yu.S. Voronina, P.K. Silaev, *Vestn.Mos.univ.Phys.Astron.* **1**, 37 (2009).

[3] Yu.S. Voronina, P.K. Silaev, *Vestn.Mos.univ.Phys.Astron.* **3**, 14 (2009).

TWO-LOOP GELL-MANN–LOW FUNCTION FOR GENERAL RENORMALIZABLE $N = 1$ SUPERSYMMETRIC THEORY, REGULARIZED BY HIGHER DERIVATIVES

Ekaterina Shevtsova [a]

Faculty of Physics, Moscow State University, 119991 Moscow, Russia

Abstract. For the general renormalizable N=1 supersymmetric Yang–Mills theory, regularized by higher covariant derivatives, a two-loop β-function is calculated. It is shown that all integrals, needed for its obtaining are integrals of total derivatives.

We calculate β-function for a general renormalizable N=1 supersymmetric Yang–Mills theory, which is described in the massless case by the action [1, 2]:

$$S = \frac{1}{2e^2}\mathrm{Re}\,\mathrm{tr}\int d^4x\,d^2\theta\,W_a C^{ab} W_b + \frac{1}{4}\int d^4x\,d^4\theta\,(\phi^*)^i (e^{2V})_i{}^j \phi_j + \\ +\Big(\frac{1}{6}\int d^4x\,d^2\theta\,\lambda^{ijk}\phi_i\phi_j\phi_k + \text{h.c.}\Big).$$

In order to regularize this, we add the following higher derivatives terms:

$$S_\Lambda = \frac{1}{2e^2}\mathrm{tr}\,\mathrm{Re}\int d^4x\,d^4\theta\,V\frac{(\boldsymbol{D}_\mu^2)^{n+1}}{\Lambda^{2n}}V + \qquad (1)$$

$$\frac{1}{8}\int d^4x\,d^4\theta\,\Big((\phi^*)^i\Big[e^{\boldsymbol{\Omega}^+}e^{2V}\frac{(\boldsymbol{D}_\alpha^2)^m}{\Lambda^{2m}}e^{\boldsymbol{\Omega}}\Big]_i{}^j\phi_j + +(\phi^*)^i\Big[e^{\boldsymbol{\Omega}^+}\frac{(\boldsymbol{D}_\alpha^2)^m}{\Lambda^{2m}}e^{2V}e^{\boldsymbol{\Omega}}\Big]_i{}^j\phi_j\Big).$$

We are interested in [3, 4] $d\Big(d^{-1}(\alpha_0,\lambda_0,\Lambda/p) - \alpha_0^{-1}\Big)\Big/d\ln\Lambda\,|_{p=0} = \beta(\alpha_0)/\alpha_0^2$. After calculation the corresponding supergraphs, we obtain the following result for the two-loop β-function [5]:

$$\beta_2(\alpha) = -\frac{3\alpha^2}{2\pi}C_2 + \alpha^2 T(R) I_0 + \alpha^3 C_2^2 I_1 + \frac{\alpha^3}{r}C(R)_i{}^j C(R)_j{}^i I_2 + \alpha^3 T(R) C_2 I_3 + \\ +\alpha^2 C(R)_i{}^j\frac{\lambda^*_{jkl}\lambda^{ikl}}{4\pi r}I_4,$$

where $I_i = I_i(0) - \sum\limits_I c_I I_i(M_I)$ for $I = 0,2,3,$ $\sum\limits_I c_I = 1,$ $\sum\limits_I c_I M_I^2 = 0,$

$$I_0(M) = 4\pi\int\frac{d^4q}{(2\pi)^4}\frac{d}{d\ln\Lambda}\frac{1}{q^2}\frac{d}{dq^2}\left[\ln\Big(P(q,M)\Big) + \frac{M^2}{P(q,M)} - \frac{2m\,q^{2m}/\Lambda^{2m}q^2 B(q,m)}{P(q,M)}\right];$$

$$I_1 = 96\pi^2\int\frac{d^4q}{(2\pi)^4}\frac{d^4k}{(2\pi)^4}\frac{d}{d\ln\Lambda}\frac{1}{k^2}\frac{d}{dk^2}\left[\frac{1}{q^2(q+k)^2 B(q,n)B(q+k,n)}\left(\frac{n+1}{B(k,n)} - \frac{n}{B^2(k,n)}\right)\right];$$

[a] e-mail: shevtsova-katya@yandex.ru

$$I_2(M) = -16\pi^2 \int \frac{d^4q}{(2\pi)^4}\frac{d^4k}{(2\pi)^4}\frac{d}{d\ln\Lambda}\frac{1}{q^2}\frac{d}{dq^2}\frac{B(q+k,m)}{P(q+k,M)k^2B(k,n)}\frac{q^2V(q,q+k)}{P(q,M)}\times$$

$$\times\left[\frac{q^4V^2(q,q+k)B^3(q,m)}{P^2(q,M)}+\frac{mq^{2m}}{\Lambda^{2m}}\left(-2B(q,m)+V(q,q+k)-\frac{2M^2V(q,q+k)}{P(q,M)}\right)\right];$$

$$I_3(M) = 4\pi^2\int\frac{d^4q}{(2\pi)^4}\frac{d^4k}{(2\pi)^4}\frac{d}{d\ln\Lambda}\left\{\frac{\partial}{\partial q_\alpha}\left[\frac{k_\alpha}{(k+q)^2B(q+k,n)}\times\right.\right.$$

$$\times\left(-\frac{V^2(q,k)B^3(k,m)B(q,m)}{P^2(k,M)P(q,M)}-\frac{m\,k^{2m}/\Lambda^{2m}V^2(q,k)B(q,m)}{k^2P(k,M)P(q,M)}+\right.$$

$$\left.\left.+\frac{2m\,k^{2m}/\Lambda^{2m}V(q,k)B(k,m)B(q,m)}{k^2P(k,M)P(q,M)}+\frac{2m\,M^2k^{2m}/\Lambda^{2m}V^2(q,k)B(q,m)}{k^2P^2(k,M)P(q,M)}\right)\right]-$$

$$\left.-\frac{1}{k^2}\frac{d}{dk^2}\left[\frac{2V^2(q,q+k)B(q,m)B(q+k,m)}{P(q,M)P(q+k,M)}\left(\frac{1}{B(k,n)}+\frac{nk^{2n}/\Lambda^{2n}}{B^2(k,n)}\right)\right]\right\};$$

$$I_4 = 64\pi^2\int\frac{d^4q}{(2\pi)^4}\frac{d^4k}{(2\pi)^4}\frac{d}{d\ln\Lambda}\frac{1}{q^2}\frac{d}{dq^2}\left[\frac{1}{k^2(q+k)^2B(k,m)}\frac{1}{B(q+k,m)}\left(\frac{1}{B(q,m)}+\frac{mq^{2m}/\Lambda^{2m}}{B^2(q,m)}\right)\right].$$

where we use the following notation: $2+q^{2m}/\Lambda^{2m}+(q+k)^{2m}/\Lambda^{2m}\equiv V(q,q+k)$, $q^2B^2(q,m)+M^2\equiv P(q,M)$, $B(q,m)\equiv 1+q^{2m}/\Lambda^{2m}$. We see that all the integrals are integrals of total derivatives and can be easily calculated. The result is

$$\beta(\alpha) = -\frac{\alpha^2}{2\pi}\Big(3C_2-T(R)\Big)+\frac{\alpha^3}{(2\pi)^2}\Big(-3C_2^2+T(R)C_2+\frac{2}{r}C(R)_i{}^jC(R)_j{}^i\Big)-$$
$$-\frac{\alpha^2C(R)_i{}^j\lambda^*_{jkl}\lambda^{ikl}}{8\pi^3r}+\ldots$$

In the two-loop approximation this agrees with the exact NSVZ β-function [6].

References

[1] A.A.Slavnov, *Nucl.Phys.* **B** 31, 301 (1971).
[2] P.West, *Nucl.Phys.* **B** 268, 113 (1986).
[3] A.A.Soloshenko, K.V.Stepanyantz, *hep-th/0304083; Theor.Math.Phys.* 140, 1264 (2004).
[4] A.B.Pimenov, E.S.Shevtsova, A.A.Soloshenko, K.V.Stepanyantz, *Russ.Phys.J.* 51, 444 (2008).
[5] A.B.Pimenov, E.S.Shevtsova, K.V.Stepanyantz, *hep-th/0912.5191.*
[6] V.Novikov, M.Shifman, A.Vainstein, V.Zakharov, *Nucl.Phys.* **B** 229, 381 (1986).

CASIMIR EFFECT WITHIN $(3+1)D$ MAXWELL-CHERN-SIMONS ELECTRODYNAMICS

Oleg Kharlanov [a], Vladimir Zhukovsky [b]
Department of Theoretical Physics, Faculty of Physics, Moscow State University, 119991 Moscow, Russia

Abstract. Within the context of electrodynamics extended with the Lorentz- and CPT-violating Chern-Simons term, we consider the photon Casimir effect between two parallel perfectly conducting plates. We find the one-photon eigenstates between the plates and then renormalize the sum over their energies using the residue theorem. As a result, we find the leading quadratic Chern-Simons correction to the Casimir force and place a constraint on the Chern-Simons coupling.

1 Introduction

In the late 1990's, the framework of the Lorentz-violating Standard Model Extension (SME) [1] was suggested to phenomenologically account for the low-energy behavior of the mysterious Planck-scale physics. Indeed, some Planck-scale theories predict the spontaneous Lorentz symmetry breaking at near-Planck energies. We work within the so-called (3+1)-dimensional Maxwell-Chern-Simons (MCS) electrodynamics [2], which is a special case of SME, after the electroweak symmetry breaking occurs in the latter theory. We focus on the photon Casimir effect with a setup of two parallel superconductor plates.

2 One-photon eigenstates

The Lagrangian of the MCS electrodynamics contains the Lorentz-violating topological Chern-Simons term in the pure-photon sector,

$$\mathcal{L} = -\frac{1}{4}F_{\mu\nu}F^{\mu\nu} + \frac{1}{2}\eta^{\mu}\epsilon_{\mu\nu\alpha\beta}A^{\nu}F^{\alpha\beta}, \tag{1}$$

where $F_{\mu\nu} = \partial_\mu A_\nu - \partial_\nu A_\mu$ is the field strength tensor, $\epsilon_{\mu\nu\alpha\beta}$ is the Levi-Cività tensor, and η^μ is a constant axial vector. We will consider the case $\eta^\mu = (\eta, \mathbf{0})$, in which we can fix the radiation gauge $A^0 = 0$, $\mathrm{div}\boldsymbol{A} = 0$. Then the equations of motion take the form

$$\Box\boldsymbol{A} = 2\eta\,\mathrm{rot}\boldsymbol{A}, \quad \mathrm{div}\boldsymbol{A} = 0. \tag{2}$$

Choose the cartesian coordinates x, y, z in such a way that the plates occupy the planes $z = \pm a$, $a = D/2$. Then, searching for the one-photon eigenstate wavefunctions in the form

$$\boldsymbol{A}(\boldsymbol{x}, t) = e^{\mp i\omega_n t + i\boldsymbol{k}\cdot\boldsymbol{x}}\boldsymbol{f}_n(z), \qquad \boldsymbol{k} = (k_x, k_y, 0), \tag{3}$$

$$\boldsymbol{k}\cdot\boldsymbol{f}_n(-z) = -\Pi\boldsymbol{k}\cdot\boldsymbol{f}_n(z), \quad \boldsymbol{k}\times\boldsymbol{f}_n(-z) = \Pi\boldsymbol{k}\times\boldsymbol{f}_n(z), \quad \Pi = \pm 1, \tag{4}$$

[a] e-mail: okharl@mail.ru
[b] e-mail: zhukovsk@phys.msu.ru

where $n = (k_x, k_y, \Pi, n_z)$ is the complete set of quantum numbers, we find the implicit expression for the energy eigenvalues ω_n

$$g_\Pi(\omega_n^2) \equiv \varphi_\Pi(\varkappa_+ a)\varphi_{-\Pi}(\varkappa_- a)\sin\theta_- + \varphi_\Pi(\varkappa_- a)\varphi_{-\Pi}(\varkappa_+ a)\sin\theta_+ = 0, \quad (5)$$

where $\sin\theta_\lambda = \varkappa_\lambda/K_\lambda$, $\varkappa_\lambda = \sqrt{K_\lambda^2 - k^2}$, $K_\lambda = -\lambda\eta + \sqrt{\omega_n^2 + \eta^2}$, $\varphi_\pm(\xi) = \left\{ {\cos\xi \atop \sin\xi} \right.$, $\lambda = \pm 1$.

3 Vacuum energy

We have shown in [3] that the Casimir force, which is a derivative of the vacuum energy with respect to the plate separation D, is gauge-invariant, although the energy-momentum tensor $T_{\mu\nu}$ cannot be made gauge-invariant [1]. Within the radiation gauge, the vacuum energy per unit plate area equals [3]

$$\frac{E_{\rm vac}}{L^2} = \frac{1}{2L^2}\sum_n \omega_n(D) = \lim_{\Lambda\to\infty}\frac{1}{2}\int_0^\infty \frac{k\,dk}{2\pi}\sum_{\Pi=\pm 1} D S_\Pi(D,\Lambda), \quad (6)$$

$$S_\Pi(D,\Lambda) = \frac{1}{D}\sum_{n_z}\omega_{k,\Pi,n_z}(D)e^{-\omega_{k,\Pi,n_z}(D)/\sqrt{k\Lambda}}, \quad (7)$$

where Λ is the cutoff regulator. We show in [3] that the zeros ω_{k,Π,n_z} of the function $g_\Pi(\omega^2)$ can be replaced by the zeros of the meromorphic function

$$\tilde{g}_\Pi(K_+) \equiv \frac{g_\Pi(\omega^2)}{\varphi_\Pi(\varkappa_+ a)\varphi_\Pi(\varkappa_- a)} = \tan^\Pi \varkappa_+ a \sin\theta_+ + \tan^\Pi \varkappa_- a \sin\theta_-, \quad (8)$$

which correspond to $K_+ \in (0,\Lambda)$. Here, K_+ is chosen as the parameter, while $\omega = \sqrt{K_+ K_-}$, $K_- = K_+ + 2\eta$. The function $\tilde{g}_\Pi$ does not depend on the sign of η, so we assume $\eta \geq 0$. Now, using the residue theorem, we transform the discrete sum S_Π into a contour integral over the complex K_+-plane:

$$S_\Pi = \frac{1}{D}\oint_C \frac{\omega\, e^{-\omega/\sqrt{k\Lambda}}dK_+}{2\pi i}\left(\frac{\partial\tilde{g}_\Pi/\partial K_+}{\tilde{g}_\Pi(K_+)} - \Psi_\Pi(K_+)\right), \quad (9)$$

where the term containing $\Psi_\Pi(K_+) = a\Pi\sum_{\lambda=\pm 1}\frac{\tan^\Pi \varkappa_\lambda a}{\sin\theta_\lambda}$ cancels the residues of the other term at the poles of $\partial\tilde{g}_\Pi/\partial K_+$. The contour $C = C_{\rm Im} + C_\Lambda$ encloses the domain $\Delta = \{|K_+| < \Lambda, {\rm Re}K_+ > 0\}$ and consists of the segment $C_{\rm Im} = [i\Lambda, -i\Lambda]$ and the semicircle C_Λ of radius Λ.

As it is common in the Maxwell electrodynamics, the renormalization of the above expression involves subtracting the two following counterterms:

$$S_\Pi(D,\Lambda) \quad\to\quad S_\Pi^{\rm ren}(D,\Lambda) = S_\Pi(D,\Lambda) - \frac{C_\Pi^{(1)}(\Lambda)}{D} - C_\Pi^{(2)}(\Lambda). \quad (10)$$

The integral over the semicircle C_Λ has the form of the second counterterm, except for the terms which vanish in the $\Lambda \to \infty$ limit, and is thus fully cancelled when renormalized [3]. After extracting the counterterms from the integral over the imaginary axis, through the consideration of its $D \to \infty$ asymptotic, we find the renormalized expression for the vacuum energy. After making some variable redefinitions, we get

$$\frac{E_{\rm vac}^{\rm ren}}{L^2} = -\frac{1}{4\pi a^3} \int\limits_0^\infty \frac{k\,dk}{2\pi} \sum_{\Pi=\pm 1} \int\limits_{-\infty}^{\infty} \frac{\omega(K_+)dK_+}{\tilde{g}_\Pi} \sum_{\lambda=\pm 1} (1 - \tanh^\Pi \varkappa_\lambda \, \tilde{\Xi}_{\Pi,\lambda}), \quad (11)$$

$$\tilde{\Xi}_{\Pi,\lambda} = 1 + (1 - \tanh^\Pi \varkappa_{-\lambda}) \frac{\cosh\theta_\lambda}{\cosh\theta_{-\lambda}} + \sum_{\lambda'=\pm 1} \frac{\lambda\lambda' \sinh^2\theta_{\lambda'} \cosh\theta_{-\lambda'}}{(\cosh\theta_+ + \cosh\theta_-)\varkappa_{\lambda'}}, \quad (12)$$

$$\tilde{g}_\Pi = \cosh\theta_+ \tanh^\Pi \varkappa_+ + \cosh\theta_- \tanh^\Pi \varkappa_-, \quad (13)$$

where $\omega(K_+) = \sqrt{K_+ K_-}$, $K_- = K_+ - 2i\eta a$, $\varkappa_\lambda = \sqrt{K_\lambda^2 + k^2}$, $\sinh\theta_\lambda = \frac{k}{K_\lambda}$ after these redefinitions. The Taylor expansion of this integral gives us the leading correction to the Casimir force per unit plate area

$$f_{\rm Casimir} = -\frac{\partial}{\partial D} \frac{E_{\rm vac}^{\rm ren}}{L^2} = -\frac{\pi^2}{240 D^4} \left(1 + \frac{25(\eta D)^2}{3\pi^2} + o((\eta D)^3)\right), \quad (14)$$

which is valid for $D \ll 1/|\eta|$. This result differs from the one found in [4] both in sign and in magnitude. Comparing this result with the experimental data on the Casimir effect [5,6], we obtain the constraint on the absolute value of η:

$$|\eta| \lesssim 2 \cdot 10^{-2} \text{ eV}. \quad (15)$$

Although the constraint we have obtained is weak compared to, e.g., astrophysical constraints on η, the effect we have considered is a signature of the quantum structure of the vacuum with broken Lorentz invariance and is thus a subject of experimental interest.

[1] D. Colladay and V. A. Kostelecký, *Phys. Rev. D* **58**, 116002 (1998).

[2] S. M. Carroll, G. B. Field, and R. Jackiw, *Phys. Rev. D* **41**, 1231 (1990).

[3] O. G. Kharlanov and V. Ch. Zhukovsky, *e-Print:* arXiv:0905.3680 [hep-th], accepted for publication in *Phys. Rev. D.*

[4] M. Frank and I. Turan, *Phys. Rev. D* **74**, 033016 (2006).

[5] B. W. Harris, F. Chen, and U. Mohideen, *Phys. Rev. A* **62**, 052109 (2000).

[6] P. Antonini, G. Bressi, and G. Carugno, et al., *New J. Phys.* **8**, 239 (2006).

CASIMIR ENERGY CALCULATIONS FOR CHERN-SIMONS SURFACES AND DIELECTRIC PLATES WITHIN THE FORMALISM OF LATTICE QUANTUM FIELD THEORY

Maxim Ulybyshev [a]

Faculty of Physics, Moscow State University, 119991 Moscow, Russia

Abstract. A new method based on the Monte-Carlo calculation on the lattice is proposed to study the Casimir effect in the noncompact lattice QED. This method can be used for Chern-Simons surfaces (thin metal films) and for dielectric plates.

1 Introduction and general motivation

During the last few years Casimir effect has attracted much attention due to the great experimental and theoretical progress in studying of this phenomenon. This macroscopic quantum effect plays crucial role in nanophysics, micro-mechanics and material science.

Analytic methods for Casimir effect calculation are very complicated and strongly dependent on the shape of a surfaces. So a lot of theoretical methods for calculation of the Casimir effect were proposed. Typically, various approximate methods (like the proximity force approximation method [1], [2]) are used in the case of curved surfaces.

Based on this arguments, it seems a very important task now to create a general method for calculation of Casimir effects which would work well for different shapes of boundary surfaces, for different fields, at non-zero temperature and density and under other external factors. It means that such a method should be formulated very generally for working in different coupling regimes and different external conditions. And we believe that direct lattice calculations in quantum field theory can meet all these requirements.

2 Lattice variables for the description of the Casimir efffect.

In our work we use the four-dimensional hyper-cubical lattice in Euclidian space-time and the following action of so-called non-compact QED:

$$S = \frac{\beta}{2} \sum_{x} \sum_{\mu\nu} \Theta_{\mu\nu}(x)\Theta_{\mu\nu}(x).$$

Here the link variable is simply $\theta_\mu(x) = gA_\mu(x)a$ (a is the step of the lattice) and plaquette variables are defined as:

$$\Theta_{\mu\nu}(x) = \theta_\mu(x) + \theta_\nu(x + \hat{\mu}) - \theta_\mu(x + \hat{\nu}) - \theta_\nu(x).$$

[a] e-mail: ulybyshev@goa.bog.msu.ru

Physical quantities are calculated in the lattice formalism by means of field configuration averaging, where the field configurations (the set of all link variables) are generated with the statistical weight e^{-S}.

Lattice variables for the description of the Casimir effect can be written as:

$$e^{-iS_{add}},$$

where S_{add} is an additional action that describes interaction of boundary surfaces or macroscopic bodies with the electromagnetic field. So we can obtain the Casimir energy of this boundaries or bodies by calculation of the configuration average:

$$\langle e^{iS_{add}} \rangle \to Ce^{-E_{cas}T},$$

where T is size of the bodies in t-direction.

The main task is to rewrite these additional actions in terms of lattice variables (links and plaquettes).

Chern-Simons action is the 3-dimentional integral on the boundary surface S:

$$S = \frac{\lambda}{2} \oint d^3s\, \varepsilon^{\sigma\mu\nu\rho} n_\sigma A_\mu(x) F_{\nu\rho}(x)$$

It can be rewritten on the lattice by the following way:

$$S_{add.CS} = \frac{1}{8}\beta \sum_{x\in\Sigma} n_\mu(x)\theta_{c,\mu}(x), \tag{1}$$

where

$$\begin{aligned}\theta_{c,\mu}(x) = \varepsilon_{\mu\nu\rho\sigma}(\theta_{l,\nu}(x) + \theta_{l,\nu}(x+\hat{\rho}) + \theta_{l,\nu}(x+\hat{\sigma}) +\\ \theta_{l,\nu}(x+\hat{\rho}+\hat{\sigma}))(\theta_{p,\rho\sigma}(x) + \theta_{p,\rho\sigma}(x+\hat{\nu})).\end{aligned} \tag{2}$$

The same operation can be performed in the case of dielectric bodies. The additional action for them can be written as follows:

$$S_{add.} = \frac{(\varepsilon - 1)}{2} \sum_{x\in V} \sum_{i=1}^{3} \Theta_{0i}^2(x).$$

So we have lattice variables for the dielectric bodies and Chern-Simons surfaces. These variables can be used for the Casimir energy calculations in the case of complicated geometry of interacting bodies.

References

[1] 1. M.Bordag, U.Mohideen and V.M.Mostepanenko, *Phys.Rept.* 353 (2001).

[2] 2. B.V.Deryagin, I.I.Abrikosova and E.M.Lifshitz, *Quart.Rev.Chem.Soc.* 10, 295 (1958).

BOUND STATES OF THE FOURTH-GENERATION QUARKS

Oleg Pavlovsky [a], Maxim Ulybyshev [b]
ITPM, Moscow State University, 119991 Moscow, Russia

Abstract. The problem of the boundary state formation of the fourth-generation quarks due to the interaction with the Higgs field was considered. Such interaction leads to the change of the vacuum expectation value of the Higgs field locally around the heavy fermion and to the non-topological soliton generation. This phenomenon plays a very important role for the fourth-generation quarks physics.

The extension of the Standard Model by taking into the account a 4th-generation quarks is a very interesting possibility for the New Physics search and the 4th-generation quarks have been taken into the theoretical considerations. The 4th-generation particles are denoted by the same symbols as the third-generation ones with an added prime (b' and t'). Experimentally, the masses of the 4th generation quarks should be more than 250 GeV. Typically, in theoretical models of the 4th generation extension of the Standard Model the b' quark mass is about 300 GeV, and t' quark mass should be more the 1 TeV.

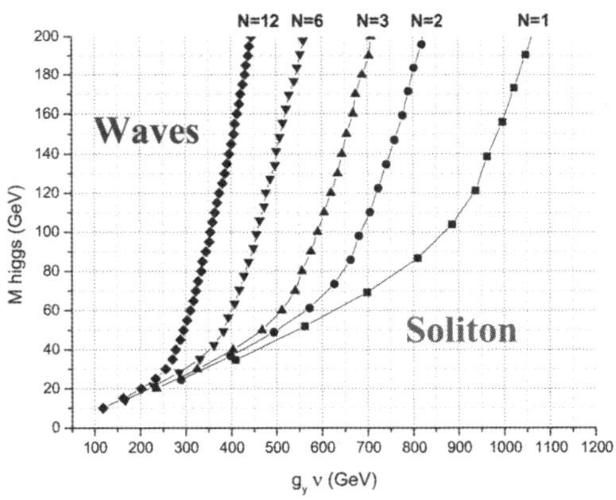

Figure 1: Wave-soliton phase diagram for the bubbles with different numbers of quarks.

Analogously to the Standard Model, let as consider the 4th-generation quarks as the massless Dirac particles been in the Higgs field. Due to the spontaneously symmetry breaking phenomenon the Higgs field has a vacuum expectation value ν and the Yukawa coupling generates the values of the mass for the quarks

[a] e-mail: ovp@goa.bog.msu.ru
[b] e-mail: ulybyshev@goa.bog.msu.ru

$m_q = g_Y \nu/\sqrt{2}$. But as we will show, this scenario of the mass generation correct only for the small coupling constant g_Y. Strong coupling leads to the deflection of Higgs mechanism steam from the fact that in case than the Fermion-Higgs interaction is strong enough it is favorable for the quasi-massless quarks to be inside a bubble-like soliton than to be free and massive.

For illustration of this idea let as consider the spherically symmetrical field configuration for the Higgs fields and the quarks. In this simplest case our problem is to integrate a system of the ordinary differential equations for potentials and finding the solution of the eigenvalue problem for the energy E of the quarks.

On Fig.1 one can find the soliton-wave phase diagram for the different numbers of fermions (N) which are bound inside the bubble (the largest N is 12 due to the Pauli exclusion principle). The generation of the soliton (and the changing of the Higgs mechanism) for one fermion bubbles take place only for very large coupling constants which can be associated with t'-quark physics only. However for multi-quark bubbles the formation of such states is possible for b'-quark too.

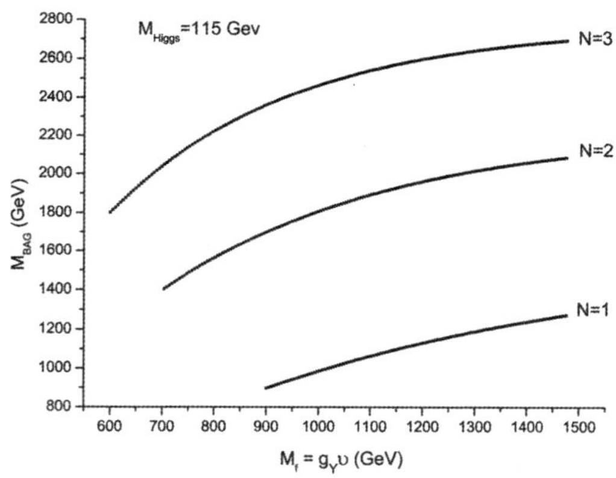

Figure 2: The mass the bubbles with N fermions.

The deflection of the Higgs mechanism in strong Yukawa coupling regime is illustrated on Fig. 2. The linear growth of the mass with coupling is changed on a saturation for any N and the Higgs mass. The physical origin of this phenomenon consists in the generation of the non-topological soliton of the Higgs field around the 4th-generation quarks. The energy of this soliton can be interpret as a binding energy of these quarks in the case misti-quarks bubbles.

This work was partially supported by President RF Grant N.Sh. 3159.2010.2

EXPONENTIAL FORM OF THE MIXING MATRIX IN THE LEPTON SECTOR OF THE STANDARD MODEL

G.Dattoli [a]

Gruppo Fisica Teorica e Matematica Applicata Unita Tecnico Scientifica Tecnologie Fisiche Avanzante (FIS-MAT), ENEA - Centro Richerche Frascati. Via Enrico Fermi 45, I-00044 Frascati, Rome, Italy

K.Zhukovsky [b]

Faculty of Physics, M.V.Lomonosov Moscow State University, Leninskie Gory, 119991, Moscow, Russia

Abstract. We propose a new unitary exponential parameterisation of the neutrino mixing matrix in the Standard Model, where the CP-violating term and the Majorana phases term are accounted for by a special term, separated from the rotational one and discuss its properties in comparison with the O(3) rotation matrix in the angle-axis form.

The mass mixing in the Standard Model (SM) is based on the fact that weak and mass eigenstates differ. Flavour states ν_e, $\nu_\mu$or ν_τ, are linear combinations of neutrino states with different masses. The flavour content of a quantum state may change as it propagates freely. The strength of flavour changes under weak interactions of neutrinos is specified by the Pontecorvo-Maki-Nakagawa-Sakata (PMNS) matrix $\boldsymbol{U}_{PMNS}$ [1]:

$$|\nu_\alpha\rangle = \sum_{i=1,2,3} \mathbf{U}^*_{PMNS\alpha\, i}\, |\nu_i\rangle, \quad \mathbf{U}_{PMNS\alpha\, i} \equiv \langle \nu_\alpha \mid \nu_i \rangle\,. \tag{1}$$

For three neutrinos the following standard parameterisation (SP) of the $\boldsymbol{U}_{PMNS}$ [2] is approved:

$$\mathbf{U}_{PMNS} = \mathbf{U}\mathbf{P}_{Mjr}, \quad \mathbf{P}_{Mjr} = diag\left(e^{i\alpha_1/2}, e^{i\alpha_2/2}, 1\right), \tag{2}$$

where matrix $\boldsymbol{U}$ is identical to that of quark mixing with three mixing angles θ_{12}, θ_{23}, θ_{13}; the CP violating phases are δ and α_i. For Majorana particles $\alpha_i \neq 0$, this has physical consequences only in lepton number violating processes. Regardless of that, α_i do not influence the neutrino oscillations. Experimentally determined values for the mixing angles read as follows [2]: $\theta_{12} \cong 33.9 \pm 2.4°$, $\theta_{23} \cong 45 \pm 7°$, $\theta_{13} \leq 13°$. The angles θ_{12} and θ_{23} are large and only θ_{13} is small in contrast with the mixing matrix for quarks. The tri-bimaximal (TBM) parameterisation [3] of the $\boldsymbol{U}_{PMNS}$ matrix is consistent with experimental data. We advocate exactly unitary exponential parameterisation for the neutrino mixing, following that we constructed for quarks [4]:

$$\tilde{\mathbf{V}} = \mathbf{P_{Rot}}\mathbf{P_{CP}}\mathbf{P_{Mjr}}. \tag{3}$$

[a] e-mail: dattoli@frascati.enea.it

[b] e-mail: zhukovsk@phys.msu.ru

The matrix $\boldsymbol{P}_{Rot}$ (3) represents the angle Φ – axis $\boldsymbol{n}$ form of a rotation in 3D:

$$\mathbf{P_{Rot}} = e^{\mathbf{A}_{Rot}} = \exp\begin{pmatrix} 0 & \lambda & \mu \\ -\lambda & 0 & -\nu \\ -\mu & \nu & 0 \end{pmatrix} = \mathbf{M} = \exp\left[\Phi\begin{pmatrix} 0 & -n_z & n_y \\ n_z & 0 & -n_x \\ -n_y & n_x & 0 \end{pmatrix}\right], \tag{4}$$

the CP violating part and the Majorana part respectively consist of the matrices

$$\begin{aligned} \mathbf{P_{CP}} &= \exp\begin{pmatrix} 0 & 0 & \mu\left(-1+e^{i\delta}\right) \\ 0 & 0 & 0 \\ \mu\left(1-e^{-i\delta}\right) & 0 & 0 \end{pmatrix}, \\ \mathbf{P_{Mjr}} &= \exp\left[i\begin{pmatrix} \alpha_1/2 & 0 & 0 \\ 0 & \alpha_2/2 & 0 \\ 0 & 0 & 0 \end{pmatrix}\right]. \end{aligned} \tag{5}$$

The rotation angles in $\tilde{V}$ coincide with those in the SP form of $\boldsymbol{U}_{PMNS}$ [2] only when δ is small, but $\boldsymbol{U}_{PMNS}$ does not contain any small parameters because of the angles θ_{12} and θ_{23} are large and δ, α_1 and α_2 are unconstrained by experiments. Thus we construct $\tilde{U}$:

$$\tilde{\mathbf{U}} = \mathbf{MP_{CP}} = \left(\ \Xi_1\cos 2\Delta + \kappa^-\Xi_3\sin 2\Delta\,,\quad \Xi_2\,,\quad \Xi_3\cos 2\Delta + \kappa^+\Xi_1\sin 2\Delta\ \right), \tag{6}$$

where

$$\Delta = \mu\sin\frac{\delta}{2},\quad \kappa^{\pm} = ie^{\pm i\frac{\delta}{2}}, \Xi_1 = \begin{pmatrix} M_x \\ M_{yx} \\ M_{zx} \end{pmatrix}, \Xi_2 = \begin{pmatrix} M_{xy} \\ M_y \\ M_{zy} \end{pmatrix}, \Xi_3 = \begin{pmatrix} M_{xz} \\ M_{yz} \\ M_z \end{pmatrix} \tag{7}$$

and the rotation matrix $\boldsymbol{M}$ in 3D has the following well known tensor form:

$$M_{ij} = (1-\cos\Phi)\, n_i n_j + \delta_{ij}\cos\Phi - \varepsilon_{ijk} n_k \sin\Phi\ ,\ i,j,k = x,y,z, \hat{\vec{\mathbf{n}}} = (n_x,\, n_y,\, n_z) \tag{8}$$

where $\boldsymbol{n}$ and Φ are the rotation vector and angle. We omit the expressions for them in terms of the mixing angles θ_{12}, θ_{23}, θ_{13} of the SP form of $\boldsymbol{U}_{PMNS}$ [2] and derive them from (4):

$$n_x = \frac{\nu}{\Phi}, n_y = \frac{\mu}{\Phi},\quad n_z = -\frac{\lambda}{\Phi},\quad \Phi = \pm\sqrt{\lambda^2+\mu^2+\nu^2}. \tag{9}$$

Thus, with account for the TBM entries [3] we obtain the values for λ, μ, ν in $\boldsymbol{P}_{Rot}$: $\lambda \cong 0.5831$, $\mu \cong -0.2415$, $\nu \cong -0.7599$. Surprisingly, the rotation axes $\boldsymbol{n}$ (9) for the neutrino and the quark mixing matrix [4], calculated from the experimental data [2], form the angle of 44.5 degrees. The product $\boldsymbol{P}_{CP}.\boldsymbol{P}_{Mjr}$ yields the following $\boldsymbol{V}_{MCP}$ matrix, responsible for the CP violation effects in the exponential parameterisation of the $\boldsymbol{U}_{PMNS}$ and its expansion in series of the Bessel functions:

$$\mathbf{V_{MCP}} = \mathbf{P_{CP}}\mathbf{P_{Mjr}} = \begin{pmatrix} \xi_1 \cos\, 2\Delta & 0 & \kappa^+ \sin 2\Delta \\ 0 & \xi_2 & 0 \\ \xi_1 \kappa^- \sin 2\Delta & 0 & \cos\, 2\Delta \end{pmatrix}, \quad \xi_{1,2} = e^{i\frac{\alpha_{1,2}}{2}}, \tag{10}$$

$$\mathbf{V_{MCP}} = \begin{pmatrix} \xi_1 \sum\limits_{n=-\infty}^{\infty} J_n\,(2\mu) \cos\left(\frac{n\delta}{2}\right) & 0 & \kappa^+ \sum\limits_{n=-\infty}^{\infty} J_n\,(2\mu) \sin\left(\frac{n\delta}{2}\right) \\ 0 & \xi_2 & 0 \\ \xi_1 \kappa^- \sum\limits_{n=-\infty}^{\infty} J_n\,(2\mu) \sin\left(\frac{n\delta}{2}\right) & 0 & \sum\limits_{n=-\infty}^{\infty} J_n\,(2\mu) \cos\left(\frac{n\delta}{2}\right) \end{pmatrix}, \tag{11}$$

where the CP non-conserving parameter δ is separated from the parameter μ in the arguments of the Bessel functions $J_n(2\mu)$. The obtained result is exact in the terms of δ and α_1, α_2. Assuming that α and δ are small[c] we have

$$\mathbf{V_{MCP}} = \mathbf{P_{CP}}\mathbf{P_{Mjr}} \cong \exp\begin{pmatrix} i\alpha_1/2 & 0 & \left(-1 + e^{i\delta}\right)\mu \\ 0 & i\alpha_2/2 & 0 \\ \left(1 - e^{-i\delta}\right)\mu & 0 & 0 \end{pmatrix} + O\left(\alpha_1 \mu \delta\right). \tag{12}$$

The $\boldsymbol{V}_{MCP}$ matrix is asymmetric with respect to $\xi_{1,2}$ and the Majorana term interplays with the δ phase. When $\alpha_1 = 0$, the symmetry in (10) is restored. For $\xi_2 = 1$, i.e. $\alpha_2 = 0$, the $\boldsymbol{V}_{MCP}$ matrix reminds the mixing matrix for just two lepton generations, acting on $\boldsymbol{e}$, τ and ν_e, ν_τ with the weights ξ_1 for the entry (1,1), κ^+ for the entry (3,1) and $\xi_1\ \kappa^-$ for the entry (1,3). For $\alpha_2 \neq 0$, the vector of the mixed neutrino states under the action of the $\boldsymbol{V}^*_{MCP}$ matrix becomes

$$|\tilde{\nu}_\alpha\rangle = \begin{pmatrix} |\nu_1\rangle\, \xi_1^* \cos 2\Delta + |\nu_3\rangle\, \kappa^{+*} \sin 2\Delta \\ |\nu_2\rangle\, \xi_2^* \\ |\nu_1\rangle\, \xi_1^* \kappa^{-*} \sin 2\Delta + |\nu_3\rangle \cos 2\Delta \end{pmatrix}. \tag{13}$$

The direct check confirms the unitarity of $\boldsymbol{V}_{MCP}$ of the exponential parameterisation $\tilde{V}$:

$$\mathbf{V_{MCP}^{-1}} \cdot \mathbf{V_{MCP}} = \mathbf{V_{MCP}^{+}} \cdot \mathbf{V_{MCP}} = \mathbf{I}, \quad \tilde{\mathbf{V}}^{-1} \cdot \tilde{\mathbf{V}} = \tilde{\mathbf{V}}^{+} \cdot \tilde{\mathbf{V}} = \mathbf{I}. \tag{14}$$

The action of $\boldsymbol{P}_{Rot}$ on the neutrino vector (13) yields the following final result:

$$\begin{pmatrix} |\bar{\nu}_e\rangle \\ |\bar{\nu}_\mu\rangle \\ |\bar{\nu}_\tau\rangle \end{pmatrix} = \mathbf{M}\,\mathbf{V^*_{MCP}} \begin{pmatrix} |\nu_1\rangle \\ |\nu_2\rangle \\ |\nu_3\rangle \end{pmatrix} = F\,|\nu_1\rangle + G\,|\nu_2\rangle + H\,|\nu_3\rangle\,, \tag{15}$$

[c]Note that $|\mu|$ is rather small, compared with 1, and hence the expansion in power series yields $\mathbf{U} \cong \mathbf{P_{Rot}}\mathbf{P_{CP}}$ when $\mu\delta \ll 1$. However, the condition $\mu\delta \ll 1$ is fulfilled only roughly here, contrary to the case of quarks, and we do not make use of this expansion.

where α denotes the type of the neutrino, $\boldsymbol{M}$ is the rotation matrix (4), (8) and

$$\begin{aligned} F &= \xi_1^* \left(\Xi_1 \cos 2\Delta + \Xi_3 \kappa^{-*} \sin 2\Delta\right), \\ G &= \xi_2^* \Xi_2 H = \xi_2^* \left(\Xi_1 \kappa^{+*} \sin 2\Delta + \Xi_3 \cos 2\Delta\right). \end{aligned} \tag{16}$$

Ξ_i, Δ, κ and ξ in (16) are given by (7),(10),(8), where the $\boldsymbol{n}$ and Φ are expressed in terms of μ, λ, ν (9) of the exponential mixing matrix. The contribution due to ν_2 neutrinos is affected by α_2 in G. The ν_1 and ν_3 neutrino enter respectively with α_1 and α_2 dependent factors F and H (16), which can be seen as a rotation in the angle 2Δ, determined by the CP phase δ with the weights Ξ and κ. When δ=0, the neutrino mixing can be viewed as the rotation around a fixed axis in 3D space in the angle Φ. For $\Phi = 0$, the neutrino mixing fades out, $\boldsymbol{U}_{PMNS}$ becomes unit matrix I. The CP-violating phases δ and α_i break this symmetry. From the experimental data and [3] we calculate $\Phi \cong 56°$ and conclude that the rotation axes for the neutrino and the quark mixing form the angle of $\approx$ 45 degrees, thus reformulating the hypothesis of equality and complementary angles for quarks and neutrinos [5]. The exponential parameterisation of the $\boldsymbol{U}_{PMNS}$ gives several advantages. Generation of new unitary parameterisations with separated CP violating part is possible. Combined effect and the interplay of δ and Majorana phases α_i in the CP violation can be studied in the exponential forms (10),(12). The Bessel series (11) separate δ and α_i terms in the coefficients. Transformation of the neutrino vector by the exponential parameterisation of the $\boldsymbol{U}_{PMNS}$ (15),(16) distinguishes the contribution of the CP violating phase for each neutrino type, providing a useful tool for the analysis of experimental data on neutrino oscillations.

Acknowledgments

We thank the Organizing Committee for the invitation to the 14th Lomonosov Conference on Elementary Particle Physics and A.V.Borisov and A.E.Lobanov for useful discussions of the results of the work.

References

[1] Z.Maki, M.Nakagawa and S.Sakata, *Progr. Theor. Phys.*, 28, 870 (1962).

[2] PARTICLE DATA GROUP W-M Yao et al., *J. Phys. G: Nucl. Part. Phys.*, 33, 1,(2006).

[3] S.F.King, *arXiv:hep-ph.* /0710.0530.

[4] G.Dattoli and K.Zhukovsky, *Eur. Phys. J.* **C**, 50, 817 (2007); *Eur. Phys. J.* **C**, 52, 591 (2007).

[5] H.Minakata, A.Yu.Smirnov, *Phys.Rev.* **D**, 70, 073009 (2004).

"THE BUNDLE FOR MEMORY" FROM AN EXOTIC QUASIPARTICLES

Vladimir Skvortsov [a], Nadeshda Vogel [b]
Moscow Institute for Physics and Technology (State University), 141700, Moscow region, Dolgoprudny, Institutive lane 9, Russia (a); University of Technology Chemnitz, 09107 Chemnitz, Germany (b)

*Abstract.*The recent results of micro fibers of diamond bundles and gyroscope creation by using an exotic quasiparticles (magnetic charges) are represented. We note that classic L. Euler's model of gyroscope can be used as theoretical model of magnetic charge vortex. A new type of an exotic (twin) quasiparticles have been identified, which are also generated and detected in laser-discharge experiments in vacuum and air.

1 Introduction

Few interesting physical effects had been observed at first during our investigation of the processes generation of and exotic quasi particles and its interaction with matter in laser-produced plasma experiments (see its schemes in [2]). Namely, we observed the discrete trajectory of moving an exotic quasiparticles (in vacuum) as was predicted at first by W. Heisenberg [1] (see, for example Fig.1 and an additional pictures in [2–5]). Here the height of open slit $= 15mm$. By using the new method for diamond synthesis [6] we were able to obtain a strongly magnetized diamond fibers (see Fig.2) which attracted by steel pincers and had specific specter of Raman scattering (Fig.3). The analogous bundles have been obtained on the surfaces of dielectric detectors from CR39 (Fig.4). We already wrote that an exotic quasiparticles have vortex-like structures [2,5] (Fig.5, height of spinning top equals $\approx 1cm$). Now we can add that sometimes a diamonds in form as "freezing" gyroscope (Fig.6) had been obtained. In this paper we try to give our explanation of mentioned above physical phenomena. Finely, we would like to remember the well forgotten scientific work of M.V. Lomonosov, who was the first who observed and described analogous an exotic quasiparticles in 1753 year [7].

2 The physical nature of observed an exotic quasi particles

Why we think that observed an exotic quasi particles are magnetic charges? As we already demonstrated an exotic quasiparticles separated in two groups near two different poles of magnets. They produces character tracks on dielectrics (for example, tubular tracks, see Fig.7, scale $X = Y = 921.3\ \mu$) predicted by L. Schiff [8], and long line tracks which at passed times try to found by E. Fermi, E. Teller and others, see (Fig.8). Besides this the super strong magnetic

[a]e-mail: skv@mail.mipt.ru
[b]e-mail: n.vogel@physik.tu-chemnitz.de

fields and anomalous plasma droplets with few miniature magnetic dipoles had been detected and investigated [9–11]. The observed bundle topology (Fig.2,4) show us that most probably the bundle theory for describing an elementary particles has a good chance for its the best descriptions. As already wrote in [12], the structure of magnetic monopole may be very similar to gyroscope described by Leonard Euler equations in case of zero main inertial moments. This is in a good agreement with our measurements of mass an exotic quasi particles [4] $m_d \approx 2 \cdot 10^{-35}$g (namely due to this fact we gave name "monopolino" to an exotic quasiparticles). The generation of an exotic quasiparticles may be connected with the development of thermomagnetic instability [5] or an explosive instability [13, 14] in laser-produced plasmas. Such instabilities are accompanying by super strong magnetic fields generation, as well as by the generation of shooting vertexes - laser "magnetoms" [5], which in turn can be "evaporated" as miniature black holes and hence the magnetic charge generation can take place [15]. In accordance with [16] classical electrodynamics may be interpreted as a theory of an infinitesimal connection in a principal fibre bundle with the structure group U(1). A similar interpretation can be given to the Yang-Mills fields and in general to all fields resulting from "gauge transformations of the second kind". In absence of external fields the simplification is possible in theory [17] and then monopoles are Abelian objects [18]. As mentioned in this paper the Abelian monopoles and the center vortices appears to be strongly correlated with each other: almost all monopoles are sitting on top of the vortices. In reality (in general case) the creation of magnetic monopoles take place in strong magnetic fields inducing in laser-produced plasma (as in our case, or in other fields which can be in concrete laboratory system), and hence the mentioned above simplified model is not correct [17]. So we regards that detected in our experiments an exotic quasiparticles are non-Abelian particles on its nature like as leptons [19]. In addition, the observed percolating properties of the monopoles and the vortices are revealing the nature of the magnetic condensation in the non-Abellian case [18]. As it had been noted in this paper in SU(2) gauge theory the magnetic monopoles system can be considered as monopole-vortex chain. Namely such system we observed in our experiments [20]. In Fig 9,10, the traces of low energy scattering of non-Abelian magnetic monopole are shown (theoretically predicted trace [21], left, and our experimental trace, center). We can note, that predicted theoretical model developed in [21] could be very interesting due to a good agreement with observed trace on CR39 (Fig.10). Figures (2,4) are the best arguments for demonstration that bundles are a generalization of the concept of Cartesian product $T * S$, where T is time axis and S is the three-dimensional space [16] (note only, that T can be also 3D, see [4] and its references). The full set of Maxwell's equations is known to follow from a simple action principle in the Kaluza-Klein theory (or one of its modifications), and one can construct a principal fibre bundle from the Kaluza-Klein space [16]. So, in principle, we are not needed in Yang-Mills theoretical models for describing the real observed monopoles. In particularly, instead Nahm's model [12] it is worth to use the classical Euler model of gyroscope to describe magnetic monopoles (vortexes).

3 The twin exotic quasiparticles

The X-ray images produced by RFR-4 [22] demonstrate us not only discrete trajectory of moving and radiative quasiparticles but also traces of double particles. It is distinctly viewed in CR39 after standard procession procedure (etching in 6.25 N suspension of NaOH during 20 minutes at temperature $(65 \pm 10)^\circ$ C, see for example, Fig.11, right). The behavior of such twin particles are different from behavior of magnetic monopoles (or monopolinos). If last of theme are "very nuclear active" [23] the twin particles are more "quiet". Now we can propose that it may be dyons. By using F. Wilczek's remarks on dyons [24] and knowing only its typical size $r_o \approx 10\ \mu$ (see Fig.11) we can estimate the mass of dyons $m_d \approx 10^{-34}$g when $q = e$ (it is easy to show that case when magnetic g and electric q charges are equals to each other is not realistic), and then $g \approx (137/4)e$, i.e in twice less then possible minimal magnetic Dirac's monopole charge. So dyon can be constructed from two pieces with total magnetic charge $g = g_d$ - minimal magnetic charge of monopole Dirac . But dyons of course could not being single possible candidate for explanation of our doublet particles observations. In addition, for example, we could proposed, that they can be neutral Higgs particles H^0 which have doublet structure [25] or two positive and negative Higgs bosons [19] but only with extra small masses. Because of, if we have such slight magnetic charges we must have very slight Higgs bosons too. Similar to monopole-vortex chains [20] were found in numerous (non-)supersymmetric non-Abelian gauge theories involving various Higgs fields (see [18] and its references). These questions will be considered in more detail in other our publications. Finely, we would like to emphasize, that most probably an exotic quasiparticles (slight magnetic monopoles) was observed by M.V. Lomonosov in 1753 year at first by using "storm machine"! He wrote ([7], p.191) that a pyramid-like sparks and red sparks moving in air produced a "hissing light". The red color is a preferably that due to Compton's scattering of light by slight magnetic charges must be in red and infrared regions [4]. Acoustic radiation effects, which we detected in our experiments [2] are similar to acoustic effects which had been listen by M.V. Lomonosov and others eyewitnesses who observed of ball lightning decomposition (see also our notices in [14, 26, 27]). When magnetic charge moves in air its Cerenkov's radiation must be in opposite direction and also has a pyramid-like form [4].

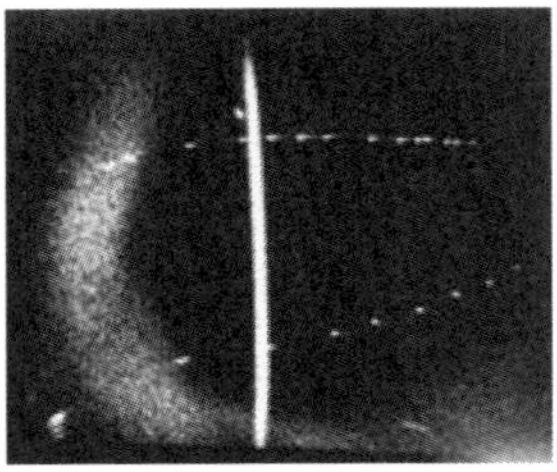

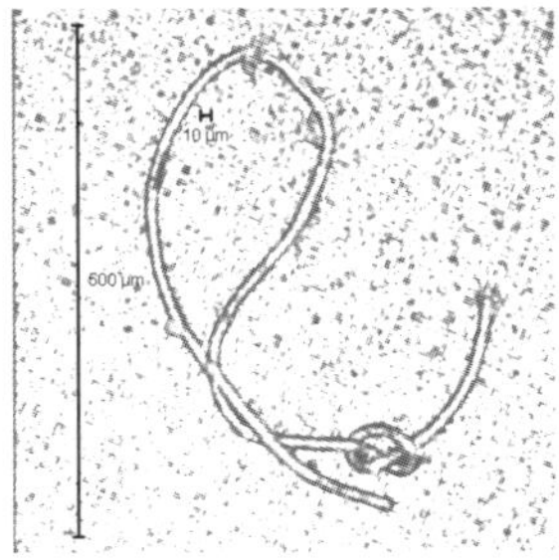

Fig. 1(left), Fig. 2(right)

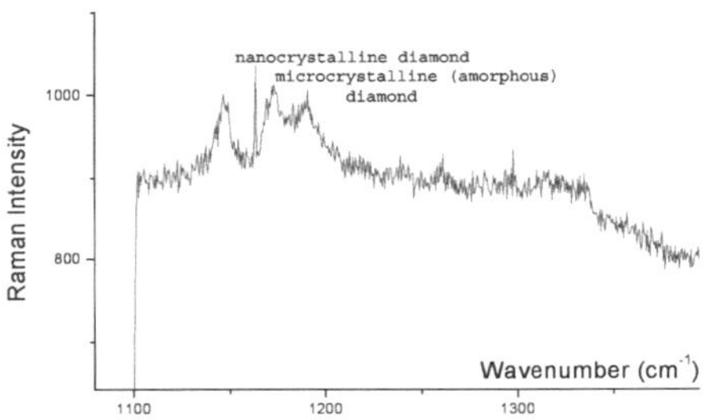

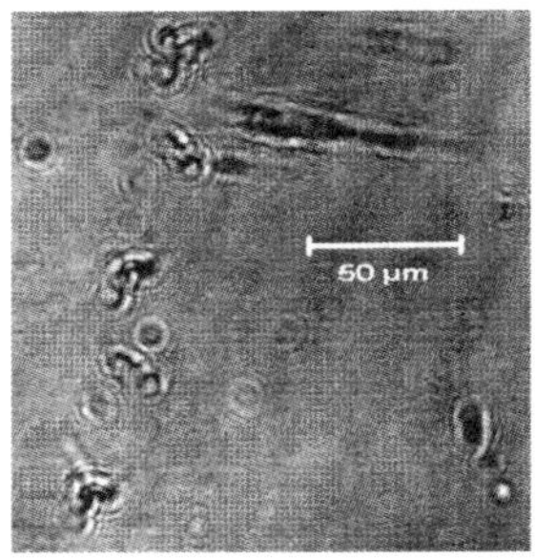

Fig. 3 (left), Fig.4 (right)

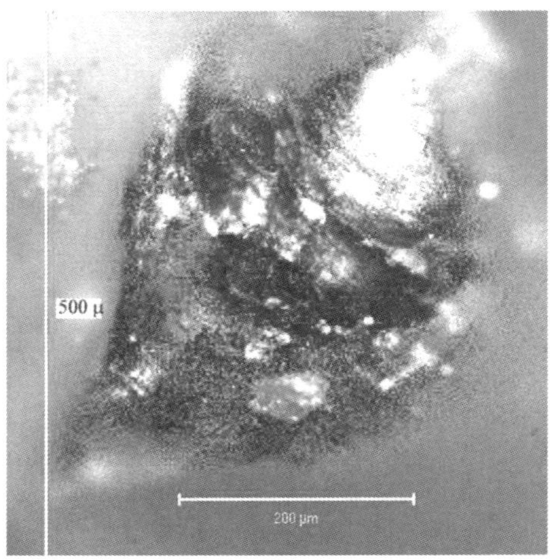

Fig. 5 (left) Fig.6 (right)

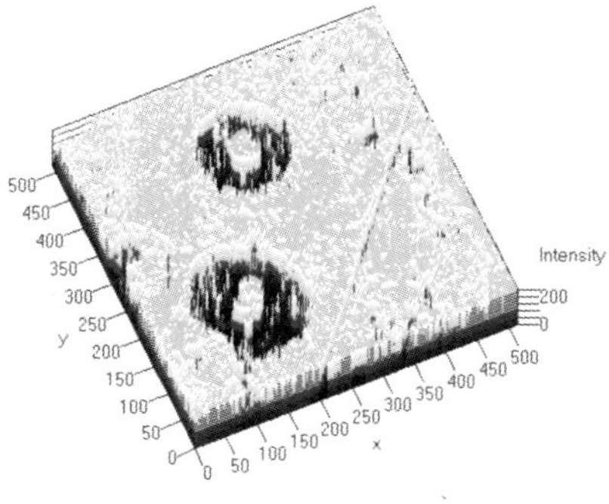

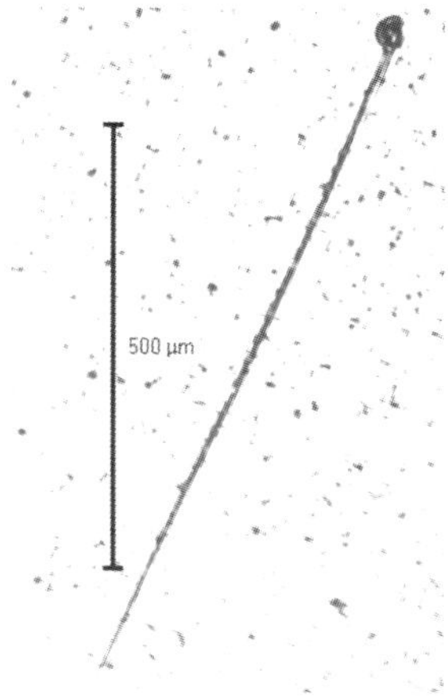

Fig.7 (left), Fig.8 (right)

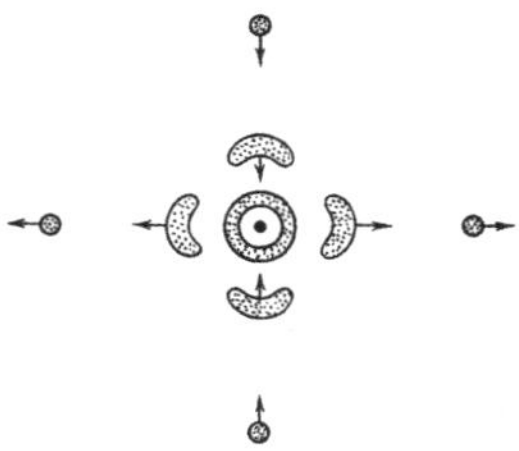

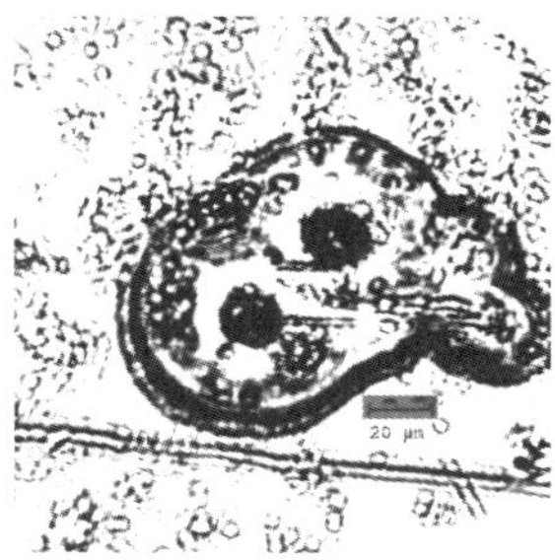

Fig.9 (left), Fig.10 (center), Fig.11 (right)

References

[1] W. Heisenberg, *ZPh*, **43**, 172-198 (1927).

[2] V. Skvortsov, N. Vogel, *Electromag. waves and el. sys.*,**7**, No.7, 64 (2002).

[3] V. Skvortsov, N. Vogel, *"The Magnetic Monopoles Generation in Laser-Induced Discharges"*,(Proc.29th EPS Conf. on Plasma Physics and Contr. Fusion, Montreux, 17-21 June), **26B**, D-5.013, 2002; and i.b.i.d., O-2.30.

[4] V. Skvortsov, N. Vogel, *"The generation of exotic quasiparticles"*,(Proc. 11-th Lomonosov Conf. on Elementary particle Physics, Moscow, 21-27 August 2003,World Sci., Singapore, 373 (2005).

[5] V. Skvortsov, *"Physics of matter under extreme states"*, Moscow, MIPT, 200 pages, 2007.

[6] V. Skvortsov, N. Vogel, *"Method for synthesising diamond with the aid of magnetic monopoles"*,(PCT/RU/2002/000422, WO2004/025002 A1, publ.25.03.2004); (Petent RU No.2293147), publ. 10.02.2007. Bull.No.4.

[7] M.V. Lomonosov,*"Transactions on Physics 1753-1765"*, M.-Leningrad,

Ac.Sci. USSR, 1952, **3**, pp. 190-192.

[8] L. Schiff, *Phys.Rev.* **160**, No. 5, 1257-1262 (1967).

[9] N. Vogel, V. Skvortsov,*"The generation of toroidal dipoles in laser plasmas"*, (Proc. XXXVI Int. Conf. on Plasma Phys. and Contr. Fusion), Zvenigorod, 9-13 Febr. 2009, 169 (see also p.296).

[10] V. Skvortsov, N.Vogel, *"The super strong magnetic fields and fine structure of anomalous plasma droplets"*,(Proc. of Int. Conf. "Physics impulse discharges in condenced media"), Nikolaev, IIPT, N.A.S. of Ukraine. 17-21 August, 2009, pp. 12-14.

[11] V. Skvortsov, N. Vogel. *"The generation of superstrong magnetic fields in plasma of laser- induced discharges"*, (Proc. Int. Conf. MEGAGAUSS-XI), Imperial College. London. 10-14 Sept. 2006. OI.4.

[12] N.J. Hitchin,*Comm. Math. Phys.*, **89**, 147-190 (1983).

[13] V. Skvortsov,*"Explosive instabilities in condenced and plasma media"*,(Proc. Int. Conf. "Phys. impulse discharges in condensed media"), Nikolaev, IIPT, N.A.S. Ukraine.17 August, 2009,3-6.

[14] V. Skvortsov, N. Vogel, *IEEE Tr. on PS*, to be published in 2010.

[15] A. Sakharov, *"Scientific Proceedings"*, (CENTRCOM, Moscow) 1995.

[16] A. Trautman. *Reports on mathematical physics*, **1**, No.1, 29-62 (1970).

[17] M.F. Atiah, *"Geometry and physics of bundles"*,(Moscow, MIR, 1995).

[18] M.N. Chernodub, V.I. Zakharov,*"Monopoles and vortices in Yang-Mills plasmas"*, (ITEP-LAT/2008-12).

[19] G. Kane, *"Modern elementary particle physics"*,(M., MIR, 1990).

[20] V. Skvortsov, N. Vogel, *"Effect of multiple generation of magnetic monopoles"*,(Proc. XXIX Conf. on Plasma Phys. and Contr. Fusion, Zvenigorod, Rus. Ac. Sci.), 186, 2002.

[21] M.Atiah, N.Hitchin,*Phil. trans. Royal Soc. London*,**A315**, 459 (1985).

[22] S. Petrov, V. Lazarchuk , V. Murugov et. al., Proc. 22 nd Int. Congr. on High-Speed Photography and Photonics. Santa Fe, USA, 1996.

[23] V. Skvortsov, N. Vogel,*"Nuclear reactions in superstrong magnetic fields"*,(Proc. Int. Conf. MEGAGAUSS-XI), (Imperial College. London. 10-14 Sept., 2006. P2.40.)

[24] F. Wilczek, *PRL*, **48**, No.17, 1146-1149 (1982).

[25] T.-P. Cheng, L.-F.Li, *"Gauge theory of elementary particle physics"*, (Clarendon Press, Oxford) 1984.

[26] V. Skvortsov, N. Vogel,*"Ball lightning synthesis of diamonds" and "Synthesis of ball lightning and diamonds by using magnetic monopoles"*, (Proc. XXX Conf. on Plasma Phys. and Contr. Fusion, 24-28 Febr., 2003, Zvenigorod, Rus. Ac.Sci., M.), 196 and 197, 2003.

[27] V. Skvortsov, N. Vogel,*"Ball lightning as an unusial dipole current generator"*, (Proc. XXXIV Conf. on Plasma Phys. and Contr. Fusion, 12-16 Febr. 2007, Zvenigorod. Rus. Ac. Sci., M.), 263, 2007.

ANGULAR MOMENTUM RADIATION OF SPIN LIGHT

Vladimir A. Bordovitsyn [a], Olga A. Konstantinova [b]
Faculty of Physics, Tomsk State University, 634026 Tomsk, Russia
Abstract. Orbital, spin and it's mixed density tensors of angular momentum radiation of electromagnetic field are constructed on the basis of technique of covariant differentiation. As an example the angular momentum of spin light of a magneton moving with a constant velocity is considered.

1 Introduction

The spin light [1] corresponds to the electromagnetic radiation of a relativistic spin particle with a proper magnetic momentum. Such radiation arises only from the electrically neutral particles (neutron [2], neutrino [3]). If the particles have also an electrical charge, the interference of radiations arises. In what follows the problem of identification of the spin light was studied in details on the basis of the semiclassical theory of relativistic radiation [1,4-6].

At the present time spin light as a special type of electromagnetic radiation is discussed in the case of synchrotron and undulator radiation of electrons; high energy neutrino radiation in matter and in gravitational fields [1, 7-9].

In this work we suggest a research of fundamental properties of the spin light radiation associated with the proper angular momentum of electromagnetic field (AMEF). For the construction of our theory we will use the well-known methods of the classical relativistic radiation theory [1, 9].

2 Proper AMEF of spin light and it's power

With respect to Teitelboim's assumption [9] the spin density tensor of AMEF is defined like

$$\mathcal{M}^{\mu\nu\lambda} = \frac{1}{c}(R^\mu \mathcal{E}^{\nu\lambda} - R^\nu \mathcal{E}^{\mu\lambda}).$$

Here $R^\mu = r^\mu + \tilde{r}^\mu$, $r^\mu = (ct, \mathbf{r})$, $\tilde{r}^\mu$ is the light-like four-dimensional vector drawn from the charge to the spectacular point; $\mathcal{E}^{\mu\nu}$ is a symmetrical stress tensor of energy of spin light. In this case the division of the full radiated AMEF into orbital and spin parts has gauge and relativistically invariance:

$$\mathcal{M}^{\mu\nu\lambda} = L^{\mu\nu\lambda} + \Pi^{\mu\nu\lambda}.$$

Here

$$L^{\mu\nu\lambda} = \frac{1}{c}(r^\mu \mathcal{E}^{\nu\lambda} - r^\nu \mathcal{E}^{\mu\lambda}), \qquad \Pi^{\mu\nu\lambda} = \frac{1}{c}(\tilde{r}^\mu \mathcal{E}^{\nu\lambda} - \tilde{r}^\nu \mathcal{E}^{\mu\lambda})$$

[a] e-mail: vabord@sibmail.com
[b] e-mail: olgakonst87@mail.ru

are the orbital and the spin density tensors AMEF correspondingly. Then we will consider AMEF located at the wave zone, where

$$\tilde{L}^{\mu\nu\lambda} = \frac{1}{c}(r^{\mu}_{(-2)}\mathcal{E}^{\nu\lambda} - r^{\nu}_{(-2)}\mathcal{E}^{\mu\lambda}), \qquad \tilde{\Pi}^{\mu\nu\lambda} = \frac{1}{c}(\tilde{r}^{\mu}_{(-3)}\mathcal{E}^{\nu\lambda} - \tilde{r}^{\nu}_{(-3)}\mathcal{E}^{\mu\lambda}).$$

The indexes (-2) and (-3) describe a degree of falling of field subject to the distance. Similarly we can construct the density tensor AMEF for mixed radiation. Especially for a particle moving with a constant it becomes in the form

$$\tilde{\Pi}^{\mu\nu\lambda}_{em} = \frac{ec^2}{4\pi}\frac{\tilde{r}^{[\mu}\overset{\circ\circ}{\Pi}{}^{\nu\rho]}}{(\tilde{r}_{\rho}\upsilon^{\rho})^5}\tilde{r}_{\rho}\tilde{r}^{\lambda}.$$

Here a square brackets denote an antisymmetrization: $a^{[\mu}b^{\nu]} = a^{\mu}b^{\nu} - a^{\nu}b^{\mu}$. $\Pi^{\mu\nu} = (\mathbf{\Phi}, \mathbf{\Pi})$ is an antisymmetric space-like tensor of the proper magnetic momentum of the particle, symbol $\circ$ means a derivative with respect to the proper time τ. From the condition of divergence $D_{\lambda}\tilde{\Pi}^{\mu\nu\lambda}_{em} = 0$ (retardation derivative is defined like $D^{\mu} = \partial/\partial R^{\mu}$) after tha integration over four-dimensional solid angle the rate of mixed AMEF per unit proper time is given by

$$\frac{d\tilde{\Pi}^{\mu\nu}_{em}}{d\tau} = \frac{2e}{3c^2}\overset{\circ\circ}{\Pi}{}^{\mu\nu}.$$

3 Conclusion

Thus, this paper we demonstrate the properties of intrinsic angular momentum of spin light radiation, a new phenomenon of nature.

This work was supported by RF President grant SS 871.2008.2, and by the Russian Science and Innovations Federal Agency under contract 02.740.11.0245.

References

[1] V.Bordovitsyn, I.Ternov, V.Bagrov, *Usp. Fiz. Nauk* 165, 1083 (1995).
[2] I.Ternov, V.Bagrov, A.Khapaev, *JETP* 48, 961 (1965).
[3] A.Lobanov, A.Studenikin, *Phys. Lett.* **B** 564, 27 (2003).
[4] A.Bondar, E.Soldin, *Nuel. Instrum. Meth.* 195, 577 (1982).
[5] I.Sokolov, I.Ternov, *Dokl. Akad. Nauk.* 153, 1052 (1963).
[6] I.Ternov, *Usp. Fiz. Nauk.* 38, 1037 (1995).
[7] A.Lobanov, *Int. J. Mod. Phys.* **A** 39, 7517 (2006).
[8] M.Dvornikov, A.Grigoriev, A.Studenikin, *Int. J. Mod. Phys.* **D** 14, 309 (2005).
[9] C.Teitelboim, D.Villarroel, Ch.G.van Weert, *Riv. Nuovo Cim.* 3, 1 (1980).

FORCE-MOMENTUM RADIATION FROM RELATIVISTIC CHARGED PARTICLES

V. A. Bordovitsyn [a], E. A. Nemchenko [b]

Faculty of Physics, Tomsk State University, 634050 Tomsk, Russia

Abstract. Two alternative definitions of density of angular momentum of electromagnetic field by Ivanenko-Sokolov and by Teitelboim et al. are discussed. It is shown that both definitions give identical integral characteristics of radiation for an arbitrary moving relativistic charge. The total powers of orbital angular momentum and spin momentum of radiation are deduced and the exactly physical interpretation obtained.

1 Introduction

In 1897 A. I. Sadowsky put forward a hypothesis that circularly polarized light must possess the proper angular momentum [1]. In 1935-1936, B. A. Beth in USA [2] and A. N. S. Holborn in England [3] experimentally proved that circularly polarized light does possess angular momentum. At the present time there is no doubt about the existence of the angular momentum of circularly polarized electromagnetic waves. However, the general definition of angular momentum of the electromagnetic field (AMEF) is argued over among physicists even nowadays [4–8].

2 Relativistic theory of the AMEF radiation

In this work two different definitions of density of intrinsic angular momentum of electromagnetic field by Ivanenko-Sokolov [9] and by Teitelboim- Villarroel-van Weert [10] are discussed. We find that in spite of the absence of the outward sign of similarity these two methods give identical results for the rate of the total power for both orbital and spin angular momenta radiation which in the wave zone have the following form:

$$\frac{d\tilde{L}^{\mu\nu}}{d\tau} = \frac{2}{3}\frac{e^2}{c^5}\omega_\rho\omega^\rho(r^\mu v^\nu - r^\nu v^\mu), \qquad \frac{d\tilde{\Pi}^{\mu\nu}}{d\tau} = \frac{2}{3}\frac{e^2}{c^3}(v^\mu\omega^\nu - v^\nu\omega^\mu),$$

where $v^\mu = dr^\mu/d\tau$ and $\omega^\mu = dv^\mu/d\tau$ are the four-vectors of velocity and acceleration correspondingly.

Our formulas have clear-cut physical interpretation. If we put the well-known charge power of the radiation into operation [11]

$$\frac{d\tilde{P}^\mu}{d\tau} = \frac{2e^2}{3c^5}\omega_\rho\omega^\rho v^\mu,$$

[a] e-mail: vabord@sibmail.com

[b] e-mail: katrin_nemchenko@mail.ru

then we can get the expression for the orbital momentum of radiation

$$\frac{d\tilde{L}^{\mu\nu}}{d\tau} = r^\mu \frac{d\tilde{P}^\nu}{d\tau} - r^\nu \frac{d\tilde{P}^\mu}{d\tau} = \frac{d}{d\tau}(r^\mu \tilde{P}^\nu - r^\nu \tilde{P}^\mu),$$

which is formally the same relation between the torque tensor of the particle and its angular momentum as in the relativistic mechanics

$$\tilde{T}^{\mu\nu} = r^\mu \tilde{F}^\nu - r^\nu \tilde{F}^\mu = \frac{d\tilde{L}^{\mu\nu}}{d\tau}.$$

Thus, the radiation of the orbital AMEF adds up to the radiation of the field momentum of the force.

It should be noted, that the power of the radiation of the spin momentum is proportional to the Thomas's precession frequency $\Omega^{\mu\nu}_{Th}$ (see e. g. [10]), which acquires now obvious dynamical interpretation.

3 Conclusion

To sum up, it can be stated that the subject of investigation designed here has discovered, in fact, a new trend in the relativistic theory of radiation and the spin properties of relativistic particles.

This work was supported by RF President grant SS 871.2008.2, and by the Russian Science and Innovations Federal Agency under contract 02.740.11.0238.

References

[1] A. Sadowsky, *Zhurn. Russ. Fiz.-Khim. Ob-va* 2, 82 (1987)(in Russian) see also *Acta et Comment. Imp. Univ. Jurien* 7, 1-2 (1999)(in French).
[2] B. Beth, *Phys. Rev.* 48, 471 (1935); 50, 115 (1936).
[3] A. Holborn, *Nature* 137, 31 (1936).
[4] H. Ohanian, *Amer. J. Phiz.* 54, 500 (1986).
[5] K. Wulfson, *Usp. Fiz. Nauk* 152, 668 (1987).
[6] I. Sokolov, *Usp. Fiz. Nauk* 161, 175 (1991).
[7] A. Barabanov, *Usp. Fiz. Nauk* 163, 75 (1993).
[8] G. Rosenberg, *Usp. Fiz. Nauk* 40, 328 (1950).
[9] D. Ivanenko, A. Sokolov *Classical Field Theory* (GITTL, Moscow-Leningrad) (1949) (in Russia) see also *Klassische Feld Theorie* (Akademie-Verlag,Berlin) (1953) (in German)
[10] C. Teitelboim, D. Villarroel, Ch. G. van Weert, *Riv. del Nuovo Cim.* 3, 1 (1980).
[11] V. Bordovitsyn, I. Ternov, V. Bagrov, *Usp. Fiz. Nauk* 165, 1083 (1995).

REDUCED GAUGE THEORY AND MULTI-MATRIX MODELS

A.O. Shishanin[a]
Russian People's Freindship University, Moscow

Abstract. For 2d Eguchi-Kawai model and two-matrix model with commutators we compare a correlation function.

Many interesting problems of quantum field theory and statistical physics can be simplified in the limit of infinite number for degrees of freedom $N \to \infty$. One could then develop a systematic approximation procedure by studying the case $N = \infty$. For particle physics the 1/N-expansion was introduced by 't Hooft [1]. In particular, in this review was observed that the only planar diagrams survive in the large N limit of $SU(N)$ gauge field theory.

It was explained by T. Eguchi and H. Kawai [2] that at $N = \infty$ field theories become equivalent to matrix models living at a single point. The feature of these Eguchi-Kawai models (or "reduced models") has got hope for better understanding of the $N = \infty$ limit for QCD [3], [4].

Continuum reduced Eguchi-Kawai model has action

$$S_{EK} = -\frac{1}{4g^2} Tr[A_\mu, A_\nu]^2, \tag{1}$$

where A_μ are $N \times N$ traceless anti-hermitian matrices.

In paper of T. Hotta, J. Nishimura and A. Tsuchiya [5] was estimated some correlators. Using scaling properties of the partition function we have

$$< \frac{1}{N} Tr(F^2) >= dN(1 - \frac{1}{N^2})g^2. \tag{2}$$

Let us now consider the $k+1$ matrix model [6] with

$$Z = \int dX dY_1 ... dY_k e^{-\frac{1}{2}TrX^2 - \frac{1}{2}\sum_i TrY_i^2 + \frac{1}{4g^2}\sum_i Tr[X,Y_i]^2}. \tag{3}$$

We can calculate gaussian integral over Y_i matrices. Here we can derive at large N limit equation for the eigenvalue distribution function $\rho(x)$

$$\frac{x}{2} = \int (\frac{1-k}{x-y} + \frac{k}{(x-y)(1+\frac{1}{4g^2}(x-y)^2)})\rho(y)dy. \tag{4}$$

Here normalization is $\int \rho(x)dx = N$.

Let us for simplicity consider the two-matrix model $k = 1$ with $g \to 0$ limit as approximation for $2d$ Eguchi-Kawai model. For this model the general solution was found by V. Kazakov, I. Kostov and N. Nekrasov [7]. For all g the solution

[a] e-mail:shishandr@rambler.ru

of field equations was expressed via elliptic functions. They demonstrated also that here a phase transition is absent.

The saddle point equation for this case is

$$\frac{x}{2} = \int \frac{\rho(y)dy}{(x-y)(1+\frac{1}{4g^2}(x-y)^2)}. \tag{5}$$

One can solve this equation taking into account the general result

$$\int_{-a}^{a} \frac{\rho(y)dy}{(x-y)(1+\frac{1}{4g^2}(x-y)^2)} = -2\pi g\rho'(x) \tag{6}$$

as $g \to 0$.

Then solution of this integral equation is parabolic distribution

$$\rho(x) = \frac{1}{8g\pi}(a^2 - x^2), \tag{7}$$

where $a = \sqrt[3]{3\pi g}$.

The correlation function also can be calculated as

$$< tr[X,Y]^2 >= 2 < (tr(X^2Y^2) - tr(XYXY) >= \int \frac{dxdy\rho(x)\rho(y)(x^2 - xy)}{(1+\frac{1}{4g^2}(x-y)^2)}, \tag{8}$$

where x and y change from $-a$ to a.

When $g \to 0$, we can obtain

$$< \frac{1}{N} tr[X,Y]^2 >= -2Ng^2 + \frac{4}{5}g^2(3\pi gN)^{\frac{2}{3}} + ... = -2\lambda + \frac{4}{5}(3\pi)^{\frac{2}{3}}\frac{\lambda^{\frac{5}{3}}}{N^{\frac{2}{3}}} + ... \tag{9}$$

Here we see that this formula coincides in the principal approximation with accuracy of sign (we deal with accepted notation). But next approximation terms are different.

References

[1] G. 't Hooft, Nucl.Phys. B72 (1974) 461

[2] T. Eguchi and H. Kawai, Phys. Rev. Lett. 48 (1982) 1063

[3] S. Das, Rev. Mod. Phys. 59 (1981) 235

[4] Yu. Makeenko, Methods of contemprorary gauge theory, Cambridge University Press, 2002, 417 p.

[5] T. Hotta, J. Nishimura and A. Tsuchiya, Nucl. Phys., **B545** (1999) 543 hep-th/9811220

[6] D. Berenstein, M. Hanada and S. Hartnoll, arXiv:0805.4658

[7] V. Kazakov, I. Kostov and N.Nekrasov, Nucl. Phys. **B557** (1999) 413, hep-th/9810035

Problems of Intelligentsia

THE INTELLIGENTSIA AND GLOBALIZATION

by John Kuhn Bleimaier[a]
15 Witherspoon Street, Princeton, New Jersey 08542 USA

Globalization is an international movement premised on the subordination of individual rights and sovereign prerogatives to the dictates of the transnational maximization of return on capital investment. As such, globalization is at odds with democracy, self-determination, individual economic welfare and the attainment of world peace. When globalization is recognized for the ideological malaise which it is, the intelligentsia will strenuously oppose it.

I

Democracy is most readily attained in small units of governance. An informed electorate and a responsive government are likely to work smoothly together when their interaction is intimate and frequent. For this reason a globalized polity is unlikely to be democratic. On the other hand, localized governmental administration is likely to be responsive and representative. The classic Greek city-states, the cantons of the Swiss Confederation and the town meetings of rural Vermont are good examples of successful self-government. On the other hand mega sized sovereigns are destined to be bureaucratic, distant and imperialistic. Witness the Ottoman Empire, an archetypical, dictatorial bureaucracy of Byzantine proportions. As a practical matter a large state entity must rule through indirection, and by use of fiat. The mega unit cannot relate directly to the individual constituent and must rely on mass media as an avenue of communication. Communication is always, of necessity, from the top down.

From 1848 the political history of the western world has been centered around the concept of self-determination. National groups, social classes, professional associations, all have sought to shape their own destiny. Self-determination is imperiled by globalization.

Globalization is first and foremost a financial system, which restricts the economic freedom of individual state actors in the interest of a unified, transnational market for capital, goods and services. Globalization is advocated in the exclusive interest of powerful, global economic actors. The determination of policy in the globalized society will, of necessity, be subjective and self-serving. After all, in the globalized world only the largest scale players can hope to influence the determination of policy. In a globalized world the individual person and his interests are virtually irrelevant. Self-determination is an obstacle in the path toward the attainment of global objectives. From the perspective of the global oligarchy democracy is never an efficient system of government.

The global system allocates the extraction of raw materials, the production of finished goods and the consumption of end products so as to maximize the return on the capital investment of large international proprietors. The objectives and needs of the individual members of society are of no consequence. Thus, globalization is inherently undemocratic. Indeed, globalization is anti-democratic because it seeks to eliminate the disturbances to the optimized

---

[a]B.A.; Master of International Affairs; Juris Doctor; member of the New York, New Jersey and US Supreme Court bars; Fellow of Mathey College, Princeton University

flow of capital, goods and services which popular self-determination potentially present.

The implement used by the global proprietors, which eradicates democratic institutions, is the international mass media. Establishing and operating the means of communication in a technological society is an inherently expensive proposition. The ownership of means of mass communication requires a sufficiently large capital investment that it is only within the grasp of governments or oligarchs. Thus, ironically the fourth estate, once the advocate of individual liberty, equality and brotherhood is transformed in the globalized society into the handmaiden of "Big Brother." Mass media becomes a true opiate of the masses, restraining aspirations, providing mindless distractions and diversions. The globalized media insures that the mass audience is preoccupied with trivial show business and yellow journalism, while guaranteeing that substantive issues are disregarded. Mass media, to the extent that it addresses fundamental issues at all, portrays them as hopelessly complex and insoluble. The mass audience is discouraged from contemplating the basic problems facing humankind and their solution.

The mass media invariably attempts to establish a conventional wisdom that is consistent with the economic interests of its proprietors. Thus mass media is the principal means used by the international capital investment oligarchy to establish a global network of financial markets and a global distribution of assets and production, which maximizes the return on capital investment.

What is the intelligentsia's reaction to this state of affairs? By definition the intelligentsia is that cadre in society, which thinks deeply, critically and independently. Membership in the intelligentsia reflects education, but first and foremost it reflects an intellectual predisposition to think outside the box. The members of the intelligentsia are the torchbearers of national culture and the theoreticians in both the arts and in science. The intelligentsia has a long and distinguished history as the conscience of society and the repository of its most dearly held values.

The intelligentsia has been deeply committed to western notions of individualism and self-determination. It is individualism, coupled with freedom of conscience, which has led the intelligentsia support democracy as a form of government. The intelligentsia has never been monolithic. However, it is the diversity of point of view of the intelligentsia which motivates the intelligentsia to endorse democratic governance. Realizing that thoughtful individuals can respectfully disagree and require the opportunity to air their differences, leads inescapably to the embrace of democratic institutions.

II

The concept of the intelligentsia as an integral component of the social fabric derives from 19th Century Russia. It was in Russia that an extraordinary cadre of intellectuals without formal organization spontaneously brought forth enormous achievements in literature, philosophy, the arts and science. The attainments of this group of individuals brought Russian culture to the forefront of attention in European civilization. The 19th Century Russian intelligentsia established itself as a moral force of overpowering significance. And the term, "intelligentsia," entered the lexicon.

Among the 19th Century Russian men of letters no one had a more profound impact than Count Leo Tolstoy. First as a consummate novelist and then as a social thinker, Tolstoy became the embodiment of the Russian intelligentsia. Tolstoy's moral indignation directed at the social injustices of his time played an enormous role in the transformation of Russian society.

At the beginning of the 20th Century Leo Tolstoy wrote a political essay "on patriotism and government" wherein he excoriated both the former and the latter. Tolstoy depicted patriotic emotion as an antiquated tribal sense of solidarity, which was being used by an unscrupulous leadership to consolidate power in the hands of the few. Tolstoy attacked government as the framework for the exercise of the power of a preeminent minority to the determent of the majority. "The ruling classescan retain their position - exceptionally advantageous in comparison with the laboring masses - thanks only to government organization, which rests on patriotism." A stern indictment, indeed.

Tolstoy was prophetic when he envisioned the devastating conflagration, which awaited Christendom in a twentieth century as a result of world wars and incessant strife. Cynical governments exploited national feelings, engendering enmity between peoples in order to strengthen state power and retain or expand concentrated economic advantage. An era of unprecedented bloodshed lined the pockets of those who financed the military industrial complex. "The ruling classes have in their hands the army, money, the schools the churches and the press."

We are justified in reading this essay by Tolstoy as a political tract. It is not a closely reasoned, dispassionate philosophical treatise, but is rather a provocative outcry against a status quo characterized by mobilizing mass armies bent on destruction. "No anarchical disorder could be worse than the position to which governments have already led their peoples." The author's shrill tenor and emotional rhetoric were apparently motivated by a premonition of impending disaster.

At the beginning of the 21st Century we are living in the backwash of the cataclysms that Tolstoy foresaw a hundred years ago. Two world wars and innumerable undeclared conflicts have left millions dead and have totally reshaped the international social fabric. However, the oligarchic financial interests, which he decried, persist to this day. They continue to manipulate the emotions of the masses and utilize the institutions of government for their economic advantage.

The tactics of the oligarchs have changed. Instead of inflaming patriotism and strengthening national governments, today the oligarchs tout a concept of anti-nationalism and seek to subordinate the sovereign state to multinational capital and finance. It is ironic that the same interests, which sought to engender patriotism and empower governments, would now denigrate nationalism and subordinate sovereign states to an international cabal of investors and bankers.

Because smaller units of governance are yet susceptible of reflecting the self-determination of peoples, the 21st Century oligarchy seeks to subordinate the sovereign nation to the globalized institution. The International Monetary Fund, for example, intrudes itself into the life of the state by first offering development capital at attractive rates of interest. Then, as it manipulates the international economic climate it can dictate policy to all debtor nations. National economic policy is thus subordinated to global financial interests. The ability of an electorate to control its economic destiny is effectively abrogated in a globalized environment.

It is not difficult to extrapolate Tolstoy's values and principles into our 21st Century world. It seems clear that Tolstoy would decry the global manipulators of public opinion and the global financial stranglehold just as he condemned cynical, artificial patriotism and selfish, despotic government.

Evil individuals put good institutions to bad use. Alas, Tolstoy's indictment of patriotism and government did not take into consideration the potential for good that is inherent in both patriotic emotion and governmental organization. Patriotism is first and foremost an extension of the feeling of attachment and affection, which is felt toward one's family, one's neighbors and one's community. Thus patriotism can motivate us to care for others and to work for the betterment of the society at large. By the same token, government is what stands between us and the Hobbsian state of nature where life is nasty, brutish and short.

Tolstoy's ringing attack at the dawn of the 20th Century would have been more productively aimed directly against the oligarchs who subvert the positive emotion of patriotism into jingoistic aggression and who contrive to make government an instrument for personal gain. Unfortunately, by helping to turn a generation of Russian intellectuals against patriotism and against the Russian state, Tolstoy unwittingly facilitated the success of the Bolshevik revolution, which brought with it more suffering than the most incompetent or venal Romanoff autocrats. Ironically the late Soviet state exploited emotions akin to patriotism in order to motivate its citizens and the government embarked on bellicose foreign adventures no less costly in blood and material that its capitalist counterparts.

In 21st Century Russia the evils of the communist state are but a distant memory. However, the new Russian Federation is in the grip of an oligarchy that consists of the literal and figurative descendents of the communist party bosses. That oligarchy suppresses national feeling and controls a puppet government. The suppression of patriotic emotion is undertaken in the interest of removing an element of social cohesion, which might unify the populace in opposition to the international economic cabal we know as globalization. By the same token the current oligarchy in the Russian Federation demands that government not be an expression of popular will but rather be an organ of social control. The present governmental structure derives its legitimacy from the mass media and both are the creatures of the international capital investors who have financed them.

The foregoing analysis of the present state of affairs is not inordinately complex. Knowing what we do of Tolstoy's idealism and Christian charity, it is obvious that, were he alive today, Tolstoy would be an outspoken opponent of the tyranny of globalization. I propose that he would support a generous spirit of patriotism, which is an extension of loving one's neighbor. He would also advocate a form of government, free of corruption, which would reflect the moral will of the people and work unselfishly for the greater common good.

We can read all of Leo Tolstoy's political writings as advocating the application of Christ's social teachings in the world of everyday societal intercourse. Because of Tolstoy's well publicized disputes with the organized Church of his day relating to ecclesiastic dogma, we often forget that this spokesman for the Russian intelligentsia was first and foremost a Christian philosopher. Tolstoy's disagreements with the clerical and monarchal powers were based on his assessment that the leadership failed to live up to Christian principles of brotherly love and charity. How much more repugnant would Tolstoy find our 21st Century social and political environment which has more or less explicitly divorced itself from the beatitudes expounded in the Sermon on the Mount. Now idealism is dismissed as na?ve, while materialism is touted as virtue. Today's globalist ideology is inconsistent with the core values of the intelligentsia.

III

Globalization is based on bogus economics. A twisted interpretation of this social science has been promulgated, premised on the notion that the interests of society are served when we exult in man's selfish nature and allow it the unfettered reign of laissez-faire. Clearly, it is the function of the intelligentsia to unmask and debunk retrograde cynicism.

The advocates of globalization argue the merits of their ideology by referring to the alleged maximization of economic efficiency when every nation produces goods and services only in the sector or sectors in which it has a natural advantage. For example it is urged that societies with an abundance of cheap labor should produce labor-intensive products. Countries with a high level of education should engage in research and build high tech products. There is a certain superficial logic to this proposition. However careful analysis proves it to be simplistic and flawed.

The globalization advocates really seek the maximum return on capital investment, not the international maximization of economic efficiency. The globalist approach to the manufacture of automobiles would dictate that the factories in Europe and North America should be shut down so that cars can be build in the less developed Asian nations, where manpower is abundant and wages are low. Thus, profit margins on the sale of motorcars can be increased and the return on investment in this industry is maximized. However, if the automobile factories in the developed countries shut down the erstwhile workers in this heavy industry will be compelled to take lower paying jobs in other fields and may no longer be able to purchase new cars. Whereas the workers in the new third world factories sole economic advantage is their willingness to work for low wages. Thus, even if the new autoworkers are fully and gainfully employed they will not be able to purchase new automobiles either. They will still be commuting to work on their bicycles. In this example the global maximization of return on investment is penny wise but pound foolish as the cost of producing cars decreases in direct proportion to the decrease in the size of the market for new automobiles.

It should be taken as axiomatic that in a healthy economy the workers should be able to afford to purchase the fruits of their labor. Thus there is no exploitation and true economic efficiency is attained. Capital and labor are prudently and effectively invested. The division of labor, which characterizes organized society, demands that all persons are simultaneously producers and consumers. When one group merely produces and another merely consumes the social organization is fundamentally flawed.

The globalist paradigm, where research is conducted in the educated countries and products are built in the cheap labor countries, fails to take into account one of the critical elements of the industrial process. Technical innovation is greatest where development and manufacture take place in close physical proximity to one another. New ideas often come from the factory floor, from the workers and managers who are doing the hands on building of a finished product. When research and development are conducted at a distant location by an alien culture the opportunities for technical advancement are minimized.

At a sociological level mankind needs to feel the satisfaction of producing a tangible product. Western man has traditionally been a farmer, a builder or a craftsman. Healthy societies take into account the emotional well being of the populous. For western man the universal service industry is incapable of providing occupational satisfaction for workers. A healthy nation is one which can take pride in its tangible products. While some are inclined for creative scientific research or the hucksterism of the marketplace, most workers are happiest producing a concrete utile good. Indeed our philology itself associates the word, "goods," with positive virtue.

An efficient economy and a stable society represent a mix of industries and occupations. Individuals are possessed of diverse gifts and attributes. A successful country presents the spectrum of economic opportunities, which address the talents of the workforce. There should be work for intellectuals, for technicians, for healers, for the physically strong and for the weak as well. This kind of optimal society cannot be achieved on a globalized planet where industries are distributed on the various continents exclusively for the maximization of return on capital investment.

Globalization also has the potential for furthering ecological catastrophe. The interests of global capital are addressed when environmental pollution can be relegated to the parts of the planet least likely to resist its onslaught. In the globalized economy dirty industries are concentrated in poor countries where environmentalist awareness is limited; where impoverished populations are unlikely to complain; where corrupt politicians can be cheaply bought off. Yet we do live on a single biologically unified planet. Spewing pollutants in a distant, poor country has the same long-term negative effect as dumping in our own backyard. Once again, the globalization approach is inefficient in the long run.

We would be remiss if we did not consider the practical, political consequences of globalization. As discussed earlier, globalization is theoretically at odds with the principles of democracy and self-determination. But at its most immediate and pragmatic level, globalization impacts the political life of the state. When global capital invests in the export of jobs to a distant land it creates a low paying job abroad and terminates a higher paying job at home. From the perspective of an American capital investor the full employment of a laborer in Mauritius has no immediate political impact. However, the unemployment of your neighbor can be politically catastrophic. Your disgruntled neighbor can figuratively and literally break your windows. The export of high paying jobs from industrialized countries creates the imminent risk of political instability. Furthermore, the creation of low paying, dangerous and environmentally hazardous employment opportunities in the Third World also has the potential for global destabilization. A new industrial proletariat developing in the poor countries of the world is a wild card for the future.

The concentration of higher education and high tech research in a small number of nations is not a positive development. First it creates a harmful "brain drain" effect whereby the most intellectually gifted individuals are drawn away from their native lands and are concentrated in an elite number of host countries. If all the fine physicians, the trained engineers and research scientists are drained off from the developing world, these poorer societies are destined to languish in poverty forever. Furthermore the concentration of the world's intellectual elite in a handful of countries spells an end to cultural identity and thus an end to international cultural diversity.

National security demands that every viable sovereign foster certain critical industries within its borders. A pure service economy is de facto at the mercy of its trading partners who supply hard goods. While trade can encourage coexistence it also has the potential to facilitate political black mail. Only a nation state which consciously supports its diversified industrial base can realistically aspire to the status of a responsible, independent actor in the international community.

Interestingly the two nation states which have the most to loose from globalization are the United States of American and the Russian Federation. Both these countries are geographically large, possessed of significant natural resources, and blessed with a highly qualified work force. Both America and Russia have the potential to "go it alone" possessed of adequate internal supply and demand to attain prosperity. Yet, with both these great nations in the grip of the same globalist oligarchic cabal, they are both presently working against their national self-interest. America has lost its industrial base and has been transformed into a service economy and international gendarme. Russia, on the other hand, has also lost her manufacturing industries. She is now an extractor of raw materials and a geopolitical buffer state. Globalization has exacted a heavy toll on both the American and Russian peoples.

It is the obligation of the intelligentsia to expose the fallacies of globalization and to defend the healthy interests of the population at large. The inherent idealism of the intelligentsia demands no less.

IV

Internationalism and globalization are opposites. The internationalism traditionally backed by the intelligentsia centered around the attainment of world peace. The establishment of international institutions and the promulgation of international law were based on the idealistic objective of lowering the risk of international conflict. The world wars of the 20th Century and the technological feasibility of world nuclear annihilation spurred the development of international cooperation and encouraged the intelligentsia to take a broader worldview.

On the other hand, globalization has a negative impact on the chances for international peace. The maximization of international return on capital investment bears a strong relation to an older international phenomenon: imperialism. The global distribution of industries and markets in response to capital accumulation on the part of a global oligarchy carries with it the seeds of international conflict. Access to raw materials, most particularly oil, has caused otherwise law-abiding nations to resort to policies of "regime change" and destabilization in relation to other sovereign states. Globalized capital investment interests can encourage one country or a single military alliance to undertake the role of international policeman without official sanction under international law. While bilateral diplomacy or international organization might have addressed contentious issues in the past, now global investors require immediate and unilateral action by puppet "peace keepers." Globalization is premised on rapidly deployed police actions whereas internationalism was committed to consensus oriented peaceful resolutions.

Globalization requires a centralized and unified source of power. For the existence of a global system there must be a single policy for the attainment of internationally maximized return on capital investment. This requires that the global policeman be under the control of the capital supplying oligarchy. It demands a single control center with coordinated strategic and tactical response. Notions of democracy and national self-determination are foreign to the optimization of return on investment.

Just as the intelligentsia conceived and advanced the cause of peaceful internationalism, it must now recognize and oppose the threat posed by globalization.

Conclusion

Globalization is simultaneously the antithesis of nationalism and of internationalism. The global maximization of return on capital investment is inconsistent with notions of democracy and self-determination. Because the intelligentsia has traditionally been the repository of national culture and of internationalist idealism, the intelligentsia must oppose globalization. The monolithic structure of a globalized society makes this threat a natural catalyst for unification of the intelligentsia in opposition to the financial oligarchy.

Programme of the 14th Lomonosov Conference on Elementary Particle Physics and the 8th International Meeting on Problems of Intelligentsia

19 August, Wednesday

09.00 – 09.15 **Opening** (Conference Hall)
A.Studenikin, MSU
V.Belokurov, Vice Rector of Moscow State University
A.Slavnov, Head of Department of Theoretical Physics, Faculty of Physics of MSU
V.Matveev, Director of Institute of Nuclear Research
A.Sisakian, Director of Joint Institute of Nuclear Research
M.Panasyuk, Director of Institute of Nuclear Physics of MSU

09.15 – 13.45 **MORNING SESSION** (Conference Hall)
Chairman: A. Slavnov
09.15 D.Shirkov (JINR) *Remembering N.N.Bogoliubov, physicist and mathematician* (15')
09.30 D.Shirkov (JINR) *60 years of broken symmetries in quantum physics (from the Bogoliubov theory of superfluidity to the Standard Model)* (25 min)
09.55 P.Wells (CERN) *Status of the ATLAS experiment* (30 min)
10.25 B.Esposito (INFN-Frascati) *Early physics with ATLAS at LHC* (25 min)
10.50 A.Sisakian (JINR) *The Nuclotron-based Ion Collider Facility (NICA): a new project of Dubna* (25 min)
11.15 – 11.40 Coffee break
Chairman: L. Stodolsky
11.40 C.Biino (INFN-Turin) *CMS: status and physics prospects* (25 min)
12.30 W.Bonivento (INFN-Milan) *Status and physics potential of the LHCb experiment* (25')
12.55 W.Gradl (Univ. of Mainz) *Recent results from BABAR* (25 min)
13.20 A.Yu.Smirnov (ICTP & INR) *Neutrinos: a phenomenological overview* (25 min)

13.45 – 15.00 Lunch

15.00 – 18.40 **AFTERNOON SESSION** (Conference Hall)
Chairman: P. Spillantini
15.00 A.Dolgov (INFN-Ferrara&ITEP) *Cosmic antimatter: models and phenomenology* (25')
15.25 P.Picozza (Univ. of Rome-II) *Cosmic ray studies with PAMELA* (25 min)
16.15 M.Panasyuk (SINP MSU) *Cosmic rays "many knees" problem for spaceborn experiments solutions* (25 min)
16.40 – 17.00 Coffee break
Chairman: P. Spillantini
17.00 A.Starobinsky (Landau Inst.) *Present observational and theoretical status of dark energy* (25 min)
17.25 R.Bernabei (INFN & Univ. of Rome-II) *Dark matter particles in the galactic halo* (25')
17.50 G.Hammond (SUPA, Univ. of Glasgow) *Gravitational Wave Astronomy* (25 min)
18.15 N.Krasnikov (INR) *"Unparticles" and fields with variable distributed mass* (25 min)

18.40 – 23.00 **SPECIAL SESSION** (40^0)
Reception banquet will be held on board of a ship that will stream along the river across the central part of Moscow; the conference buses to the ship will depart from the entrance to the Faculty of Physics at **18.40**

20 August, Thursday

09.00 – 13.35 **MORNING SESSION** (Conference Hall)

Chairman: A. Smirnov

09.00 V.Gavrin (INR) *Solar neutrinos* (25 min)

09.25 A.Bertolin (INFN-Padua) *Neutrino physics with OPERA* (25 min)

09.50 T.Nakaya (Kyoto Univ.) *New results from the FNAL SciBooNE neutrino experiment* (25 min)

10.15 J.Hartnell (Univ. of Sussex) *Neutrino oscillations with MINOS* (25 min)

10.40 M.Shibata (KEK) *Status of the T2K experiment* (25 min)

11.05 – 11.30 Coffee break

Chairman: N. Fornengo

11.30 V.Aynutdinov (INR) *Baikal Neutrino Experiment: from NT200 to NT1000* (25 min)

11.55 S.Sukhotin (RRC"Kurchatov Institute") *The status of the Double-CHOOZ experiment* (20 min)

12.15 N.Titov (INR) *KATRIN Experiment preparation status - Systematic improvement* (25')

12.40 A.Garfagnini (Univ. of Padua) *Search for neutrinoless double beta decay of Ge-76 with the GERmanium Detector Array "GERDA"* (20 min)

13.00 S. Di Domizio (INFN-Genova) *Cuore: neutrinoless double beta decay with bolometers* (15 min)

13.15 R.Dvornicky (Comenius Univ.) *Endpoint spectra of tritium and rhenium beta decays for massive neutrinos* (20 min)

13.35 – 15.00 Lunch

15.00 – 19.20 **AFTERNOON SESSION** (Conference Hall)

Chairman: P. Di Bary

15.00 L.Stodolsky (MPI Munich) *On "Practical Applications" of neutrinos* (25 min)

15.25 C.Giunti (INFN-Turin) *The GSI time anomaly: facts and fiction* (25 min)

15.50 A.Studenikin (MSU) *Electromagnetic interaction of neutrino* (15 min)

16.05 C.Salvioni (Univ. of Insubria) *Searching for neutrinoless double beta decay with the Cuore and Cuoricino experiments* (20 min)

16.25 A.Borisov (MSU),A.Ali (DESY), D.Zhuridov (National Tsing Hua Univ.) *Neutrinoless double beta decay: searching for new physics with comparison of different nuclei* (15 min)

16.40 E.Litvinovich (RRC "Kurchatov Institute") *Recent results of the Borexino experiment* (20 min)

17.00 – 17.30 Tea break

Chairman: T. Nakaya

17.30 M.Circella (INFN-Bari) *Neutrino telescopes in the deep sea* (25 min)

17.55 D.Zaborov (ITEP) *The ANTARES experiment in the Mediterranean sea: overview and first results* (20 min)

18.15 O.Ryazhskaya (INR) *On one group of experimental results related to the search for neutrino radiation from SN1987A: commentary* (20 min)

18.35 M.Dvornikov (USM, Chile & IZMIRAN) *Propagation of neutrinos in rapidly rotating neutron stars* (15 min)

18.50 A.Okrugin (Yaroslavl State Univ.) *Dirac-neutrino magnetic moment and the dynamics of a supernovae explosion* (15 min)

19.05 K.Zhukovsky (MSU) *Exponential form of the mixing matrix in the lepton sector of the SM* (15 min)

19.30 – 22.00 **Sight-seeing bus excursion in Moscow**

21 August, Friday

09.00 – 13.45 MORNING SESSION (Conference Hall)

Chairman: F.Ricci

09.00 C.Cecchi (Univ. of Perugian & INFN) *The Fermi/LAT mission: scientific results from the first year* (25 min)

09.25 A.Foerster (MPI-K Heidelberg) *The Gamma-Ray Sky seen with H.E.S.S.* (25 min)

09.50 A.Chekhtman (George Mason Univ.) *FERMI electron results* (20 min)

10.10 M.Pravdin (ICRA-SB-RAS) *The cosmic ray spectrum at ultrahigh energies* (20 min)

10.30 A.Panov (SINP MSU) *Systematics and fine structure in the cosmic ray electron spectrum measured by ATIC* (15 min)

10.45 D.Galtsov (MSU) *No black holes at LHC* (15 min)

Chairman: D.Galtsov

11.30 N.Fornengo (Univ. of Turin & INFN - Turin) *Advances in the exploration of particle dark matter signals* (25 min)

11.55 V.Berezinsky (LNGS), V.Dokuchaev (INR), Y.Eroshenko (INR) *Dark matter annihilation in the galaxy* (15 min)

12.10 D.Gorbunov (INR) *Minimal extension of the Standard Model of particle physics* (20')

12.30 R.Cerulli (LNGS-INFN) *New results on rare processes by DAMA at Gran Sasso* (15')

12.45 O.Miranda (Cinvestav-Mexico) *Constraining nonstandard neutrino interactions* (20')

13.05 V.Semikoz (IZMIRAN) *Neutrino asymmetry and growth of hypermagnetic fields in hot plasma of early Universe* (15 min)

13.20 P. Di Bari (Univ. of Southampton) *Leptogenesis* (25 min)

15.00 – 19.00 AFTERNOON SESSION (Conference Hall)

Chairman: P.Patel

15.00 S.Kulagin (INR) *Nuclear effects in the searching of high-energy neutrino* (15 min)

15.15 J-M.Levy (CNRS) *Recent results of the NOMAD experiment* (15 min)

15.30 D.Naumov (JINR) *Precision measurements by NOMAD experiment* (15 min)

15.45 A.Kataev (INR) *The perturbative relation between pole and running masses of leptons and the hypothesis of "maximal transcendality"* (15 min)

16.00 A.Nesterenko(JINR)*Hadronic effects in low-energy QCD:Adler function and tau-decay* (15')

16.15 O.Teryaev (JINR) *Axial anomaly and strange quarks* (15 min)

16.30 M.Tokarev (JINR) *Self-similarity of jet production and QCD* (15 min)

16.45 M.Osipenko (SINP MSU & INFN-Genoa) *Semi-inclusive production of charged pions with CLAS* (15 min)

17.00 D.Peshekhonov (JINR) *Helicity quark distributions from DIS and SIDIS measured in COMPASS* (15 min)

Chairman: A.Kataev

17.40 B.Brzozowska (Univ. of Warsaw) *Scaled momentum spectra in the current region of the Breit frame at HERA* (15 min)

17.55 V.Braguta (IHEP) *eta_b ->J/\Psi J/\Psi decay within light cone formalism* (10 min)

18.05 B.Mangazeev (Irkutsk State Univ.) *Electromagnetic structure functions of nucleons in the region of very small x* (10 min)

18.15 L.Laperashvili (ITEP) *New bound states of heavy quarks at LHC and TEVATRON* (15')

18.30 M.Nekrasov (IHEP) *Unstable-particles pair production in MPT approach in NNLO* (10 min)

18.40 V.Kuksa (SouthernFederalUniv.)*Bozon-pair production in the model of unstable particles* (10')

18.50 V.Beylin (Southern Federal Univ.) *Radiative and hadronic decays of vector mesons in the gauge model of quark-meson interactions* (10 min)

22 August, Saturday

09.00 – 13.35 **MORNING SESSION** (Conference Hall)

Chairman: K. Urbanovski

09.00 F.Ricci (Univ. of Rome-I) *Present and future of the VIRGO gravitational wave experiment* (25 min)

09.25 A.Zakharov (ITEP) *Gravitational lensing: from micro to nano* (20 min)

09.45 A.Burinskii (Nuclear Safety Inst.) *Holographic structure of the Kerr-Schild black-holes, beamlike excitations and fluctuating horizon* (20 min)

10.05 A.L.Smirnov (INR) *On the global geometry of Brane Universe models* (15 min)

10.20 A.Starinets (Oxford Univ.) *Transport coefficients in strongly interacting quantum field theories from gauge-gravity duality* (20 min)

10.40 M.Passera (INFN-Padua) *Review on G-2 of charge leptons* (25 min)

11.05 – 11.30 Tea break

Chairman: A. Starinets

11.30 D.Kazakov (JINR) *Infrared-finite observables in N=4 SYM theory* (20 min)

11.50 K.Stepanyantz (MSU) *Revealing structure of quantum corrections in N=1 supersymmetric theories using the Schwinger-Dyson equations* (15 min)

12.05 S.Demidov (INR) *Production of kink-antikink pair in collisions of high energy particles* (15 min)

12.20 D.Kirpichnikov (INR) *Quasilocalization in high-energy collisions* (15 min)

12.35 D.Levkov (INR) *Long quantum transitions due to unstable semiclassical dynamics* (15 min)

12.50 T.Rashba (MPI Munich) *Search for axion-like particles in astrophysical observations* (20 min)

13.10 N.Mankoc (Univ. of Ljubljana) *Is the fifth family, predicted by the approach unifying spins and charges, the candidate for forming the dark matter clusters?* (25 min)

13.35 – 15.00 Lunch

15.00 – 18.15 **AFTERNOON SESSION** (Conference Hall)

Chairman: A. Borisov

15.00 A.Bakulev (JINR) *Fractional analytic perturbation theory* (20 min)

15.20 M.Katanaev (Steklov Math.Inst.) *Geometric theory of defects in solids* (20 min)

15.40 K.Urbanowski (Univ. of Zielona Gora) *Energy of unstable states at long times* (20 min)

16.15 – 16.40 Tea break

Chairman: P. Pronin

16.40 A.Nikishov (Lebedev Inst.) *Atoms as rods and clocks in a gravitational field* (10 min)

16.50 O.Kharlanov (MSU), V.Zhukovsky (MSU) *Casimir effect within (3+1)D Maxwell-Chern-Simons electrodynamics* (15 min)

17.05 M.Fil'chenkov (Inst. of Gravitation and Cosmology) *Nonhydrogen-like graviatom radiation* (15 min)

17.20 E.Nemchenko (Tomsk State Univ.) *Radiation of force-momentum from relativistic charged particles* (15 min)

17.35 O.Konctantinova (Tomsk State Univ.) *Angular momentum of spin light* (15 min)

17.50 H.Yamamoto (Tohoku Univ.) *Recent physics results from Belle* (25 min)

23 August, Sunday

9.00 – 19.00 Sergiev Posad

24 August, Monday

09.00 – 13.35 **MORNING SESSION** (Conference Hall)
Chairman: D. Krofcheck
09.00 A. De Santis (INFN-Rome) *Recent KLOE results* (25 min)
09.25 S.Serednyakov (INP Novosibirsk) *ISR physics at BABAR* (20 min)
09.45 P.Patel (McGill Univ.) *Bottomonium spectroscopy at BABAR* (20 min)
10.05 Yu.Kolomensky (UC Berkeley & LBNL) *Search for exotics in bottomonium decay with BABAR* (20 min)
10.25 M.Sigamani (Univ. of London) *Semileptonic B decays at BABAR* (20 min)
10.45 M.Giorgi (Univ. of Pisa) *Search for lepton flavor violation in tau decays with BABAR* (20 min)
11.05 – 11.40 Tea break
Chairman: M.Giorgi
11.40 H.Fisk (Fermilab) *CP violation and mixing in b-decays* (20 min)
12.00 A.Askew (Rutgers Univ.) *Recent electroweak measurements at the TEVATRON* (20 min)
12.20 N.Serra (NIKHEF) *Prospects for rare B-decays at LHCb* (20 min)
12.40 T.du Pree (NIKHEF) *Prospects for CP Violation Studies at LHCb* (20 min)
13.00 V.Palladino (INFN & Univ. of Naples) *Searches for new physics at NA62* (15 min)
13.15 C.Biino (INFN-Turin) *QCD tests with kaon decays* (20 min)

13.35 – 15.00 Lunch

15.00 – 18.40 **AFTERNOON SESSION** (Conference Hall)
Chairman: V. Savrin
15.00 D.Pallin (CNRS/IN2P3) *Top physics with the ATLAS detector at LHC* (20 min)
15.20 S.Eckweiler (University of Mainz) *QCD studies at LHC with the ATLAS detector* (20 min)
15.40 C.Weiser (Univ. of Freiburg) *ATLAS discovery potential of the SM Higgs* (20 min)
16.00 M.Biglietti (Univ. & INFN - Rome-I) *SUSY searches with ATLAS at LHC* (20 min)
16.20 D.Krofcheck (Univ. of Auckland) *Signatures of AdS/CFT using the CMS experiment at the LHC* (15 min)
16.35 N.Ilina (ITEP) *Study of jet transverse structure with CMS experiment at* $\sqrt{s}$ *= 10 TeV* (15 min)
16.50 S.Petrushanko (SINP MSU) *Elliptic flow studies in heavy-ion collisions using the CMS detector at LHC* (15 min)
17.05 L.Sarycheva (MSU) *Ion physics in CMS experiment at LHC* (15 min)
17.20 – 17.40 Tea break
Chairman: A. Grigoriev
17.40 S.Senyukov (University of Eastern Piedmont, INFN Turin) *Open charm study in ALICE at the LHC* (20 min)
18.00 S.Bitykov (IHEP) *Systematic uncertainties in experiments at LHC* (15 min)
18.15 E.Shabalin (ITEP) *How the interaction of pi-mesons in the final state changes a correlation between K-2pi and K-3pi amplitudes* (15 min)
18.30 O.Pavlovsky (MSU-ITPM) *Is multiple t-quarks bound states possible due to Higgs exchange?* (10 min)

25 August, Tuesday

09.00 – 13.25 **MORNING SESSION** (Conference Hall)

Chairman: H.Fisk

09.00 U.Schneekloth (DESY) *Electroweak Physics and Searches for New Physics at HERA* (25 min)

09.25 V.Chekelian (MPI Munich) *Proton structure measurements at HERA* (20 min)

09.45 A.Shchukin (IHEP) *Searches for physics beyond standard model* (25 min)

10.10 E.Varnes (Univ. of Arizona) *Testing the Standard Model with top quarks* (25 min)

10.35 M.Vysotsky (ITEP) *Bounds on new light particles from high-energy and very small momentum transfer np elastic scattering data* (20 min)

10.55 V.Skvortsov (MIPT-Moscow), N.Vogel (Univ. of Tech.- Chemnitz) *"The bundle for memory" from an exotic quasi particles* (15 min)

11.10 – 11.50 Tea break

Chairman: H.Fisk

11.50 M.Polikarpov (ITEP) *Chiral symmetry breaking and Chiral Magnetic Effect in QCD with strong magnetic field* (25 min)

13.00 – 15.00 Lunch

ROUND TABLE DISCUSSION
"The Year of Astronomy:
Astroparticle Physics and Frontiers of the Universe"

Chairman: A. Studenikin

15.00 L.Stodolsky (MPI Munich) *The Year of Astronomy and the heritage of Galilei* (25 min)

15.25 A.Kappes (Univ. of Erlangen) *High-energy neutrinos from Galactic sources* (25 min)

15.50 P.Spillantini (INFN-Florence) *Radiation exposure and mission strategies for interplanetary manned missions and interplanetary habitats* (25 min)

16.15 Discussion

16.40 – 17.00 Tea break

EIGHTH INTERNATIONAL MEETING ON
PROBLEMS OF INTELLIGENTSIA:
"Intelligentsia and World Crisis"

Chairman: A. Studenikin

17.00 – 17.50 **Conference Hall**

17.00 S.Kapitza (Inst.of Phys.Problems, RAS) *Demography and the origin of crisis* (25 min)

17.25 J.Bleimaier (Princeton) *Governance and the international crisis* (25 min)

17.50 Discussion and conclusion

Closing of the 14th Lomonosov Conference on Elementary Particle Physics and the 8th International Meeting on Problems of Intelligentsia

SPECIAL SESSION (40^0)

POSTER SESSION

I.Balantsev (MSU) *Massive neutrino in rotating media: new neutrino trapping mechanism in neutron stars*
V.Bordovitsyn, O.Konctantinova (Tomsk State Univ.) *Angular momentum of spin light*
V.Bordovitsyn, E.Nemchenko (Tomsk State University) *Radiation of force-momentum from relativistic charged particles*
V.Boyarkin (INR), O.Ryazhskaya (INR) *Sodium Chloride as a Target for Supernovae Neutrinos*
V.Bytev (JINR) *Differential reduction of generalized hypergeometric functions in application to Feynman diagrams*
K.Heeger (Univ. of Wisconsin) *Reactor neutrino experiments: recent results and future prospects*
A.Khmialeuski (Joint Inst. for Power and Nucl.Research, Turov) *Scaling behavior of factorial moments of the multiplicity difference and phase transition QGP to hadrons*
A.Lokhov (MSU) *Radiative neutrino decay in dense matter*
A.Mikhailov (MSU) *Vacuumless defects in the Randall-Sundrum models*
I.Narodetskiy (ITEP), Yu.Simonov (ITEP), and A.Veselov (ITEP) *Diquark and triquark correlations in the deconfined phase of QCD*
A.Okrugin (Yaroslavl State Univ.) *Reexamination of a bound on the dirac neutrino magnetic moment from the supernova neutrino luminosity*
E.Prokopiev (ITEP) *Opportunities of positron diagnostics for research of dust space plasma*
E.Shevtsova (MSU) *Two-loop Gell-Mann-Low function for general renormalizable N=1 supersymmetric theory, regularized by higher derivatives*
Yu.Voronina (ITPM-MSU) *Casimir pressure regularization and renormalization in two-dimensional scalar field model*
M.Ulybyshev (MSU) *Casimir effect calculations within the formalism of lattice quantum field theory*

List of participants of the 14th Lomonosov Conference on Elementary Particle Physics and the 8th International Meeting on Problems of Intelligentsia

Ali Ahmed	DESY	ahmed.ali@desy.de
Askew Andrew	Rutgers Univ.	askew@fnal.gov
Aynutdinov Vladimir	INR	aynutdin@pcbai10.inr.ruhep.ru
Balantsev Ilya	MSU	balantsev@physics.msu.ru
Bakulev Alexander	JINR	bakulev@theor.jinr.ru
Belokurov Vladimir	MSU	belokur@rector.msu.ru
Berezinsky Veniamin	LNGS	venya.berezinsky@lngs.infn.it
Bernabei Rita	DAMA	rita.bernabei@roma2.infn.it
Bertolin Alessandro	INFN-Padua	bertolin@pd.infn.it
Beylin Vitaly	Southern Federal Univ.	vbey@rambler.ru
Biino Cristina	INFN, Turin	cristina.biino@cern.ch
Biglietti Michela	Univ. & INFN - Rome-I	Michela.Biglietti@cern.ch
Bityukov Sergey	IHEP	Serguei.Bitioukov@cern.ch
Bleimaier John	Princeton	bleimaier@aol.com
Bonivento Walter	CERN	walter.bonivento@cern.ch
Bordovitsyn Vladimir	Tomsk State University	vabord@sibmail.com
Borisov Anatoly	MSU	borisov@ave.phys.msu.su
Bytev Vladimir	JINR	bvv@jinr.ru
Braguta Victor	IHEP	braguta@mail.ru
Brzozowska Beata	Univ. of Warsaw	beata.brzozowska@fuw.edu.pl
Burinskii Alexander	Nuclear Safety Inst.	bur@ibrae.ac.ru
Cecchi Claudia	Univ. of Perugian & INFN	claudia.cecchi@pg.infn.it
Cerulli Riccardo	LNGS-INFN	riccardo.cerulli@lngs.infn.it
Chekelian Vladimir	MPI Munich	shekeln@mppmu.mpg.de
Chekhtman Alexandre	George Mason Univ.	achekhtm@gmu.edu
Cherepashchuk Anatoly	SAI	cher@sai.msu.ru
Circella Marco	INFN-Bari	marco.circella@ba.infn.it
De Santis Antonio	INFN-Rome	antonio.desantis@roma1.infn.it
Demidov Sergei	INR	demidov@ms2.inr.ac.ru
Di Bari Pasquale	Univ. of Southampton	dibari@pd.infn.it
Di Domizio Sergio	INFN-Genova	sergio.didomizio@ge.infn.it
Di Ruzza Benedetto	Univ. of Trieste & INFN Trieste	benedetto.diruzza@ts.infn.it
Dokuchaev Vladislav	INR	dokuchaev@inr.npd.ac.ru
Dolgov Alexander	INFN-Ferrara & ITEP	dolgov@fe.infn.it
Dvornicky Rastislav	Comenius Univ.	dvornicky@teller.dnp.fmph.uniba.sk
Dvornikov Maxim	USM, Chile & IZMIRAN	maxim.dvornikov@usm.cl
du Pree Tristan	NIKHEF	tdupree@nikhef.nl
Eroshenko Yury	INR	erosh@inr.npd.ac.ru
Eckweiler Sebastian	University of Mainz	sebastian.eckweiler@uni-mainz.de

Esposito Bellisario	INFN Frascati	bellisario.esposito@lnf.infn.it
Fil'chenkov Michael	Peoples' Friendship Univ. of Russia	fmichael@mail.ru
Fisk Gene	Fermilab	hefisk@fnal.gov
Foerster Andreas	MPI-K Heidelberg	andreas.foerster@mpi-hd.mpg.de
Fornengo Nicolao	Univ. of Turin & INFN - Turin	fornengo@to.infn.it
Galtsov Dmitry	MSU	galtsov@phys.msu.ru
Garfagnini Alberto	Univ. of Padua	alberto.garfagnini@pd.infn.it
Gavrin Vladimir	INR	gavrin@dionis.iasnet.ru
Gavryuseva Elena	Inst. of Astrophys. and Space Research, Arcetri	elena.gavryuseva@gmail.com
Giunti Carlo	INFN-Turin	giunti@to.infn.it
Gorbunov Dmitry	INR	gorby@ms2.inr.ac.ru
Gradl Wolfgang	Univ. of Mainz	wgradl@slac.stanford.edu
Giorgi Marcello	Univ. of Pisa	marcello.giorgi@pi.infn.it
Grats Yuri	MSU	grats@string.phys.msu.ru
Grigoriev Alexander	MSU	ax.grigoriev@mail.ru
Hammond Giles	SUPA, Univ. of Glasgow	g.hammond@physics.gla.ac.uk
Hartnell Jeff	Univ. of Sussex	j.j.hartnell@sussex.ac.uk
Heeger Karsten	Univ. of Wisconsin	heeger@wisc.edu
Ilina Natalia	ITEP	Natalia.Ilina@cern.ch
Kadyshevsky Vladimir	JINR	kadyshev@jinr.ru
Kapitza Sergey	Inst.of Phys.Problems, RAS	Sergey@kapitza.ras.ru
Kappes Alexander	Univ. of Erlangen	kappes@physik.uni-erlangen.de
Kataev Andrei	INR	kataev@ms2.inr.ac.ru
Katanaev Mikhail	Steklov Math.Inst.	katanaev@mi.ras.ru
Kazakov Dmitri	JINR	kazakovdi@mail.ru
Kirpichnikov Dmitriy	INR	kirp@ms2.inr.ac.ru
Kharlanov Oleg	MSU	okharl@mail.ru
Khmialeuski Anatoly	Joint Inst. for Power and Nucl.Research, Turov	akhmelevskiy@sosny.bas-net.by
Kolomensky Yury	UC Berkeley & LBNL	ygkolomensky@lbl.gov
Konstantinova Olga	Tomsk State University	olgakonst87@mail.ru
Krasnikov Nikolay	INR	krasniko@ms2.inr.ac.ru
Krofcheck David	Univ. of Auckland	d.krofcheck@auckland.ac.nz
Kruse Mark	CDF & ATLAS, Duke Univ.	mkruse@phy.duke.edu
Kuksa Vladimir	Southern Federal Univ.	kuksa@list.ru
Kulagin Sergey	INR	kulagin@ms2.inr.ac.ru
Laperashvili Larisa	ITEP	laper@itep.ru
Lokhov Alexey	MSU	lokhov.alex@gmail.com
Levkov Dmitry	INR	levkov@ms2.inr.ac.ru
Levy Jean-Michel	CNRS	jmlevy@lpnhe.in2p3.fr
Litvinovich Evgeny	RRC "Kurchatov Institute"	e.litvinovich@kiae.ru
Lobanov Andrei	MSU	lobanovmsu@yandex.ru
Madigozhin Dmitry	JINR	madigo@sunse.jinr.ru
Mangazeev Boris	Irkutsk State Univ.	MangazeevBV@yandex.ru
Mankok Borstnik Norma	Univ. of Ljubljana	norma.mankoc@fmf.uni-lj.si

Matthiae Giorgio	Univ. of Rome-II	Giorgio.Matthiae@roma2.infn.it
Matveev Victor	INR	matveev@ms2.inr.ac.ru
Miranda Omar	Cinvestav-Mexico	omr@fis.cinvestav.mx
Murchikova Elena	MSU	murchikova@yahoo.com
Nakaya Tsuyoshi	Kyoto Univ.	t.nakaya@scphys.kyoto-u.ac.jp
Narodetskii Ilya	ITEP	naro@heron.itep.ru
Naumov Dmitry	JINR	naumov@nusun.jinr.ru
Nekrasov Maksim	IHEP	Maksim.Nekrasov@ihep.ru
Nemchenko Ekaterina	Tomsk State Univ.	katusha77@sibmail.com
Nesterenko Alexander	JINR	nesterav@theor.jinr.ru
Nikishov Anatoly	Lebedev Inst.	nikishov@lpi.ru
Okrugin Alexander	Yaroslavl State Univ.	okrugin@uniyar.ac.ru
Okun Lev	ITEP	okun@itep.ru
Osipenko Mikhail	SINP MSU & INFN-Genoa	osipenko@ge.infn.it
Palladino Vito	INFN & Univ. of Naples	vito.palladino@cern.ch
Pallin Dominique	CNRS/IN2P3	pallin@clermont.in2p3.fr
Panasyuk Mikhail	SINP & MSU	panasyuk@sinp.msu.ru
Panov Alexander	SINP MSU	panov@dec1.sinp.msu.ru
Passera Massimo	INFN-Padua	passera@pd.infn.it
Patel Popat	McGill Univ.	patel@hep.physics.mcgill.ca
Pavlovsky Oleg	MSU	ovp@goa.bog.msu.ru
Peshekhonov Dmitry	JINR	pdv@lhep.jinr.ru
Petrushanko Sergey	SINP MSU	sergeant@mail.cern.ch
Picozza Piergiorgio	Univ. of Rome-II	picozza@roma2.infn.it
Polikarpov Mikhail	ITEP	polykarp@itep.ru
Pravdin Mikhail	ICRA-SB-RAS	m.i.pravdin@ikfia.ysn.ru
Prokopiev Eugene	ITEP	epprokopiev@mail.ru
Rashba Timur	MPI Munich	timur@mppmu.mpg.de
Ricci Fulvio	Univ. of Rome-I	fulvio.ricci@roma1.infn.it
Ros Eduardo	IFIC-Valencia	Eduardo.Ros@cern.ch
Royzen Ilya	Lebedev Inst.	royzen@lpi.ru
Ryazhskaya Olga	INR	ryazhskaya@vaxmw.tower.ras.r u
Salvioni Chiara	Univ. of Insubria	chiara.salvioni@mib.infn.it
Sarycheva Lyudmila	MSU	lis@alex.sinp.msu.ru
Semikoz Victor	IZMIRAN	semikoz@yandex.ru
Serednyakov Sergey	INP Novosibirsk	seredn@inp.nsk.su
Serra Nicola	NIKHEF	nicola.serra@cern.ch
Shabalin Evgeny	ITEP	shabalin@heron.itep.ru
Shevtsova Ekaterina	MSU	vrednuk@gmail.com
Shibata Masahiro	KEK	mshibata@neutrino.kek.jp
Shirkov Dmitry	JINR	shirkovd@thsun1.jinr.ru
Shchukin Andrey	IHEP	schukin@fnal.gov
Shmatov Sergei	JINR	shmatov@cern.ch
Schneekloth Uwe	DESY	uwe.schneekloth@desy.de
Senyukov Serhiy	University of Eastern Piedmont, INFN Turin	serhiy.senyukov@cern.ch
Sigamani Michael	Univ. of London	m.sigamani@qmul.ac.uk,

		sigamani@slac.stanford.edu
Simonov Yuri	ITEP	simonov@itep.ru
Sisakian Alexey	JINR	sisakian@jinr.ru
Skvortsov Vladimir	MIPT, Univ. of Tech. Chemnitz	sva@gol.ru
Slavnov Andrey	Steklov Math. Inst & MSU	slavnov@mi.ras.ru
Smirnov Alexei Yu.	ICTP & INR	smirnov@ictp.it
Smirnov Alexey L.	INR	smirnov@ms2.inr.ac.ru
Spillantini Piero	INFN-Florence	spillantini@fi.infn.it
Starobinsky Alexei	LITP	alstar@landau.ac.ru
Starinets Andrei	Oxford Univ.	andrei.starinets@physics.ox.ac.u k
Starostin Alexander	ITEP	starostin@itep.ru
Stepanyantz Konstantin	MSU	stepan@m9com.ru
Stodolsky Leo	MPI Munich	les@mppmu.mpg.de
Studenikin Alexander	MSU	studenik@srd.sinp.msu.ru
Sukhotin Sergey	RRC "Kurchatov Institute"	sukhotin@kiae.ru
Teryaev Oleg	JINR	teryaev@theor.jinr.ru
Titov Nikita	INR	titov@al20.inr.ru
Tokarev Mikhail	JINR	tokarev@sunhe.jinr.ru
Trukhin Vladimir	Faculty of Physics, MSU	dean@phys.msu.ru
Ulybyshev Maxim	MSU	oulibishev@mtu-net.ru
Urbanowski Krzysztof	Univ. of Zielona Gora	K.Urbanowski@proton.if.uz. zgora.pl
Varnes Erich	Univ. of Arizona	varnes@physics.arizona.edu
Voronina Yuliya	ITPM-MSU	voronina-yulya@yandex.ru
Vysotsky Mikhail	ITEP	vysotsky@itep.ru
Weiser Christian	Univ. of Freiburg	Christian.Weiser@cern.ch
Wells Phillippa	CERN	Pippa.Wells@cern.ch
Yamamoto Hitoshi	Tohoku Univ.	yhitoshi@awa.tohoku.ac.jp
Zaborov Dmitry	ITEP	Dmitry.Zaborov@itep.ru
Zakharov Alexander	ITEP	zakharov@itep.ru
Zhukovsky Konstantin	MSU	vchzh@mail.ru
Zhukovsky Vladimir	MSU	zhukovsk@phys.msu.ru
Zhuridov Dmitri	National Tsing Hua Univ.	jouridov@mail.ru